Industrial Polymers, Specialty Polymers, and Their Applications

PLASTICS ENGINEERING

Founding Editor

Donald E. Hudgin

Professor
Clemson University
Clemson, South Carolina

Industrial Polymers, Specialty Polymers, and Their Applications

Manas Chanda
Salil K. Roy

CRC Press
Taylor & Francis Group
Boca Raton London New York

CRC Press is an imprint of the
Taylor & Francis Group, an **informa** business

The material was previously published in *Plastics Technology Handbook, Fourth Edition* © Taylor & Francis 2007.

CRC Press
Taylor & Francis Group
6000 Broken Sound Parkway NW, Suite 300
Boca Raton, FL 33487-2742

First issued in paperback 2019

© 2009 by Taylor & Francis Group, LLC
CRC Press is an imprint of Taylor & Francis Group, an Informa business

No claim to original U.S. Government works

ISBN-13: 978-1-4200-8058-2 (hbk)
ISBN-13: 978-0-367-38715-0 (pbk)

Library of Congress Cataloging-in-Publication Data

Chanda, Manas, 1940-
    Industrial polymers, specialty polymers, and their applications / Manas Chanda and Salil K. Roy.
        p. cm. -- (Plastics engineering ; 73)
    Includes bibliographical references and index.
    ISBN 978-1-4200-8058-2 (alk. paper)
    1. Polymers. 2. Polymers--Industrial applications. I. Roy, Salil K., 1939- II. Title. III. Series.

TP1087.C43 2008
668.9--dc22
                                                      2008011622

Visit the Taylor & Francis Web site at
http://www.taylorandfrancis.com

and the CRC Press Web site at
http://www.crcpress.com

Contents

Preface

The polymer industry consumes more than half of the total output of the organic chemical industry, but only a few polymers account for most of it. For example, the combined output of polypropylene, high-density polyethylene, and linear low-density polyethylene, all produced by polymerization with transition-metal catalysts, was responsible for 42% of the 41 million metric tons of thermoplastic resins produced in the United States in the year 2000, while another 40% represented a handful of polymers that are produced using a free-radical mechanism, namely, low-density polyethylene, poly (vinyl chloride), polystyrene, poly(vinyl alcohol), poly(methyl methacrylate), and poly(vinyl acetate). Most of the remaining 18% corresponded to polyurethanes and a few condensation polymers, such as poly(ethylene terephthalate), poly(butylene terephtahalate), polyamides, and polycarbonates. Even the period from 2000 to 2005, which witnessed considerable stress and change in the U.S. chemical industry, did not cause any significant change in the landscape (Villa, C. M. 2007. *Ind. Eng. Chem. Res.*, 46, 5815–5823). In the present book, all these polymers are prominently discussed in Chapter 1, which also includes other less widely used polymer types such as acrylics, ether polymers, cellulosics, sulfide polymers, silicones, polysulfones, polyether ether ketones, and polybenzimidazoles. Polyblends and interpenetrating network polymers are also included in this chapter.

Besides the aforesaid industrial polymers, most of which we often encounter in everyday life, there are hundreds of other polymers, polymer derivatives, and polymeric combinations that play special and often critical roles in diverse fields of human activities. Chapter 2 deals with such polymers. Some of these *specialty polymers* possess, inherently, one or more special properties that make them indispensable for specific applications, while there are many others that are tailor-made or made by modification of the aforesaid industrial polymers to meet specific and critical needs. For a systematic discussion, these polymers can be placed in different groups according to their properties and/or areas of uses, such as high-temperature and fire-resistant polymers, liquid crystal polymers, electroactive polymers, electrolytic polymers, photoresist polymers, ionic and ion-exchange polymers, packaging polymers, biodegradable polymers, adhesive polymers, polymers in optical information storage, polymeric sensors, conductive fiber fillers, polymeric optical fibers, polymer electrolyte membranes, biodegradable polymer scaffolds, scavenger resins, permselective polymer membranes, hydrogels, smart polymers, dendritic polymers, shape memory polymers, microencapsulation polymers, polymer nanocomposites, wood–polymer composites, and polymerization-filled composites.

As we live in a plastic age, the diverse fields of plastics technology are also undergoing rapid change both qualitatively and quantitatively with many newer applications of common polymers and specialty polymers coming to light. While there are continuous improvements in the established uses of polymers, new uses are being developed in such diverse areas as the automotive and aerospace industries, packaging, agriculture, horticulture, domestic and sports appliances, office equipment, communication, electronics and electrical technology, and biomedical applications. Chapter 3 presents a comprehensive overview of new developments in polymer uses in all these areas.

The material in this book is included in our well-known *Plastics Technology Handbook* and it focuses on a wide range of polymers, both of common and special types, and their myriad applications. For readers who are not quite familiar with polymers and their characteristics, it would be advisable to read *Plastics Fundamentals, Properties, and Testing*, before taking up the present book. We thank Allison Shatkin, Materials Science and Chemical Engineering Editor at CRC Press/Taylor & Francis, who first conceived the idea of this book and took the initiative in publishing it.

Manas Chanda
Salil K. Roy

Authors

Manas Chanda has been a professor and is presently an emeritus professor in the Department of Chemical Engineering, Indian Institute of Science, Bangalore, India. He also worked as a summer-term visiting professor at the University of Waterloo, Ontario, Canada with regular summer visits from 1980 to 2000. A five-time recipient of the International Scientific Exchange Award from the Natural Sciences and Engineering Research Council, Canada, Dr. Chanda is the author or coauthor of nearly 100 scientific papers, articles, and books, including *Introduction to Polymer Science and Chemistry* (CRC Press/Taylor & Francis). A fellow of the Indian National Academy of Engineers and a member of the Indian Plastics Institute, he received a BS (1959) and MSc (1962) from Calcutta University, and a PhD (1966) from the Indian Institute of Science, Bangalore, India.

Salil K. Roy is a professor in the Postgraduate Program in Civil Engineering of the Petra Christian University, Surabaya, Indonesia. Earlier he worked as lecturer, senior lecturer, and associate professor at the National University of Singapore. Prior to that he was a research scientist at American Standard, Piscataway, New Jersey.

Dr. Roy is a fellow of the Institution of Diagnostic Engineers, U.K., and has published over 250 technical papers in professional journals and conference proceedings; he also holds several U.S. Patents. He received a BSc (1958) and MSc (Tech.) (1961) from the University of Calcutta, India, and a ScD (1966) from the Massachusetts Institute of Technology, Cambridge, Massachusetts. Dr. Roy is a subject of biographical record in the prestigious *Great Minds of the 21st Century* published by the American Biographical Institute, *Who's Who in the World* published by the Marquis Who's Who in the World, and *2000 Outstanding Intellectuals of the 21st Century* published by the International Biographical Centre, Cambridge, England.

1

Industrial Polymers

1.1 Introduction

The first completely synthetic plastic, phenol-formaldehyde, was introduced by L. H. Baekeland in 1909, nearly four decades after J. W. Hyatt had developed a semisynthetic plastic—cellulose nitrate. Both Hyatt and Baekeland invented their plastics by trial and error. Thus the step from the idea of macromolecules to the reality of producing them at will was still not made. It had to wait till the pioneering work of Hermann Staudinger, who, in 1924, proposed linear molecular structures for polystyrene and natural rubber. His work brought recognition to the fact that the macromolecules really are linear polymers. After this it did not take long for other materials to arrive. In 1927 poly(vinyl chloride) (PVC) and cellulose acetate were developed, and 1929 saw the introduction of urea-formaldehyde (UF) resins.

The production of nylon-6,6 (first synthesized by W. H. Carothers in 1935) was started by Du Pont in 1938, and the production of nylon-6 (perlon) by I. G. Farben began in 1938, using the caprolactam route to nylon developed by P. Schlock. The latter was the first example of ring-opening polymerization. The years prior to World War II saw the rapid commercial development of many important plastics, such as acrylics and poly(vinyl acetate) in 1936, polystyrene in 1938, melamine–formaldehyde (formica) in 1939, and polyethylene and polyester in 1941. The amazing scope of wartime applications accelerated the development and growth of polymers to meet the diverse needs of special materials in different fields of activity.

The development of new polymeric materials proceeded at an even faster pace after the war. Epoxies were developed in 1947, and acrylonitrile–butadiene–styrene (ABS) terpolymer in 1948. The polyurethanes, introduced in Germany in 1937, saw rapid development in the United States as the technology became available after the war. The discovery of Ziegler–Natta catalysts in the 1950s brought about the development of linear polyethylene and stereoregular polypropylene. These years also saw the emergence of acetal, polyethylene terephthalate, polycarbonate, and a host of new copolymers. The next two decades saw the commercial development of a number of highly temperature-resistant materials, which included poly(phenylene oxide) (PPO), polysulfones, polyimides, polyamide-imides, and polybenzimidazoles.

Numerous plastics and fibers are produced from synthetic polymers: containers from polypropylene, coating materials from PVC, packaging film from polyethylene, experimental apparatus from Teflon, organic glasses from poly(methyl methacrylate), stockings from nylon fiber—there are simply too many to mention them all. The reason why plastics materials are popular is that they may offer such advantages as transparency, self-lubrication, lightweight, flexibility, economy in fabricating, and decorating.

Properties of plastics can be modified through the use of fillers, reinforcing agents, and chemical additives. Plastics have thus found many engineering applications, such as mechanical units under stress, low-friction components, heat- and chemical-resistant units, electrical parts, high-light-transmission applications, housing, building construction functions, and many others. Although it is true that in these applications plastics have been used in a proper manner according to our needs, other ways of utilizing both natural and synthetic polymers may still remain. To investigate these further possibilities, active research has been initiated in a field called *specialty polymers*. This field relates to synthesis of new polymers with high additional value and specific functions.

Many of the synthetic plastic materials have found established uses in a number of important areas of engineering involving mechanical, electrical, telecommunication, aerospace, chemical, biochemical, and biomedical applications. There is, however, no single satisfactory definition of engineering plastics. According to one definition, engineering plastics are those which possess physical properties enabling them to perform for prolonged use in structural applications, over a wide temperature range, under mechanical stress and in difficult chemical and physical environments. In the most general sense, however, all polymers are engineering materials, in that they offer specific properties which we judge quantitatively in the design of end-use applications.

For the purpose of this discussion, we will classify polymers into three broad groups: addition polymers, condensation polymers, and special polymers. By convention, polymers whose main chains consist entirely of C–C bond are *addition polymers,* whereas those in which hetero atoms (e.g., O, N, S, Si) are present in the polymer backbone are considered to be *condensation polymers.* Grouped as *special polymers* are those products which have special properties, such as temperature and fire resistance, photosensitivity, electrical conductivity, and piezoelectric properties, or which possess specific reactivities to serve as functional polymers.

Further classification of polymers in the groups of additional polymers and condensation polymers has been on monomer composition, because this provides an orderly approach, whereas classification based on polymer uses, such as plastics, elastomers, fibers, coatings, etc. would result in too much overlap. For example, polyamides are used not only as synthetic fibers but also as thermoplastics molding compounds and polypropylene, which is used as a thermoplastic molding compound has also found uses as a fiber-forming material.

All vinyl polymers are addition polymers. To differentiate them, the homopolymers have been classified by the substituents attached to one carbon atom of the double bone. If the substituent is hydrogen, alkyl or aryl, the homopolymers are listed under polyolefins. Olefin homopolymers with other substituents are described under *polyvinyl* compounds, except where the substituent is a nitrile, a carboxylic acid, or a carboxylic acid ester or amide. The monomers in the latter cases being derivatives of acrylic acid, the derived polymers are listed under *acrylics.* Under olefin copolymers are listed products which are produced by copolymerization of two or more monomers.

Condensation polymers are classified as polyesters, polyamides, polyurethanes, and ether polymers, based on the internal functional group being ester (–COO–), amide (–CONH–), urethane (–OCONH–), or ether (–O–). Another group of condensation polymers derived by condensation reactions with formaldehyde is described under formaldehyde resins. Polymers with special properties have been classified into three groups: heat-resistant polymers, silicones and other inorganic polymer, and functional polymers. Discussions in all cases are centered on important properties and main applications of polymers.

1.2   Part I: Addition Polymers

Addition polymers are produced in largest tonnages among industrial polymers. The most important monomers are ethylene, propylene, and butadiene. They are based on low-cost petrochemicals or natural gas and are produced by cracking or refining of crude oil. Polyethylene, polypropylene, poly(vinyl chloride), and polystyrene are the four major addition polymers and are by far the least-expensive industrial polymers on the market. In addition to these four products, a wide variety of other addition polymers are commercially available.

For addition polymers four types of polymerization processes are known: free-radical-initiated chain polymerization, anionic polymerization, cationic polymerization, and coordination polymerization (with Ziegler–Natta catalysts). By far the most extensively used process is the free-radical-initiated chain polymerization. However, the more recent development of stereo regular polymers using certain organometallic coordination compounds called Ziegler–Natta catalysts, which has added a new dimension to polymerization processes, is expected to play a more important role in coming years. The production of linear low-density polyethylene (LLDPE) is a good example. Ionic polymerization is used to a lesser extent. Thus, anionic polymerization is used mainly in the copolymerization of olefins, such as the production of styrene–butadiene elastomers, and cationic polymerization is used exclusively in the production of butyl rubber.

Different processes are used in industry for the manufacture of polymers by free-radical chain polymerization. Among them *homogeneous bulk polymerization* is economically the most attractive and yields products of higher purity and clarity. But it has problems associated with the heat of polymerization, increases in viscosity, and removal of unreacted monomer. This method is nevertheless used for the manufacture of PVC, polystyrene, and poly(methyl methacrylate). More common processes are *homogeneous solution polymerization* and *heterogeneous suspension polymerization*.

Solution polymerization is used for the manufacture of polyethylene, polypropylene, and polystyrene, but by far the most widely used process for polystyrene and PVC is suspension polymerization. In the latter process (also known as *bead, pearl,* or *granular polymerization* because of the form in which the final products may be obtained), the monomer is dispersed as droplets (0.01–0.05 cm in diameter) in water by mechanical agitation. Various types of stabilizers, which include water-soluble organic polymers, electrolytes, and water-insoluble inorganic compounds, are added to prevent agglomeration of the monomer droplets. Each monomer droplet in the suspension constitutes a small bulk polymerization system and is transformed finally into a solid bead. Heat of polymerization is quickly dissipated by continuously stirring the suspension medium, which makes temperature control relatively easy.

1.2.1 Polyolefins

1.2.1.1 Polyethylene

$$+ CH_2 - CH_2 \, +_n$$

Monomer	Polymerization	Major Uses
Ethylene	LDPE: free-radical-initiated chain polymerization	LDPE: film and sheet (55%), housewares and toys (16%), wire and cable coating (5%)
	HDPE: Ziegler–Natta or metal-oxide catalyzed chain polymerization	HDPE: bottles (40%), housewares, containers, toys (35%), pipe and fittings (10%), film and sheet (5%)

Polyethylene is the most widely used thermoplastic material and is composed of ethylene. The two main types are LDPE and high-density polyethylene (HDPE) [1].

1.2.1.1.1 Manufacturing Processes

LDPE is manufactured by polymerization of ethylene under high pressures (15,000–50,000 psi, i.e., 103–345 MPa) and elevated temperatures (200–350°C) in the presence of oxygen (0.03–0.1%) as free-radical initiator. Ethylene is a supercritical fluid with density 0.4–0.5 g/cm^3 under these conditions. Polyethylene remains dissolved in ethylene at high pressures and temperatures but separates in the lower ranges. Branch polyethylene is produced due to chain transfer to polymer. The type and extent of branching depends on the local reaction temperature and concentrations of monomer and polymer. The molecular weight distributions and the frequencies of long and short branches on polymer chains depend strongly on reactor geometry and operation. The branched products (LDPE) are less crystalline and rigid than higher density species (HDPE) made by low pressure coordination polymerization.

Linear polyethylenes are produced in solution, slurry, and increasingly, gas-phase low-pressure processes. The Phillips process developed during the mid 1950s used supported chromium trioxide catalysts in a continuous slurry process (or particle-form process) carried out in loop reactors. Earlier, Standard Oil of Indiana patented a process using a supported molybdenum oxide catalyst. The polyethylenes made by both these processes are HDPE with densities of 0.950–0.965 g/cm^3 and they are linear with very few side-chain branches and have a high degree of crystallinity.

During the late 1970s, Union Carbide developed a low-pressure polymerization process (Unipol process) capable of producing polyethylene in the gas phase that required no solvents. The process employed a chromium based catalyst. In this process (Figure 1.1) ethylene gas and solid catalysts are fed continuously to a fluidized bed reactor. The fluidized material is polyethylene powder which is produced as a result of polymerization of the ethylene on the catalyst. The ethylene, which is recycled, supplies monomer for the reaction, fluidizes the solid, and serves as a heat-removal medium. The reaction is exothermic and is normally run at temperatures 25–50°C below the softening temperatures of the polyethylene powder in the bed. This operation requires very good heat transfer to avoid hot spots and means that the gas distribution and fluidization must be uniform.

The keys to the process are active catalysts. These are special organochromium compounds on particular supports. The catalysts yield up to about 10^6 kg of polymer per kilogram of metallic chromium. Branching is controlled by the use of comonomers like propylene or 1-butene, and hydrogen is used as a chain transfer agent. The catalyst is so efficient that its concentration in the final product is negligible. The absence of a solvent and a catalyst removal step makes the process less expensive. The products marketed as linear LLDPE can be considered as linear polyethylenes having a significant number of branches (pendant alkyl groups). The linearity imparts strength, the branches impart toughness.

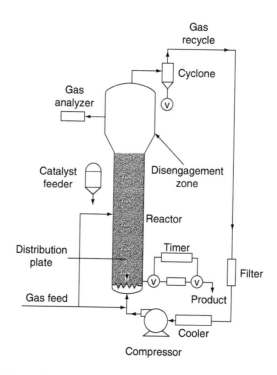

FIGURE 1.1 Union Carbide gas phase process for the production of polyethylene.

TABLE 1.1 Types of Polyethylene

Material	Chain Structure	Density (g/cm^3)	Crystallinity (%)	Process
LDPE	Branched	0.912–0.94	50	High pressure
LLDPE	Linear/less branched	0.92–0.94	50	Low pressure
HDPE	Linear	0.958	90	Low pressure

During the late 1970s, the Dow Chemical Co. also began producing polyethylene using a proprietary solution process based on Ziegler–Natta-type catalysts. Resins are made at low pressures and with lower densities in a system derived essentially from high-density resin technology (Ziegler–Natta). Higher boiling comonomers (1-hexene and 1-octene) are used to produce LLDPE having superior mechanical properties and film-drawing tendencies. A difference of the Dow solution products from gas-phase products is that they are produced in standard pellet form with any needed additives incorporated into the pellet.

Polyethylene is partially amorphous and partially crystalline. Linearity of polymer chains affords more efficient packing of molecules and hence a higher degree of crystallinity. On the other hand, side-chain branching reduces the degree of crystallinity. Increasing crystallinity increases density, stiffness, hardness, tensile strength, heat and chemical resistance, creep resistance, barrier properties, and opacity, but it reduces stress-crack resistance, permeability, and impact strength. Table 1.1 shows a comparison of three types of polyethylene.

Polyethylene has excellent chemical resistance and is not attacked by acids, bases, or salts. (It is, however, attacked by strong oxidizing agents.) The other characteristics of polyethylene which have led to its widespread use are low cost, easy process ability, excellent electrical insulation properties, toughness and flexibility even at low temperatures, freedom from odor and toxicity, reasonable clarity of thin films, and sufficiently low permeability to water vapor for many packaging, building, and agricultural applications.

Major markets for LDPE are in packaging and sheeting, whereas HDPE is used mainly in blow-molded products (milk bottles, household and cosmetic bottles, fuel tanks), and in pipe, wire, and cable applications. Ultra-high-molecular-weight polyethylene materials (see later), which are the toughest of plastics, are doing an unusual job in the textile machinery field.

1.2.1.1.2 Chlorinated Polyethylene

Low chlorination of polyethylene, causing random substitution, reduces chain order and thereby also the crystallinity. The low chlorine products (22–26% chlorine) of polyethylene are softer, more rubber-like, and more compatible and soluble than the original polyethylene. However, much of the market of such materials has been taken up by chlorosulfonated polyethylene (Hypalon, Du Pont), produced by chlorination of polyethylene in the presence of sulfur dioxide, which introduces chlorosulfonyl groups in the chain.

Chlorosulfonated LDPE containing about 27% chlorine and 1.5% sulfur has the highest elongation. Chlorosulfonated polyethylene rubbers, designated as CSM rubbers, have very good heat, ozone, and weathering resistance together with a good resistance to oils and a wide spectrum of chemicals. The bulk of the output is used for fabric coating, film sheeting, and pit liner systems in the construction industry, and as sheathing for nuclear power cables, for offshore oil rig cables, and in diesel electric locomotives.

1.2.1.1.3 Cross-Linked Polyethylene

Cross-linking polyethylene enhances its heat resistance (in terms of resistance to melt flow) since the network persists even about the crystalline melting point of the uncross-linked material. Cross-linked polyethylene thus finds application in the cable industry as a dielectric and as a sheathing material. Three main approaches used for cross-linking polyethylene are (1) radiation cross-linking, (2) peroxide cross-linking, and (3) vinyl silane cross-linking.

Radiation cross-linking is most suitable for thin sections. The technique, however, requires expensive equipment and protective measure against radiation. Equipment requirements for peroxide curing are simpler, but the method requires close control. The peroxide molecules break up at elevated temperatures, producing free radicals which then abstract hydrogen from the polymer chain to produce a polymer free radical. Two such radicals can combine and thus cross-link the two chains. It is important, however, that the peroxide be sufficiently stable thermally so that premature cross-linking does not take place during compounding and shaping operations. Dicumyl peroxide is often used for LDPE but more stable peroxides are necessary for HDPE. For cross-linking polyethylene in cable coverings, high curing temperatures, using high-pressure steam in a long curing tube set into the extrusion line, are normally employed.

Copolymers of ethylene with a small amount of vinyl acetate are often preferred for peroxide cross-linking because the latter promotes the cross-linking process. Large amounts of carbon black may be incorporated into polyethylene that is to be cross-linked. The carbon black is believed to take part in the cross-linking process, and the mechanical properties of the resulting product are superior to those of the unfilled material.

In the vinyl silane cross-linking process (*Sioplas process*) developed by Dow, an easily hydrolysable trialkoxy vinyl silane, $CH_2=CHSi(OR)_3$, is grafted onto the polyethylene chain, the site activation having been achieved with the aid of a small amount of peroxide. The material is then extruded onto the wire. When exposed to hot water or low-pressure steam, the alkoxy groups hydrolyze and then condense to form a siloxane cross-link:

$$(-Si-O-Si-)$$

The cross-linking reaction is facilitated by the use of a cross-linking catalyst, which is typically an organotin compound. There are several variations of the silane cross-linking process. In one process, compounding, grafting, and extrusion onto wire are carried out in the same extruder.

Cross-linked LDPE foam (see 'Polyolefin Foams' in Chapter 1 of *Plastics Fabrication and Recycling*) has been produced by using either chemical cross-linking or radiation cross-linking. These materials have been used in the automotive industry for carpeting, boot mats, sound deadening, and pipe insulation, and as flotation media for oil-carrying and dredging hose.

1.2.1.1.4 Linear Low-Density Polyethylene

Chemically, LLDPE can be described as linear polyethylene copolymers with alpha-olefin comonomers in the ethylene chain. They are produced primarily at low pressures and temperatures by the copolymerization of ethylene with various alpha-olefins such as butene, hexane, octane, etc., in the presence of suitable catalysts. Either gas-phase fluidized-bed reactors or liquid-phase solution-process reactors are used. (In contrast, LDPE is produced at very high pressures and temperatures either in autoclaves or tubular reactors.)

Polymer properties such as molecular weight, molecular-weight distribution (MWD), crystallinity, and density are controlled through catalyst selection and control of reactor conditions. Among the LLDPE processes, the gas-phase process has shown the greatest flexibility to produce resins over the full commercial range.

The molecular structure of LLDPE differs significantly from that of LDPE: LDPE has a highly branched structure, but LLDPE has the linear molecular structure of HDPE, though it has less crystallinity and density than the latter (see Table 1.1).

The stress-crack resistance of LLDPE is considerably higher than that of LDPE with the same melt index and density. Similar comparisons can be made with regard to puncture resistance, tensile strength, tensile elongation, and low- and high-temperature toughness. Thus LLDPE allows the processor to make a stronger product at the same gauge or an equivalent product at a reduced gauge.

LLDPE is now replacing conventional LDPE in many architectures because of the combination of favorable production economics and product performance characteristics. For many architectures (blow molding, injection molding, rotational molding, etc.) existing equipment for processing LDPE can be used to process LLDPE. LLDPE film can be treated, printed, and sealed by using the same equipment used for LDPE. Heat-sealing may, however, require slightly higher temperatures.

LLDPE films provide superior puncture resistance, high tensile strength, high impact strength, and outstanding low-temperature properties. The resins can be drawn down to thicknesses below 0.5 mil without bubble breaks. Slot-cast films combine high clarity and gloss with toughness. LLDPE films are being increasingly used in food packaging for such markets as ice bags and retail merchandise bags, and as industrial liners and garment bags.

Good flex properties and environmental stress-crack resistance combined with good low-temperature impact strength and low warp age make LLDPE suitable for injection-molded parts for housewares, closures, and lids. Extruded pipe and tubing made from LLDPE exhibit good stress-crack resistance and good bursting strength.

Blow-molded LLDPE parts such as toys, bottles, and drum liners provide high strength, flex life, and stress-crack resistance. Light-weight parts and faster blow-molding cycle times can be achieved. The combination of good high- and low-temperature properties, toughness, environmental stress-crack resistance, and good dielectric properties suit LLDPE for wire and cable insulation and jacketing applications.

A new class of linear polyethylene copolymers with densities ranging between 0.890 and 0.915 g/cm^3, known as *very low-density polyethylene* (VLDPE), was introduced commercially in late 1984 by Union Carbide. These resins are produced by copolymerization of ethylene and alpha-olefins in the presence of a catalyst.

VLDPE provides flexibility previously available only in lower-strength materials, such as ethylene–vinyl acetate (EVA) copolymer (see later) and plasticized PVC, together with the toughness and broader operating temperature range of LLDPE. Its unique combination of properties makes VLDPE suited for a wide range of applications.

Generally, it is expected that VLDPE will be widely used as an impact modifier. Tests suggest that it is suited as a blending resin for polypropylene and in HDPE films for improved tear strength.

On its own, VLDPE should find use in applications requiring impact strength, puncture resistance, and dart drop resistance combined with flexibility. The drawdown characteristics of VLDPE allow for very thin films to be formed without pinholing. Soft flexible films for disposable gloves, furniture films, and high-performance stretch and shrink film are potential markets.

1.2.1.1.5 *High-Molecular-Weight High-Density Polyethylene*

High-molecular-weight high-density polyethylene (HMW-HDPE) is defined as a linear homo-polymers or copolymer with a weight-average molecular weight ($\bar{M}_w$) in the range of approximately 200,000–500,000. HMW-HDPE resins are manufactured using predominantly two basic catalyst systems: Ziegler-type catalysts and chromium oxide-based catalysts. These catalysts produce linear polymers which can be either homopolymers when higher-density products are required or copolymers with lower density. Typical comonomers used in the latter type of products are butene, hexane, and octenes.

HMW-HDPE resins have high viscosity because of their high molecular weight. This presents problems in processing and, consequently, these resins are normally produced with broad MWD.

The combination of high molecular weight and high density imparts the HMW-HDPE good stiffness characteristics together with above-average abrasion resistance and chemical resistance. Because of the relatively high melting temperature, it is imperative that HMW-HDPE resins be specially stabilized with antioxidant and processing stabilizers. HMW-HDPE products are normally manufactured by the extrusion process; injection molding is seldom used.

The principal applications of HMW-HDPE are in film, pressure pipe, large blow-molded articles, and extruded sheet. HMW-HDPE film now finds application in T-shirt grocery sacks (with 0.6–0.9-mil thick sacks capable of carrying 30 lb of produce), trash bags, industrial liners, and specialty roll stock. Sheets 20–100 mils thick and 18–20 ft. wide are available that can be welded in situ for pond and tank liners.

HMW-HDPE piping is used extensively in gas distribution, water collection and supply, irrigation pipe, industrial effluent discharge, and cable conduit. The availability of pipe materials with significantly higher hydrostatic design stress (800 psi compared with 630 psi of the original HDPE resins) has given added impetus to their use. Large-diameter HMW-HDPE piping has found increasing use in sewer relining.

Large-blow-molded articles, such as 55-gal shipping containers, are produced. Equipment is now available to blow mold very large containers, such as 200- and 500-gal capacity industrial trash receptacles and 250-gal vessels to transport hazardous chemicals.

1.2.1.1.6 *Ultrahigh-Molecular-Weight Polyethylene*

Ultrahigh-molecular-weight polyethylene (UHMWPE) is defined by ASTM as "polyethylene with molecular weight over three million (weight average)." The resin is made by a special Ziegler-type polymerization.

Being chemically similar to HDPE, UHMWPE shows the typical polyethylene characteristics of chemical inertness, lubricity, and electrical resistance, while its very long substantially linear chains provide greater impact strength, abrasion resistance, toughness, and freedom from stress cracking. However, this very high molecular weight also makes it difficult to process the polymer by standard molding and extrusion techniques. Compression molding of sheets and ram extrusion of profiles are the normal manufacturing techniques.

Forms produced by compression molding or specialty extrusion can be made into final form by machining, sintering, or forging Standard wood-working techniques are employed for machining; sharp tools, low pressures, and good cooling are used. Forging can be accomplished by pressing a perform and billet and then forging to then final shape. Parts are also formed from compression-molded or skived sheets by heating them above 300°F ($\sim$150°C) and stamping them in typical metal-stamping equipment. Such UHMPWE items have better abrasion resistance than do steel or polyurethanes. They also have high impact strength even at very low temperatures, high resistance to cyclic fatigue and stress cracking, low coefficient friction, good corrosion and chemical resistance, good resistance to nuclear radiation, and resistance to boiling water.

Fillers such as graphite, talc, glass beads or fibers, mica, and powdered metals can be incorporated to improve stiffness or to reduce deformation and deflection under load. Resistance to abrasion and deformation can be increased by peroxide cross-linking, described earlier.

UHMWPE was first used in the textile machinery field picker blocks and throw sticks, for example. Wear strips, timing wheels, and gears made of the UHMW polymer are used in material handling, assembly, and packaging lines. Chemical resistance and lubricity of the polymer are important in its applications in chemical, food, beverage, mining, mineral processing, and paper industries. All sorts of self-unloading containers use UHMWPE liners to reduce wear, prevent sticking, and speed up the unloading cycles. The polymer provides slippery surfaces that facilitate unloading even when the product is wet or frozen.

The polymer finds applications in transportation, recreation, lumbering, and general manufacturing. Metal equipment parts in some cases are coated or replaced with UHMWPE parts to reduce wear and prevent corrosion. Sewage plants have used this polymer to replace cast-iron wear shoes and rails, bearings, and sprockets. There is even an effort to use UHMW polymer chain to replace metal chain, which is corroded by such environments.

Porous UHMW polymer is made by sintering to produce articles of varied porosity. It has found growing use for controlled-porosity battery separators. Patents have been issued on the production of ultrahigh strength, very lightweight fibers from UHMW polymer by gel spinning.

1.2.1.2 Polypropylene

$$+CH_2 - CH +_n$$
$$|$$
$$CH_3$$

Monomer	Polymerization	Major Uses
Propylene	Ziegler–Natta catalyzed chain polymerization	Fiber products (30%), housewares and toys (15%), automotive parts (15%), appliance parts (5%)

Since its conception in the late 1950s, the propylene polymerization is being revolutionized in terms of both manufacturing hardware and, more importantly, catalyst technology [2–4]. New catalyst technologies, in conjunction with state-of-the-art polymerization, have established polypropylene as the kingpin in the field of polyolefins. A brief survey of manufacturing processes and catalyst technologies are presented below.

1.2.1.2.1 *Manufacturing Processes*

Commercial production of crystalline polypropylene (PP) was first put on stream in late 1959 by Hercules in the United States, by Montecatini in Italy, and by Farbenwerke Hoechst AG in Germany. The workhorse process for commercial production of PP has been slurry polymerizations in liquid hydrocarbon diluent, for example, hexane or heptane. These are carried out either in stirred batch or continuous reactors.

High purity ($>99.5\%$) propylene is fed to the reactor containing diluent, as a suspension of solid (Ziegler–Natta) catalyst particles is metered in. The reaction is carried out at 50–80°C and 5–20 atmospheric pressure. The crystalline polymer produced is insoluble and forms a finely divided granular solid enveloping the solid catalyst particles. Monomer addition is continued until the slurry reaches 20–40% solids. Residence time varies from minutes to several hours, depending on the catalyst concentration and activity, as well as the specific reaction conditions. Molecular weight is controlled preferentially by the addition of hydrogen as the chain transfer agent.

The reactor slurry of PP is discharged to a stripping unit where the unreacted monomer flashes out for recycling. The catalyst is then deactivated and solubilized by the addition of alcohol. The bulk of the diluent, solubilized catalysts, and atactic polypropylene is solution are removed at this point by centrifuging. The crystalline polymer is purified by steam distillation and/or by water washing with surface active agents, followed by filtration and centrifuging and then drying. The dried polymer can be stored, transported, or premixed with stabilizers to be used with or without pelletization.

The efficiency of the Ziegler–Natta catalysts is of the order of 1500 g polymer formed per gram of transition metal. Residual catalyst has adverse effects on the corrosiveness, color, and light stability of the polymer, and extraction processes must be used to remove it from the product. However, by utilizing state-of-the-art, high-mileage (supported) catalyst systems, polymer yields are obtained which are several orders of magnitude higher than those obtained with first generation, Ziegler catalysts. Typically the high-mileage catalysts produce about 300,000 g of polymer per gram of transition metal. Since there are only about 3 ppm of residual metal in the polymer, catalyst removal is unnecessary. The expensive catalyst removal (deashing) steps required for the products made by earlier Ziegler catalyst systems are thus eliminated.

Propylene is readily polymerized in bulk; that is, in the liquid monomer itself. Arco, El Paso, Phillips, and Shell are practitioners of bulk processing in stirred or loop reactor systems. In either case, liquid propylene (and ethylene, if random copolymer is desired) is continuously metered to the polymerization reactor along with a high-activity/high-stereospecificity catalyst system. Polymerization temperatures are normally in the range of 45–80°C with pressures sufficient to maintain propylene in the liquid phase

(250–500 psi, that is, 1.7–3.5 MPa). Hydrogen is used for molecular weight control. The polymer slurry (approximately 30–50% solids in liquid propylene) is continuously discharged from loop reactors through a series of sequence valves into a zone maintained essentially at atmospheric pressure an containing terminating agents. Technologies also exist for production of propylene/ethylene block copolymers via bulk polymerization employing stirred or loop reactors [5].

Modern vapor-phase polymerization is represented in one form by the stirred gas-phase process originally developed by BASF and licensed by Norchem in the United States [6]. BASF process reactors contain a spiral or double-helical agitator to stir the polymer bed. Cooling of the bed is maintained by continuous injection of fresh, high-purity propylene in a liquid or partly liquefied state into the reaction zone. The unreacted propylene is removed from the top of the reactor during polymerization, condensed, and reinjected with fresh propylene. Evaporation of the unreacted propylene absorbs the heat of polymerization and also brings about intense mixing of the solid polymer particles with the gas phase. Energy costs of the process are economically attractive [7]. The diluent-free BASF process provides sufficiently high yield of polymer per unit of catalyst so that deashing is not required. Although products made in this way contain relatively high levels of titanium and aluminum residues, a unique finishing step during extrusion palletizing reduces active chlorides to an innocuous level [6].

With dramatic improvements in Ziegler–Natta catalyst technology, the *Spheripol process,* first developed by Montedison and Mitsui Petrochemical with simplified bulk (liquid propylene) process technology operating with loop reactors, is capable of directly producing a relatively large round bead with suitable density to eliminate the need for pelletizing for many applications [5]. Subsequently, Montedison and Hercules, Inc., formed a joint venture worldwide company, Himont, Inc., which assumed responsibility for all polypropylene operations and technology of the parent companies.

The Himont spheripol loop reactor process is initiated by injecting specially prepared supported catalyst and cocatalyst into liquid propylene circulated in a relatively simply high L/D ratio loop reactor, followed by monomer removal (Figure 1.2). The homopolymers so produced can be circulated through ethylene and ethylene/propylene gas phase reactors for insertion of copolymer fractions before final monomer stripping.

The Unipol low-pressure gas-phase fluidized-bed process, which was introduced by Union Carbide in 1977 for LLDPE, has also been adapted to the production of PP homopolymers and block copolymers using Shell Chemicals high-activity (Ziegler-type) catalyst technology. The Spheripol and Unipol

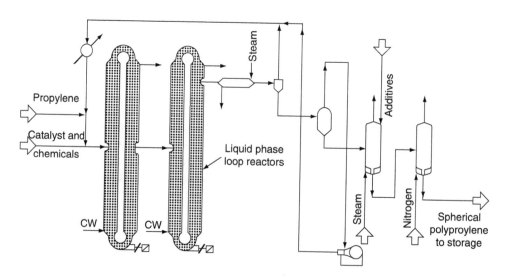

FIGURE 1.2   A simplified flow diagram of the Himont spherical loop reactor process.

processes are capable of producing polymer in crumb bead or granular forms, with the potential for direct marketing without any pelletizing finishing operation.

Regardless of the polymerization process used, the PP homo- and co-polymer must be stabilized to some degree to prevent oxidative degradation. The general practice is to incorporate a small quantity of stabilizer in the polymer prior to the first exposure to elevated temperatures of a drying operation or long-term storage. Inert-gas (nitrogen) blanketing is also used in some storage/transfer systems. Additional stabilizers, up to 1%, are added to the polymer during pelletizing. Most commercial PP compositions contain mixtures of hindered phenols and hydroperoxide decomposers or various phosphates (see Chapter 1 of *Plastics Fundamentals, Properties, and Testing*).

1.2.1.2.2 *Catalyst Technology*

The origin of sophisticated catalyst used today are to be found in the early work of Karl Ziegler (1950) and Guilio Natta (1954). In the last five decades, several distinct "generations" of catalyst technologies have emerged. The earliest commercial catalysts (*first generation*) were essentially titanium trichloride, simply prepared by reducing $TiCl_4$ with alkylaluminums to yield brown (β) $TiCl_3$, which was subsequently heated to convert it to the stereospecific purple (γ) form.

In the 1970s, improved or *second-generation catalysts* were developed. The essence of the improvement was that catalyst poisons $AlCl_3$ or $AlEtCl_2$, which are cocrystallized with or absorbed onto the $TiCl_3$ catalyst, were removed by using dialkyl ethers (especially di-*n*-butyl ether and di-isoamyl ether).

The 1980s heralded the widespread commercial implementation of supported catalysts. These *third-generation catalysts* comprise of $TiCl_4$ on a specially prepared $MgCl_2$ support. Commercially available $MgCl_2$ is converted to "active $MgCl_2$" by treating with "activating agents," which are electron donors (Lewis bases) such as ethyl benzoate, diisobutylphthalate, and phenyl triethoxy silane. These are also used in conjunction with the cocatalyst (trialkylaluminum) as a "selectivity control agent."

While the first-generation catalysts were suitable for slurry process, in which polymerization occurs in a paraffinic solvent, third-generation supported catalysts with dramatically higher activity (typically 1500 kg PP/g Ti compared to 15 kg/g Ti for first generation $TiCl_3$ catalyst) and stereo-specificity not only allowed the full exploitation of the advantages of a solventless polymerization, but also made substantial simplification of slurry process possible through elimination of atactic removal and catalyst de-ashing. These third-generation catalysts can be used in both the bulk and simplified slurry processes. They are, however, unsuitable for a gas-phase process.

The major advance offered by the later (*fourth generation*) supported catalysts is their controlled morphology, which rendered them suitable for all commercial polymerization processes. These are the only catalysts suitable for the full exploitation of the advantages of the solvent-less polymerization including the gas-phase process. These catalysts come in a variety of regular shapes, such as spherical, cubical, or cylindrical, as single particles or clusters of several particles, and are characterized by sufficiently narrow particle size distributions with the minimum of fines or coarse particles.

During polymerization, replication of the catalyst shape occurs, and hence spherical PP beads are obtained directly from the reactor, if the catalyst is spherical. By producing catalysts with a dense, spherical, and uniform particle shape it is thus possible to generate PP particles that, for many applications, can be shipped and used without a granulation or extrusion step to generate the "nibs" desired by most customers. This concept is already part of the Himont Spheripol process mentioned earlier.

While attractive in terms of economy, since the energy consuming extrusion step is obviated in the above process, there are various disadvantages such as the difficulty of homogeneously administrating additives/stabilizers, the loss of the options for molecular weight modifications via extruder operation, and the fact that the process yields particles of smaller size than what is preferred by the industry simply because such large-size particles cannot be kept in suspension during polymerization.

Montecatini has developed spherical morphology $MgCl_2$ supported catalysts for ethylene and propylene polymerization. The name of this process is "spherilene." The major advantage in such

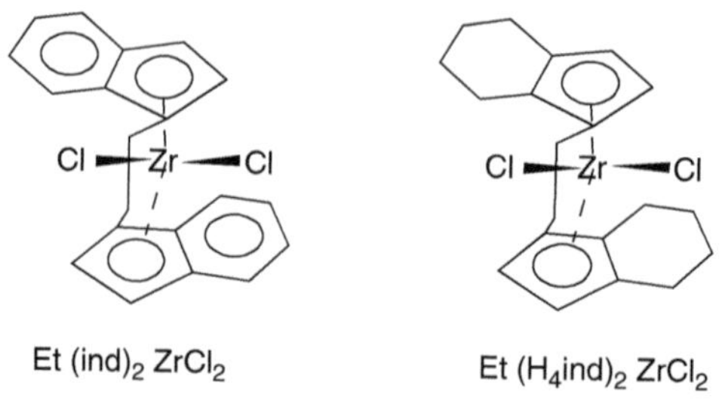

Et (ind)$_2$ ZrCl$_2$ Et (H$_4$ind)$_2$ ZrCl$_2$

FIGURE 1.3   Structures of the Brintzinger catalysts.

system is the preparation of marketable powders directly from polymerization plants and the catalyst is suitable for slurry, bulk, and gas-phase processes.

In the so-called reactor granule technology developed by Montecatini, the growing particle itself serves as the reactor within which the polymerization takes place. Thus alloys and blends previously possible by mixing and melt extrusion of polyolefins can be made with this technology directly in the reactor from the individual monomers, dramatically decreasing the producing cost by eliminating the need for compounding equipment.

A variety of specialty polyolefins and polyolefin alloys can now be made directly in the reactor taking advantage of the new technology. Examples are the *catalloy* materials from Himont, which are polyolefin alloys made by synthesis and not by the conventional route of compounding. Hivalloy is a polypropylene/polystyrene alloy made by synthesis and combines the properties of both crystalline and amorphous engineering polymers. Such materials could challenge the established positions of several thermoplastic elastomers.

Despite the domination of supported catalysts in polyolefin production, industrial research work has shifted towards new generation, *single-site,* homogeneous, Ziegler–Natta catalyst systems, the main compound of which is the Group IVB transition metallocenes (titanocene, zirconocenes, and hafnocenes) [8]. The molecular structure of the two famous Brintzinger catalysts, ethylenebis (indenyl) zirconium dichloride Et(ind)$_2$ZrCl$_2$ and ethylenebis (tetrahydroindenyl) zirconium dichloride Et(H$_4$Ind)$_2$ZrCl$_2$ are depicted in Figure 1.3. Methylalumoxane (MAO) is the most important co-catalyst that activates the group IVB metallocenes in homogeneous Ziegler–Natta polymerization.

Before the discovery of the MAO cocatalyst, the homogeneous Ziegler–Natta catalyst Cp$_2$TiCl$_2$ (Cp = cyclopentadiene) was activated with alkyl aluminum chloride, which led to poor catalyst activity. The use of MAO cocatalyst raised the catalyst activity by several orders of magnitude. Possible structures of MAO are shown in Figure 1.4. Using the chiral zirconocene catalyst in combination with the MAO cocatalyst, polypropylene can be obtained in high purity and high yield [9] even at relatively high temperatures (room temperature and above).

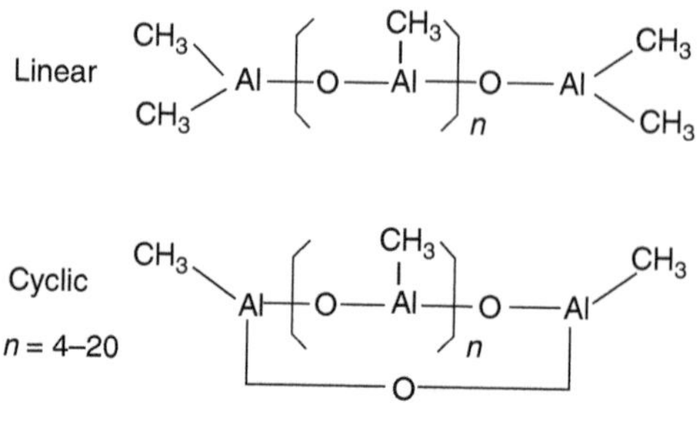

FIGURE 1.4   Possible structures of methylalumoxane. (After Kaminsky, W., Sinn, H., and Woldt, R. 1983. *Macromol. Chem. Rapid Commun.,* 4, 417.)

The zirconium catalysts are three orders of magnitude more active than their titanium counterparts. Unlike the heterogeneous Ziegler–Natta catalysts in which usually several types of active catalyst sites are present, metallocene catalysts are *single-site* in nature, that is, they have a single active catalyst site on the catalyst structure and thus they make one type of polymer, ensuring a high degree of purity.

The very high selectivity of the zirconocene single-site catalysts and their high activity, which approaches that of the $MgCl_2$-supported catalysts described in the previous section, makes them a serious contender for future processes. It can be expected [4] that the relative ease of tailoring of homogeneous catalysts compared to complicated heterogeneous systems will enable these catalysts to be further improved and exploited in terms of activity and selectivity (isotacticity, molecular weight, and distribution).

It is worthwhile to mention that homogeneous Ziegler–Natta catalysts other than metallocene-based catalysts have also been developing rapidly in recent years The catalyst precursors of these systems are nonmetallocene organometallic compounds, such as monocyclopentadienyl derivatives. A representative of monocyclopentadienyl catalysts is the *constrained geometry (CG) catalyst*. This new type of homogeneous catalyst was developed by Dow Plastics [10]. The catalyst system is based on group IVB transition metals such as Ti, covalently bonded to a cyclopentadienyl group bridged with a heteroatom such as nitrogen. The components are linked in such a way that a constrained cyclic structure is formed with Ti at the center. The bond angle between the monocyclopentadienyl group, Ti center, and heteroatom is less than 115°. The catalyst is activated by strong Lewis acid systems to a highly efficient cationic form.

The CG catalysts produce highly processable polyolefins with a unique combination of narrow MWD and long chain branches. Ethylene–octene copolymers produced with CG catalysts have useful properties across a range of densities and melting indexes. These novel copolymer families are called *polyolefin plastomers* (POP) and *polyolefin elastomers* (POE). POPs possess plastic and elastic properties while POEs containing greater than 20 wt% octene comonomer units have higher elasticity.

The CG catalyst technology is not limited to the typical selection of C_2–C_8 α-olefins, but can include higher α-olefins. The open structure of the CG catalyst significantly increases the flexibility to insert higher α-olefin comonomers into the polymer structure. This technology also allows addition of vinyl-ended polymer chains to produce long chain branching.

1.2.1.2.3 *Properties*

"Polypropylene" is not one or even 100 products. Rather it is a multidimensional range of products with properties and characteristics interdependent on the type of polymer (homopolymers, random, or block copolymer), molecular weight and molecular weight distribution, morphology and crystalline structure, additives, fillers and reinforcing fillers, and fabrication techniques.

Commercial homopolymers are usually about 90–95% isotactic, the other structures being atactic and syndiotactic (a rough measure of isotacticity is provided by the "isotactic index"—the percentage of polymer insoluble in heptane): the greater the degree of isotacticity the greater the crystallinity and hence the greater the softening point, stiffness, tensile strength, modulus, and hardness.

Although very similar to HDPE, PP has a lower density (0.90 g/cm^3) and a higher softening point, which enables it to withstand boiling water and many steam sterilizing operations. It has a higher brittle point and appears to be free from environmental stress-cracking problems, except with concentrated sulfuric acid, chromic acid, and aqua regia. However, because of the presence of tertiary carbon atoms occurring alternately on the chain backbone, PP is more susceptible to UV radiation and oxidation at elevated temperatures. Whereas PE cross-links on oxidation, PP undergoes degradation to form lower-molecular-weight products. Substantial improvement can be made by the inclusion of antioxidants, and such additives are used in all commercial PP compounds. The electrical properties of PP are very similar to those of HDPE.

Because of its reasonable cost and good combination of the foregoing properties, PP has found many applications, ranging from fibers and filament to films and extrusion coatings. A significant portion of the PP produced is used in moldings, which include luggage, stacking chairs, hospital sterilizable equipment, toilet cisterns, washing machine parts, and various auto parts, such as accelerator pedals, battery cases, dome lights, kick panels, and door frames.

Although commercial PP is a highly crystalline polymer, PP moldings are less opaque when unpigmented than are corresponding HDPE moldings, because the differences between amorphous and crystal densities are less with PP (0.85 and 0.94 g/cm^3, respectively) than with polyethylene (0.84 and 1.01 g/cm^3, respectively).

A particularly useful property of PP is the excellent resistance of thin sections to continued flexing. This has led to the production of one-piece moldings for boxes, cases, and accelerator pedals in which the hinge is an integral part of the molding.

Monoaxially oriented polypropylene film tapes have been widely used for carpet backing and for woven sacks (replacing those made from jute). Combining strength and lightness, oriented PP straps have gained rapid and widespread acceptance for packaging.

Nonoriented PP film, which is glass clear, is used mainly for textile packaging. However, biaxially oriented PP film is more important because of its greater clarity, impact strength, and barrier properties. Coated grades of this material are used for packaging potato crisps, for wrapping bread and biscuits, and for capacitor dielectrics. In these applications PP has largely replaced regenerated cellulose. (The high degree of clarity of biaxially oriented PP is caused by layering of the crystalline structures. Layering reduces the variations in refractive index across the thickness of the film, which thus reduces the amount of light scattering.)

Polypropylene, produced by an oriented extrusion process, has been uniquely successful as a fiber. Its excellent wear, inertness to water, and microorganisms, and its comparatively low cost have made it extensively used in functional applications, such as carpet backing, upholstery fabrics, and interior trim for automobiles.

Random ethylene–propylene copolymers, another important variety of polypropylene, are noted for high clarity, a lower and broader melting range than homopolymers grades, reduced flexural modulus, and higher melt strengths. They are produced by the random addition of ethylene to a polypropylene chain as it grows. The melt-flow rate of random copolymers ranges from 1 g/10 min for a blow-molding grade to 35 g/10 min for an injection-molding grade. The density is about 0.90 g/cm^3 and the notched Izod impact strength of the materials ranges from under 1 to more than 5 ft.-lb/in.

The blow-molded bottles capitalize on the good clarity provided by the random copolymer. The high gloss and very broad heat sealing range of this resin is useful in such cast-film applications as trading cards and document protectors. Polypropylene copolymers with a melt-flow rate of 35 g/10 min or above find applications in thin wall parts, usually for injection molded food packaging such as delicatessen containers or yogurt cups. Such containers have walls with a length-to-thickness ratio as high as 400:1; yet they retain the properties of top-load strength, impact resistance, and recyclability that are typical of polypropylene.

Block copolymers, preferably with ethylene, are classed as having medium, high, or extra-high impact resistance with particular respect to subzero temperatures. Block copolymers consist of a crystalline PP matrix containing segments of EPR-type elastomer and/or crystalline PE for energy impact absorption in the rubber phase [5]. The level of the ethylene comonomer as well as the size of these segments has an important bearing on the physical properties of the final block copolymer.

1.2.1.2.4 Use Pattern

Few materials are as compatible with as many processing techniques or are used in as many commercial applications as polypropylene. It is found in everything from flexible and rigid packaging to fibers and large molded parts for automotive and consumer products. Largely conforming to this diversity of applications is the fact that the material can be processed by most methods, including extrusion, extrusion coating, blown and cast film, blow molding, injection molding, and thermoforming.

Polypropylene fibers and filaments form the largest market area, which is comprised of several segments with carpeting applications being the largest; these include primary and secondary woven and nonwoven uses, carpet backging face yarns, indoor/outdoor constructions, automotive interior mats and trunk linings, and synthetic turf.

The good wickability of PP is utilized in such nonwoven applications as disposable diaper. Other applications include clothing inner liners, drapes and gowns, sleeping bags, wall coverings, wiping cloths, and tea bags. Furniture and automotive upholstery fabrics are produced from both continuous monofilament and staple fibers.

The second largest PP market is film, both oriented (OPP) and cast. Large users of OPP films are packaging for snack foods, bakery products, dry foods, candy, gum, cheese, tobacco products, and electrical capacitors. Cast (unoriented) film is used for packaging textile soft goods, cheese, snack foods, and bakery products.

Largest users of injection molded PP are in transportation, particularly automotive and truck batter cases. PP copolymers have secured about 90% of this market as a result of a drive by automotive manufacturers to reduce weight and cost. In addition to being lightweight, PP also provides outstanding resistance to creep and fatigue, high temperature rigidity, impact strength, and resistance to corrosion.

The next largest molded product market for PP is packaging, especially closures and containers. Child-resistant, tamperproof, linerless features are important design factors as also inherent chemical resistance, stress-crack resistance, and high productivity at low cost. Housewares utilize random copolymers for refrigerator and shelf-storage containers and lids. Medium-impact copolymers are used for hot/cold thermos containers, lunch boxes, coolers, and picnic ware.

Medical applications of PP such as disposable syringes, hospital trays, and labware are contingent on sterilizability, either autoclaving or radiation. Disposable syringes that are sterilized by radiation require special formulations to prevent discoloration (yellowing) or brittleness as a consequence of degradation and cross-linking.

PP finds highly successful uses in both major and small appliances. Washing machines, dish washers, tub liners, agitators, bleach and detergent dispensing units, valve and control assemblies, drain tubes, pump housings, door liners, coffee makers, hair dryers, vacuum cleaners, can openers, knife sharpeners, room humidifiers and dehumidifiers, floor and ceiling fans, and window air-conditioner units are some examples.

1.2.1.3 Polyallomer

$$\left(\!\!\begin{array}{c} CH_2-CH \\ | \\ CH_3 \end{array}\!\!\right)_{\!m} \quad \left(\!\!\begin{array}{c} CH_2-CH_2 \end{array}\!\!\right)_{\!n}$$

A proprietary polymerization process, developed in the mid 1960s by staff researchers of Eastman Chemical Products, produces copolymers of 1-olefins that give a degree of crystallinity normally obtained only with homopolymers. The term polyallomer was coined to identify the polymers manufactured by this process and to distinguish them from conventional copolymers. The polyallomer materials available today are based on block copolymers of propylene and ethylene.

Polyallomers combine the most desirable properties of both crystalline polypropylene and high-density polyethylene (HDPE) and can offer impact strengths three or four times that of polypropylene. Resistance to heat distortion is better than that of HDPE but not quite as good as that of polypropylene. Polyallomer has better abrasion resistance than polypropylene and comparable hinge-forming characteristics.

Polyallomer lightweight cases can be molded entirely in one piece. Back, front, hinges, handles, and snap clasps can be molded in at the same time in a wide range of colors. Polyallomer is thus used in such injection molded items as fishing tackle boxes, typewriter cases, gas-mask cases, and bowling-ball bags.

Shoe toes molded of polyallomer resist cracking and denting under repeated hammer blows. They withstand temperatures from $-40°C$ to $150°C$ and can withstand up to 300 pounds of force (1335 N).

Polyallomers can be processed easily on conventional molding and extruding equipment. Polypropylene color concentrates can be used to color polyallomer, since these two polymers are compatible.

1.2.1.4 Poly(Vinyl Chloride)

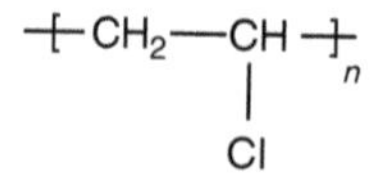

$$\left[CH_2 - \underset{\underset{Cl}{|}}{CH} \right]_n$$

Monomer	Polymerization	Major Uses
Vinyl chloride	Free-radical-initiated chain polymerization	Pipe and fittings (35%), film and sheet (15%), flooring materials (10%), wire and cable insulation (5%), automotive parts (5%), adhesives and coatings (5%)

PVC is produced by polymerization of vinyl chloride by free-radical mechanisms, mainly in suspension and emulsion, but bulk and solution processes are also employed to some extent [11–14]. (The control of vinyl chloride monomer escaping into the atmosphere in the PVC production plant has become important because cases of angiosarcoma, a rare type of liver cancer, were found among workers exposed to the monomer. This led to setting of stringent standards by governments and modification of manufacturing processes by the producers to comply with the standards.)

At processing temperatures used in practice (150–200°C), sufficient degradation may take place to render the product useless. Evidence points to the fact that dehydrochlorination occurs at an early stage in the degradation process and produces polyene structures:

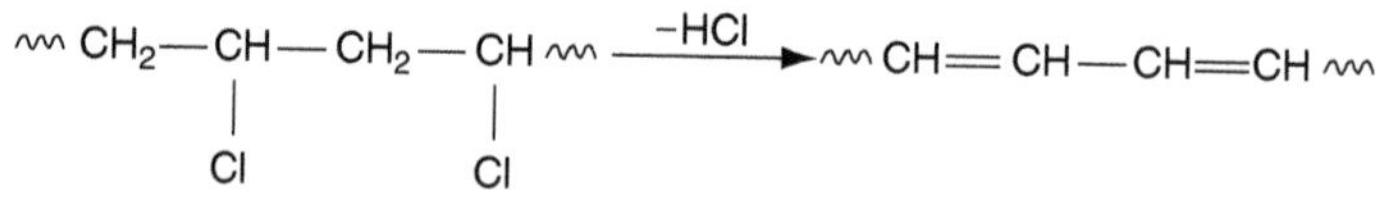

$$\sim CH_2 - \underset{\underset{Cl}{|}}{CH} - CH_2 - \underset{\underset{Cl}{|}}{CH} \sim \quad \xrightarrow{-HCl} \quad \sim CH = CH - CH = CH \sim$$

It is believed that the liberated hydrogen chloride can accelerate further decomposition and that oxygen also has an effect on the reaction. However, incorporation of certain materials known as *stabilizers* retards or moderates the degradation reaction so that useful processed materials can be obtained. Many stabilizers are also useful in improving the resistance of PVC to weathering, particularly against degradation by UV radiation.

1.2.1.4.1 *Characterization of Commercial Resins*

Commercial PVC polymers are largely amorphous with molecular weights in the range $\bar{M}_w = 100,000$–$200,000$ and $\bar{M}_n = 45,000$–$64,000$, although values may be as low as 40,000 and as high as 480,000 for $\bar{M}_w$. In practice, the ISO viscosity number is often used to characterize the molecular weight of a PVC polymer. Table 1.2 compares typical correlations between number and weight average molecular weights with ISO numbers. Most general purpose polymer for use in plasticized PVC compounds have ISO numbers of about 125. Because of processing problems the polymer used for unplasticized PVC compounds have lower molecular weights, typical ISO numbers being 105 for pipe, 85–95 for rigid sheet, and as low as 70 for injection molding compounds [14].

With commercial polymers the major differences are in the characteristics of the particle, i.e., its shape, size, size distribution, and porosity. Such differences considerably affect the processing behavior of a polymer.

Considerable effort has been expended to develop suitable process to control porosity, surface area, and diffusivity of PVC particles and this has led to great improvements over the years in the processability of PVC.

TABLE 1.2   Molecular Weight Characterization of PVC

Average Molecular Weight		ISO/R174-1961(E): Viscosity Number
Weight	Number	
54,000	26,000	57
70,000	36,000	70
1,00,000	45,500	87
1,40,000	55,000	105
2,00,000	64,000	125
2,60,000	73,000	145
3,40,000	82,000	165

Source: Matthews, G. A. R. 1972. Vinyl and Allied Polymers, Vol. 1. *Vinyl Chloride and Vinyl Acetate Polymers*, Iliffe, London, UK.

If PVC polymer particles are mixed, at room temperature, with plasticizers, the immediate product may take one of two forms. If the plasticizer quantity is insufficient to fill all the gaps between the particles, a mush will be produced. If all the voids are filled then the particles will become suspended in the excess plasticizer and a paste will be formed.

The viscosity of a PVC paste (see Plastisol Casting in Chapter 1 of *Plastics Fabrication and Recycling*) made from a fixed polymer–plasticizer ratio depends to a great extent on the particle size and size distribution. To obtain a low viscosity paste the amount of plasticizer required to fill the voids between particles should be low so that more plasticizer is available to act as a lubricant for the particles, facilitating their general mobility in suspension. Thus in general PVC pastes in which the polymer has a wide particle-size distribution (but within limits set by problems of significant plasticizer absorption even at room temperature by very small particles and settling caused by large particles) so that particles pack efficiently and leave less voids (see Figure 1.5a) will have lower viscosity than those of constant particle size (Figure 1.5b).

The use of "filler" polymers in increasing quantities in PVC paste technology is an extension of this principle. These filler polymers are made by suspension (granular, dispersion) polymerization and by themselves the particles are too large to make stable pastes. However, in the presence of much smaller paste polymer particles they remain in stable suspension. As shown in Figure 1.6, the replacement in space of a mixture of paste-polymer particles and plasticizer by a large granular polymer particle releases plasticizer which then acts as a lubricant, i.e., a viscosity depressant.

PVC pastes exhibit complex rheological behavior with the viscosities showing dependence on the shear rate and on the time of shear. A paste viscosity may increase with shear rate (*dilatancy*) or decrease (*shear thinning* or *pseudoplasticity*). Some pastes may show dilatant tendencies over one range of shear rates but

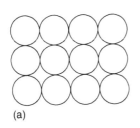
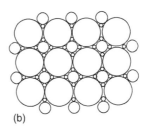

(a)   (b)

FIGURE 1.5   (a) PVC paste polymer particles with homogeneous particle size—less efficient packing. (b) PVC paste polymer particles with distribution of size—efficient packing. (After Brydson, J. A. 1982. *Plastics Materials*. Butterworth Scientific, London, UK.)

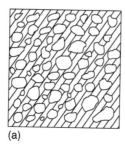
(a)

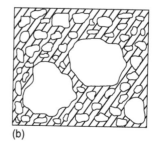
(b)

FIGURE 1.6   (a) PVC paste polymer suspended in plasticizer. (b) PVC paste containing filler polymer. Less plasticizer is required to fill voids in unit volume. (After Brydson, J. A. 1982. *Plastics Materials.* Butterworth Scientific, London, UK.)

be shear thinning over another range. The viscosities may also decrease with time of stirring (*thixotropy*) or increase with it (*rheopexy*) [14].

It has been observed that spherical particles with distribution of size giving a high degree of packing are closest to Newtonian liquids in their behavior. Spherical particles of homogeneous size, however, give shear thinning pastes. This may be due to the fact that these particles tend to aggregate at rest while shearing causes disaggregation and hence easier movement of particles. Very coarse and lumpy uneven granules do not slide past each other in pastes and tend to become more entangled as shear rate increases. Such pastes commonly show dilatant behavior.

1.2.1.4.2   Compounding Ingredients

PVC is a colorless rigid material with limited heat stability and with a tendency to adhere to metallic surfaces when heated. For these and other reasons it is necessary to compound the polymer with other ingredients to make useful products. It is possible in this way to make a wide range of products including rigid piping and soft elastic cellular materials.

A PVC compound may contain, besides the polymer, the following ingredients: stabilizers, plasticizers, extenders, lubricants, fillers, pigments, and polymeric processing aids. Other ingredients also used occasionally include impact modifiers, fire retardants, optical bleaches, and blowing agents.

1.2.1.4.3   Stabilizers

The most important class of stabilizers are the lead compounds which form lead chloride on reaction with the hydrogen chloride evolved during decomposition. Basic lead carbonate (white lead), which has a low weight cost, is more commonly used. A disadvantage of lead carbonate is that it may decompose with the evolution of carbon dioxide at higher processing temperatures and lead to a porous product. For this reason, tribasic lead sulfate, which gives PVC products with better electrical insulation properties than lead carbonate, is often used despite its somewhat higher weight cost.

Other lead stabilizers are of much more specific applications. For example, dibasic lead phthalate, which is an excellent heat stabilizer, is used in heat-resistant insulation compounds (e.g., in 150°C wire), in high-fidelity gramophone records, in PVC coatings for steel, and in expanded PVC formulation.

The use of lead compounds as stabilizers has been subjected to regulation because of its toxicity. Generally, lead stabilizers are not allowed in food-packaging PVC materials, but in most countries they are allowed in PVC pipes for conveying drinking water, with reduction in the level of use of such stabilizers.

Today the compounds of cadmium, barium, calcium, and zinc have gained prominence as PVC stabilizers. A modern stabilizing system may contain a large number of components. A typical cadmium–barium packaged stabilizer may have the following composition: cadmium–barium phenate 2–3 parts, epoxidized oils 3–5 parts, trisnonyl phenyl phosphite 1 part, stearic acid 0.5–1 part, and zinc octoate 0.5 part by weight. For flooring compositions, calcium–barium, magnesium–barium, and copper–barium compounds are sometimes used in conjunction with pentaerythritol (which has the function of reducing color by chelating iron present in asbestos).

Another group of stabilizers are the organotin compounds. Development of materials with low toxicity, excellent stabilizing performance, and improving relative price situation has led to considerable growth in the organotin market during the last decade. Though the level of toxicity of butyltins is not sufficiently low for application in contact with foodstuffs, many of the octylins, such as dioctyltin dilaurate and dioctyltin octylthiog–lycolate, meet stringent requirements for use in contact with foodstuffs. Further additions to the class of organotins include the estertins characterized by low toxicity, odor, and volatility, and the methyltins having higher efficiency per unit weight compared with the more common organotins.

Organotin compounds that are salts of alkyltin oxides with carboxylic acids (e.g., dioctyltin dilaurate) are usually called organotin carboxylates. Organotin compounds with at least one tin-sulfur bond (e.g., dioctyltin octylthioglycolates) are generally called organotin mercaptides. The latter are considered to be the most efficient and most universal heat stabilizers. The important products which are on the market have the following structures:

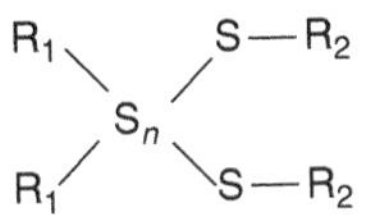

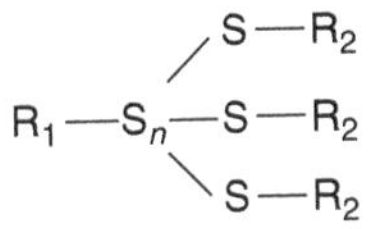

$$
\begin{matrix}
R_1 \diagdown & \diagup S-R_2 \\
& Sn \\
R_1 \diagup & \diagdown S-R_2
\end{matrix}
\qquad
R_1-Sn_n
\begin{matrix}
\diagup S-R_2 \\
-S-R_2 \\
\diagdown S-R_2
\end{matrix}
$$

Where R_1 is H_3C–, n-C_4H_9–, n-C_8H_{17}–, n-$C_{12}H_{25}$ or alkyl–O–CO–CH$_2$–CH$_2$–, and R_2 is –CH$_2$–CO–O– alkyl, –CH$_2$– CH$_2$–CO–O–alkyl, –CH$_2$–CH$_2$–O–CO–alkyl or –alkyl.

One of the most important properties—not only of the sulfur-containing tin stabilizers but also of the whole group of organotin stabilizers—is absolute crystal clarity, which can be achieved by means of a proper formulation. Clarity is required for bottles, containers, all kinds of packaging films, corrugated sheets, light panels, and also for hose, profiles, swinging doors, and transparent top coats of floor or wall coverings made from plasticized PVC.

1.2.1.4.4 *Plasticizers*

In addition to resin and stabilizers, a PVC compounds may contain ingredients such as plasticizers, extenders, lubricants, fillers, pigments, polymeric processing aids, and impact modifiers.

Plasticizers (see also Chapter 1 of *Plastics Fundamentals, Properties, and Testing*) are essentially nonvolatile solvents for PVC. At the processing temperature of about 150°C, molecular mixing occurs in a short period of time to give products of greater flexibility. Phthalates prepared from alcohols with about eight carbon atoms are by far the most important class and constitute more than 70% of plasticizers used. For economic reasons, diisooctyl phthalate (DIOP), di-2-ethylhexyl phthalate (DEHP or DOP), and the phthalate ester of the C7–C9 oxo-alcohol, often known as dialphanyl phthalate (DAP) because of the ICI trade name "Alphanol-79" for the C7-C9 alcohols, are used. DIOP has somewhat less odor, whereas DAP has the greatest heat stability. Dibutyl phthalate and diisobutyl phthalate are also efficient plasticizers and continue to be used in PVC (except in thin sheets) despite their high volatility and water extractability.

Phosphate plasticizers such as tritolyl phosphate and trixylyl phosphate are generally used where good flame resistance is required, such as in insulation and mine belting. These materials, however, are toxic and give products with poor low-temperature resistance, i.e., with a high cold flex temperature (typically, −5°C).

For applications where it is important to have a compound with good low-temperature resistance, aliphatic ester plasticizers are of great value. Dibutyl sebacate, dioctyl sebacate, and, more commonly, cheaper esters of similar effect derived from mixed acids produced by the petrochemical industry are used. These plasticizers give PVC products with a cold flex temperature of −42°C.

Esters based on allyl alcohol, such as diallyl phthalate and various polyunsaturated acrylates, have proved useful in improving adhesion of PVC to metal. They may be considered as polymerizable plasticizers. In PVC pastes they can be made to cross-link by the action of peroxides or perbenzoates when the paste is spread on to metal, giving a cured coating with a high degree of adhesion [14]. The high adhesion of these rather complex compounds has led to their development as metal-to-metal adhesives used, for example, in car manufacture. Metal coatings may also be provided from plasticized powders containing polymerizable plasticizers by means of fluidized bed or electrostatic spraying techniques.

1.2.1.4.5 Extenders

In the formulation of PVC compounds it is not uncommon to replace some of the plasticizer with an extender, a material that is not in itself a plasticizer because of its very low compatibility but that can be used in conjunction with a true plasticizer. Commercial extenders are cheaper than plasticizers and can often be used to replace up to one-third of the plasticizer without seriously affecting the properties of the compound. Three commonly employed types of extenders are chlorinated paraffin waxes, chlorinated liquid paraffinic fractions, and oil extracts [14].

1.2.1.4.6 Lubricants

In plasticized PVC it is common practice to incorporate a lubricant whose main function is to prevent sticking of the compound to processing equipment [14]. The material used should have limited compatibility such that it will sweat out during processing to form a film between the bulk of the compound and the metal surfaces of the processing equipment. The additives used for such a purpose are known as *external lubricants*.

In the United States normal lead stearate is commonly used. This material melts during processing and lubricates like wax. Also used is dibasic lead stearate, which does not melt but lubricates like graphite and improves flow properties. In Britain, stearic acid is mostly used with transparent products, calcium stearate with nontransparent products.

An unplasticized PVC formulation usually contains at least one other lubricant, which is mainly intended to improve the flow of the melt, i.e., to reduce the apparent melt viscosity. Such materials are known as *internal lubricants*. Unlike external lubricants they are reasonably compatible with the polymer and are more like plasticizers in their behavior at processing temperatures, whereas at room temperature this effect is negligible. Among materials usually classified as internal lubricants are montan wax derivatives, glyceryl monostearate, and long-chain esters such as cetyl palmitate.

1.2.1.4.7 Fillers

Fillers are commonly employed in opaque PVC compounds to reduce cost and to improve electrical insulation properties, to improve heat deformation resistance of cables, to increase the hardness of a flooring compound, and to reduce tackiness of highly plasticized compounds. Various calcium carbonates (such as whiting, ground limestone, precipitated calcium carbonate) are used for general-purpose work, china clay is commonly employed for electrical insulation, and asbestos for flooring applications. Also employed occasionally are the silicas and silicates, talc, light magnesium carbonate, and barytes (barium sulfate).

1.2.1.4.8 Pigments

Many pigments are now available commercially for use with PVC. Pigment selection should be based on the pigments ability to withstand process conditions, its effect on stabilizer and lubricant, and its effect on end-use properties, such as electrical insulation.

1.2.1.4.9 Impact Modifiers and Processing Aids

Unplasticized PVC present some processing difficulties due to its high melt viscosity; in addition, the finished product is too brittle for some applications. To overcome these problems and to produce toughening, certain polymeric additives are usually added to the PVC. These materials, known as *impact modifiers*, are generally semicompatible and often some what rubbery in nature [14]. Among the most important impact modifiers in use today are butadiene–acrylonitrile copolymers (nitrile rubber), acrylonitrile–butadiene–styrene (ABS) graft terpolymers, methacrylate–butadiene–styrene (MBS) terpolymers, chlorinated polyethylene, and some polyacrylates.

ABS materials are widely used as impact modifiers, but they cause opacity and have only moderate aging characteristics. Many grades also show severe *stress whitening*, A phenomenon advantageously employed in labeling tapes, such as Dymotape. MBS modifiers have been used such as where tough PVC materials of high clarity are desired (e.g., bottles and film). Chlorinated polyethylene has been widely used as an impact modifier where good aging properties are required.

A number of polymeric additives are also added to PVC as *processing aids*. They are more compatible with PVC and are included mainly to ensure more uniform flow and thus improve the surface finish. In chemical constitution they are similar to impact modifiers and include ABS, MBS, acrylate–methacrylate copolymers, and chlorinated polyethylene.

Typical formulations of several PVC compounds for different applications are given in Appendix A3 of *Plastics Fabrication and Recycling*.

1.2.1.4.10 *Properties and Applications*

PVC is one of the most versatile of plastics and its usage ranges widely from building construction to toys and footwear. PVC compounds are made in a wide range of formulations, which makes it difficult to make generalizations about their properties. Mechanical properties are considerably affected by the type and amount of plasticizer. Table 1.3 illustrates differences in some properties of three distinct types of compound. To a lesser extent, fillers also affect the physical properties.

Unplasticized PVC (UPVC) is a rigid material, whereas the plasticized material is tough, flexible, and even rubbery at high plasticizer loadings. Relatively high plasticizer loadings are necessary to achieve many significant improvement in impact strength. Thus incorporation of less than 20% plasticizer does not give compounds with impact strength higher than that of unplasticized grades. Lightly plasticized grades are therefore used when the ease of processing is more important than achieving good impact strength.

PVC is resistant to most aqueous solutions, including those of alkalis and dilute mineral acids. The polymer also has a good resistance to hydrocarbons. The only effective solvents appear to be those which are capable of some form of interaction with the polymer. These include cyclohexanone and tetrahydrofuran.

At ordinary temperatures, PVC compounds are reasonably good electrical insulators over a wide range of frequencies, but above the glass transition temperature their value as an insulator is limited to low-frequency applications. The volume resistivity decreases as the amount of plasticizer increases.

PVC has the advantage over other thermoplastic polyolefins of built-in fire retardancy because of its 57% chlorine content.

Copolymers of vinyl chloride with vinyl acetate have lower softening points, easier processing, and better vacuum-forming characteristics than the homopolymers. They are soluble in ketones, esters, and certain chlorinated hydrocarbons, and have generally inferior long-term heat stability.

About 90% of the PVC produced is used in the form of homopolymers, the other 10% as copolymers and terpolymers. The largest application of homopolymers PVC compounds, particularly unplasticized grades, is for rigid pipes and fittings, most commonly as suspension homopolymers of high bulk density compounded as *powder blends*.

TABLE 1.3 Properties of Three Types of PVC Compounds

Property	Unplasticized PVC	PVC + DIOP (50 Parts Per 100 Resin)	Vinyl Chloride–Vinyl Acetate Copolymer (Sheet)
Specific gravity	1.4	1.31	1.35
Tensile strength			
lbf/in.^{2}	8500	2700	7000
MPa	58	19	48
Elongation at break (%)	5	300	5
Vicat softening (°C)	80	Flexible at room temperature	70

Source: Brydtson, J. A. 1982. *Plastics Materials*, Butterworth Scientific, London, UK.

In addition to the afore said properties, UPVC has an excellent resistance to weathering. Moreover, when the cost of installation is taken into account, the material frequently turns out to be cheaper. UPVC is therefore becoming used increasingly in place of traditional materials. Important uses include translucent roof sheathing with good flame-retarding properties, window frames, and piping that neither corrodes nor rots.

As a pipe material PVC is widely used in soil pipes and for drainage and above-ground applications. Piping with diameters of up to 60 cm is not uncommon.

UPVC is now being increasingly used as a wood replacement due to its more favorable economics, taking into account both initial cost and installation. Specific applications include bench-type seating at sports stadia, window fittings, wall-cladding, and fencing. UPVC bottles have better clarity, oil resistance, and barrier properties than those made from polyethylene. Compared with glass they are also lighter, less brittle, and possess greater design flexibility. These products have thus made extensive penetration into the packaging market for fruit juices and beverages, as well as bathroom toiletry. Sacks made entirely of PVC enable fertilizers and other products to be stored outdoors.

The largest applications of plasticized PVC are wire and cable insulation and as film and sheet. PVC is of great value as an insulator for direct-current and low-frequency alternating-current carriers. It has almost completely replaced rubber in wire insulation. PVC is widely used in cable sheathing where polyethylene is employed as the insulator.

Other major outlets of plasticized PVC include floor coverings, leathercloth, tubes and profiles, injection moldings, laminates, and paste processes.

When a thin layer of plasticized PVC is laminated to a metal sheet, the bond may be strong enough that the laminate can be punched, cut, or shaped without parting the two layers. A pattern may be printed or embossed on the plastic before such fabrication. Typewriter cases and appliance cabinets have been produced with such materials.

PVC leathercloth has been widely used for many years in upholstery and trim in car applications, house furnishings, and personal apparel. The large-scale replacement of leather by PVC initiated in the 1950s and 1960s was primarily due to the greater abrasion resistance, flex resistance, and washability of PVC. Ladies handbags are frequently made from PVC leathercloth. House furnishing applications include kitchen upholstery, printed sheets, and bathroom curtains. Washable wallpapers are obtained by treating paper with PVC compounds.

Special grades of PVC are used in metal-finishing applications, for example, in stacking chairs. Calendered plasticized PVC sheet is used in making plastic rainwear and baby pants by the high-frequency welding technique. The application of PVC in mine belting is still important in terms of the actual tonnage of material consumption. All-PVC shoes are useful as beachwear and standard footwear. PVC has also proved to be an excellent abrasion-resistant material for shoe soles. PVC adhesives, generally containing a polymerizable plasticizer, are useful in many industries.

The two main applications of vinyl chloride–vinyl acetate copolymers are phonograph records and vinyl floor tiles. The copolymers contain an average of about 13% of vinyl acetate. They may be processed at lower temperatures than those used for the homopolymers. Phonograph records contain only a stabilizer, lubricant, pigment, and, possibly, an antistatic agent; there are no fillers. Preformed resin biscuits are normally molded in compression presses at about 130–140°C. The press is a flash mold that resembles a waffle iron. The faces of the mold may be nickel negatives of an original disc recording that have been made by electrodeposition.

Floor tiles contain about 30–40 parts plasticizer per 100 parts copolymer and about 400 parts filler (usually a mixture of asbestos and chalk). Processing involves mixing in an internal mixer at about 130°C, followed by calendering at 110–120°C.

1.2.1.4.11 Pastes

A PVC paste is obtained when the voids between the polymer particles in a powder are completely filled with plasticizer so that the particles are suspended in it. To ensure a stable paste, there is an upper limit and a lower limit to the order of particle size. PVC paste polymers have an average particle size of about

TABLE 1.4 Typical Formulations[a] of Three Types of PVC Pastes

Ingredient	Plastisol	Organosol	Plastigel
PVC paste polymer	100	100	100
Plasticizer (e.g., DOP)	80	30	80
Filler (e.g., china clay)	10	10	10
Stabilizer (e.g., white lead)	4	4	4
Naphtha	—	50	—
Aluminum stearate	—	—	4

[a] Parts by weight.

0.2–1.5 μm. The distribution of particle sizes also has significant influences on the flow and fluxing characteristics of the paste.

The main types of PVC pastes are plastisols, organosols, plastisols incorporating filler polymers (including the rigisols), plastigels, hot-melt compounds, and compounds for producing cellular products. Typical formulations of the first three types are shown in Table 1.4. The processing methods were described in Chapter 1 of *Plastics Fabrication and Recycling*.

Plastisols are of considerable importance commercially. They are converted into tough, rubbery products by heating at about 160°C (*gelation*). *Organosols* are characterized by the presence of a volatile organic diluent whose sole function is to reduce the paste viscosity. The diluent is removed after application and before gelling the paste.

Another method of reducing paste viscosity is to use a *filler polymer* to replace a part of the PVC paste polymer. The filler polymer particles are too large to make stable pastes by themselves, but in the presence of pastepolymer particles they remain in stable suspension. Being very much larger than paste-polymer particles and having a low plasticizer absorption, the take up large volumes in the paste and make more plasticizer available for particle lubrication, thus reducing paste viscosity. The use of filler polymers has increased considerably in recent years. Pastes prepared using filler polymers and only small quantities of plasticizer (approximately 20 parts per 100 parts of polymer) are termed *rigisols*.

The incorporation of such materials as aluminum stearate, fumed silicas, or certain bentonites gives a paste that shows pronounced Bingham Body behavior (i.e., it only flows on application of shearing stress above a certain value). Such putty-like materials (called *pastigels*), which are usually thixotropic may be hand-shaped and subsequently gelled (see 'Plastisol Casting' in Chapter 1 of *Plastics Fabrication and Recycling*).

Plastigels are often compared with *hot-melt PVC compounds*. These later materials are prepared by fluxing polymer with large quantities of plasticizers and extenders. They melt at elevated temperatures and become very fluid, so they may be poured. These compounds are extensively used for casting and prototype work.

Sigma-blade trough mixers are most commonly used for mixing PVC pastes. It is common practice to mix the dry ingredients initially with part of the plasticizer so that the shearing stresses are high enough to break down the aggregates. The remainder of the plasticizer is then added to dilute the product. The mix is preferably deaerated to remove air bubbles before final processing.

A large proportion of PVC paste is used in the manufacture of leathercloth by a *spreading technique*. A layer of paste is smeared on the cloth by drawing the latter between a roller or endless belt and a doctor blade against which there is a rolling bank of paste. The paste is gelled by passing through a heated tunnel or under infrared heaters. Embossing operations may be carried out by using patterned rollers when the gelled paste is still hot. The leathercloth is then cooled and wound up. Where it is desired that the paste should enter the interstices of the cloth, a shear-thinning (pseudo-plastic) paste is employed. Conversely, where strike-through should be minimized, a dilatant paste (viscosity increases with shear rate) is employed.

Numerous methods exist for producing cellular products (see Chapter 1 of *Plastics Fabrication and Recycling*) from PVC pastes. Closed-cell products can be made if a blowing agent such as azodiisobutyronitrile is incorporated into the paste. The paste is then heated in a mold to cause the blowing agent to decompose and the compound to gel. Since the mold is full, expansion does not take place at this stage. The unexpanded block is removed after thoroughly cooling the mold and is heated in an oven at about 100°C to produce uniform expansion.

One method of producing a flexible, substantially open-cell product is to blend the paste with carbon dioxide (either as dry ice or under pressure). The mixture is heated to volatilize the carbon dioxide to produce a foam, which is then gelled at a higher temperature.

1.2.1.4.12 Chlorinated PVC

Closely related to PVC, but with distinct properties of its own, is chlorinated poly(vinyl chloride) (CPVC), a polymer produced by postchlorination of PVC. The effect of adding more chlorine to the PVC molecule is to raise the T_g of the base resin to 115–135°C (239–275°F) range and the heat deflection temperature under load to around 115°C (239°F). CPVC also has higher tensile strength, higher modulus, and greater resistance to combustion and smoke generation.

The compounding process for CPVC is similar to that used in PVC compounding, but is more complex. Processing of CPVC is done by the traditional thermoplastic operations of extrusion, calendaring, and injection molding. However, because of the high temperature of CPVC polymer melt (205–230°C), extrusion of the resin requires chrome-plated or stainless steel dies. Injection molding of CPVC requires low-compression screws with good exit depth and molds should be stainless steel or chrome- or nickel-plated.

Traditional applications of CPVC compounds are hot and cold water distribution piping, fittings, and valves that can handle industrial liquids and chemicals. The increasing popularity of CPVC in its application in hot and cold water pipes in residential units stems from its continuous-use rating of 80°C (176°F) and 100 psi, its approval for potable water by the National Sanitation Foundation (U.S.A.), and its low heat loss along with lack of sweating and scale buildup.

The high-heat capability, low combustion ratings, and resistance to grease- and oil-induced cracking have made CPVC a strong contender for applications in automotive interiors. With its combination of excellent properties of PVC and the added ability to perform at elevated temperatures, CPVC is beginning to penetrate markets formerly dominated by metals or the more expensive engineering polymers.

1.2.1.5 Poly(Vinylidene Chloride)

$$\left[\mathrm{CH_2-\underset{\underset{Cl}{|}}{\overset{\overset{Cl}{|}}{C}}}\right]_n$$

Monomer	Polymerization	Major Uses
Vinylidene chloride	Free-radical-initiated chain polymerization	Film and sheeting for food packaging

The polymer may be prepared readily by free-radical mechanisms in bulk, emulsion, and suspension; the latter technique is usually preferred on an industrial scale. Copolymers of vinylidene chloride with vinyl chloride, acrylates, and acrylonitrile are also produced.

Since the poly(vinylidene chloride) molecule has an extremely regular structure (and the question of tacticity does not arise), the polymer is capable of crystallization. Because of the resultant close packaging and the presence of heavy chloride atoms the polymer has a high specific gravity (1.875) and a low permeability to vapors and gases. The chlorine present gives a self-extinguishing polymer. Vinylidene chloride–vinyl chloride copolymers are also self-extinguishing and possess very good resistance to a wide range of chemicals, including acids and alkalies. Because of a high degree of crystallization, even in the copolymers, high strengths are attained even though the products have relatively low molecular weights ($\sim$ 20,000–50,000).

Both poly(vinylidene chloride) and copolymers containing vinylidene chloride are used to produce flexible films and coatings. Flexible films are used extensively for food packaging because of their superior barrier resistance to water and oxygen. The coating resins are used for cellophane, polyethylene, paper, fabric, and container liner applications. Dow's trade name for a copolymer of vinylidene chloride (87%) and vinyl chloride (13%) is *Saran*. Biaxially stretched Saran film is a useful, though expensive, packaging material possessing exceptional clarity, brilliance, toughness, and impermeability to water and gases.

Vinylidene chloride–vinyl chloride copolymers are used in the manufacture of filaments. The filaments have high toughness, flexibility, durability, and chemical resistance. They find use in car upholstery, deck-chair fabrics, decorative radio grilles, doll hair, filter presses, and other applications. A flame-resisting fiber said to be a 50:50 vinylidene chloride–acrylonitrile copolymer is marketed by Courtaulds with the name Teklan.

1.2.1.6 Polytetrafluoroethylene and Other Fluoropolymers

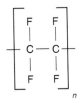

Monomer	Polymerization	Major Uses
Tetrafluoroethylene, hexafluoropropylene, chlorotrifluoroethylene, vinyl fluoride, vinylidene fluoride, perfluoroalkyl vinyl ether	Free-radical-initiated chain polymerization	Coatings for chemical process equipment, cable insulation, electrical components, nonsticking surfaces for cookware

Polytetrafluoroethylene (PTFE) was discovered in 1947. Today PTFE probably accounts for at least 85% of the fluorinated polymers and, in spite of its high cost, has a great diversity of applications [16–18]. It is produced by the free-radical chain polymerization of tetrafluoroethylene.

With a linear molecular structure of repeating –CF_2–CF_2–units, PTFE is a highly crystalline polymer with a melting point of 327°C. Density is 2.13–2.19 g/cm^3.

Commercially PTFE is made by two major processes—one leading to the so-called granular polymer, the second to a dispersion of polymers of much finer particle size and lower molecular weight.

Since the carbon–fluorine bond is very stable and since the only other bond present in PTFE is the stable C–C bond, the polymer has a high stability, even when heated above its melting point. Its upper-use temperature is given as 260°C. It is reported to give ductile rather than brittle failures at temperatures just above absolute zero, signifying a useful temperature range of more than 500°C. In many instances PTFE has been used satisfactorily as a totally enclosed gasket for considerable periods of time at temperatures well above the recommended upper-use temperature.

Because of its high crystallinity ($>$90%) and incapability of specific interaction, PTFE has exceptional chemical resistance and is *insoluble in all organic solvents*. (PTFE dissolves in certain fluorinated liquids such as perfluorinated kerosenes at temperatures approaching the melting point of the polymer.) At room temperature it is attacked only by alkali metals and, in some cases, by fluorine. Treatment with a solution of sodium metal in liquid ammonia sufficiently alters the surface of a PTFE sample to enable it to be cemented to other materials by using epoxide resin adhesives.

PTFE has good weathering resistance. The polymer is not wetted by water and has negligible water absorption (0.005%). The permeability of gases is low—the rate of transmission of water vapor is approximately half that of poly(ethylene terephthalate) and low-density polyethylene. PTFE is however degraded by high-energy radiation. For example, the tensile strength of a given sample may be halved by exposure to a dosage of 70 Mrad.

PTFE is a tough, flexible material of moderate tensile strength (2500–3800 psi, i.e., 17–21 MPa) at 23°C. Temperature has a considerable effect on its properties. It remains ductile in compression at temperatures as low as 4 K (-269°C). The creep resistance is low in comparison to other engineering plastics. Thus, even at 20°C unfilled PTFE has a measurable creep with compression loads as low as 300 psi (2.1 MPa).

The coefficient of friction is unusually low and is lower than almost any other material. The values reported in the literature are usually in the range 0.02–0.10 for polymer to polymer and 0.09–0.12 for polymer to metal. The polymer has a high oxygen index and will not support combustion.

PTFE has outstanding insulation properties over a wide range of temperatures and frequencies. The volume resistivity exceeds 10^{20} ohm-m. The power factor (<0.003 at 60 Hz and <0.0003 at 10^6 Hz) is negligible in the temperature range -60°C to $+250$°C. PTFEs low dielectric constant (2.1) is unaffected by frequency. The dielectric strength of the polymer is 16–20 kV/mm (short time on 2-mm thick sheet).

1.2.1.6.1 *Processing*

PTFE is commonly available in three forms [14] (a) granular polymers with average particle size of 300–600 μm; (b) dispersion polymers (obtained by coagulation of a dispersion) consisting of agglomerates with an average diameter of 450 μm, which are made up of primary particles 0.1 μm in diameter; and (c) dispersion (lattices) containing about 60% polymer in the form of particles with an average diameter of about 0.16 μm.

ASTM Standard Specification D1457 covers the forms sold as powders and defines seven type of PTFE molding and extrusion materials ranging from general-purpose granular resin (Type I) to presintered resin (Type VII). ASTM D4441 describes eight types of aqueous dispersion that differ, primarily, in the solids content and the amount of surfactant.

PTFE cannot be processed by the usual thermoplastics processing techniques because of its exceptionally high melt viscosity ($\sim 10^{10}$–10^{11} P at 350°C). Granular polymers are processed by press and sinter methods used in powder metallurgy. In principle, these methods involve performing the sieved powder by compressing in a mold at 1.0–3.5 tonf/in.^{2} (16–54 MPa) usually at room temperature or at 100°C, followed by sintering at a temperature above the melting point (typically at about 370°C), and then cooling.

Free-sintering of the perform in an oven at about 380°C is also satisfactory. The sintering period depends on the thickness of the sample. For example, a 0.5-in. (1.25-cm)-thick sample will need sintering for 3.5 h. Granular polymers may also be extruded, though at very low rates (2.5–16 cm/min), by screw and ram extruders. The extrudates are reasonably free of voids.

Billets in sizes varying from less than a kilogram up 700 kg (among the largest moldings made of any plastic material) are made by perform sintering. The powder is placed in a mold at or slightly above room temperature and compressed at pressures from 2000–5000 psi (14–34 MPa). After being removed from the mold, the perform is sintered by heating it unconfined in an oven at temperatures in the range of 360–380°C for times ranging from a few hours to several days. The time-temperature schedule depends on the size and shape of the billet.

The billets are often in the form of cylinder, which are then mounted on a mandrel. Sheeting is prepared by skiving, much like plywood is cut from large logs. The sheeting is cut in the thickness range of 0.025 mm (0.001 in.) to 2.5 mm (0.1 in.).

With sheet molding, the procedures are very similar to those used as described above for billet molding except for the shape and size of the molding. This process is used for sheeting above 2.5 mm (0.1 in.) up to large blocks. The latter are used for a block method skiving operation similar to that used traditionally for cellulose nitrate.

With ram extrusion, the PTFE powder is fed into a cavity at one end of a heated tube. A reciprocating ram compacts the powder and forces in into the tube. While it is being transported down the length of the tube, the PTFE is melted and coalesced. Continuous lengths of sintered rod or shapes comes out of the other end of the extruder.

PTFE moldings and extrudates may be machined without difficulty. Continuous film may be obtained by peeling a pressure-sintered ring and welding it to a similar film by heat sealing under pressure at about 350°C.

A PTFE dispersion polymer leads to products with improved tensile strength and flex life. Preforms are made by mixing the polymer with 15–25% of a lubricant and extruded. This step is followed by lubricant removal and sintering. In a typical process a mixture of PTFE dispersion polymer (83 parts) and petroleum ether (17 parts) with a 100–120°C boiling range is compacted into a perform billet which is then extruded by a vertical ram extruder. The extrudate is heated in an oven at about 105°C to remove the lubricant and is then sintered at about 380°C. Because of the need to remove the lubricant, only thin sections can be produced by this process. Thin-walled tubes with excellent fatigue resistance can be produced, or wire can be coated with very thin coatings of PTFE.

Tapes also may be made by a similar process. However, in this case the lubricant used in a nonvolatile oil. The perform is extruded in the shape of a rod, which is then passed between a pair of calendared rolls at about 60–80°C. The unsintered tape finds an important application in pipe-thread sealing. If sintered tape is required, the calendered product is first degreased by passing through boiling trichloroethylene and then sintered by passing through a salt bath. The tape made in this way is superior to that obtained by machining granular polymer molding [14].

A convenient way to apply PTFE dispersions to surfaces is by use of aqueous dispersion coatings. For this a dispersion as provided by the manufacturer may be used. There are also many coating formulations that can be purchased for use in such applications as cookware and bakeware. Coating is done as dip coating and, with thickened dispersions, as roller coating.

The dispersion should have a concentration of 45–50% PTFE and 6–9% of a wetting agent, based on the amount of PTFE. Usually a nonionic agent is used, such as Triton X-100 from Rohm and Haas or similar materials from other suppliers. The amount of PTFE deposited on each coat must be restricted so as to prevent mud cracking when the coating is dried. For drying the coating, infrared lamps or forced-convention ovens at 85–95°C are usually used. Multiple coats are used to obtain thicker films.

Drying is followed by baking and sintering to remove the wetting agent and then coalesce the PTFE. Temperatures of 260–315°C are used for the baking and 360–400°C for the sintering. The time required for each of these steps varies from several seconds to a few minutes, depending on the shape of the surface and thickness of the coating. To obtain good homogeneity, multiple dips are used, with baking and sintering after each.

Casting is used to make thin films for use in such applications as heart–lung machines, special electronic equipment and various specialty applications. Often it is done by allowing the dispersion to flow onto a support surface, usually a polished stainless-steel belt, that caries the film through the successive steps of coating, drying, baking, sintering, recoating, and finally, stripping from the belt.

Glass-coated PTFE laminates may be produced by piling up layers of glass cloth impregnated with PTFE dispersions and pressing at about 330°C. Asbestos–PTFE laminates may be produced in a similar way. The dispersions can also be used for producing filled PTFE molding material. The process typically involves stirring fillers into the dispersion, coagulating with acetone, drying at 280–290°C, and disintegrating the resulting cake of material.

1.2.1.6.2 *Applications*

The exceptional properties of PTFE make it highly useful. It is selected for a wide range of applications that affect every person. The applications fall in the five areas that require one or more of its chemical, mechanical, electrical, thermal, and surface properties. In essentially every instance, a successful application employs at least two of the outstanding properties of this polymer. However, because of its high volume cost, PTFE is not generally used to produce large objects.

In many cases it is possible to coat a metal object with a layer of PTFE to meet the particular requirement. Nonstick home cookware is perhaps the best-known example. PTFE is used for lining chutes and coating other metal objects where low coefficients of friction, chemical inertness, nonadhesive characteristics are required. The same properties make PTFE useful for coverings on rollers in food processing equipment, xerographic copiers and saw blades, coatings on snow shovels, and many other similar applications. The Alaskan oil pipeline, for example, rests on PTFE-coated steel plates. Most new bridges and tunnels use similar supports.

Because of its exceptional chemical resistance over a wide temperature range, PTFE is used in a variety of seals, gaskets, packings, valve and pump parts, and laboratory equipment.

Its excellent electrical insulation properties and heat resistance lead to its use in high-temperature wire and cable insulation, molded electrical components, insulated transformers, hermetic seals for condensers, laminates for printed circuitry, and many other electrical applications.

Reinforced PTFE applications include bushings and seals in compressor hydraulic applications, automotive applications, and pipe liners. A variety of moldings are used in aircraft and missiles and also in other applications where use at elevated temperatures is required.

An important application for aqueous dispersions of PTFE is the architectural fabric market. This product consists of fiberglass fabric coated with special forms of the aqueous dispersion. The resulting material is used as roofs in a wide variety of buildings, especially where a large area must be covered with minimum support. Notable examples of such use are the Pontiac "Silver Dome," the airport terminal in Jeddah, Saudi Arabia (where 105 acres are enclosed), and many college and university stadiums or union buildings.

Copolymers of tetrafluoroethylene were developed in attempts to provide materials with the general properties of PTFE and the melt process-ability of the more conventional thermoplastics. Two such copolymers are tetrafluoroethylene–hexafluoropropylene (TFE–HFP) copolymers (Teflon FEP resins by Du Pont; FEP stands for fluorinated ethylene propylene) with a melting point of 290°C and tetrafluoroethylene–ethylene (ETFE) copolymers (Tefzel by Du Pont) with a melting point of 270°C. These products are melt processable. A number of other fluorine containing melt processable polymers have been introduced.

Polychlorotrifluoroethylene (PCTFE) was the first fluorinated polymer to be produced on an experimental scale and was used in the United States early in World War II. It was also used in the handling of corrosive materials, such as uranium hexafluoride, during the development of the atomic bomb.

PCTFE is a crystalline polymer with a melting point of 218°C and density of 2.13 g/cm^3. The polymer is inert to most reactive chemicals at room temperature. However, above 100°C a few solvents dissolve the polymer, and a few, especially chlorinated types, swell it. The polymer is melt processable, but processing is difficult because of its high melt viscosity and its tendency to degrade, resulting in deterioration of its properties.

PCTFE is marketed by Hoechst as Hostaflon C2 and in the United States by Minnesota Mining and Manufacturing (3M) as Kel-F and by Allied Chemical as Halon. The film is sold by Allied Corp. as Aclar.

PCTFE is used in chemical processing equipment and cryogenic and electrical applications. Major applications include wafer boats, gaskets, O-rings, seals, and electrical components. PCTFE has outstanding barrier properties to gases, the PCTFE film has the lowest water-vapor transmission of any transparent plastic film. It is used in pharmaceutical packaging and other applications for its vapor barrier properties, including electroluminescent lamps.

Other melt-processable fluoroplastics include ethylene–chlorotrifluoroethylene (ECTFE) copolymer (melting point 240°C), polyvinylidene fluoride (PVDF) (melting point 170°C), and polyvinyl fluoride (PVF), which is commercially available only as film.

PVF is tough and flexible, has good abrasion and staining resistance, and has outstanding weathering resistance. It maintains useful properties over a temperature range of −70°C to 110°C. PVF can be laminated to plywood, hardboard, vinyl, reinforced polyesters, metal foils, and galvanized steel. These laminates are used in aircraft interior panels, lighting panels, wall coverings, and a variety of building applications. PVF is also used as glazing in solar energy collectors. PVF film is marketed by Du Pont as Tedlar.

PVDF is correctly named poly(1,1-difluoroethylene) and represented by $(-CF_2CH_2-)_n$. It is a hard, tough thermoplastic fluoropolymer. PVDF is prepared by free-radical initiated polymerization, either in suspension or (usually) in emulsion systems. The basic raw material for PVDF is vinylidene fluoride $(CH_2=CF_2)$, a preferred synthesis of which is dehydrochlorination of chlorodifluoroethane.

PVDF has the lowest melting point of any of the commercial fluoropolymers. As a result, its upper use temperature is limited to about 150°C, compared to values of 200°C for FEP and 260°C for PTFE. At temperatures in its useful range, however, PVDF maintains its stiffness and toughness very well.

PVDF exhibits the excellent resistance to harsh environments, characteristic of fluoropolymers. It is widely used in the chemical processing industry, in piping systems, vales, tanks (both molded and lined), and other areas where its combination of excellent mechanical properties and superb resistance to most chemicals make it an ideal material for fluid handling equipment. Increasingly important is use of PVDF as the base resin for long-life, exterior coatings on aluminum, steel, masonry, wood, and plastics.

The high dielectric loss and high dielectric constant of PVDF (8–9) both restrict its use in some electrical applications and provide superior performance in others. PVDF has very unusual piezoelectric and pyroelectric properties which are opening up many new applications with a very high value in use (see Chapter 2).

PVDFs cost, about the lowest of the melt-processible fluoropolymers, is an important advantage. Essentially all the common procedures available for thermoplastic polymers can be used with PVDF. Pennwalt Corp. is the leading producer and markets a full line of PVDF resins under the trade name of Kynar.

Perfluoroalkoxy (PFA) resins represent another class of commercially available class of melt-processable fluoroplastics. Their general chemical structure is

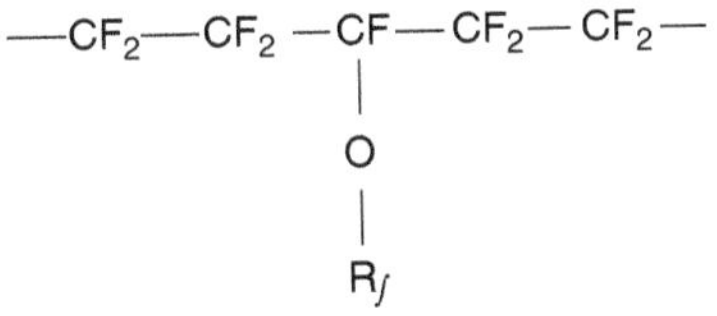

$$-CF_2-CF_2-CF-CF_2-CF_2-$$
$$|$$
$$O$$
$$|$$
$$R_f$$

where $R_f=-C_nF_{2n+1}$.

PFA resin has somewhat better mechanical properties than FEP above 150°C and can be used up to 260°C. In chemical resistance it is about equal to PTFE. PFA resin is sold by Du Pont under the Teflon trademark.

1.2.1.7 Polyisobutylene

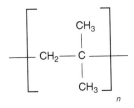

Monomer	Polymerization	Major Uses
Isobutylene	Cationic-initiated chain polymerization	Lubricating oils, sealants

High-molecular-weight polyisobutylene (PIB) is produced by cationic chain polymerization in methyl chloride solution at −70°C using aluminum chloride as the catalyst. Such polymers are currently available from Esso (Vistanex) and BASF (Oppanol).

PIB finds variety of uses. It is used as a motor oil additive to improve viscosity characteristics, as a blending agent for polyethylene to improve its impact strength and environmental stress-cracking resistance, as a base for chewing gum, as a tackifier for greases; it is also used in caulking compounds and tank linings. Because of its high cold flow, it has little use as a rubber in itself, but copolymers containing about 2% isoprene to in troduce unsaturation for cross-linking are widely used (butyl rubber; see later).

1.2.1.8   Polystyrene

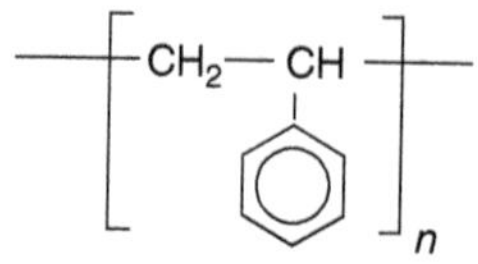

Monomer	Polymerization	Major Uses
Styrene	Free-radical-initiated chain polymerization	Packaging and containers (35%), housewares, toys and recreational equipment (25%), appliance parts (10%), disposable food containers (10%)

Polystyrene is made by bulk or suspension polymerization of styrene. Polystyrene is very low cost and is extensively used where price alone dictates. Its major characteristics [19,20] include rigidity, transparency, high refractive index, no taste, odor, or toxicity, good electrical insulation characteristics, low water absorption, and ease of coloring and processing. Polystyrene has excellent organic acid, alkali, salts, and lower alcohol resistance. It is, however, attacked by hydrocarbons, esters, ketones, and essential oils.

A more serious limitation of polystyrene in many applications is its brittleness. This limitation led to the development of rubber modified polystyrenes (containing usually 5–15% rubber), the so-called *high impact polystyrenes* (HIPS). The most commonly used are styrenebutadiene rubber and *cis*-1,4-polybutadiene.

The method of mixing the polystyrene and rubber has a profound effect on the properties of the product. Thus, much better results are obtained if the material is prepared by polymerization of styrene in the presence of the rubber rather than by simply blending the two polymers. The product of the former method contains not only polystyrene and straight rubber but also a graft copolymer in which polystyrene side chains are attached to the rubber.

Compared to straight or general-purpose polystyrenes, high-impact polystyrene materials have much greater toughness and impact strength, but clarity, softening point, and tensile strength are not as good. Expanded or foamed polystyrene (see Chapter 1 of *Plastics Fabrication and Recycling*), which has become very important as a thermal insulating material, has a low density, has a low weight cost, is less brittle, and can be made fire retarding.

End uses for all types of polystyrene are packaging, toys, and recreational products, housewares, bottles, lenses, novelties, electronic appliances, capacitor dielectrics, low-cost insulators, musical instrument reeds, light-duty industrial components, furniture, refrigeration, and building and construction uses (insulation).

Packaging is by far the largest outlet: bottle caps, small jars and other injection-molded containers, blow-molded containers, toughened polystyrene liners (vacuum formed) for boxed goods, and oriented polystyrene film for foodstuffs are some of its uses.

The second important outlet is refrigeration equipment, including door liners and inner liners (made from toughened polystyrene sheet), molding for refrigerator furnishings, such as flip lids and trays, and expanded polystyrene for thermal insulation.

Expanded polystyrene products have widely increased the market for polystyrene resin (see the section on polystyrene foams in Chapter 1 of *Plastics Fabrication and Recycling*). With as light a weight as 2 lb/ft^3 (0.032 g/cm^3), the thermal conductivity of expanded polystyrene is very low, and its cushioning value is high. It is an ideal insulation and packaging material. Common applications include ice buckets, water coolers, wall panels, and general thermal insulation applications.

Packaging uses of expanded polystyrene range from thermoformed egg boxes and individually designed shipping packages for delicate equipment, such as cameras and electronic equipment, to individual beads (which may be about 1 cm in diameter and up to 5 cm long) for use as a loose fill material in packages.

1.2.1.9 Polybutadiene (Butadiene Rubber)

$$+CH_2-CH=CH-CH_2+_n$$

Monomer	Polymerization	Major Uses
Butadiene	Ziegler–Natta-catalyzed chain polymerization	Tires and tire products (90%)

Polybutadiene is made by solution polymerization of butadiene using Ziegler–Natta catalysts. Slight changes in catalyst composition produce drastic changes in the stereoregularity of the polymer. For example, polymers containing 97–98% of *trans*-1,4 structure can be produced by using Et$_3$Al/VCl$_3$ catalyst, those with 93–94% *cis*-1,4 structure by using Et$_2$AlCl/COCl$_2$, and those with 90% 1,2-polybutadiene by using Et$_3$Al/Ti(OBu)$_4$. The stereochemical composition of polybutadiene is important if the product is to be used as a base polymer for further grafting. For example, a polybutadiene with 60% *trans*-1,4, 20% *cis*-1,4, and 20% 1,2 configuration is used in the manufacture of ABS resin.

Polybutadiene rubbers generally have a higher resilience than natural rubbers at room temperature, which is important in rubber applications. On the other hand, these rubbers have poor tear resistance, poor tack, and poor tensile strength. For this reason polybutadiene rubbers are usually used in conjunction with other materials for optimum combination of properties. For example, they are blended with natural rubber in the manufacture of truck tires and with styrene–butadiene rubber (SBR) in the manufacture of automobile tires.

Polybutadiene is also produced in low volume as specialty products. These include low-molecular-weight, liquid 1,2-polybutadienes (60–80%, 1,2 content) used as potting compounds for transformers and submersible electric motors and pumps, liquid *trans*-1,4-polybutadienes used in protective coatings inside metal cans, and hydroxy-terminated polybutadiene liquid resins for use as a binder and in polyurethane and epoxy resin formulations.

1.2.1.10 Polyisoprene

$$\left[CH_2-C=CH-CH_2\right]_n$$
$$\qquad\quad |$$
$$\qquad\ CH_3$$

Monomer	Polymerization	Major Uses
Isoprene	Ziegler–Natta-catalyzed chain polymerization	Car tires (55%), mechanical goods, sporting goods, footwear, sealants, and caulking compounds

Polyisoprene is produced by solution polymerization using Ziegler–Natta catalysts. The *cis*-1,4-polyisoprene is a synthetic equivalent of natural rubber. However, the synthetic polyisoprenes have cis contents of only about 92–96%; consequently, these rubbers differ from natural rubber in several ways. The raw synthetic polyisoprene is softer than raw natural rubber (due to a reduced tendency for a stress-induced crystallization because of the lower cis content) and is therefore more difficult to mill. On the other hand, the unvulcanized synthetic material flows more readily; this feature makes it easier to injection mold. The synthetic product is somewhat more expensive than natural rubber.

Polyisoprene rubbers are used in the construction of automobile tire carcasses and inner liners and truck and bus tire treads. Other important applications are in mechanical goods, sporting goods, footwear, sealants, and caulking compounds.

1.2.1.11 Polychloroprene

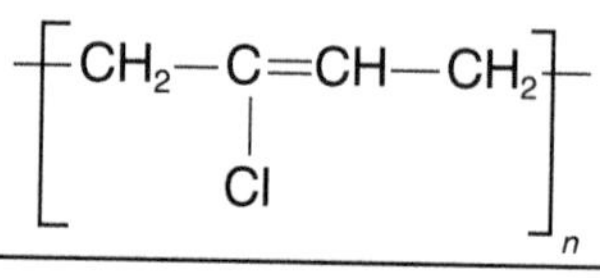

$$\left[CH_2-\underset{\underset{Cl}{|}}{C}=CH-CH_2 \right]_n$$

Monomer	Polymerization	Major Uses
Chloroprene (2-chlorobuta-1,3-diene)	Free-radical-initiated chain polymerization (mostly emulsion polymerization)	Conveyor belts, hose, seals and gaskets, wire and cable sheathing

The polychloroprenes were first marketed by Du Pont in 1931. Today these materials are among the leading special-purpose or non-tire rubbers and are well known under such commercial names as Neoprene (Du Pont), Baypren (Bayer), and Butachlor (Distagul).

A comparison of polychloroprene and natural rubber or polyisoprene molecular structures shows close similarities. However, while the methyl groups activates the double bond in the polyisoprene molecule, the chlorine atom has the opposite effect in polychloroprene. Thus polychloroprene is less prone to oxygen and ozone attack than natural rubber is. At the same time accelerated sulfur vulcanization is also not a feasible proposition, and alternative vulcanization or curing systems are necessary.

Vulcanization of polychloroprene rubbers is achieved with a combination of zinc and magnesium oxide and added accelerators and antioxidants. The vulcanizates are broadly similar to those of natural rubber in physical strength and elasticity. However, the polychloroprene vulcanizates show much better heat resistance and have a high order of oil and solvent resistance (though less resistant than those of nitrile rubber). Aliphatic solvents have little effect, although aromatic and chlorinated solvents cause some swelling. Because of chlorine chloroprene rubber is generally self-extinguishing.

Because of their greater overall durability, chloroprene rubbers are used chiefly where a combination of deteriorating effects exists. Products commonly made of chloroprene rubber include conveyor belts, V-belts, diaphragms, hoses, seals, gaskets, and weather strips. Some important construction uses are highway joint seals, pipe gaskets, and bridge mounts and expansion joints. Latexes are used in gloves, balloons, foams, adhesives, and corrosion-resistant coatings.

1.2.2 Olefin Copolymers

1.2.2.1 Styrene–Butadiene Rubber

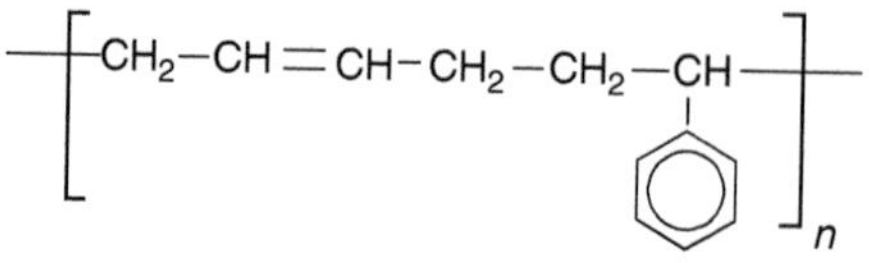

$$\left[CH_2-CH=CH-CH_2-CH_2-\underset{\underset{\bighexagon}{|}}{CH} \right]_n$$

Monomer	Polymerization	Major Uses
Styrene, butadiene	Free-radical-initiated chain polymerization (mostly emulsion polymerization)	Tires and tread (65%), mechanical goods (15%), latex (10%), automotive mechanical good (5%)

In tonnage terms SBR is the world's most important rubber [21,22]. Its market dominance is primarily due to three factors: low cost, good abrasion resistance, and a higher level of product uniformity than can be achieved with natural rubber.

Reinforcement of SBR with carbon black leads to vulcanizates which resemble those of natural rubber, and the two products are interchangeable in most applications. As with natural rubber, accelerated sulfur systems consisting of sulfur and an *activator* comprising a metal oxide (usually zinc oxide) and a fatty acid (commonly stearic acid) are used. A conventional curing system for SBR consists of 2.0 parts sulfur, 5.0 parts zinc oxide, 2.0 parts stearic acid, and 1.0 part *N-t*-butylbenzothiazole-2-sulfenide (TBBS) per 100 parts polymers.

The most important application of SBR is in car tires and tire products, but there is also widespread use of the rubber in mechanical and industrial goods. SBR latexes, which are emulsions of styrene–butadiene copolymers (containing about 23–25% styrene), are used for the manufacture of foam rubber backing for carpets and for adhesive and molded foam applications.

1.2.2.2 Nitrile Rubber

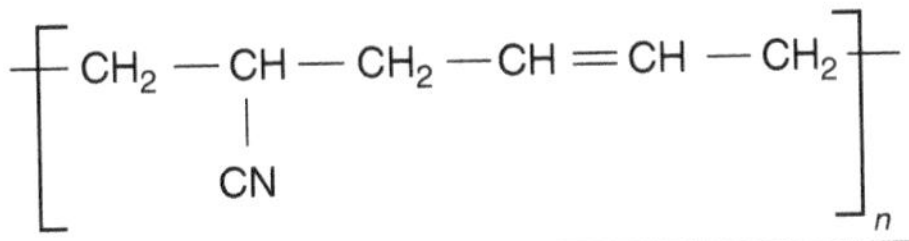

$$\left[CH_2 - CH - CH_2 - CH = CH - CH_2 \right]_n$$
$$\overset{|}{CN}$$

Monomer	Polymerization	Major Uses
Acrylonitrile, butadiene	Free-radical-initiated chain polymerization (mostly emulsion polymerization)	Gasoline hose, seals, gaskets, printing tools, adhesive, footwear

Nitrile rubber (acrylonitrile–butadiene copolymer) is a unique elastomer. The acrylonitrile content of the commercial elastomers ranges from 25% to 50% with 34% being a typical value. This non-hydrocarbon monomer imparts to the copolymer very good hydrocarbon oil and gasoline resistance. The oil resistance increases with increasing amounts of acrylonitrile in the copolymer. Nitrile rubber is also noted for its high strength and excellent resistance to abrasion, water, alcohols, and heat. Its drawbacks are poor dielectric properties and poor resistance to ozone.

Because of the diene component, nitrile rubbers can be vulcanized with sulfur. A conventional curing system consists of 2.5 parts sulfur, 5.0 parts zinc oxide, 2.0 parts stearic acid, and 0.6 parts *N-t*-butylbenzothiazole-2-sulfenamide (TBBS) per 100 parts polymer.

Nitrile rubbers (vulcanized) are used almost invariably because of their resistance to hydrocarbon oil and gasoline. They are, however, swollen by aromatic hydrocarbons and polar solvents such as chlorinated hydrocarbons, esters, and ketones.

1.2.2.3 Ethylene–Propylene–Elastomer

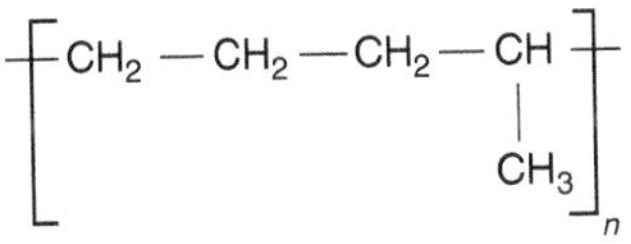

$$\left[CH_2 - CH_2 - CH_2 - CH \right]_n$$
$$\overset{|}{CH_3}$$

Monomer	Polymerization	Major Uses
Ethylene, propylene	Ziegler–Natta-catalyzed chain polymerization	Automotive parts, radiator and heater hoses, seals

Two types of ethylene–propylene elastomers are currently being produced; ethylene–propylene binary copolymers (EPM rubbers) and ethylene–propylene-diene ternary copolymers (EPDM rubbers).

Because of their saturated structure, EPM rubbers cannot be vulcanized by using accelerated sulfur systems, and the less convenient vulcanization with free-radical generators (peroxide) is required. In contrast, EPDM rubbers are produced by polymerizing ethylene and propylene with a small amount (3–8%) of a diene monomer, which provides a cross-link site for accelerated vulcanization with sulfur.

A typical vulcanization system for EPDM rubber consists of 1.5 parts sulfur, 5.0 parts zinc oxide, 1.0 part stearic acid, 1.5 parts 2-mercaptobenzothiazole (MBT), and 0.5 part tetramethylthiuram disulfide (TMTD) per 100 parts polymer.

The EPDM rubbers, though hydrocarbon, differ significantly from the diene hydrocarbon rubbers considered earlier in that the level of unsaturation in the former is much lower, giving rubbers much better heat, oxygen, and ozone resistances. Dienes commonly used in EPDM rubbers include dicyclopentadiene, ethylidene norbornene, and hexa-1,4-diene (Table 1.5). Therefore the double bonds in the polymer are either on a side chain or as part of a ring in the main chain. Hence, should the double bond become broken, the main chain will remain substantially intact, which also accounts for the greater stability of the product.

The use of EPDM rubbers for the manufacture of automobile and truck tires has not been successful, mainly because of poor tire cord adhesion and poor compatibility with most other rubbers. However, EPDM rubbers have become widely accepted as a moderately heat-resisting material with good weathering, oxygen, and ozone resistance. They find extensive use in nontire automobile applications, including body and chassis parts, car bumpers, radiator and heater hoses, weatherstrips, seals, and mats. Other applications include wire and cable insulation, appliance parts, hoses, gaskets and seals, and coated fabrics.

TABLE 1.5 Principal Diene Monomers Used in EPDM Manufacture

Monomer	Predominant Structure Present in Terpolymer
Dicyclopentadiene 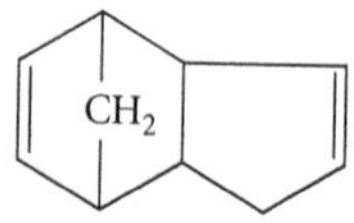	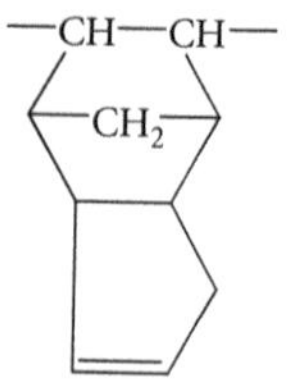
4-Ethylidenenorborn-2-ene 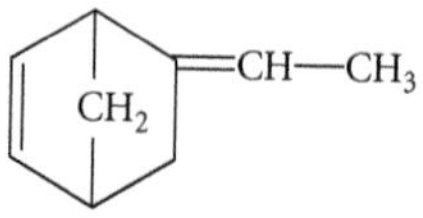	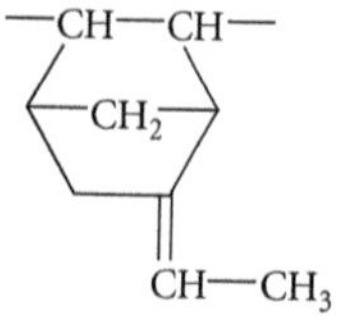
Hexa-1,4-diene $CH_2=CH-CH_2-CH=CH-CH_3$	

These rubbers are now also being blended on a large scale with polyolefin plastics, particularly polypropylene, to produce an array of materials ranging from tough plastics at one end to the thermoplastic polyolefin rubbers (see later) at the other.

1.2.2.4 Butyl Rubber

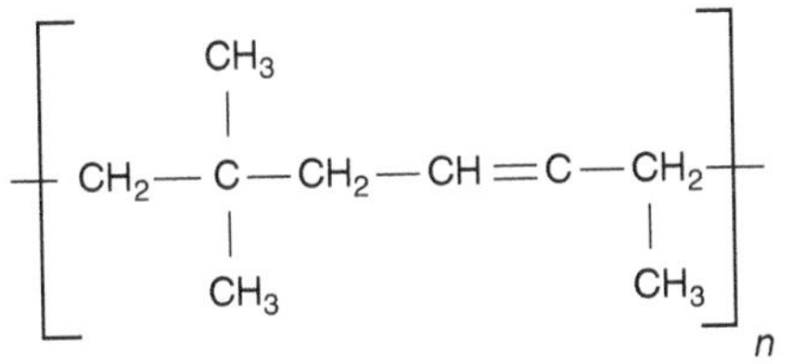

Monomer	Polymerization	Major Uses
Isobutylene, isoprene	Cationic chain polymerization of isobutylene with 0.5–2.5 mol% of isoprene	Tire inner tubes and inner liners of tubeless tires (70%), inflatable sporting goods

Although polyisobutylene described earlier is a nonrubbery polymer exhibiting high cold flow, the copolymer containing about 2% isoprene can be vulcanized with a powerful accelerated sulfur system to give rubbery polymers. Being almost saturated, they are broadly similar to the EPDM rubbers in many properties.

The most outstanding property of butyl rubber is its very low air permeability, which has led to its extensive use in tire inner tubes and liners. A major disadvantage is its lack of compatibility with SBR, polybutadiene, and natural rubber. An ozone resistant copolymer of isobutylene and cyclopentadiene has also been marketed.

1.2.2.5 Thermoplastic Elastomers

Monomer	Polymerization	Major Uses
Butadiene, isoprene, styrene	Anionic block polymerization	Footwear, automotive parts, hot-melt adhesives

Conventional rubbers are vulcanized, that is, cross-linked by primary valence bonding. For this reason vulcanized rubbers cannot dissolve or melt unless the network structure is irreversibly destroyed. These products cannot therefore be reprocessed like thermoplastics. Hence, if a polymer could be developed which showed rubbery properties at normal service temperatures but could be reprocessed like thermoplastics, it would be of great interest.

During the past three decades a few groups of materials have been developed that could be considered as being in this category. Designated as *thermoplastic elastomers,* they include (1) styrene–diene–styrene triblock copolymers; (2) thermoplastic polyester elastomers and thermoplastic polyurethane elastomers; and (3) thermoplastic polyolefin rubbers (polyolefin blends).

1.2.2.5.1 *Styrene–Diene–Styrene Triblock Elastomers*

The styrene–diene–styrene triblocks consist of a block of diene units joined at each end to a block of styrene units and are made by sequential anionic polymerization of styrene and a diene. In this way two important triblock copolymers have been produced—the styrene–butadiene–styrene (SBS) and styrene–isoprene–styrene (SIS) materials, developed by Shell.

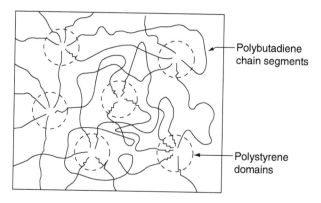

FIGURE 1.7 Schematic representation of polystyrene domain structure in styrene–butadiene–styrene triblock copolymers.

The commercial thermoplastic rubbers, Clarifex TR (Shell), are produced by joining styrene–butadiene or styrene–isoprene diblocks at the active ends by using a difunctional coupling agent. Similarly, copolymer molecules in the shape of a T, X, or star have been produced (e.g., Solprene by Phillips) by using coupling agents of higher functionality.

The outstanding behavior of these rubbers arises from the natural tendency of two polymer species to separate. However, this separation is restrained in these polymers since the blocks are covalently linked to each other. In a typical commercial SBS triblock copolymer with about 30% styrene content, the styrene blocks congregate into rigid, glassy domains which act effectively to link the butadiene segments into a network (Figure 1.7) analogous to that of cross-linked rubber.

As the SBS elastomers is heated above the glass transition temperature (T_g) of the polystyrene, the glass domains disappear and the polymer begins to flow like a thermoplastic. However, when the molten material is cooled (below T_g), the domains reharden and the material once again exhibits properties similar to those of a cross-linked rubber.

Below T_g of polystyrene the glassy domains also fulfill another useful role by acting like a reinforcing particulate filler. It is also an apparent consequence of this role that SBS polymers behave like carbon-black-reinforced elastomers with respect to tensile strength.

The styrene–diene–styrene triblock copolymers are not used extensively in traditional rubber applications because they show a high level of creep. The block copolymers can, however, be blended with many conventional thermoplastics such as polystyrene, polyethylene, and polypropylene, to obtain improved properties. A major area of use is in footwear, where blends of SBS and polystyrene have been used with remarkable success for crepe soles.

Other important uses are adhesives and coatings. A wide variety of resins, plasticizers, fillers, and other ingredients commonly used in adhesives and coatings can be used with styrene–diene–styrene triblock copolymers. With these ingredients properties such as tack, stiffness, softening temperatures, and cohesive strength can be varied over a wide range. With aliphatic resin additives the block copolymers are used for permanently tacky pressure-sensitive adhesives, and in conjunction with aromatic resins they are used for contact adhesives. The copolymers can be compounded into these adhesives by solution or hot-melt techniques.

The block copolymers are also used in a wide variety of sealants, including construction, industrial, and consumer-grade products. They are unique in that they can be formulated to produce a clear, water white product. Other applications include bookbinding and product assembly and chemical milling coatings.

1.2.2.5.2 Thermoplastic Polyester Elastomers

Because of the relatively low T_g of the short polystyrene blocks, the styrene–diene–styrene triblock elastomers have very limited heat resistance. One way to overcome this problem is to use a block copolymer in which one of the blocks is capable of crystallization and has a melting temperature well above room temperature. This approach coupled with polyester technology has led to the development of thermoplastic polyester elastomers (Hytrel by Du Pont and Arnitel by Akzo). A typical such polymer consists of relatively long sequences of *tetra*-methylene terephthalate (which segregate into rigid domains of high melting point) and softer segments of polyether (see Section 1.3.4.1.2 for more details).

Being polar polymers, these rubbers have good oil and gasoline resistance. They have a wider service temperature range than many general-purpose rubbers, and they also exhibit a high resilience, good flex fatigue resistance, and mechanical abuse resistance. These rubbers have therefore become widely accepted in such applications as seals, belting, water hose, etc.

1.2.2.5.3 Thermoplastic Polyurethane Elastomers

Closely related to the polyether–ester thermoplastic elastomers are thermoplastic polyurethane elastomers, which consist of polyurethane or urethane terminated polyurea hard blocks, with T_g above normal ambient temperature, separated by soft blocks of polyol, which in the mass are rubbery in nature (see Section 1.3.4.1.3 for more details). The main uses of thermoplastic rubbers (e.g., *Estane* by Goodrich) are for seals, bushes, convoluted bellows, and bearings.

One particular form of thermoplastic polyurethane elastomer is the elastic fiber known as Spandex. Several commercial materials of this type have been introduced, which include Lycra (Du Pont), Dorlastan (Bayer) Spanzelle (Courtaulds), and Vyrene (U.S. Rubber). Spandex fibers have higher modulus, tensile strength, and resistance to oxidation, and are able to produce finer deniers than natural rubber. They have enabled lighter-weight garments to be produced. Staple fiber blends of Spandex fiber with non-elastic fibers have also been introduced.

1.2.2.5.4 Thermoplastic Polyolefin Elastomers

Blends of EPDM rubbers with polypropylene in suitable ratios have been marketed a thermoplastic polyolefin rubbers. Their recoverable high elasticity is believed to be due to short propylene blocks in the EPDM rubber co-crystallizing with segments of the polypropylene molecules so that these crystalline domains act like cross-linking agents. Having good weathering properties, negligible toxicity hazards, and easy processability, these rubbers have received rapid acceptance for use in a large variety of nontire automotive applications such as bumper covers, headlight frames, radiator grilles, door gaskets, and other auto parts. They have also found use in cable insulation.

1.2.2.5.5 Ionic Elastomers

Ionic elastomers have been obtained using sulfonated EPDM. In one case an EPDM terpolymer consisting of 55% ethylene units, 40% propylene units, and 5% ethylidene norbornene units is sulfonated to introduce about 1 mol sulfonate groups (appended to some of the unsaturated groups of the EPDM). The sulfonic acid group is then neutralized with zinc acetate to form the zinc salt. The ionized sulfonic groups create inoic cross-links in the intermolecular structure (Figure 1.8), giving properties normally associated with a cross-linked elastomer. However, being a thermoplastic material, it can be processed in conventional molding machines. This rubber, however, has a very high melt viscosity, which must be reduced by using a polar flow promoter, such as zinc stearate, at levels of 9.5–19%.

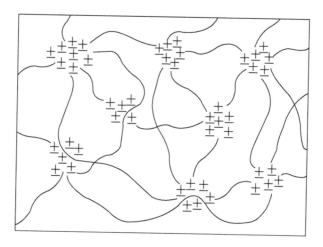

FIGURE 1.8   Schematic representation of domain structure in ionic elastomers.

1.2.2.6   Fluoroelastomers

Monomers	Polymerization	Major Uses
Vinylidene fluoride ($CH_2=CF_2$), chlorotrifluoroethylene ($CF_2=CFCl$), tetrafluoroethylene ($CF_2=CF_2$), hexafluoropropylene ($CF_3CF=CF_2$), perfluoromethyl vinyl ether (CF_2CFOCF_3)	Free-radical-initiated chain polymerization	Aerospace industry (20%), industrial equipment, wire and cable jacketing, and other insulation applications

The fluoroelastomers are a general family of fluorinated olefin copolymers. To be rubbery, the copolymer must have a flexible backbone and be sufficiently irregular in structure to be noncrystalline. A number of important fluororubbers are based on vinylidene fluoride ($CH_2=CF_2$). Several common products are listed in Table 1.6.

TABLE 1.6   Commercial Fluoroelastomers

Composition	Trade Name (manufacturer)	Remarks
Vinylidene fluoride–hexafluoro-propylene copolymer	Viton A (Du Pont)	60–85% VF. Largest tonnage production among fluororubbers
Vinylidene fluoride–hexafluoro-propylene-tetrafluoroethylene terpolymer	Fluorel (MMM) Viton B (Du Pont)	Superior resistance to heat, chemical, and solvent
Vinylidene fluoride–chlorotri-fluoroethylene copolymer	Daiel G-501 (Daikin Kogyo) Kel-F 3700 (MMM)	Superior resistance to oxidizing acids
Vinylidene fluoride-1-hydro-pentafluoropropylene–tetrafluoroethylene terpolymer	Kel-F 5500 (MMM) Tecnoflon T (Montecatini)	Superior resistance to oil, chemical, and solvent
Tetrafluoroethylene–perfluoro-(methyl vinyl ether) + cure site monomer terpolymer	Kalrez (Du Pont)	Excellent air oxidation resistance to 315°C
Tetrafluoroethylene–propylene + cure site monomer[a] terpolymer	Aflas (Asahi Glass)	Cross-linked by peroxides. Resistant to inorganic acids and bases. Cheaper alternative to Kalrez

[a] Suggested as triallyl cyanurate.

The most important of the above products are the copolymers of vinylidene fluoride and hexafluoropropylene (VF$_2$–HFP), as typified by the Du Pont product Viton A. The terpolymer of these two monmers together with tetrafluoroethylene (VF$_2$–HFP–TFE) is also of importance (e.g., Du Pont product Viton B). This terpolymer is the best among oil-resistant rubbers in its resistance to heat aging, although its actual strengths are lower than for some other rubbers. The copolymers of vinylidene fluoride and chlorotrifluoroethylene (VF$_2$–CTFE) are notable for their superior resistance to oxidizing acids such as fuming nitric acid.

Fluoroelastomers with no C–H groups will be expected to exhibit a higher thermal stability. Du Pont thus developed a terpolymer of tetrafluoroethylene, perfluoro(methyl vinyl ether) and, in small amounts, a cure site monomer of undisclosed composition. This product, marketed as Kalrez, has excellent air-oxidation resistance up to 315°C and exhibits extremely low swelling in a wide range of solvents, which is unmatched by any other commercial fluroelastomer. Table 1.7 lists a number of commercial elastomers with their main properties and applications.

1.2.2.7 Styrene–Acrylonitrile Copolymer

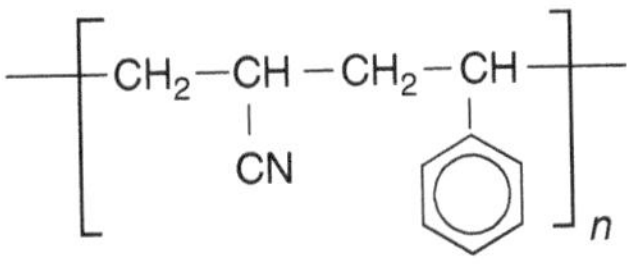

Monomer	Polymerization	Major Uses
Acrylonitrile, styrene	Free-radical-initiated chain polymerization	Components of domestic appliances, electrical equipment and car equipment, picnic ware, housewares

Because of the polar nature of the acrylonitrile molecule, styrene–acrylonitrile (SAN) copolymers have better resistance to hydrocarbons, oils, and greases than polystyrene. These copolymers have a higher softening point, a much better resistance to stress cracking and crazing, and a higher impact strength than the homopolymer polystyrene, yet they retain the transparency of the latter. The toughness and chemical resistance of the copolymer increases with the acrylonitrile content but so do the difficulty in molding and the yellowness of the resin. Commercially available SAN copolymers have 20–30% acrylonitrile content. They are produced by emulsion, suspension, or continuous polymerization.

Due to their rigidity, transparency, and thermal stability, SAN resins have found applications for dials, knobs, and covers for domestic appliances, electrical equipment, car equipment, dishwasher-safe housewares, such as refrigerator meat and vegetable drawers, blender bowls, vacuum cleaner parts, humidifier parts, plus other industrial and domestic applications with requirements more stringent than can be met by polystyrene.

SAN resins are also reinforced with glass to make dashboard components and battery cases. Over 35% of the total SAN production is used in the manufacture of ABS blends.

1.2.2.8 Acrylonitrile–Butadiene–Styrene Terpolymer

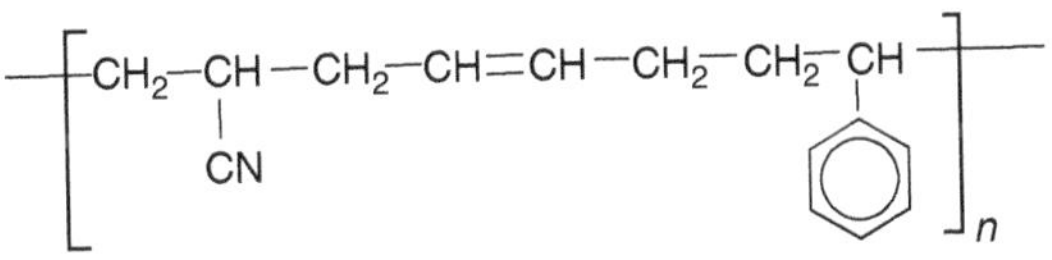

TABLE 1.7   Commercial Elastomer Products

Type	Properties	Major Uses
Natural rubber	Excellent properties of vulcanizates under conditions not demanding high levels of heat, oil, and chemical resistance	Tires, bushings, couplings, seals, footwear and belting; second place in global tonnage
Styrene–butadiene (SBR)	Reinforcement with carbon black leads to vulcanizates which resemble those of natural rubber; more effectively stabilized by antioxidants than natural rubber	Tires, tire products, footwear, wire and cable covering, adhesives; highest global tonnage
Polybutadiene	Higher resilience than similar natural rubber compounds, good low-temperature behavior and adhesion to metals, but poor tear resistance, poor tack, and poor tensile strength	Blends with natural rubber and SBR; manufacture of high-impact polystyrene
Polyisoprene	Similar to natural rubber, but excellent flow characteristics during molding	Tires, belting, footwear, flooring
Butyl	Outstanding air-retention property, but low resiliency and poor resistance to oils and fuels	Tire inner tubes, inner liners, seals, coated fabrics
Ethylene–propylene	Outstanding resistance to oxygen and ozone, poor fatigue resistance, poor tire-cord adhesion	Nontire automotive parts, radiator and heater hoses, wire and cable insulation
Nitrile	Excellent resistance to oils and solvents, poor low-temperature flexibility and poor resistance to weathering	Hoses, seals, gaskets, footwear
Chloroprene	High order of oil and solvent resistance (but less than nitrile rubber), good resistance to most chemicals, oxygen and ozone, good heat resistance, high strength but difficult to process	Mechanical automotive goods, conveyor belts, disphragms, hose, seals and gaskets
Silicone	Outstanding electrical and high-temperature properties, retention of elasticity at low temperature, poor tear and abrasion resistance, relatively high price	Gaskets and sealing rings for jet engines, ducting, sealing strips, vibration dampers and insulation equipment in aircraft, cable insulation in naval craft, potting and encepsulation
Polyurethane	High tensile strength, tear, abrasion and oil resistance, relatively high price	Oil seals, shoe soles and heels, forklift truck tires, disaphragms, fiber coatings resistant to dry cleaning, variety of mechanical goods
Chlorosulfonated polyethylene	Very good heat, ozone, and weathering resistance, good resistance to oil and a wide range of chemicals, high elasticity, good abrasion resistance	Wire and cable coating, chemical plant hose, fabric coating, film sheeting, footwear, pond liners
Polysulfide	Excellent oil, solvent, and water resistance, high impermeability to gases, low strength, unpleasant odor (particularly during processing)	Adhesive, sealants, binders, hose
Epichlorohydrin	Low air permeability, low resilience, excellent ozone resistance, good heat resistance, flame resistance, and weathering resistance	Seals, gaskets, wire and cable coating
Fluoroelastomers	Outstanding heat resistance, superior oil, chemical, and solvent resistance, highest-priced elastomer	Aerospace applications, high quality seals, and gaskets

Monomers	Polymerization	Major Uses
Acrylonitrile, butadiene styrene	Free-radical-initiated chain polymerization	Pipe and fittings (30%), automotive and appliance (15%), telephones and business machine housings

A range of materials popularly referred to as ABS polymers first became available in the early 1950s. They are formed basically from three different monomers: acrylonitrile, butadiene, and styrene, Acrylonitrile contributes chemical resistance, heat resistance, and high strength; butadiene contributes toughness, impact strength, and low-temperature property retention; styrene contributes rigidity, surface appearance (gloss), and processability. Not only may the ratios of the monomers be varied, but the way in which they can be assembled into the final polymer can also be the subject of considerable variations. The range of possible ABS-type polymers is therefore very large [23].

The two most important ways of producing ABS polymers are (1) blends of styrene–acrylonitrile copolymers with butadiene–acrylonitrile rubber, and (2) interpolymers of polybutadiene with styrene and acrylonitrile, which is now the most important type. A typical blend would consist of 70 parts styrene–acrylonitrile (70:30) copolymer and 40 parts butadiene–acrylonitrile (65:35) rubber.

Interpolymers are produced by copolymerizing styrene and acrylonitrile in the presence of polybutadiene rubber (latex) by using batch or continuous emulsion polymerization. The resultant materials are a mixture of polybutadiene, SAN copolymer, and polybutadiene grafted with styrene and acrylonitrile. The mixture is made up of three phases: a continuous matrix of SAN, a dispersed phase of polybutadiene, and a boundary layer of SAN graft.

ABS polymers are processable by all techniques commonly used with thermoplastics. They are slightly hygroscopic and should be dried 2–4 h at 180–200°F (82–93°C) just prior to processing. A dehumidifying circulating air-hopper dryer is recommended. ABS can be hot stamped, painted, printed, vacuum metallized, electroplates, and embossed. Common fabrication techniques are applicable, including sawing, drilling, punching, riveting, bonding, and incorporating metal inserts and threaded and non-threaded fasteners. The machining characteristics of ABS are similar to those of nonferrous metals.

ABS materials are superior to the ordinary styrene products and are commonly described as tough, hard, and rigid. This combination is unusual for thermoplastics. Moreover, the molded specimens generally have a very good surface finish, and this property is particularly marked with the interpolymer type ABS polymers. Light weight and the ability to economically achieve a one-step finished appearance part have contributed to large-volume applications of ABS.

Adequate chemical resistance is present in the ABS materials for ordinary applications. They are affected little by water, alkalis, weak acids, and inorganic salts. Alcohol and hydrocarbon may affect the surfaces. ABS has poor resistance to outdoor UV light; significant changes in appearance and mechanical properties will result after exposure. Protective coatings can be applied to improve resistance to UV light.

ABS materials are employed in thousand of applications, such as house-hold appliances, business machine and camera housings, telephone handsets, electrical hand tools (such as drill housings), handles, knobs, cams, bearings, wheels, gears, pump impellers, automotive trim and hardware, bathtubs, refrigerator liners, pipe and fittings, shower heads, and sporting goods. Business machines, consumer electronics, and telecommunications applications represent the fastest-growing areas for ABS. Painted, electroplated, and vacuum-metallized parts are used throughout the automotive, business machine, and electronics markets.

Multilayered laminates with an ABS outer layer can be produced by coextrusion. In this process two or three different polymers may be combined into a multilayered film or sheet. Adhesion is enhanced by cooling the extruded laminate directly from the melt rather than in a separate operation after the components of the sheet have been formed and cooled separately. In one process flows from individual extruders are combined in a *flow block* and then conveyed to a single manifold die. All the polymer streams should have approximately the same viscosity so that laminar flow can be maintained.

Multilayered films and sheets have the advantage that a chemically resistant sheet can be combined with a good barrier to oxygen and water diffusion, or a decorative glossy sheet can be placed over a tough, strong material. One commercial example is an ABS-high-impact-polystyrene sheet, which can be thermoformed (see Chapter 1 of *Plastics Fabrication and Recycling*) to make the inside door and food compartment of a refrigerator. Another example is a four-layered sheet comprising ABS, polyethylene, polystyrene, and rubber-modified polystyrene for butter and margarine packages. However, not all combinations adhere equally well, so there are limits to the design of such structures.

A variety of special ABS grades have been developed. These include high-temperature-resistant grades have been developed. (for automotive instrument panels, power tool housings), fire-retardant grades (for appliance housings, business machines, television cabinets), electroplating grades (for automotive grilles and exterior decorative trim), high-gloss, low-gloss, and matte-finish grades (for molding and extrusion applications), clear ABS grades (using methyl methacrylate as the fourth monomer), and structural foam grades (for molded parts with high strength-to-weight ratio). The structural foam grades are available for general purpose and flame-retardant applications. The cellular structure can be produced by injecting nitrogen gas into the melt just prior to entering the mold or by using chemical blowing agents in the resin (see also the section on foaming processes in Chapter 1 of *Plastics Fabrication and Recycling*).

1.2.2.9 Ethylene–Methacrylic Acid Copolymers (Ionomers)

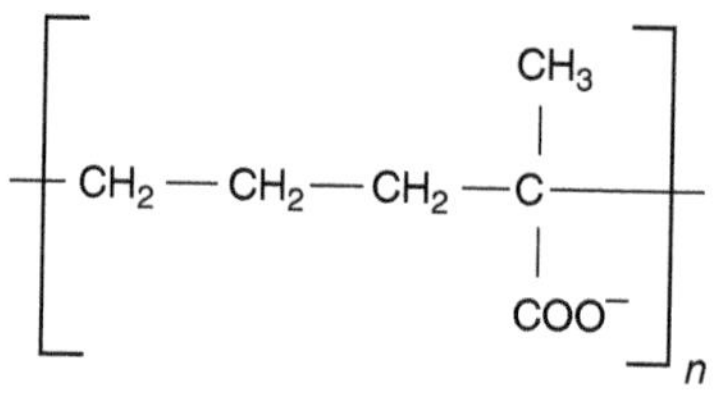

$$\left[CH_2 - CH_2 - CH_2 - \overset{\displaystyle CH_3}{\underset{\displaystyle COO^-}{C}} \right]_n$$

Monomers	Polymerization	Major Uses
Ethylene, methacrylic acid	Free-radical-initiated chain polymerization	Packaging film, golf ball covers, automotive parts, footwear

1.2.2.9.1 *Ionomers*

Ionomer is a generic name for polymers containing interchain ionic bonding [24]. Introduced in 1964 by Du Pont, they have the characteristics of both thermoplastics and thermosetting materials and are derived by copolymerizing ethylene with a small amount (1–10% in the basic patent) of an unsaturated acid, such as methacrylic acid, using the high-pressure process. The carboxyl groups in the copolymer are then neutralized by monovalent and divalent cations, resulting in some form of ionic cross-links (see Figure 1.8) which are stable at normal ambient temperatures but which reversibly break down on heating. These materials thus process the advantage of cross-linking, such as enhanced toughness and stiffness, at ambient temperatures, but they behave as linear polymers at elevated temperatures, so they may be processed and even reprocessed without undue difficulty.

Copolymerization used in making ionomers has had the effect of depressing crystallinity, although not completely eliminating it, so the materials are also transparent. Ionomers also have excellent oil and grease resistance, excellent resistance to stress cracking, and a higher water vapor permeability than does polyethylene.

The principal uses of ionomers are for film lamination and coextrusion for composite food packaging. The ionomer resin provides an outer layer with good sealability and significantly greater puncture resistance than an LDPE film. Sporting goods utilize the high-impact toughness of ionomers. Most major golf ball manufacturers use covers of durable ionomer. Such covers are virtually cut proof in normal use and retain a greater resiliency over a wider temperature range; they are superior to synthetic *trans*-polyisoprene in these respects.

Automotive uses (bumper pads and bumper guards) are based on impact toughness and paintability. In footwear applications, resilience and flex toughness of ionomers are advantages in box toes, counters, and shoe soles. Ski boot and ice skate manufacturers produce light weight outer shells of ionomers. Sheet and foamed sheet products include carpet mats, furniture tops, ski lift seat pads, boat bumpers, and wrestling mats.

Ionomers should be differentiated from polyelectrolytes and ion-exchange resins, which also contain ionic groups. Polyelectrolytes show ionic dissociation in water and are used, among other things, as thickening agents. Common examples are sodium polyacrylate, ammonium poly-methacrylate (both anionic polyacrylate) and poly(N-butyl-4-vinyl–pyridinium bromide), a cationic polyelectrolyte. Ion-exchange resins used in water softening, in chromatography, and for various industrial purposes, are cross-linked polymers containing ionic groups.

Polyelectrolytes and ion-exchange resins are, in general, intractable materials and not processable on conventional plastics machinery. In ionomers, however, the amount of ionic bonding is limited to yield useful and tractable plastics. Using this principle, manufacturers can produce rubbers which undergo ionic cross-linking to give the effect of vulcanization as they cool on emergence from an extruder or in the mold of an injection-molding machine (see the section on thermoplastic polyolefin rubbers).

1.2.3 Acrylics

Acrylic polymers may be considered structurally as derivatives of acrylic acid and its homologues. The family of acrylics includes a range of commercial polymers based on acrylic acid, methacrylic acid, esters of acrylic acid and of methacrylic acid, acrylonitrile, acrylamide, and copolymers of these compounds. By far the best known applications of acrylics are acrylic fibers and acrylonitrile copolymers such as NBR, SAN, and ABS.

1.2.3.1 Polyacrylonitrile

$$\left[CH_2 - CH \atop \underset{CN}{|} \right]_n$$

Monomer	Polymerization	Major Uses
Acrylonitrile	Free-radical-initiated chain polymerization	Fibers in apparel (70%) and house furnishings (30%)

Polyacrylonitrile and closely related copolymers have found wide use as fibers [25]. The development of acrylic fibers started in the early 1930s in Germany. In the United States they were first produced commercially about 1950 by Du Pont (Orlon) and Monsanto (Acrilan).

In polyacrylonitrile appreciable electrostatic forces occur between the dipoles of adjacent nitrile groups on the same polymer molecule. This restricts the bond rotation and leads to a stiff, rodlike structure of the polymer chain. As a result, polyacrylonitrile has a very high crystalline melting point (317°C) and is soluble in only a few solvents, such as dimethylformamide and dimethylacetamide, and in concentrated aqueous solutions of inorganic salts, such as calcium thiocyanate, sodium perchlorate, and zinc chloride. Polyacrylonitrile cannot be melt processed because its decomposition temperature is close to the melting point. Fibers are therefore spun from solution by either wet or dry spinning (see Chapter 1 of *Plastics Fabrication and Recycling*).

Fibers prepared from straight polyacrylonitrile are difficult to dye. To improve dyeability, manufacturers invariably add to monomer feed minor amounts of one or two comonomers, such as methyl acrylate, methyl methacrylate, vinyl acetate, and 2-vinyl–pyridine. Small amounts of ionic monomers (sodium styrene sulfonate) are often included for better dyeability. Modacrylic fibers are composed of 35–85% acrylonitrile and contain comonomers, such as vinyl chloride, to improve fire retardancy.

Acrylic fibers are more durable than cotton, and they are the best alternative to wool for sweaters. A major portion of the acrylic fibers produced are used in apparel (primarily hosiery). Other uses include pile fabrics (for simulated fur), craft yarns, blankets, draperies, carpets, and rugs.

1.2.3.2   Polyacrylates

$$\left[\!\!\begin{array}{c} CH_2 - CH \\ | \\ COOR \end{array}\!\!\right]_n$$

Monomer	Polymerization	Major Uses
Acrylic acid esters	Free-radical-initiated chain polymerization	Fiber modification, coatings, adhesives, paints

The properties of acrylic ester polymers depend largely on the type of alcohol from which the acrylic acid ester is prepared [26]. Solubility in oils and hydrocarbons increases as the length of the side chain increases. The lowest member of the series, poly(methyl acrylate), has poor low-temperature properties and is water sensitive. It is therefore restricted to such applications as textile sizes and leather finishes. Poly(ethyl acrylate) is used in fiber modifications and in coatings; and poly(butyl acrylate) and poly (2-ethylhexyl acrylate) are used in the formulation of paints and adhesives.

The original acrylate rubbers first introduced in 1948 by B. F. Goodrich and marketed as Hycar 4021 were a copolymer of ethyl acrylate with about 5% of 2-chloroethyl vinyl either (CH_2=CH–O–CH_2–CH_2Cl) acting as a cure site monomer. Such polymers are vulcanized through the chlorine atoms by amines (such as triethylenetetramine), which are, however, not easy to handle. Therefore, 2-chloroethyl vinyl ether has been replaced with other cure site monomers, such as vinyl and allyl chloroacetates; the increased reactivity of the chlorine in these monomers permits vulcanization with ammonium benzoate (which decomposes on heating to produce ammonia, the actual cross-linking agent) rather than amines.

Acrylate rubbers have good oil resistance. In heat resistance they are superior to most rubbers, exceptions being the fluororubbers, the fluororubbers, the silicones, and the fluorosilicones. It is these properties which account for the major use of acrylate rubbers, i.e., in oil seals for automobiles. They are, however, inferior in low-temperature properties.

A few acrylate rubbers (such as Hycar 2121X38) are based on butyl acrylate. These materials are generally copolymers of butyl acrylate and acrylonitrile ($\sim$10%) and may be vulcanized with amines. They have improved low-temperature flexibility compared to ethyl acrylate copolymers but swell more in aromatic oils.

Ethylene copolymers with acrylates represent a significant segment of the ethylene copolymer market, as many LDPE producers use copolymerization as a strategy to obtain products more resistant to displacement by HDPE and LLDPE. Ethylene copolymers with methyl methacrylate and ethyl, butyl, and methyl acrylates are similar to EVA copolymers in properties (discussed later) but have improved thermal stability during extrusion and increased low-temperature flexibility.

The commercial products in this category generally contain 15–30% of the acrylate or methacrylate comonomer. Applications include medical packaging, disposable gloves, hoses, tubing, gaskets, cable insulation, and squeeze toys. Use of ethylene–ethyl acrylate copolymers for making vacuum cleaner hoses demonstrates the increased flexibility and long flex life that is possible with such materials.

Copolymers in which the acrylate monomer is the major component are useful as ethylene–acrylate elastomers (trade name: Vamac). These are terpolymers containing a small amount of an alkenoic acid to introduce sites (C=C) for subsequent cross-linking via reaction with primary diamines [see Equation 1.34 in Chapter 1 of *Plastics Fundamentals, Properties, and Testing*]. These elastomers have excellent oil resistance and stability over a wide temperature range ($-50°C$ to $200°C$), being superior to chloroprene and nitrile rubbers. Although not superior to silicone and fluoroelastomers, they are less costly; uses include automotive (hydraulic systems seals, hoses) and wire and cable insulation.

1.2.3.3 Polymethacrylates

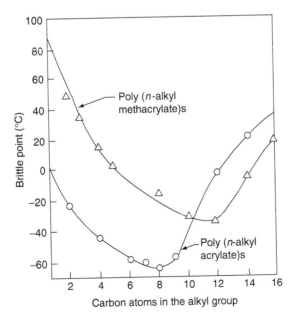

A large number of alkyl methacrylates, which may be considered as esters of poly(methacrylic acid), have been prepared [26]. By far the most important of these polymers is poly(methyl methacrylate), which is an established major plastics material. As with other linear polymers, the mechanical and thermal properties of polymethacrylates are largely determined by the intermolecular attraction, spatial symmetry, and chain stiffness.

As the size of the ester alkyl group increases in a series of poly(n-alkyl methacrylate)s, the polymer molecules become spaced further apart and the intermolecular attraction is reduced. Thus, as the length of the side chain increases, the softening point decreases, and the polymers become rubbery at progressively lower temperatures (Figure 1.9). However, when the number of carbon atoms in the side chain exceeds 12, the polymers become less rubbery, and the softening point, brittle point, and other

FIGURE 1.9 Brittle points of poly(n-alkyl acrylate)s and poly(n-alkyl methacrylate)s. (After Rehberg, C. E. and Fisher, C. H. 1948. *Ind. Eng. Chem.*, 40, 1431.)

TABLE 1.8 Vicat Softening Points of Polymethacrylates Derived from Monomers of Type $CH_2=C(CH_3)COOR$

R–	Softening Point (°C)
$CH_3–$	119
$CH_3–CH_2–$	81
$CH_3–CH_2–CH_2–$	55
$CH_3–CH_2–CH_2–CH_2–$	30
$CH_3–CH_2–CH_2–CH_2–CH_2–$	—[a]
$(CH_3)_2CH–$	88
$(CH_3)_2CH–CH_2–$	67
$(CH_3)_3C–$	104
$(CH_3)_2CH–CH_2–CH_2$	46
$(CH_3)_3C–CH_2–$	115
$(CH_3)_3C–CH–$ 　　　　　$\mid$ 　　　　CH_3	119

[a] Too rubbery for testing.

properties related to the glass transition temperature rise with an increase in chain length (Table 1.8 and Table 1.9). As with the polyolefins, this effect is due to side-chain crystallization.

Poly(alkyl methacrylate)s in which the alkyl group is branched have higher softening points (see Table 1.8) and are harder than their unbranched isomers. This effect is not simply due to the better packing possible with the branched isomers. The lumpy branched structures impede rotation about the carbon–carbon bond on the main chain, thus contributing to stiffness of the molecule and consequently a higher transition temperature. Similarly, since the α-methyl group in polymethacrylates reduces chain flexibility, the lower polymethacrylates have higher softening points than the corresponding polyacrylates do.

Poly(methyl methacrylate (PMMA) is by far the predominant polymethacrylate used in rigid applications because it has crystal clear transparency, excellent weatherability (better than most other plastics), and a useful combination of stiffness, density, and moderate toughness. The glass transition temperature of the polymer is 105°C (221°F), and the heat deflection temperatures range from 75 to 100°C (167–212°F). The mechanical properties of PMMA can be further improved by orientation of heat-cast sheets.

PMMA is widely used for signs, glazing, lighting, fixtures, sanitary wares, solar panels, and automotive tail and stoplight lenses. The low index of refraction (1.49) and high degree of uniformity make PMMA an excellent lens material for optical applications.

Methyl methacrylate has been copolymerized with a wide variety of other monomers, such as acrylates, acrylonitrile, styrene, and butadiene. Copolymerization with styrene gives a material with improved melt-flow characteristics. Copolymerization with either butadiene or acrylonitrile, or blending PMMA with SBR, improves impact resistance. Butadiene–methyl methacrylate copolymer has been used in paper and board finishes.

TABLE 1.9 Glass Transition Temperatures of Polymethacrylates

Ester Group	T_g (°C)
Methyl	105
Ethyl	65
n-Butyl	20
n-Decyl	−70
n-Hexadecyl	−9

Higher *n*-alkyl methacrylate polymers have commercial applications. Poly(*n*-butyl-), poly(*n*-octyl-) and poly(*n*-nonyl methacrylate)s are used as leather finishes; poly(lauryl methacrylate) is used to depress the pour point and improve the viscosity-temperature property of lubricating oils.

Mention may also be made here of the 2-hydroxyethyl ester of methacrylic acid, which is the monomer used for soft contact lenses. Copolymerization with ethylene glycol dimethacrylate produces a hydrophilic network polymer (a *hydrogel*). Hydrogel polymers are brittle and glassy when dry but become soft and plastic on swelling in water.

Terpolymers based on methyl methacrylate, butadiene, and styrene (MBS) are being increasingly used as tough transparent plastics and as additives for PVC.

1.2.3.4 Polyacrylamide

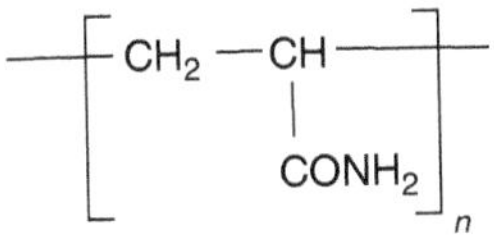

$$\left[\begin{array}{c} CH_2 - CH \\ | \\ CONH_2 \end{array} \right]_n$$

Monomer	Polymerization	Major Uses
Acrylamide	Free-radical-initiated chain polymerization	Flocculant, adhesives, paper treatment, water treatment, coatings

Polyacrylamide exhibits strong hydrogen bonding and water solubility. Most of the interest in this polymer is associated with this property. Polymerization of acrylamide monomer is usually conducted in an aqueous solution, using free-radical initiators and transfer agents.

Copolymerization with other water-soluble monomers is also carried out in a similar manner. Cationic polyacrylamides are obtained by copolymerizing with ionic monomers such as dimethylaminoethyl methacrylate, dialkyldimethylammonium chloride, and vinylbenzyltrimethylammonium chloride. These impart a positive charge to the molecule. Anionic character can be imparted by copolymerizing with monomers such as acrylic acid, methacrylic acid, 2-acrylamido-2-methylpropanesulfonic acid, and sodium styrene sulfonate. Partial hydrolysis of polyacrylamide, which converts some of the amide groups to carboxylate ion, also results in anionic polyacrylamides.

Polyacrylamides have several properties which lead to a multitude of uses in diverse industries. Table 1.10 lists the main functions of the polymers and their uses in various industries.

Polyacrylamides are used as primary flocculants or coagulant aids in water clarification and mining application. They are effective for clarification of raw river water. The capacity of water clarifiers can be

TABLE 1.10 Applications of Polyacrylamide

Function	Application	Industry
Flocculation	Water clarification	General
	Waste removal	Sewage
	Solids recovery	Mining
	Retention aid	Paper
	Drainage aid	Paper
Adhesion	Dry strength	Paper
	Wallboard cementing	Construction
Rheology control	Waterflooding	Petroleum
	Viscous drag reduction	Petroleum
		Fire fighting
		Irrigation pumping

increased when the polymer is used as a secondary coagulant in conjunction with lime and ferric chloride. Polyacrylamides, and especially cationic polyacrylamides, are used for conditioning municipal and industries sludges for dewatering by porous and empty sand beds, vacuum filters, centrifuges, and other mechanical devices.

Certain anionic polyacrylamides are approved by the U.S. Environmental Protection Agency for clarification of potable water. Polymer treatment also allows filters to operate at higher hydraulic rates. The function of clarification is not explained by a simple mechanism. The long-chain linear polymer apparently functions to encompass a number of individual fine particles of the dispersed material in water, attaching itself to the particles at various sites by chemical bonds, electrostatic attraction, or other attractive forces. Relatively stable aggregates are thus produced, which be removed by filtration, settling, or other convenient means.

Polyacrylamides are useful in the paper industry as processing aids, in compounding and formulating, and as filler-retention aids. Polyacrylamides and copolymers of acrylamide and acrylic acid are used to increase the dry strength of paper.

Polyacrylamides are used as flooding aids in secondary oil recovery from the producing oil well. Water, being of low viscosity, tends to finger ahead of the more viscous oil. However, addition of as little as 0.05% polyacrylamide to the waterflood reduces oil bypass and give significantly higher oil to water ratios at the producing wellhead. Greatly increased yields of oil result from adding polymer to waterflooding.

Solutions containing polyacrylamide are very slippery and can be used for water-based lubrication. Small amounts of polymer, when added to an aqueous solution, can significantly reduce the friction in pipes, thereby increasing the throughput or reducing the power consumption.

Other applications include additives in coatings and adhesives and binders for pigments.

Lightly cross-linked polyacrylamide is used to make superabsorbents of water. A starch-g-polyacryl-amide/clay superabsorbent composite has been synthesized [27] by graft copolymerization reaction of acrylamide, potato starch, and kaolinite micropowder (<1 μm) followed by hydrolysis with sodium hydroxide. Such a superabsorbent of compositin: 20% kaolinite, 20% potato starch, 60% acrylamide, 2% initiator (ceric amonium nitrate), and 0.04% cross-linker (N,N-methylenebisacrylamide) is found to absorb 2250 g H_2O/g at room temperature at swelling equilibrium.

1.2.3.5 Poly(Acrylic Acid) and Poly(Methacrylic Acid)

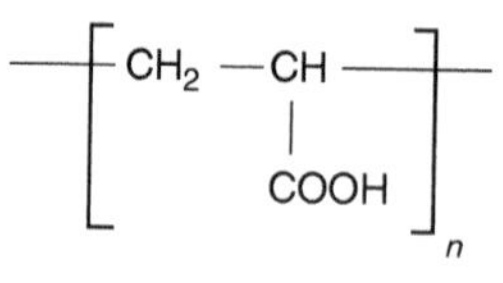

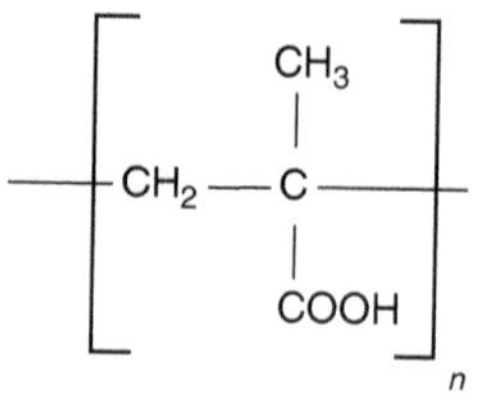

Monomers	Polymerization	Major Uses
Acrylic acid and methacrylic acid	Free-radical-initiated chain polymerization	Sodium and ammonium-salts as polyelectrolytes, thickening agents

Poly(acrylic acid) and poly(methacrylic acid) may be prepared by direct polymerization of the appropriate monomer, namely, acrylic acid or methacrylic acid, by conventional free-radical techniques, with potassium persulfate used as the initiator and water as the solvent (in which the polymers are soluble); or if a solid polymer is required, a solvent such as benzene, in which the polymer is insoluble, can be used, with benzoyl peroxide as a suitable initiator.

The multitude of applications of poly(acrylic acid) and poly(methacrylic acid) is reminiscent of the fable of the man who blew on his hands to warm them and blew on his porridge to cool it. They are used as adhesives and release agents, as flocculants and dispersants, as thickeners and fluidizers, as reaction inhibitors and promoters, as permanent coatings and removable coatings, etc. Such uses are the direct result of their varied physical properties and their reactivity. Many of the applications thus depend on the ability of these polymers to form complexes and to bond to substrates. Monovalent metal and ammonium salts of these polymers are generally soluble in water. These materials behave as anionic polyelectrolytes and are used for a variety of purposes, such as thickening agents, particularly for rubber latex.

Superabsorbent polymers are loosely cross-linked networks of partially neutralized acrylic acid polymers capable of absorbing large amounts of water and retaining the absorbed water under pressure. Although superabsorbent polymers have been abundantly used in the disposable diaper industry for the past 30 years, their applications are still being expanded to many fields, including agriculture and horticulture, sealing composites, artificial snow, drilling fluid additives, drug delivery systems, and so on.

The swelling of a hydrophilic polymer is dependent on the rubbery elasticity, ionic osmotic pressure, and affinity of the polymer toward water. Although superabsorbent polymers have the greatest absorbency in water, the addition of an inorganic salt or organic solvent will reduce the absorbency.

Clay and mineral fillers have been used for reducing production costs and improving the comprehensive water absorbing properties of superabsorbent materials For example, a poly(acrylic acid)/mica superabsorbent has been synthesized with water absorbency higher than 1100 g H_2O/g. In a typical method of preparation, acrylic acid monomer is neutralized at ambient temperature with an amount of aqueous sodium hydroxide solution to achieve 65% neutralization (optimum). Dry ultrafine (<0.2 µm) mica powder (10 wt%) is added, followed by cross-linker N,N-methylene-bisacrylamide (0.10 wt%) and radical initiator, potassium persulfate. The mixture is heated to 60–70°C in a water bath for 4 h. The product is washed, dried under vacuum at 50°C, and screened.

1.2.3.6 Acrylic Adhesives

Acrylic adhesives are essentially acrylic monomers which achieve excellent bonding upon polymerization. Typical examples are cyanoacrylates and ethylene glycol dimethacrylates. Cyanoacrylates [28] are obtained by depolymerization of a condensation polymer derived from a malonic acid derivative and formaldehyde.

Cyanoacrylates are marketed as contact adhesives. Often popularly known as *superglue*, they have found numerous applications. In dry air and in the presence of polymerization inhibitors, methyl- and ethyl-2-cyanoacrylates have a storage life of many months. As with many acrylic monomers, air can inhibit or severely retard polymerization of cyanoacrylates. These monomers are, however, prone to anionic polymerization, and even a very weak base such as water can bring about rapid polymerization.

In practice, a trace of moisture occurring on a substrate is adequate to cause polymerization of the cyanoacrylate monomer to provide strong bonding within a few seconds of closing the joint and excluding air. Cyanoacrylate adhesives are particularly valuable because of their speed of action, which obviates the need for complex jibs and fixtures. The amount of monomer applied should be minimal to obtain a strong joint. Larger amounts only reduce the strength. Notable uses of cyanoacrylates include surgical glue and dental sealants; morticians use them to seal eyes and lips.

Dimethacrylates, such as tetramethylene glycol dimethacrylate, are used as *anaerobic adhesives*. Air inhibition of polymerization of acrylic monomers is used to advantage in this application because the monomers are supplied along with a curing system (comprising a peroxide and an amine) as part of a one-part pack. When this adhesive is placed between mild steel surfaces, air inhibition is prevented since the air is excluded, and polymerization can take place. Though the metal on the surface acts as a polymerization promoter, it may be necessary to use a primer such as cobalt naphthenate to expedite the polymerization. The anaerobic adhesives are widely used for sealing nuts and bolts and for miscellaneous engineering purposes.

Dimethacrylates form highly cross-linked and, therefore, brittle polymers. To overcome brittleness, manufacturers often blend dimethacrylates with polyurethanes or other polymers such as low-molecular-weight vinyl-terminated butadiene–acrylonitrile copolymers and chlorosulfonated polyethylene. The modified dimethacrylate systems provide tough adhesives with excellent properties. These can be formulated as two-component adhesives, the catalyst component being added just prior to use or applied separately to the surface to be bonded. One-component systems also have been formulated which can be conveniently cured by ultraviolet radiation.

1.2.4 Vinyl Polymers

If the R substituent in an olefin monomer (CH_2=CHR) is either hydrogen, alkyl, aryl, or halogen, the corresponding polymer in the present discussion is grouped under polyolefins. If the R is a cyanide group or a carboxylic group or its ester or amide, the substance is an *acrylic* polymer. *Vinyl* polymers include those polyolefins in which the R substituent in the olefin monomer is bonded to the unsaturated carbon through an oxygen atom (vinyl esters, vinyl ethers) or a nitrogen atom (vinyl pyrrolidone, vinyl carbazole).

Vinyl polymers constitute an important segment of the plastics industry. Depending on the specific physical and chemical properties, these polymers find use in adhesives, in treatments for paper and textiles, and in special applications. The commonly used commercial vinyl polymers are described next.

1.2.4.1 Poly(Vinyl Acetate)

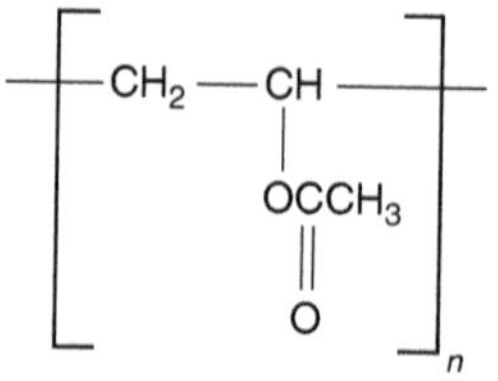

$$\left[CH_2 - CH \atop \underset{\underset{O}{\overset{\|}{OCCH_3}}}{} \right]_n$$

Monomer	Polymerization	Major Uses
Vinyl acetate	Free-radical-initiated chain Polymerization (mainly emulsion polymerization)	Emulsion paints, adhesives, sizing

Since poly(vinyl acetate) is usually used in an emulsion form, it is manufactured primarily by free-radical-initiated emulsion polymerization. The polymer is too soft and shows excessive cold flow, which precludes its use in molded plastics. The reason is that the glass transition temperature of 28°C is either slightly above or (at various times) below the ambient temperatures.

Vinyl acetate polymers are extensively used in emulsion paints, as adhesives for textiles, paper, and wood, as a sizing material, and as a permanent starch. A number of commercial grades are available which differ in molecular weight and in the nature of comonomers (e.g., acrylate, maleate, fumarate) which are often used. Two vinyl acetate copolymers of particular interest to the plastics industry are EVA and vinyl chloride–vinyl acetate copolymers.

EVA copolymers represent the largest-volume segment of ethylene copolymer market and are the products of low-density polyethylene (LDPE) technology. Commercial preparation of EVA copolymer is based on the same process as LDPE with the addition of controlled comonomer stream into the reactor. EVA copolymers are thermoplastic materials consisting of an ethylene chain incorporating 5–20 mol% vinyl acetate (VA), in general. The VA produces a copolymer with lower crystallinity than conventional ethylene homopolymer.

These lower crystallinity resins have lower melting points and heat seal temperatures, along with reduced stiffness, tensile strength, and hardness. EVAs have greater clarity, low-temperature flexibility, stress-crack resistance, and impact strength than LDPE. EVA resins are more permeable to oxygen, water vapor, and carbon dioxide. Chemical resistance is similar to that of LDPE, with somewhat better resistance to oil and grease for EVA resins of higher VA content. The VA groups contribute to improved adhesion in extrusions or hot-melt adhesive formulations.

The outdoor stability of EVA resins is superior to that of LDPE by virtue of their greater flexibility. Addition of UV stabilizers can extend the outdoor life of clear compounds to three to five years, depending on the degree of exposure. Outdoor life expectancy is also enhanced by the addition of carbon black.

In addition to specialty applications involving film and adhesives production, EVA are used in a variety of molding, compounds, and extrusion applications. Some typical end uses include flexible hose and tubing, footwear components, toys and athletic goods, wire and cable compounding, extruded gaskets, molded automotive parts (such as energy-absorbing bumper components), cap and closure seals, and color concentrates. In footwear applications, EVA resins are used in canvas box toes and flocked or fabric-laminated contours. Foamed and cross-linked EVA is used in athletic or leisure shoe midsoles and in sandals.

With increasing VA content, EVA resin properties range from those of LDPE to those of highly plasticized PVC. High VA resins are soft and flexible, with excellent toughness and good stress-crack resistance. With EVA resins these properties are permanent and do not dissipate with time because of the loss of the liquid plasticizer. EVAs also have exceptional low temperature toughness and flexibility.

The EVA copolymers are slightly less flexible than normal rubber compounds but have the advantage of simpler processing since no vulcanization is necessary. The materials have thus been largely used in injection molding in place of plasticized PVC or vulcanized rubber. Typical applications include turntable mats, based pads for small items of office equipment, buttons, car door protection strips, and for other parts where a soft product of good appearance is required.

A substantial use of EVA copolymers is as was additives and additives for hot-melt coatings and adhesives. Cellular cross-linked EVA copolymers are used in shoe parts.

EVA copolymers with only a small vinyl acetate content ($\sim$3 mol%) are best considered as a modification of low-density polyethylene. These copolymers have less crystallinity and greater flexibility, softness, and, in case of film, surface gloss.

EVA can be processed by all standard plastics processing techniques, including injection and blow molding, thermoforming, and extrusion into sheet and shapes. They accommodate high loadings of fillers, pigments, and carbon blacks. They are compatible with other thermoplastics, and thus are frequently used for impact modification and improvement of stress-crack resistance. This combination of properties makes EVA highly adaptable vehicles for color concentrates. EVA resins can be formulated with blowing agents and cross-linking to produce low density foams via compression molding.

Hydrolysis of EVA copolymers yields ethylene–vinyl alcohol copolymers (EVOH). EVOH has exceptional gas barrier properties as well as oil and organic solvent resistance. The poor moisture resistance of EVOH is overcome by coating, coextrusion, and lamination with other substrates.

Applications include containers for food (ketchup, jelly, mayonnaise) as well as chemicals and solvents.

1.2.4.2　Poly(Vinyl Alcohol)

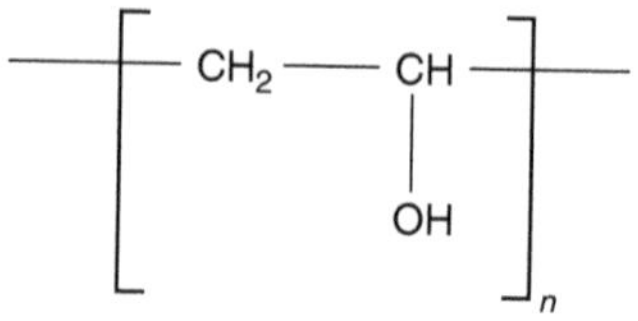

$$\left[\!\!\begin{array}{c} CH_2 \!-\! CH \\[1ex] | \\ OH \end{array}\!\!\right]_n$$

Manufacture	Major Uses
Alcoholysis of poly(vinyl acetate)	Paper sizing, textile sizing, cosmetics

Poly(vinyl alcohol) (PVA) is produced by alcoholysis of poly(vinyl acetate), because vinyl alcohol monomer does not exist in the free state [29]. (The term hydrolysis is often used incorrectly to describe this process.) Either acid or base catalysts may be employed for alcoholysis. Alkalien catalysts such as sodium hydroxide or sodium methoxide give more rapid alcoholysis. The degree of alcoholysis, and hence the residual acetate content, is controlled by varying the catalyst concentration.

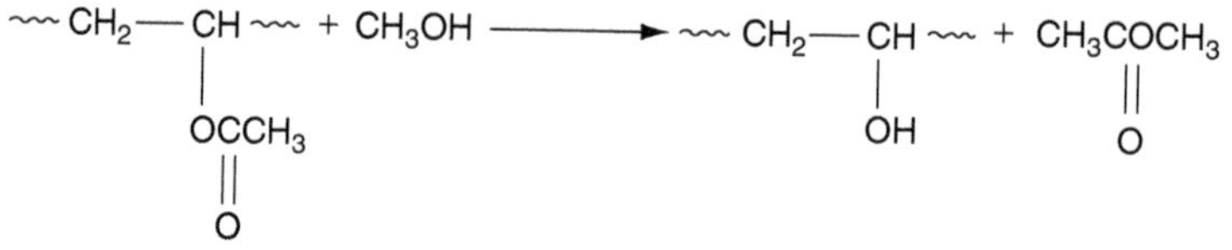

$$\sim\!\!\sim CH_2\!-\!CH\!\sim\!\!\sim + CH_3OH \longrightarrow \sim\!\!\sim CH_2\!-\!CH\!\sim\!\!\sim + CH_3COCH_3$$

The presence of hydroxyl groups attached to the main chain renders the polymer hydrophilic. PVA therefore dissolves in water to a greater or lesser extent according to the degree of hydrolysis. Polymers with a degree of hydrolysis in the range of 87–90% readily dissolve in cold water. Solubility decreases with an increase in the degree of hydrolysis, and fully hydrolyzed polymers are water soluble only at higher temperatures ($>85°C$). This apparently anomalous behavior is due to the higher degree of crystallinity and the greater extent of hydrogen bonding in the completely hydrolyzed polymers.

Commercial PVA is available in a number grades which differ in molecular weight and degree of hydrolysis. The polymer finds a variety of uses. It functions as a nonionic surface active agent and is used in suspension polymerization as a protective colloid. It also serves as a binder and thickener and is widely used in adhesives, paper coatings, paper sizing, textile sizing, ceramics, and cosmetics.

Completely hydrolyzed grades of PVA find use in quick-setting, water-resistant adhesives. Combinations of fully hydrolyzed PVA and starch are used as a quick-setting adhesive for paper converting. Borated PVA, commonly called "tackified," are combined with clay and used in adhesive applications requiring a high degree of wet tack. They are used extensively to glue two or more plies of paper together to form a variety of shapes such as tubes, cans, and cores.

Since PVA film has little tendency to adhere to other plastics, it can be used to prevent sticking to mold. Films cast from aqueous solution of PVA are used as release agents in the manufacture of reinforced plastics.

Partially hydrolyzed grades have been developed for making tubular blown film (similar to that with polyethylene) for packages for bleaches, bath salts, insecticides, and disinfectants. Use of water-soluble PVA film for packaging preweighed quantities of such materials permits their addition to aqueous systems without breaking the package or removing the contents, thereby saving time and reducing material losses. Film made from PVA may be used for hospital laundry bags that are added directly to the washing machine.

A process has been developed in Japan for producing fibers from poly(vinyl alcohol). The polymer is wet spun from a warm aqueous solution into a concentrated aqueous solution of sodium sulfate containing sulfuric acid and formaldehyde, which insolubilizes the alcohol by formation of *formal* groups (see below). These fibers are generally known as *vinal* of *vinylon* fibers.

1.2.4.3 Poly(Vinyl Acetals)

Poly(vinyl acetals) are produced by treating poly(vinyl alcohol) with aldehydes. (They may also be made directly from poly(vinyl acetate) without separating the alcohol.) Since the reaction with aldehyde involves a pair of neighboring hydroxyl groups on the polymer chain and the reaction occurs at random, some hydroxyl groups become isolated and remain unreacted. A poly(vinyl acetal) molecule will thus contain acetal groups and residual hydroxyl groups. In addition, there will be residual acetate groups due to incomplete hydrolysis of poly(vinyl acetate) to the poly(vinyl alcohol) used in the acetalization reaction. The relative proportions of these three types of groups may have a significant effect on specific properties of the polymer.

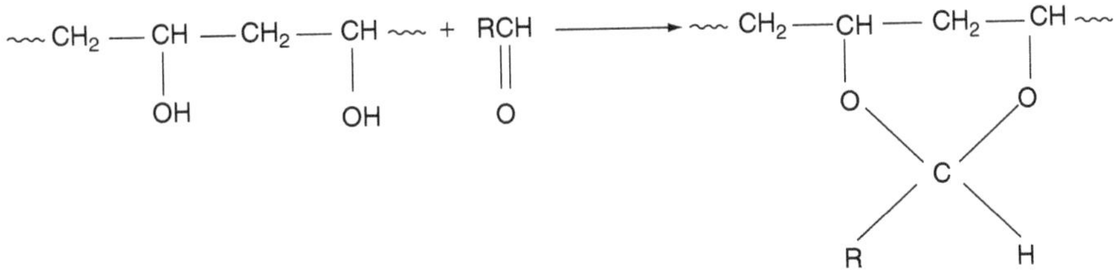

When the aldehyde in this reaction is formaldehyde, the product is poly(vinyl formal). This polymer is, however, made directly from poly(vinyl acetate) and formaldehyde without separating the alcohol. The product with low hydroxyl (5–6%) and acetate (9.5–13%) content (the balance being *formal*) is used in wire enamel and in structural adhesives (e.g., Redux). In both applications the polymer is used in conjunction with phenolic resins and is heat cured.

When the aldehyde in the acetalization reaction is butyraldehyde, i.e., $R=CH_3CH_2CH_2-$, the product is poly(vinyl butyral). Sulfuric acid is the catalyst in this reaction. Poly(vinyl butyral) is characterized by high adhesion to glass, toughness, light stability, clarity, and moisture insensitivity. It is therefore extensively used as an adhesive interlayer between glass plates in the manufacture of laminate safety glass and bullet-proof composition.

1.2.4.4 Poly(Vinyl Cinnamate)

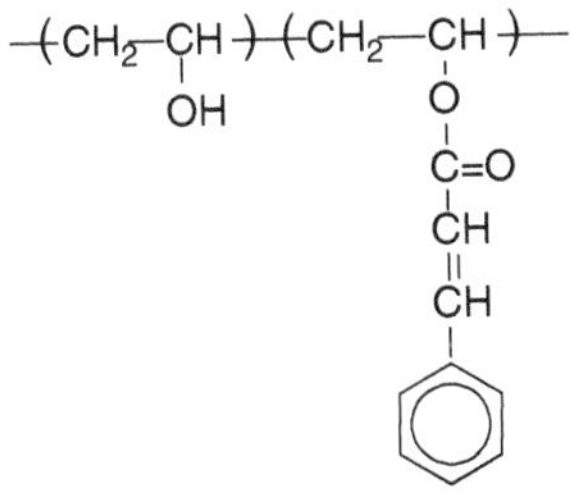

Poly(vinyl cinnamate) is conveniently made by the Schotten–Baumann reaction using poly(vinyl alcohol) in sodium or potassium hydroxide solution and cinnamoyl chloride in methyl ethyl ketone. The product is, in effect, a copolymer of vinyl alcohol and vinyl cinnamate, as shown. The polymer has the ability to cross-link on exposure to light, which has led to its important applications in photography, lithography, and related fields as a *photoresist* (see also Chapter 2).

1.2.4.5   Poly(Vinyl Ethers)

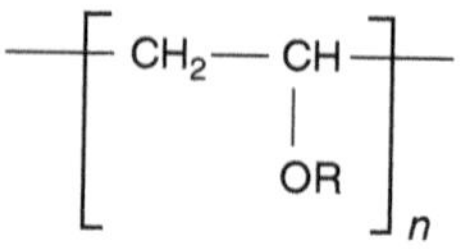

$$\left[\begin{array}{c} CH_2 - CH \\ | \\ OR \end{array} \right]_n$$

Commercial uses have developed for several poly(vinyl ethers) in which R is methyl, ethyl, and isobutyl. The vinyl alkyl ether monomers are produced from acetylene and the corresponding alcohols, and the polymerization is usually conducted by cationic initiation using Friedel–Craft-type catalysts.

Poly(vinyl methyl ether) is a water-soluble viscous liquid which has found application in the adhesive and rubber industries. One particular applications has been as a heat sensitizer in the manufacture of rubber-latex dipped goods.

Ethyl and butyl derivatives have found uses as adhesives. *Pressure-sensitive adhesive tapes* made from poly(vinyl ethyl either) incorporating antioxidants are said to have twice the shelf life of similar tapes made from natural rubber. Copolymers of vinyl isobutyl ether with vinyl chloride, vinyl acetate and ethyl acrylate are also produced.

1.2.4.6   Poly(Vinyl Pyrrolidone)

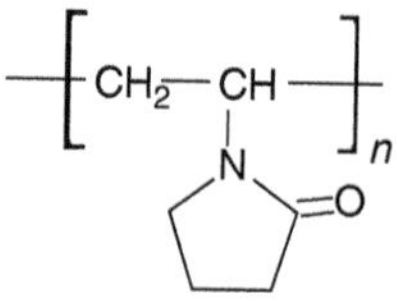

$$\left[\begin{array}{c} CH_2 - CH \\ | \\ N \end{array} \right]_n$$

Poly(vinyl pyrrolidone) is produced by free-radical-initiated chain polymerization of *N*-vinyl pyrrolidone. Polymerization is usually carried out in aqueous solution to produce a solution containing 30% polymer. The material is marketed in this form or spray dried to give a fine powder.

Poly(vinyl pyrrolidone) is a water-soluble polymer. Its main value is due to its ability to form loose addition compounds with many substances. It is thus used in cosmetics. The polymer has found several applications in textile treatment because of its affinity for dyestuffs. In an emergency it is used as a blood plasma substitute. Also, about 7% polymer added to whole blood allows it to be frozen, stored at liquid nitrogen temperatures for years, and thawed out without destroying blood cells.

1.2.4.7   Poly(Vinyl Carbazole)

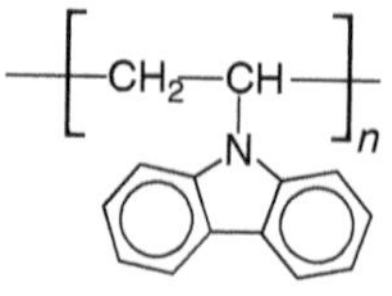

$$\left[\begin{array}{c} CH_2 - CH \\ | \\ N \end{array} \right]_n$$

Poly(vinyl carbazole) is produced by polymerization of vinyl carbazole using free-radial initiation or Ziegler–Natta catalysis.

Poly(vinyl carbazole) has a high softening point, excellent electrical insulating properties, and good photoconductivity, which has led to its application in xerography.

1.3   Part II: Condensation Polymers

According to the original classification of Carothers, condensation polymers are formed from bi- or polyfunctional monomers by reactions which involve elimination of some smaller molecule. A condensation polymer, according to this definition, is one in which the repeating unit lacks certain atoms which were present in the monomers(s) from which the polymer was formed.

With the development of polymer science and synthesis of newer polymers, this definition of condensation polymer was found to be inadequate. For example, in polyurethanes, which are classified as condensation polymers, the repeat unit has the same net composition as the two monomers—that is, a diol and a diisocyanate, which react without the elimination of any small molecule. Similarly the polymers produced by the ring-opening polymerization of cyclic monomers, such as cyclic ethers and amides, are generally classified as condensation polymers based on the presence of functional groups, such as the ether and amide linkages, in the polymer chains, even though the polymerization occurs without elimination of any small molecule.

To overcome such problems, an alternative definition has been introduced. According to this definition, polymers whose main chains consist entirely of C–C bonds are classified as *addition polymers*, whereas those in which heteratoms (O, N, S, Si) are present in the polymer backbone are considered to be condensation polymers. A polymer which satisfies both the original definition (of Carothers) and the alternative definition or either of them, is classified as a condensation polymer. Phenol–formaldehyde condensation polymers, for example, satisfy the first definition but not the second.

Condensation polymers described in Part II are classified as polyesters, polyamides, formaldehyde resins, polyurethanes, and ether polymers.

1.3.1   Polyesters

Polyesters were historically the first synthetic condensation polymers studied by Carothers in his pioneering work in the early 1930s. Commercial polyesters [30] were manufactured by polycondensation reactions, the methods commonly used being melt polymerization of diacid and diol, ester interchange of diester and diol, and interfacial polymerization (Schotten–Baumann reaction) of diacid chloride and diol. In a polycondensation reaction a by-product is generated which has to be removed as the reaction progresses.

Thermoplastic saturated polyesters are widely used in synthetic fibers and also in films and molding applications. The production of polyester fibers accounts for nearly 30% of the total amount of synthetic fibers. Unsaturated polyesters are mainly used in glass-fiber reinforced plastic products.

1.3.1.1   Poly(Ethylene Terephthalate)

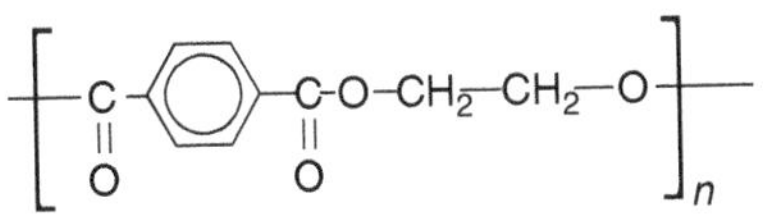

Monomers	Polymerization	Major Uses
Dimethyl terephthalate or terephthalic acid, ethylene glycol	Bulk polycondensation	Apparel (61%), home furnishings (18%), tire cord (10%)

Whinfield and Dixon, in UK, developed polyethylene terephthalate fibers (Dacron, Terylene). This first Dacron polyester plant went into operation in 1953. *Ester interchange* (also known as *ester exchange* or *alcoholysis*) was once the preferred method for making polyethylene terephthalate (PET) because dimethyl terephthalate can be readily purified to the high quality necessary for the production of the polymer. The process is carried out in two steps.

Dimethyl terephthalate (DMT) is reacted with excess ethylene glycol (mole ratio 1:2.1–2.2) at 150°C and 100 kPa (1 atm = 101 kPa). The output of the process is bis(hydroxyethyl) terephthalate (BHET). The pre-polymerization step (250–280°C, 2–3 kPa) follows in which BHET is polymerized to a degree of poymerization (DP) of up to 30. The next step is the polycondensation process where the DP is further increased to 100 by heating under vacuum, the process conditions being 280–290°C and 50–100 Pa. Up to this stage, PET is suitable for applications that do not require high molecular weight or high intrinsic viscosity $[\eta]$, such as fibers and sheets. Solid-state polymerization is used to further increase the DP to 150. The operating conditions are 200–240°C at 100 kPa and 5–25 h. Bottle-grade PET that has an $[\eta]$ of 0.73–0.81 dl g^{-1} is normally produced by solid-state polymerization at 210°C for around 15–20 h [31].

In recent years methods have been developed to produce terephthalic acid with satisfactory purity, and direct polycondensation reaction with ethylene glycol is now the preferred route to this polymer.

Virgin PET is produced at different specifications because different applications require different properties [32]. Examples of intrinsic viscosity $[\eta]$ for different applications are recording tape 0.60, carbonated drink bottles 0.73–0.81, and industrial tire cord 0.85 dl g^{-1}. PET granules can be processed in many ways depending on application and final product requirements.

PET is widely used in synthetic fibers designed to simulate wool cotton, or rayon, depending on the processing conditions. They have good wash-and-wear properties and resistance to wrinkling. In the production of fiber the molten polymer is extruded through spinnerets and rapidly cooled in air. The filaments thus formed are, however, largely amorphous and weak. They are therefore drawn at a temperature (80°C) above T_g and finally heated at 190°C under tension, whereby maximum molecular orientation, crystallinity, and dimensional stability are achieved. The melting point of highly crystalline PET is 271°C.

Crystalline PET has good resistance to water and dilute mineral acids but is degraded by concentrated nitric and sulfuric acids. It is soluble at normal temperatures only in proton donors which are capable of interaction with the ester group, such as chlorinated and fluorinated acetic acids, phenols, and anhydrous hydrofluoric acid.

PET is also used in film form (Melinex, Mylar) and as a molding material. The manufacture of PET film closely resembles the manufacture of fiber. The film is produced by quenching extruded sheet to the amorphous state and then reheating and stretching the sheet approximately threefold in the axial and transverse directions at 80–100°C. To stabilize the biaxially oriented film, it is annealed under restraint at 180–210°C. This operation increases the crystallinity of PET film and reduces its tendency to shrink on heating. The strength of PET in its oriented from is outstanding.

The principal uses of biaxially oriented PET film are in capacitors, in slot liners for motors, and for magnetic tape. Although a polar polymer, its electrical insulation properties at room temperature are good (even at high frequencies) because at room temperature, which is well below T_g (69°C), dipole orientation is severely restricted.

The high strength and dimensional stability of the polyester film have also led to its use for x-ray and photographic film and to a number of graphic art and drafting applications. The film is also used in food packaging, including boil-in-bag food pouches. Metallized polyester films have many uses as a decorative material.

Because of its rather high glass transition temperature, only a limited amount of crystallization can occur during cooling after injection molding of PET. The idea of molding PET was thus for many years not a technical proposition. Toward the end of the 1970s Du Pont introduced Rynite, which is a PET nucleated with an ionomer, containing a plasticizer and only available in glass-fiber-filled form (at 30, 45, and 55% fill levels). The material is very rigid, exceeding that of polysulfone, is less water sensitive than an unfilled polymer, and has a high heat-deflection temperature (227°C at 264 psi).

In the late 1970s the benefits of biaxial stretching PET were extended from film to bottle manufacture. Producing carbonated beverages PET bottles by blow molding has gained prominence (particularly in the United States) because PET has low permeability to carbon dioxide. The process has been extended, particularly in Europe, to produce bottles for other purposes, such as fruit juice concentrates and sauces, wide-necked jars for coffee, and other materials. Because of its excellent thermal stability, PET is also used material for microwave and conventional ovens.

Virgin PET manufacturers have tended in recent years to produce PET copolymer, such as isphthalic acid modified PET, rather than homopolymer PET. PET bottles are normally made from copolymer PET because of its lower crystallinity, improved ductitlity, better process ability, and better clarity. Some of the most improtant PET copolymers are shown in Figure 1.10.

1.3.1.2   Poly(Butylene Terephthalate)

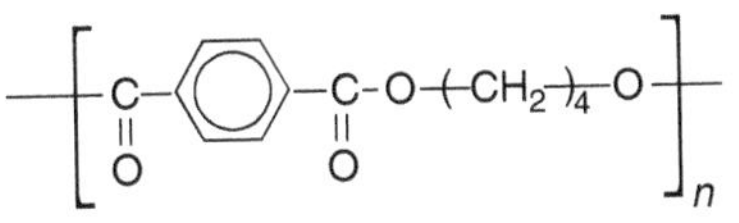

Monomers	Polymerization	Major Uses
Dimethyl terephthalate or terephthalic acid, butanediol	Bulk polycondensaiton	Machine parts, electrical applications, small appliances

Poly(butylenes terephthalate), often abbreviated to PBT or PBTP, is manufactured by condensation polymerization of dimethyl terephthalate and butane-1,4-diol in the presence of tetrabutyl titanate. The polymer is also known as poly(tetramethylene terephthalate), PTMT in short. Some trade names for this engineering thermoplastic are Tenite PTMT (Eastman Kodak), Valox (General Electric), Celanex (Celanese) in America and Arnite PBTP (Akzo), Ultradur (BASF), Pocan (Bayer), and Crastin (Ciba-Geigy) in Europe.

Because of the longer sequence of methylene groups in the repeating unit poly(butylenes terephthalate) chains are both more flexible and less polar than poly(ethylene terephthalate). This leads to lower values for melting point (about 224°C) and glass transition temperature (22–43°C). The low glass

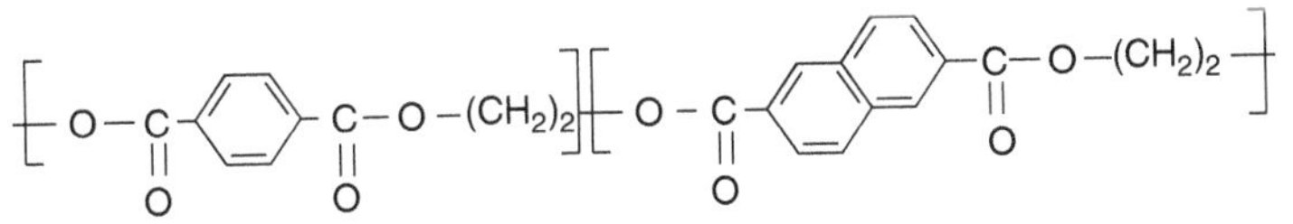

Poly[(ethylene terephthalate)-co-(ethylene 2,6-naphthalate)][PET/PEN]

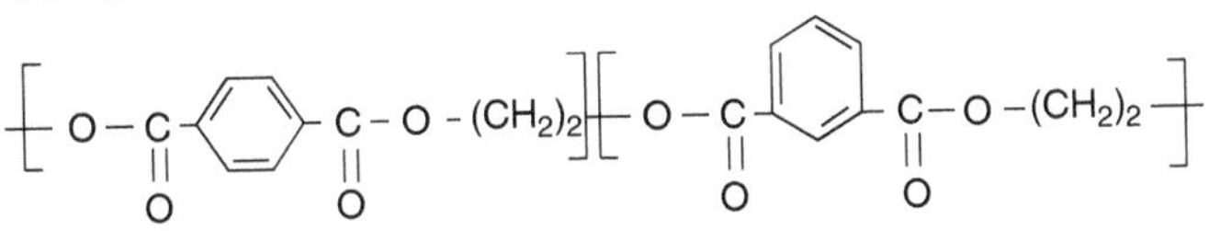

Poly[(ethylene terephthalate)-co-(ethylene isophthalate)][PET/PEI]

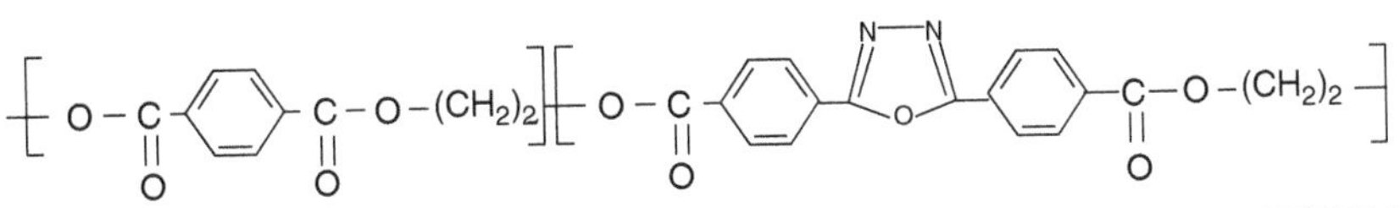

Poly[(ethylene terephthalate)-co-(ethylene 2,5-bis(4-carboxyphenyl)1,3,4,-oxadiazole)[PET/PEOD]

FIGURE 1.10   Some of the most important PET copolymers. (After Awaja, F., and Pavel, D. 2005. *Eur. Polymer J.*, 41, 1453. With permission.)

transition temperature facilitates rapid crystallization when cooling in the mold, and this allows short injection-molding cycles and high injection speeds.

PBT finds use as an engineering material due to its dimensional stability, particularly in water, and its resistance to hydrocarbon oils without showing stress cracking. PBT also has high mechanical strength and excellent electrical properties but a relatively low heat-deflection temperature 130°F (54°C) at 264 psi (1.8 MPa). The low water absorption of PBT—less than 0.1% after 24-h immersion—is outstanding. Both dimensional stability and electrical properties are retained under conditions of high humidity. The lubricity of the resin results in outstanding wear resistance.

As with PET, there is particular interest in glass-filled grades of PBT. The glass has a profound effect on such properties as tensile strength, flexural modulus, and impact strength, as can be seen from the values of these properties for unfilled and 30% glass-filled PBT: 8200 vs. 17,000 psi (56 vs. 117 MPa), 340,000 vs. $1.1–1.2 \times 10^6$ psi (2350 vs. 7580–8270 MPa) and 0.8–1.0 vs. 1.3–1.6 ft.-lbf/in.^{2} (Izod), respectively. Reinforcing with glass fiber also results in an increase in heat-deflection temperature to over 400°F (204°C) at 264 psi (1.8 MPa).

Typical applications of PBT include pump housings, impellers, bearing bushings, gear wheels, automotive exterior and under-the-hood parts, and electrical parts such as connectors and fuse cases.

1.3.1.3   Poly(Dihydroxymethylcyclohexyl Terephthalate)

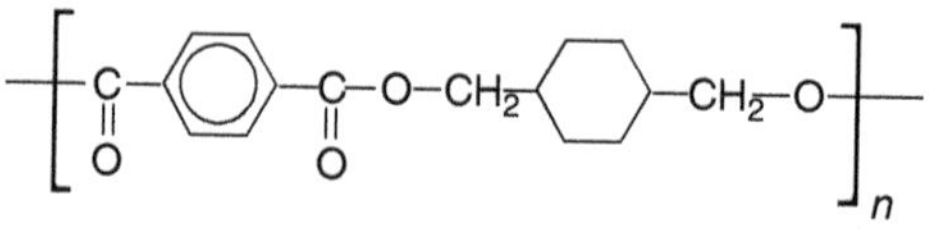

In 1958, Eastman Kodak introduced a more hydrophobic polyester fiber under the trade name Kodel. The raw material for this polyester is dimethyl terephthalate. Reduction leads to 1,4-cyclohexylene glycol, which is used with dimethyl terephthalate in the polycondensation (ester exchange) reaction.

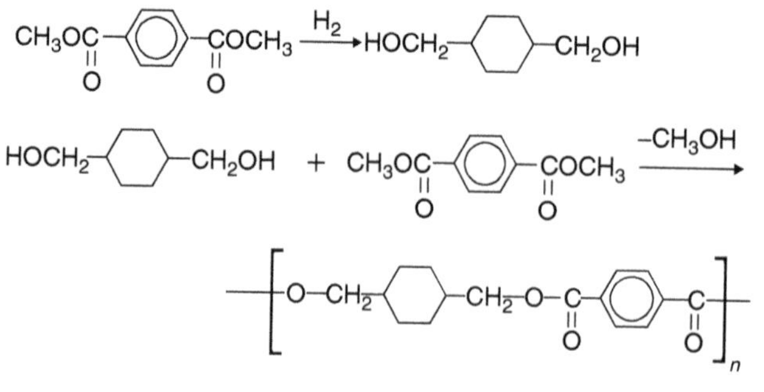

Eastman Kodak also introduced in 1972 a copolyester based on 1,4-cyclohexylene glycol and a mixture of terephthalic and isophthalic acids. The product is sold as Kodar PETG. Being irregular in structure, the polymer is amorphous and gives products of brilliant clarity.

In spite of the presence of the heterocyclic ring, the deflection temperature under load is as low as that of the poly(butylenes terephthalate)s, and the polymer can be thermoformed at draw ratios as high as 4:1 without blustering or embrittlement. Because of its good melt strength and low molding shrinkage, the material performs well in extrusion blow molding and in injection molding. The primary use for the copolymer is extrusion into film and sheeting for packaging.

Ethylene glycol-modified polyesters of the Kodel type are used in blow-molding applications to produce bottles for packaging liquid detergents, shampoos, and similar products. One such product is Kodar PETG 6703 in which one acid (terephthalic acid) is reacted with a mixture of glycols (ethylene glycol and 1,4-cyclohexylene glycol). A related glass-reinforced grade (Ektar PCTG) has also been offered.

The principle of formation of segmented or block copolymers (see Section 1.3.4.1.2) has also been applied to polyesters, with the "hard" segment formed from butanediol and terephthalic acid, and the "soft" segment provided by a hydroxyl-terminated polyether [polytetramethylene either glycol (PTMEG)] with molecular weight 600–3000.

In a typical preparation, dimethyl terephthalate is transesterified with a mixture of PTMEG and a 50% excess of butane-1,4-diol in the presence of an ester exchange catalyst. The stoichiometry is such that relatively long sequences of tetramethylene terephthalate (TMT) are produced which, unlike the polyether segments, are crystalline and have a high melting point. Since the sequences of TMT segregate into rigid domains, they are referred to as "hard" segments, and the softer polyether terephthalate (PE/T) segments are referred to as "soft" segments (Figure 1.11).

Due Pont markets this polyester elastomer under the trade name Hytrel. These elastomers are available in a range of stiffnesses. The harder grades have up to 84% TMT segments and a melting point of 214°C, and the softest grades contain as little as 33% TMT units and have a melting point of 163°C.

Processing of these thermoplastic rubbers is quite straightforward. The high crystallization rates of the hard segments facilitate injection molding, while the low viscosity at low shear rates facilitates a low shear process, such as rotational molding.

These materials are superior to conventional rubbers in a number of properties (see the section on thermoplastic rubbers). Consequently, in spite of their relatively high price they have become widely accepted as engineering rubbers in many applications.

1.3.1.4   Unsaturated Polyesters

Monomers	Polymerization	Major Uses
Phthalic anhydride, maleic anhydride, fumaric acid, isophthalic acid, ethylene glycol, propylene glycol, diethylene glycol, styrene	Bulk polycondensation followed by free-radical-initiated chain polymerization	Construction, automotive applications, marine applications

Unsaturated polyester laminating resins [14,34] are viscous materials of a low degree of polymerization (i.e., oligomers) with molecular weights of 1500–3000. They are produced by condensing a glycol with both an unsaturated dicarboxylic acid (maleic acid) and a saturated carboxylic acid (phthalic or isophthalic acid). The viscous polyesters are dissolved in styrene monomer (30–50% concentration) to reduce the viscosity. Addition of glass fibers and curing with peroxide initiators produces a cross-linked polymer (solid) consisting of the original polyester oligomers, which are now interconnected with polystyrene chains (Figure 1.12b). The unsaturated acid residues in the initial polyester oligomer provide a site for cross-linking in this curing step, while the saturated acid reduces the brittleness of the final cross-linked product by reducing the frequency of cross-links.

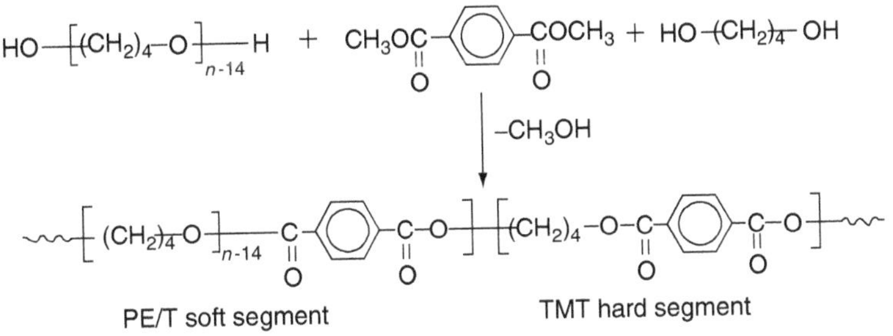

PE/T soft segment   TMT hard segment

FIGURE 1.11   Formation of polyester-type segmented or block copolymer.

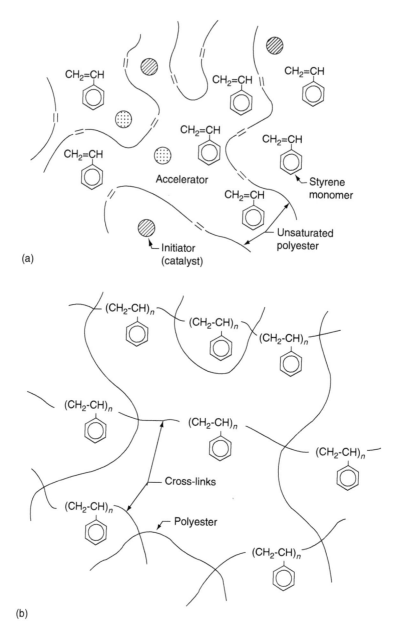

FIGURE 1.12 Curing of unsaturated polyesters. (a) Species in polyester resin ready for laminating. (b) Structures present in cured polyester resin. Cross-linking takes place via an addition copolymerization reaction. The value of $n \sim 2$–3 on average in general-purpose resins. (After Brydson, J. A. 1982. *Plastics Materials*. Butterworth Scientific, London, UK.)

In practice, the peroxide curing system is blended into the resin before applying the resin to the reinforcement, which is usually glass fiber (E type), as perform, cloth, mat, or rovings, but sisal or more conventional fabrics may also be sued. The curing system may be so varied (in both composition and quality) that curing times may range from a few minutes to several hours, and the cure may be arranged to proceed either at ambient or elevated temperatures.

The two most important peroxy materials used for room temperature curing of polyester resins are methyl ethyl ketone peroxide (MEKP) and cyclohexanone peroxide. These are used in conjunction with a cobalt compound such as a naphthenate, octoate, or other organic-solvent-soluble soap. The peroxides are referred to as *catalysts* (though, strictly speaking, these are polymerization initiators) and the cobalt compound is referred to as an *accelerator*.

In room-temperature curing it is obviously necessary to add the resin to the reinforcement as soon as possible after the curing system has been blended and before gelation can occur. Benzoyl peroxide is most commonly used for elevated-temperature curing. It is generally supplied as a paste ($\sim 50\%$) in a liquid such as dimethyl phthalate to reduce explosion hazards and to facilitate mixing.

Since the cross-linking of polyester–styrene system occurs by a free-radical chain-reaction mechanism across the double bonds in the polyesters with styrene providing the cross-links, the curing reaction does not give rise to volatile by-products (unlike phenolic and amino resins) and it is thus possible to cure without applying pressure. This fact as well as that room temperature cures are also possible makes unsaturated polyesters most useful in the manufacture of large structures such as boats and car bodies.

Unsaturated polyesters find applications mainly in two ways: polyester–glass-fiber laminates and polyester molding compositions (discussed later).

1.3.1.4.1 *Raw Materials and Resin Preparation*

General purpose resins generally employ either maleic acid (usually as the anhydride) or its *trans*-isomer fumaric acid as the unsaturated acid:

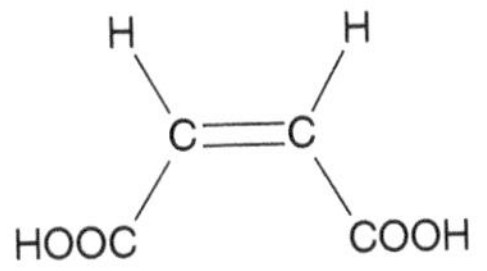

Maleic acid 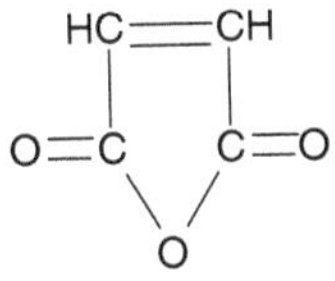   Maleic anhydride 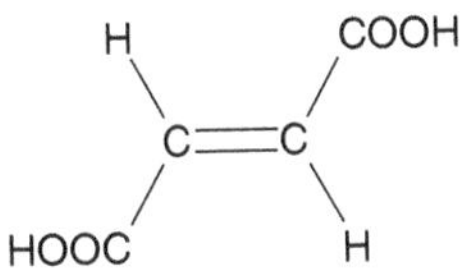   Fumaric acid

Maleic anhydride or fumaric acid confers the fundamental unsaturation to the polyester which provides the reactivity with coreactant monomers such as styrene. Maleic anhydride is a crystalline solid melting at 52.6°C (the acid melts at 130°C), while fumaric acid is a solid melting at 284°C. The latter is sometimes preferred to maleic anhydride because it is less corrosive, tends to give lighter colored products, higher impact strength, and slightly greater heat resistance.

Phthalic anhydride (melting point 131°C) is most commonly used to play the role of the saturated acid because it provides an inflexible link and maintains the rigidity in the cured resin. It is also preferred because its low price enables cheaper resins to be made. Use of isophthalic acid (melting point 347°C) in place of phthalic anhydride yields resins having higher heat distortion temperature and flexural moduli, better craze resistance, and often better water and alkali resistance. These resins are also useful in the preparation of resilient gel coats. Where a flexible resin is required, adipic acid may be used since, unlike the phthalic acids which give a rigid link, adipic acid gives highly flexible link and hence flexibility in the cured resin. Flexible resins are of value in gel coats.

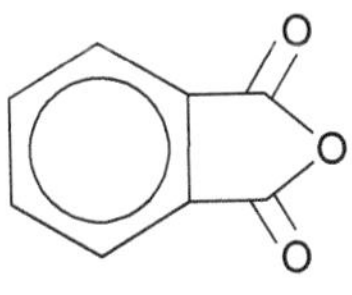

Phthalic anhydride 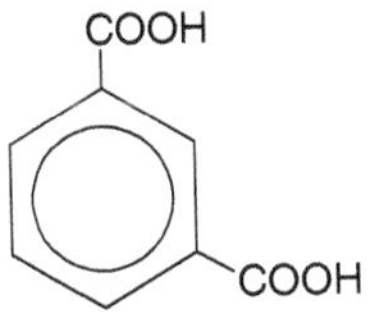   Isophthalic acid   Adipic acid

Polyester laminating resins are manufactured by heating the component acids and glycols for several hours at 150–200°C in steel reactor vessels with moderate agitation to prevent local overheating and overhead condenser systems to collect the aqueous by-products, initially under ordinary pressure and, in the last phase, under vacuum. To prevent discoloration and premature gelation caused by oxygen, the reactor is continuously purged with nitrogen or carbon dioxide. Reaction systems are fabricated from 304 or 316 stainless steels. Copper and brass valves and fittings are avoided because dissolved copper salts can affect the curing characteristics of the final resins.

A typical charge for a general purpose resin would be propylene glycol 170 parts, maleic anhydride 132 parts, and phthalic anhydride 100 parts, corresponding to molar ratios of 1.1: 0.67: 0.33. (The slight excess of glycol is primarily to allow for evaporation losses.) The glycols and dibasic acid can be substituted by other components of similar functionality, e.g., propylene glycol by ethylene and diethylene glycols, maleic anhydride by fumaric acid, and phthalic anhydride by isophthalic acid. A fusion-type reator process, also known as *fusion-melt process*, is commonly used for condensing liquid glycols with dibasic acids of low melting temperature, including phthalic anhydride and aliphatic dibasic acids. The components are added in liquid form to facilitate loading, that is, glycol followed by molten anhydrides, thus saving the heating time needed when solid anhydrides are used. Heat is supplied by hot oil circulating through internal coils.

The polycondensation reaction occurs according to third-order kinetics with the progressive development of higher molecular weight polymers aproaching an asymptotic limit. This requires extended reaction periods, e.g., 15 h at 190°C for phthalic resins to attain a satisfactory molecular weight. Steps must be taken to free the viscous melt of water, as the latter retards the reaction.

At 190°C, some glycol is vaporized and lost with the water of condensation. The reactors may be equipped with fractionating condenser systems to prevent this glycol loss. While the reaction rate is enhanced by acid catalysts, the latter also promote the formation of volatile ethers that are lost as by-products. A 5% excess glycol added to the initial charge compensates for glycol losses that may occur during the course of the reaction. Addition of glycol during the final stages is, however, not recommended.

The long reaction time in conventional fusion melt reactors, usually at temperatures above 180°C, causes the maleate ester to isomerize to the corresponding fumarate. This isomerization is of fundamental importance because the fumarate polymers display reactivities almost 20 times more than those of the maleate reaction products in subsequent polymerization reactions with styrene. The rate and extent of isomerization can be promoted at lower temperatures by a cycloaliphatic amine catalyst, e.g., morpholine.

The polyesterification reaction is followed by measuring the *acid number* (defined as the number of milligrams of potassium hydroxide equivalent to the acidity present in one gram of resin) of small samples periodically removed from the reactor. Where there are equimolar proportions of glycol and acid in the initial charge, the number average molecular weight is given by 5600/acid number. When the acid number value between 25 and 50 is reached, the heaters are switched off, the reactor is cooled to 150°C, and the contents transferred under vigorous agitation to a blend tank containing suitably inhibited styrene monomer at 30°C. The final temperature of the blend reaches about 80°C. At this temperature, even an inhibited styrenated resin polymerizes in several hours unless the blend is rapidly cooled to ambient temperature.

A mixture of inhibitors is commonly employed for the styrene diluent in order to obtain a balance of properties in respect of color, storage stability, and gelation rate of catalyzed resin. Thus, a typical composition of the diluent based on the above polyester formulation would be: styrene 172 parts, benzyltrimethylammonium chloride 0.44 part, hydroquinone 0.06 part, and quinone 0.006 part. After cooling to the ambient temperature, the resin is transferred into drums for storage and shipping.

Isophthalic acid is widely used as a substitute for phthalic anhydride since, as mentioned earlier, it improves certain properties of the cross-linked polymer. However, isophthalic acid is insoluble in the initial melt charge of maleic anhydride and propylene glycol and it also reacts more slowly with glycols than maleic anhydride. Thus while soluble components in the melt, namely, propylene glycol and maleic anhydride, react to form propylene glycol maleate polymers at lower temperatures (<190°C), isophthalic acid remains inert and requires heating subsequently to temperatures of 240°C to dissolve and react over prolonged periods, thus giving to rise to undesirable discoloration of the resin. Since maleic anhydride has a preference to form esters with the more reactive primary hydroxyl group on the propylene glycol molecule, producing propylene glycol maleate esters with a preponderance of terminal secondary hydroxyl groups, subsequent condensation with isophthalic acid proceeds much more slowly than if the available hydroxyl functionality was primary, thus contributing to the slowness of transesterification reactions. Furthermore, sufficient unreacted isophthalic acid may be present in the final polymer, thus leading to precipitation upon blending with styrene and hence to a hazy resin product.

To reduce reaction time and eliminate color problems of one-step processes, as explained above, a two-step process is used to produce isophthalic resins. In the first stage, only the isophthalic acid is reacted with the glycol at a relatively high temperature, which may be elevated rapidly to over 220°C without concern for discoloration. Since ethylene glycol and propylene glycol boil at lower temperatures under ordinary pressure, the reactors must be provided with fractionating condensers or operated under pressure to prevent glycol loss. At 220°C, a clear melt is obtained from isophthalic acid and propylene glycol in about 8 h. However, esterification catalysts, such as tetrabutyl titanate, stannous oxalate, and dibutyl tin oxide can be used to accelerate the reaction.

At the end of the first stage, the melt is cooled to 150°C, maleic anhydride is added, and the temperature is raised to 180°C to re-start the process of condensation. The progress of the reaction is monitored by the measurement of carboxylic functionality and viscosity. As the reaction proceeds rapidly, the temperature only needs to be elevated to 210°C to drive the reaction. The stability of isophthalic acid esters at higher temperatures allows the development of polymers with higher molecular weight. (In contrast, high-molecular weight phthalic resins cannot be produced by the fusion melt process, since phthalic anhydride sublimes above 200°C.)

1.3.1.4.2 *Polyester–Glass–Fiber Laminates (GRP, FRP)*

Methods of producing FRP laminates with polyesters have been described in Chapter 1 of *Plastics Fabrication and Recycling*. The major process today is the hand layup technique in which the resin is brushed or rolled into the glass mat (or cloth) by hand (see Figure 1.44 of *Plastics Fabrication and Recycling*). Since unsaturated polyesters are susceptible to polymerization inhibition by air, surfaces of the hand layup laminates may remain under-cured, soft, and, in some cases, tacky if freely exposed to air during the curing. A common way of avoiding this difficulty is to blend a small amount of paraffin wax (or other incompatible material) in with the resin. This blooms out on the surface and forms a protective layer over the resin during cure.

For mass production purposes, matched metal molding techniques involving higher temperatures and pressures are employed (see Figure 1.45 of *Plastics Fabrication and Recycling*). A number of intermediate techniques also exist involving vacuum bag, pressure bag, pultrusion, and filament winding (see Figure 1.46 through Figure 1.48 of *Plastics Fabrication and Recycling*).

Glass fibers are the preferred form of reinforcement for polyester resins. Glass fibers are available in a number of forms, such as glass cloth, chopped strands, mats, or rovings (see "Glass Fibers" in Chapter 1 of *Plastics Fabrication and Recycling*). Some typical properties of polyester–glass laminates with different forms of glass reinforcements are given in Table 1.11. it may be seen that laminates can have very high tensile strengths.

Being relatively cheaper, polyesters are preferred to epoxide and furan resins for general-purpose laminates. Polyesters thus account for no less than 95% of the low-pressure laminates produced. The largest single outlet is in sheeting for roofing and building insulation. For the greatest transparency of the laminate the refractive indices of glass-cured resins and binder should be identical.

The second major outlet is in land transport. Polyester–glass laminates are used in the building of sports car bodies, translucent roofing panel in lorries, and in public transport vehicles. In such applications the ability to construction large polyester–glass moldings without complicated equipment is used to advantage. Polyester resins in conjunction with glass cloth or mat are widely used in the manufacture of boat hulls up to 153 ft. (∼46 m) in length. Such hulls are competitive in price and are easier to maintain and to repair [14].

The high strength-to-weight ratio, microwave transparency, and corrosion resistance of the laminates have led to their use in air transport applications as in aircraft radomes, ducting, spinners, and other parts. Land, sea, and air transport applications account for nearly half the polyester resin produced.

TABLE 1.11   Typical Properties of Polyester–Glass Laminates

Property	Mat Laminate (hand layup)	Mat Laminate (press formed)	Fine Square Woven Cloth Laminate	Rod from Rovings
Specific gravity	1.4–1.5	1.5–1.8	2.0	2.19
Tensile strength				
10^3 lbf/in.^{2}	8–17	18–25	30–45	150
MPa	55–117	124–173	210–310	1030
Flexural strength				
10^3 lbf/in.^{2}	10–20	20–27	40–55	155
MPa	69–138	138–190	267–380	1100
Flexural modulus				
10^5 lbf/in.^{2}	5	6	10–20	66
MPa	3440	4150	6890–1380	45,500
Dielectric constant (10^6 Hz)	3.2–4.5	3.2–4.5	3.6–4.2	—
Power factor (10^6 Hz)	0.02–0.08	0.02–0.08	0.02–0.05	Z—
Water absorption (%)	0.2–0.8	0.2–0.8	0.2–0.8	—

Source: From Brydson, J. A. 1982. *Plastics Materials.* Butterworth Scientific, London, UK.

Other applications include such diverse items as chemical storage vessels, chemical plant components, swimming pools, stacking chairs, trays, and sports equipment.

1.3.1.4.3 *Polyester Molding Compositions*

Four types of polyester molding compounds may be recognized [14]: (1) dough-molding compound (DMC), (2) sheet-molding compound (SMC), (3) alkyd-molding compositions, sometimes referred to as *polyester alkyds,* (4) diallyl phthalate (DAP) and diallyl isophthalate (DAIP) compounds.

Dough-molding compounds of puttylike consistency are prepared by blending resins, catalyst, powdered mineral filler, reinforcing fiber (chopped strand), pigment, and lubricant in a dough mixer, usually of the Z-blade type. Formulations for three typical DMC grades are given in Table 1.12.

The tendency of thick sections of DMC structural parts to crack has been overcome by using low-profile polyester resins (or low-shrink resins). These are prepared by making a blend of a thermoplastic (e.g., acrylic) polymer–styrene system with a polyester–styrene system. Moldings of this blend cured at elevated temperatures exhibit negligible shrinkage and minimal warpage and have very smooth surface, to which paint may be applied with very little pretreatment.

A wide spectrum of properties may be obtained by varying the ratios of thermoplastics, polyester, and styrene in the blend. Among the thermoplastics quoted in the literature for such blending are poly (methyl methacrylate), polystyrene, PVC, and polyethylene. High-gloss DMCs using low-shrink resins have found uses in kitchen appliances such as toaster end plates, steam iron bases, and casings for electric heaters.

TABLE 1.12   Typical Formulations of DMC Grades

Ingredients	Low-Cost General-Purpose	High-Grade Mechanical	High-Grade Electrical
Polyester resin	100	100	100
E glass (1/4-in. length)	20	—	90
E glass (1/2-in. length)	—	85	—
Sisal	40	—	—
Calcium carbonate	240	150	—
Benzoyl peroxide	1	1	1
Pigment	2	2	2
Calcium stearate	2	2	2

Source: From Brydson, J. A. 1982. *Plastics Materials.* Butterworth Scientific, London, UK.

Since the manufacture of DMC involves intensive shear that causes extensive damage to fibers, DMC moldings have less strength than GRP laminates. This problem is largely avoided with the sheet-molding compounds (SMC). In the SMC process, unsaturated polyester resin, curing systems, filler thickening agents, and lubricant are blended together and coated onto two polyethylene films. Chopped-glass rovings are supplied between the resin layers, which are then sandwiched together and compacted as shown in Figure 1.13. Thickening occurs by the reaction of free carboxyl end groups with magnesium oxide. This converts the soft, sticky mass to a handleable sheet, which takes usually a day or two. A typical formulation consists of 30% chopped-glass fiber, 30% ground limestone and resin. For molding, the sheet may be easily cut to the appropriate weight and shape and placed between the halves of the heated mold. The main applications of SMCs are in car parts, baths, and doors.

The polyester "alkyd" molding compositions are also based on a polyester resin similar to those used for laminating. (The term alkyd is derived from *alcohol* and *acid*.) They are prepared by blending the resin with cellulose pulp, mineral filler, pigments, lubricants, and peroxide curing systems on hot mills to the desired flow properties. The mix is then removed, cooled, crushed, and ground.

Diallyl phthalate (DAP) is a diester of phthalic acid and allyl alcohol and contains two double bonds.

$$\text{(benzene ring)}\begin{array}{l}-\overset{\overset{\displaystyle O}{\|}}{C}-O-CH_2-CH=CH_2 \\ -\underset{\underset{\displaystyle O}{\|}}{C}-O-CH_2-CH=CH_2 \end{array}$$

On heating with a peroxide, DAP therefore polymerizes and eventually cross-links, forming an insoluble network polymer. However, it is possible to heat the DAP monomer under carefully controlled conditions, to give a soluble and stable partial polymer in the form of a white powder. The powder may then be blended with peroxide catalysts, fillers, and other ingredients to form a molding powder in the same manner as polyester alkyds. Similar products can be obtained from diallyl isophthalate (DAIP).

Both DAP and DAIP alkyd moldings are superior to the phenolics in their tracking resistance and in their availability in a wide range of colors; however, they tend to show a higher shrinkage on cure. The

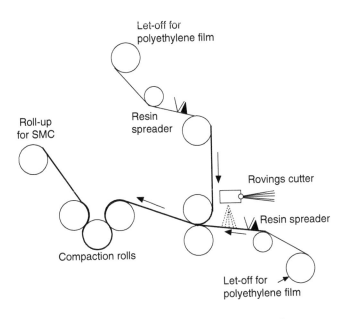

FIGURE 1.13 Schematic outline of machine used for sheet-molding compounds.

DAIP materials are more expensive than DAP but have better heat resistance. They are supposed to be able to withstand temperatures as high as 220°C for long periods.

The polyester alkyd resins are cheaper than the DAP alkyd resins but are mechanically weaker and do not maintain their electrical properties as well under severe humid conditions. Some pertinent properties of the polyester molding composition as compared in Table 1.13 along with those of a GP phenolic composition. The alkyd molding compositions are used almost entirely in electrical applications where the cheaper phenolic and amino resins are not suitable.

1.3.1.5 Aromatic Polyesters

Monomers	Polymerization	Major Uses
p-Hydroxy benzoic acid, bisphenol A, diphenyl isophthalate	Bulk polycondensation	High-temperature engineering thermoplastics, plasma coatings, abradable seals

In the 1960s the Carborundum Company introduced the homopolymer of p-hydroxybenzoic acid under the trade name Ekonol [35]. It is used in plasma coating. This wholly aromatic homopolyester is produced in practice by the self-ester exchange of the phenyl ester of p-hydroxybenzoic acid.

$$HO-\langle C_6H_4 \rangle-\overset{O}{\underset{||}{C}}-O-C_6H_5 \xrightarrow{-C_6H_5OH} \left[O-\langle C_6H_4 \rangle-\overset{O}{\underset{||}{C}} \right]_n$$

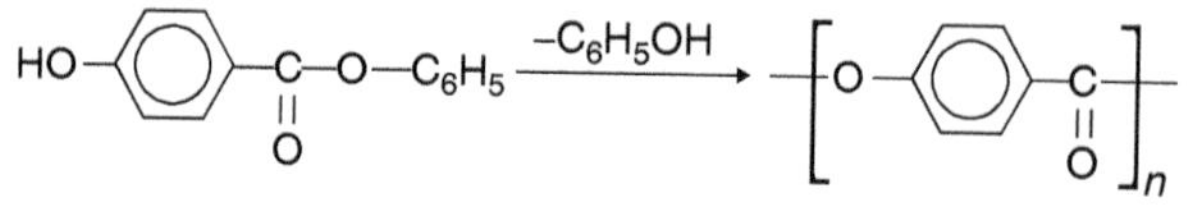

The homopolyester (mol. wt. 8000–12,000) is insoluble in dilute acids and bases and all solvents up to their boiling points. It melts at about 500°C and is difficult to fabricate. It can be shaped only by hammering (like a metal), by impact molding and by pressure sintering (420°C at 35 MPa). The difficulty in fabrication has severely limited the wider application of these polymers.

The homopolyester is available as a finely divided powder in several grades, based on particle size. The average particle size ranges from 35 to 80 μm. The material can be blended with various powdered

TABLE 1.13 Properties of Thermosetting Polyester Moldings

Property	Phenolic (GP)	DMC (GP)	Polyester Alkyd	DAP Alkyd	DAIP Alkyd
Molding temperature (°C)	150–170	140–165	140–165	150–165	150–165
Cure time (cup flow test) (sec)	60–70	25–40	20–30	60–90	60–90
Shrinkage (cm/cm)	0.007	0.004	0.009	0.009	0.006
Specific gravity	1.3	2.0	1.7	1.6	1.8
Impact strength (ft.-lb)	0.12–0.2	2.0–4.0	0.13–0.18	0.12–0.18	0.09–0.13
Volume resistivity (ohm-m)	10^{12}–10^{14}	10^{16}	10^{16}	10^{16}	10^{16}
Dielectric constant (10^6 Hz)	4.5–5.5	5.6–6.0	4.5–5.0	3.5–5.0	4.0–6.0
Dielectric strength (90°C) (kV/cm)	39–97	78–117	94–135	117–156	117–156
Power factor (10^6 Hz)	0.03–0.05	0.01–0.03	0.02–0.04	0.02–0.04	0.04–0.06
Water absorption (mg)	45–65	15–30	40–70	5–15	10–20

Source: Brydson, J. A. 1982. *Plastics Materials.* Butterworth Scientific, London, UK.

metals, such as bronze, aluminum, and nickel-chrome, and is used in flame-spray compounds. Plasma-sprayed coatings are thermally stable, self-lubricating, and wear and corrosion resistant. Applications include abradable seals for jet aircraft engine parts.

The polymer can also be blended up to 25% with PTFE. Such blends have good temperature and wear resistance and are self-lubricating. Applications include seals, bearings, and rotors.

Copolymeric aromatic polyesters, though possessing a somewhat lower level of heat resistance are easier to fabricate than are the wholly aromatic polymers; they also possess many properties that make them of interest as high-temperature materials. These materials, called *polyarylates*, are copolyester of terephthalic acid, and bisphenol A in the ratio of 1:1:2.

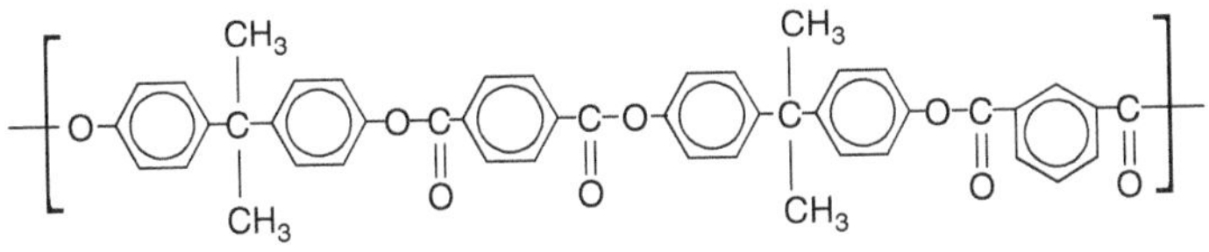

The use of two isomeric acids leads to an irregular chain which inhibits crystallization. This allows the polymer to be processed at much lower temperatures than would be possible with a crystalline homopolymer. Nevertheless the high aromatic content of these polyesters ensures a high T_g ($\sim 90°C$). The polymer is self-extinguishing with a limiting oxygen index of 34 and a self-ignition temperature of 545°C. The heat-deflection temperature under load (1.8 MPa) is about 175°C.

Among other distinctive properties of the polyarylate are its good optical properties (luminous light transmission 84–88% with 1–2% haze, refractive index 1.61), high impact strength between that polycarbonate and polysulfone, exceptionally high level of recovery after deformation (important in applications such as clips and snap fasteners), good toughness at both elevated and low temperatures with very little notch sensitivity, and high abrasion resistance which is superior to that of polycarbonates.

Polyarylates weatherability and flammability (high oxygen index, low flame spread) are inherent and are achieved without additives. The weatherability properties therefore do not deteriorate significantly with time. (Tests show that over 5000 h of accelerated weathering results in virtually no change in performance with respect to luminous light transmittance, haze, gloss, yellowness, and impact.) Having no flame-retardant additives, the combustion products of polyarylate are only carbon dioxide, carbon monoxide, and water, with no formation of toxic gas.

Several companies have marketed polyarylates under the trade names: U-polymer (Unitika of Japan), Arylef (Solvay of Belgium), Ardel (Union Carbide), and Arylon (Du Pont). These are noncrystallizing copolymers of mixed phthalic acids with a bisphenol and have repeat units of the type shown above. They are melt processable with T_g and heat distortion temperatures in the range of 150–200°C and have similar mechanical properties to polycarbonate and polyethersulfones (see later).

The polymers are useful for electrical and mechanical components that require good heat resistance and for lighting fixtures and consumer goods that operate at elevated temperatures, such as microwave ovens and hair dryers. Potential uses of the somewhat cheaper arylon type polyesters include exterior car parts, such as body panels and bumpers. Typical properties of aromatic polyesters mentioned above are shown in Table 1.14.

A different approach to obtaining polymers with good melt processability coupled with high softening point has led to another type of aromatic copolyester, the so-called *liquid crystalline polymers* (LCPs) (see Chapter 2). In these polymers, marketed under the trade names Vectra (Celanese) and Xydar (Dartco Manufacturing), the retention of liquid crystalline order in the melt gives lower melt viscosities than would otherwise be achieved. Heat distortion temperatures are also in the high range of 180–240°C (Table 1.14). LCPs have thus heralded a new era of readily molded engineering and electrical parts for high temperature use.

1.3.1.6 Wholly Aromatic Copolyester

A high-performance, wholly aromatic copolyester suitble for injection molding was commercial introduced in late 1984 by Dartco Manufacturing under the trade name Xydar. Xydar injection-molding resins are based on terephthalic acid, *p,p'*-dihydroxybiphenyl, and *p*-hydroxybenzoic acid.

Polymers of this class contain long relatively rigid chains which are thought to undergo parallel ordering in the melt, resulting in low melt viscosity and good injection-molding characteristics, although at relatively high melt temperatures—750°F to 806°F (400–430°F) (400–430°C). The melt solidifies to form tightly packed fibrous chains in the molded parts, which give rise to exceptional physical properties. The tensile modulus of the molded unfilled resin is 2.4×10^6 psi (16,500 MPa) at room temperature and 1.2×10^6 psi (8300 MPa) at 575°F (300°C). Tensile strength is about 20,000 psi (138 MPa), compressive strength is 6000 psi (41 MPa), and elongation is approximately 5%. Mechanical properties are claimed to improve at subzero temperatures.

The wholly aromatic copolyester is reported to have outstanding thermal oxidative stability, with a decomposition temperature in air of 1040 F (560°C) and 1053°F (567°C) in a nitrogen atmosphere. The resin is inherently flame retardant and does not sustain combustion. Its oxygen index is 42, and smoke generation is extremely low.

The resin is extremely inert, resists attack by virtually all chemicals, including acids, solvents, boiling water, and hydrocarbons. It is attacked by concentrated, boiling caustic but is unaffected by 30 days of immersion in 10% sodium hydroxide solution at 127°F (53°C). It withstands a high level of UV radiation and is transparent to microwaves.

The wholly aromatic copolyester for injection molding is available in filled and unfilled grades. It can be molded into thin-wall components at high speeds. The high melt flow also enables it to be molded into heavy-wall parts. No mold release is required because of the inherent lubricity and nonstick properties. No post-curing is necessary because the material is completely thermoplastic in nature. The material is expected to have many applications because of its moldability and its resistance to high temperatures, fire, and chemicals.

TABLE 1.14   Properties of Unfilled Aromatic Polyesters

Property	Ekonol	Arylef or Ardel	Arylon	Vectra (Range for Various Grades)
Tensile strength				
10^3 lbf/in.^{2}	11	10	10	20–35
MPa	74 (flexural)	70 (yield)	70	140–240
Elongation at break (%)	—	50	25	1.6–7
Tensile modulus				
10^5 lbf/in.^{2}	—	3.0	2.9	14–58
GPa	—	2.1	2.0	10–40
Flexural modulus				
10^5 lbf/in.^{2}	10	2.9	3.0	14–51
GPa	7.1	2.0	2.1	10–35
Heat distortion temperature (°C)	>550	175	155	180–240
Impact strength, notched Izod				
ft.-lbf/in.	—	2.8–4.7	5.4	1–10
J/m	—	150–250	288	53–530
Limiting oxygen index (%)	—	34	26	35–50

1.3.1.7 Polycarbonates

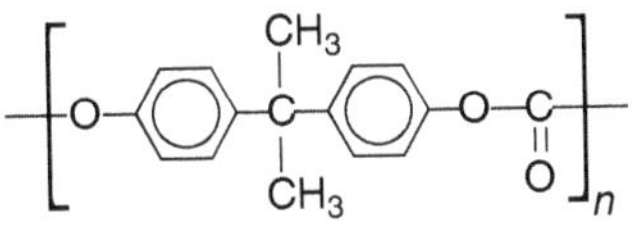

Monomers	Polymerization	Major Uses
Bisphenol A, phosgene	Interfacial polycondensation, solution polycondensation, transesterification	Glazing (37%), electrical and electronics (15%, appliances (15%), compact discs

The major processes for polycarbonate manufacture include (1) transesterification of bisphenol A with diphenyl carbonate [36,37]:

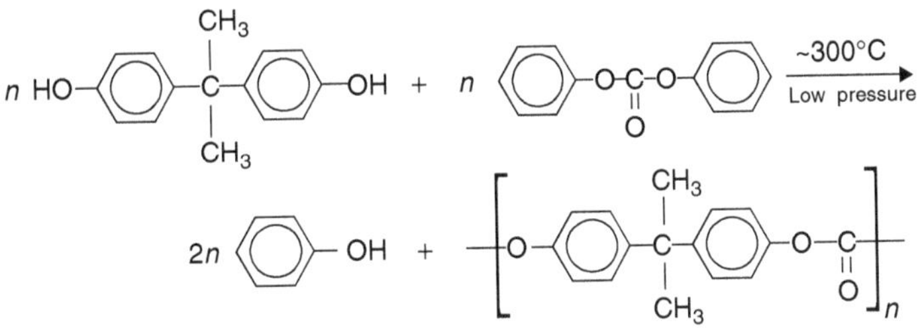

(2) solution phosgenation in the presence of an acid acceptor such as pyridine:

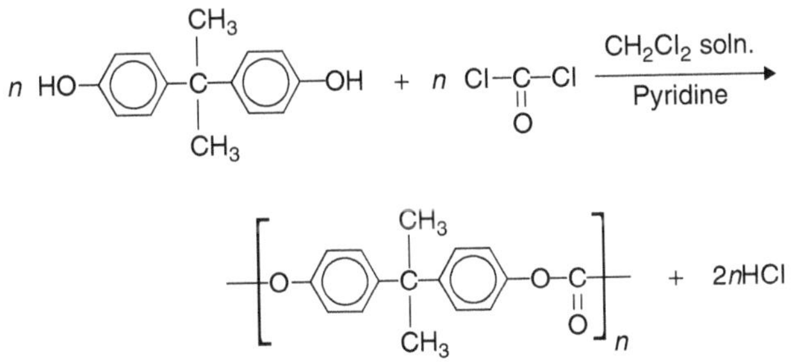

and (3) interfacial phosgenation in which the basic reaction is the same as in solution phosgenation, but it occurs at the interface of an aqueous phase and an organic phase. Here the acid acceptor is aqueous phase and an organic phase. Here the acid acceptor is aqueous sodium hydroxide, which dissolves the bisphenol A and a monohydric phenol used for molecular-weight control (without which very high-molecular-weight polymers of little commercial value will be obtained), and the organic phase is a solvent for phosgene and the polymer formed. A mixture of methylene chloride and chlorobenzene is a suitable solvent. The interfacial polycondensation method is the most important process at present for the production of polycarbonate. Interestingly, polycarbonate represents the first commercial application of interfacial polycondensation. Fire-retardant grades of polycarbonates are produced by using tetrabromobisphenol A as comonomer.

Polycarbonate resin is easily processed by all thermoplastic-molding methods. Although it is most often injection molded or extruded into flat sheets, other options include blow molding, profile extrusion, and structural foam molding. Polycarbonate sheet can be readily thermoformed. The resin should be dried to less than 0.02% moisture before processing to prevent hydrolytic degradation at the high temperatures necessary for processing.

The chemical resistance of polyester materials is generally limited due to the comparative ease of hydrolysis of the ester groups, but the bisphenol A polycarbonates are somewhat more resistant. This resistance may be attributed to the shielding of the carbonate group by the hydrophobic benzene rings on either side. The resin thus shows resistance to dilute mineral acids; however, it has poor resistance to alkali and to aromatic and chlorinated hydrocarbons.

Polycarbonates have an unusual combination of high impact strength (12–16 ft.-lbf per inch notch for 1/2-in. × 1/8-in. bar), heat-distortion temperature (132°C), transparency, very good electrical insulation characteristics, virtually self-extinguishing nature, and physiological inertness. As an illustration of the toughness of polycarbonate resins, it is claimed that an 1/8-in.-thick molded disc will stop a 22 caliber bullet, causing denting but not cracking. In creep resistance, polycarbonates are markedly superior to acetal and polyamide thermoplastics.

Because of a small dipole polarization effect, the dielectric constant of polycarbonates (e.g., 3.0 at 10^3 Hz) is somewhat higher than that for PTFE and the polyolefins (2.1–2.5 at 103 Hz). The dielectric constant is also almost unaffected by frequency changes up to 10^6 Hz and temperature changes over the normal range of operations. (Note that for satisfactory performance electrical insulating materials should have a *low* dielectric constant for *low* dissipation factor but *high* dielectric strength. For dielectrics used in capacitors, however, a high dielectric constant is desirable.)

At low frequencies (60 Hz) and in the ordinary temperature range (20–100°C), the power factor of polycarbonates ($\sim$0.0009) is remarkably low for a polar polymer. It increases, however, at higher frequencies, reaching a value of 0.010 at 10^6 Hz. The polycarbonates have a high volume resistivity (2.1 × 10^{20} ohm-cm at 23°C) and a high dielectric strength (400 kV/in., 1/8-in. sample). Because of the low water absorption, these properties are affected little by humidity. Polycarbonates, however, do have a poor resistance to tracking.

Although the electrical properties of polycarbonates are not as impressive as those observed with polyethylene, they are adequate for many purposes. These properties, coupled with the high impact strength, heat and flame resistance, transparency, and toughness have led to the extensive use of these resins in electronics and electrical engineering, which remains the largest single field of their application. Polycarbonate is the only material that can provide such a combination of properties, at least at a reasonable cost.

Known for many years, epoxy oligomers made from tetrabromobisphenol A are still used as the flame retardant in polycarbonates because they minimally affect the heat distortion temperature and even show a positive effect on impact strength. About 6–9 wt% of the epoxy oligomer is required for achieving V-0 rating and a thermotropic liquid crystal polyester helps to improve melt flow, so that thin-walled parts can be molded [38]. Antimony trioxide is not normally used in combination with halogen-containing additives in PC, because it causes loss of clarity.

At General Electric, it was found that very low additions (<1 wt%) of alkali or alkaline earth metal salts of certain arylsulfonates provide self-extinguishing performance to PC. Potassium diphenylsulfone sulfonate, sodium trichlorobenzene sulfonate, and potassium perfluorobutane sulfonate are effective in PC at one-tenth of a per cent level and these salts are used on a commercial scale. The salts are mostly active in the condensed phase where they strongly destabilize PC upon heating, thus promoting fast decomposition and melt flow which removes heat.

Phosphate esters are rarely used in plain PC because of partial loss of clarity, tendency to stress-cracking, and somewhat reduced hydrolytic stability. However, aromatic phosphates are currently the products of choice for flame-retarding PC-based blends [39]. Triphenyl phosphate and mixed tri(*t*-butylphenyl phenyl) phosphate are reasonably effective in PC/ABS blends and are used commercially, though they have the disadvantage of relatively high volatility. However, bridged aromatic diphenyl phosphates, especially resorcinol bis(diphenyl phosphate) and bisphenol A bis(diphenyl phosphate), have found much broader application than monophosphates because of good thermal stability, high efficiency, and low volatility. Nano-scale inorganic materials have been shown to improve fire-retardant performance of aromatic phosphates and provide enhanced thermal dimensional stability for PC/ABS blends.

Polycarbonate covers for time switches, batteries, and relays utilize the god electrical insulation characteristics in conjunction with transparency, toughness, and flame resistance of the polymer. Its combination of properties also accounts for its wide use in making coil formers. Many other electrical and electronic applications include moldings for computers, calculating machines and magnetic disc pack housing, contact strips, switch plates, and starter enclosures for fluorescent lamps. Polycarbonate films of high molecular weight are used in the manufacture of capacitors.

Traditional applications of polycarbonate in the medical market, such as filter housings, tubing connectors, and surgical staplers, have relied on the materials unique combination of strength, purity, transparency, and ability to stand all sterilization methods (steam, ethylene oxide gas, and gamma radiation). Polycarbonate-based blends blends and copolymers have further extended the materials usefulness to medical applications.

Recent years have seen a continuing growth of the market for polycarbonate glazing and light transmission units. Applications here include lenses and protective domes as well as glazing. The toughness and transparency of polycarbonates have led to many successful glazing applications of the polymer, such as bus shelters, telephone kiosks, gymnasium windows, lamp housings for street lighting, traffic lights, and automobiles, strip-lighting covers at ground level, safety goggles, riot-squad helmets, armor, and machine guards.

The limited scratch and weathering resistance of the polycarbonates is a serious drawback in these applications, and much effort is being directed at overcoming these problems. One approach is to coat the polycarbonate sheet with a glasslike composition by using a suitable priming material (e.g., *Margard*, marketed by the General Electric Company) to ensure good adhesion between coating and the base plastic.

Polycarbonates modified with ABS (acrylonitrile and styrene grafted onto polybutadiene) and MBS (methyl methacrylate and styrene grafted onto polybutadiene) resins have been available for many years. Usually used to the extent of 2–9%, the styrene-based terpolymers are claimed to reduce the notch sensitivity of the polycarbonate and to improve its resistance to environmental stress cracking while retaining from some grades the high impact strength of the unmodified polycarbonate. These materials find use in the electrical industry, in the automotive industry (instrument panels and glove compartment flaps), and for household appliances (coffee machine housings, hair drier housings, and steam handles). Elastomer modified polycarbonates have been used for automobile front ends and bumpers (e.g., 1982 Ford Sierra).

Polycarbonate is used for making compact audio discs, which are based on digital recording and playback technology and can store millions of bits of information in the form of minute pits in an area only which is read by the laser, Each "track" comprising a spiral of these its is laid in polycarbonate which is backed with reflective aluminum and coated with a protective acrylic layer.

The processability of the polycarbonate, or any other material used as the substrate, is crucial in the manufacture of all optical discs. Bayers polycarbonate grade Makrolon CD-2000 has been specially developed to fit such requirements.

1.3.2 Polyamides

The early development of polyamides started with the work of W.H. Carothers and his colleagues, who, in 1935, first synthesized nylon-6,6—a polyamide of hexamethylene diamine and adipic acid—after extensive and classical researches into condensation polymerization.

Commercial production of nylon-6,6 and its conversion into fibers was started by the Du Pont Company in 1939. In a parallel development in Germany, Schlack developed polyamides by ring-opening polymerization of cyclic lactams, and nylon-6 derived from caprolactam was introduced in 1939. Today nylon-6,6 and nylon-6 account for nearly all of the polyamides produced for fiber applications.

Nylon-6,6 and nylon-6 are also used for plastics applications. Besides these two polyamides, very many other aliphatic polyamides, have been prepared in the laboratory, and a few of them (nylon-11, nylon-12, and nylon-6,10 in particular) have attracted specialized interest as plastics materials. However, only about 10% of the nylons produced are used for plastics production. Virtually all of the rest goes for the production of fibers where the market is shared, roughly equally, between nylon-6 and nylon-6,6.

(Nylon is the trade name for the polyamides from unsubstituted, non-branched aliphatic monomers. A polyamide made from either an amino acid or a lactam is called nylon-*x*, where *x* is the number of carbon atoms in the repeating unit. A nylon made from a diamine and a dibasic acid is designated by two numbers, in which the first represents the number of carbons in the diamine chain and the second the number of carbons in the dibasic acid.)

For a variety of technical reasons the development of aromatic polyamides was much slower in comparison. Commercially introduced in 1961, the aromatic polyamides have expanded the maximum temperature well above 200°C. High-tenacity, high-modulus polyamide fibers (aramid fibers) have provided new levels of properties ideally suited for tire reinforcement. More recently there has been considerable interest in some new aromatic glassy polymers, in thermoplastic polyamide elastomers, and in a variety of other novel materials.

1.3.2.1 Aliphatic Polyamides

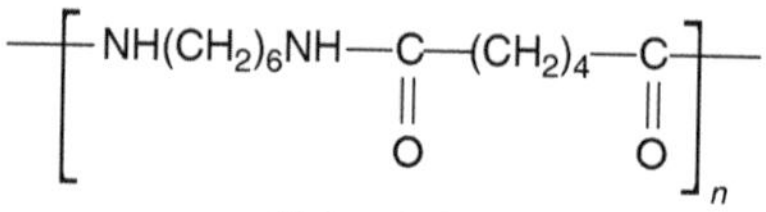

$$\left[-NH(CH_2)_6NH-\underset{\underset{O}{\|}}{C}-(CH_2)_4-\underset{\underset{O}{\|}}{C} \right]_n$$

Nylon-6,6

$$\left[-NH(CH_2)_5-\underset{\underset{O}{\|}}{C} \right]_n$$

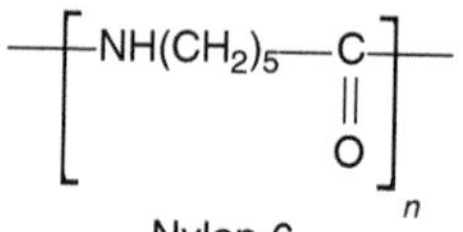

Nylon-6

Monomers	Polymerization	Major Uses
Adipic acid, hexamethylenediamine, caprolactam	Bulk polycondensation	Home furnishings, apparel, tire cord

Aliphatic polyamides are produced commercially by condensation of diamines with dibasic acids, by self-condensation of an amino acid, or by self-condensation of an amino acid, or by ring-opening polymerization of a lactam [14,40,41]. To obtain polymers of high molecular weight, there should be stoichiometric equivalence of amine and acid groups of the monomers. For amino acids and lactams the stoichiometric balance is ensured by the use of pure monomers; for diamines and dibasic acids this is readily obtained by the preliminary formation of a 1:1 ammonium salt, often referred to as a *nylon salt.* Small quantities of monofunctional compounds are often used to control the molecular weight.

The nylon-6,6 salt (melting point 190–191°C) is prepared by reacting hexamethylenediamine and adipic acid in boiling methanol, so that the comparatively insoluble salt precipitates out. A 60% aqueous slurry of the salt together with a trace of acetic acid to limit the molecular weight to the desired level (9000–15,000) is heated under a nitrogen blanket at about 220°C in a closed autoclave under a pressure of about 20 atmospheres (atm). The polymerization proceeds to approximately 80–90% without removal of by-product water. The autoclave temperature is then raised to 270–300°C, and the steam is continuously driven off to drive the polymerization to completion.

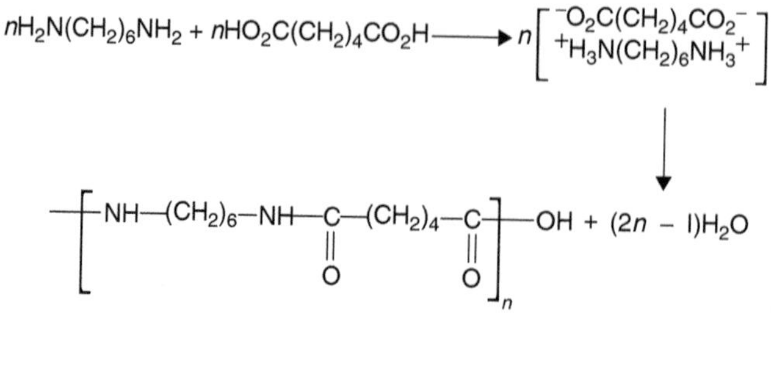

$$nH_2N(CH_2)_6NH_2 + nHO_2C(CH_2)_4CO_2H \longrightarrow n\left[\begin{array}{c} ^-O_2C(CH_2)_4CO_2^- \\ ^+H_3N(CH_2)_6NH_3^+ \end{array} \right]$$

$$\left[-NH-(CH_2)_6-NH-\underset{\underset{O}{\|}}{C}-(CH_2)_4-\underset{\underset{O}{\|}}{C} \right]_n -OH + (2n - 1)H_2O$$

The later stages of polymerization reaction constitute a melt *polycondensation*, since the reaction temperature is above the melting point of the polyamide. The molten polymer is extruded by nitrogen pressure on to a water-cooled casting wheel to form a ribbon which is subsequently disintegrated. In a continuous process for the production of nylon-6,6 similar reaction conditions are used, but the reaction mixture moves slowly through various zones of a reactor.

Nylon-6,10 is prepared from the salt (melting point 170°C) of hexamethylenediamine and sebacic acid by a similar technique. Nylon-6,9 uses azelaic acid. Decane-1,10-dicarboxylic acid is used for nylon-6,12.

In a typical batch process for the production of nylon-6 by ring-opening polymerization, a mixture of caprolactam, water (5–10% by weight), which acts as a catalyst, and a molecular-weight regulator [e.g., acetic acid ($\sim$0.1%)] is heated in a reactor under a nitrogen blanket at 250°C for above 12 h, a pressure of about 15 atm being maintained by venting off steam. The product consists of high-molecular-weight polymer (about 90%) and low-molecular-weight material (about 10%), which is mainly monomer. To obtain the best physical properties, the low-molecular-weight materials may be removed by leaching and/or by vacuum distillation.

In the continuous process, similar reaction conditions are used. In one process a mixture of molten caprolactam, water, and acetic acid is fed continuously to a reactor operating at about 260°C. Residence time is 18–20 h.

A simpler technique for the preparation of nylon-6 is the *polymerization casting* of caprolactam in situ in the mold. In this process rapid formation of polymer is achieved by anionic polymerization, initiated by strong bases such as metal amides, metal hydrides, and alkali metals. However, the anionic polymerization of lactams by strong bases alone is relatively slow because it is associated with an *induction period* due to a slow step in the initiation sequence leading to an *N*-acyl lactam which participates in the propagation reaction. The induction period may, however, be eliminated by adding along with the strong base a preformed *N*-acyl lactam or related compound at the start of the reaction.

A typical system for polymerization casting of caprolactam thus uses as a catalyst 0.1–1 mol% *N*-acetyl caprolactam and 0.15–0.50 mol% of the sodium salt of caprolactam. The reaction temperature is initially about 150°C, but during polymerization it rises to about 200°C. The technique is especially applicable to the production of large, complex shapes that could not be made by the more conventional plastics processing techniques.

Important advantages of the process are the low heats and low pressures involved. Although the polymerization process is exothermic, the relatively low heat of polymerization of caprolactam, coupled with its low melting point, makes the process easy to control and simplifies the heat transfer problem generally associated with the production of massive parts. Moldings of cast nylon-6 up to 1 ton are claimed to have been produced by these techniques.

Nylon parts made by polymerization casting of caprolactam exhibit higher molecular weights and a highly crystalline structure and are, therefore, slightly harder and stiffer than conventionally molded nylon-6.

Applications for cast nylon-6 include hug gears (e.g., a 150-kg nylon gear for driving a large steel drum drier) and bearings, gasoline and fuel tanks, buckets, building shutters, and various components for paper production machinery and mining and construction equipment. Later development has centered on adding reinforcing materials to the monomer before polymerization to produce parts with higher heat distortion temperature, impact strength and tensile strength.

Nylon-12 is produced by the ring-opening polymerization of laurolactam (dodecyl lactam) such as by heating the lactam at about 300°C in the presence of aqueous phosphoric acid. Unlike the polymerization of caprolactam, the polymerization of dodecyl lactam does not involve an equilibrium reaction. Hence, an almost quantitative yield of nylon-12 polymer is obtained by the reaction, and the removal of low-molecular-weight material is unnecessary.

Nylon-11 is produced by the condensation polymerization of ω-aminoundecanoic acid at 200–220°C with continuous removal of water. The latter stages of the reaction are conducted under reduced pressure to drive the polymerization to completion.

Nylon copolymers can be obtained by heating a blend of two or more different nylons above the melting point so that amide interchange occurs. Initially, block copolymers are formed, but prolonged reaction leads to random copolymers. For example, a blend of nylon-6,6 and nylon-6,10 heated for 2 h gives a random copolymer (nylon-6,6–nylon-6,10) which is identical with a copolymer prepared directly from the mixed monomers. Other copolymers of this type are available commercially.

1.3.2.1.1 Properties

Aliphatic polyamides are linear polymers containing polar –CONH– groups spaced at regular intervals by aliphatic chain segments. The principal structural difference between the various types of nylon is in the length of aliphatic chain segments separating the adjacent amide groups. The polar amid groups give rise to high interchain attraction in the crystalline zones, and the aliphatic segments impart a measure of chain flexibility in the amorphous zones. This combination of properties yields polymers which are tough above their glass transition temperatures.

The high intermolecular attraction also accounts for high melting points of nylons, which are usually more than 200°C. The melting point, however, decreases (which facilitates processing) as the length of the aliphatic segment in the chain increases, as indicated in Table 1.15.

Because of the high cohesive energy and their crystalline state, the nylons are resistant to most solvents. They have exceptionally good resistance to hydrocarbons and are affected little by esters, alkyl halides, and glycols. There are only a few solvents for the nylons, of which the most common are formic acid, glacial acetic acid, phenols, and cresols. Alcohols generally have some swelling action and may dissolve some copolymers (e.g., nylon-6,6, nylon-6,10, nylon-6). Nylons have very good resistance to alkalis at room temperature. Mineral acids attack nylons, built the rate of attack depends on the nature and concentration of acids and the type of nylon. Nitric acid is generally active at all concentrations.

Because of the presence of amide groups, the nylons absorb water. Figure 1.14 shows how the equilibrium water absorption of different nylons varies with humidity at room temperature, and Figure 1.15 shows how the rate of moisture absorption of nylon-6,6 is affected by the environmental conditions. Since dimensional changes may occur as a result of water absorption this effect should be considered when dimensional accuracy is required in a specific application. Manufacturers commonly supply data on the dimensional changes of their products with ambient humidity.

The various types of nylon have generally similar physical properties, being characterized by high toughness, impact strength, and flexibility (Table 1.16). Mechanical properties of nylons are affected significantly by the amount of crystallization in the test piece, ambient temperature (Figure 1.16), and humidity (Figure 1.17), and it is necessary to control these factors carefully in the determination of comparative properties. Moisture has a profound plasticizing influence on the modulus. For example, the Youngs modulus values for nylon-6,6 and nylon-6 decreases by about 40% with the absorption of 2% moisture.

Nylons have extremely good abrasion resistance. This property can be further enhanced by addition of external lubricants and by providing a highly crystalline hard surface to the bearings. The surface

TABLE 1.15 Melt Temperatures of Aliphatic Polyamides

Polyamide	T_m (°C)
Nylon-6,6	265
Nylon-6,8	240
Nylon-6,10	225
Nylon-6,12	212
Nylon-6	230
Nylon-7	223
Nylon-11	188
Nylon-12	180

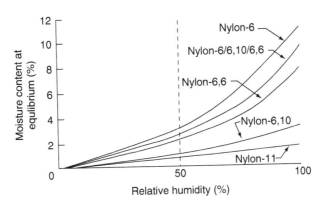

FIGURE 1.14   Effect of relative humidity on the equilibrium moisture absorption of the nylons.

crystallinity can be developed by the use of hot injection molds and by annealing in a nonoxidizing fluid at an elevated temperature (e.g., 150–200°C for nylon-6,6).

The coefficient of friction of nylon-6,6 is lower than mild steel but is higher than the acetal resins. The fractional heat buildup, which determines the upper working limits for bearing applications, is related to the coefficient of friction under working conditions. The upper working limits measured by the maximum LS value (the product of load L in psi on the projected bearing area and the peripheral speed S in ft./min) are 500–1000 for continuous operation of unlubricated nylon-6,6. For intermittent operation initially oiled nylon bearings can be used at LS values of 8000. Higher LS values can be employed with continuously lubricated bearings.

The electrical insulation properties of the nylons are reasonably good at room temperature, under conditions of low humidity, and at low frequencies. Because of the presence of polar amide groups, they are not good insulators for high-frequency work, and since they absorb water, the electrical insulation properties deteriorate as the humidity increases (see Figure 1.18).

The properties of nylons are considerably affected by the amount of crystallization and by the size of morphological structures, such as spherulites, which in turn, are generally influenced by the processing conditions. Thus, a molding of nylon-6, slowly cooled and subsequently annealed, may be 50–60% crystalline, whereas a rapidly cooled thin-walled molding may be only 10% crystalline.

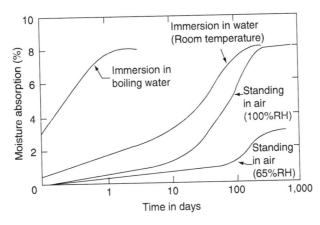

FIGURE 1.15   Effect of environmental conditions on rate of moisture absorption of nylon-6,6 (1/8-in.- thick specimens).

TABLE 1.16 Comparative Properties[a] of Typical Commercial Grades of Nylon

Property	6,6	6	6,10	11	12	6,6/6,10/6 (40:30:30)
Specific gravity	1.14	1.13	1.09	1.04	1.02	1.09
Tensile stress at yield						
10^3 lbf/in.^{2}	11.5	11.0	8.5	5.5	6.6	—
MPa	80	76	55	38	45	—
Elongation at break (%)	80–100	100–200	100–150	300	200	300
Tension modulus						
10^5 lbf/in.^{2}	4.3	4	3	2	2	2
10^2 MPa	30	28	21	14	14	14
Impact strength						
ft.-lbf/1/2-in. notch	1.0–1.5	1.5–3.0	1.6–2.0	1.8	1.9	—
Rockwell hardness	R118	R112	R111	R108	R107	R83
Heat distortion temperature (264 lbf/in.^{2}) (°C)	75	60	55	55	51	30
Coefficient of linear expansion 10^{-5} cm/cm/°C	10	9.5	15	15	12	—
Volume resistivity						
ohm-m (dry)	$>10^{17}$	$>10^{17}$	$>10^{17}$	—	—	—
ohm-m (50% RH)	10^{15}	—	10^{16}	—	—	10^{15}
Dielectric constant (10^3 Hz dry)	3.6–6.0	3.6–6.0	3.6–6.0	—	—	—
Power factor (10^3 Hz dry)	0.04	0.02–0.06	0.02	—	—	—
Dielectric strength (kV/cm) (25°C, 50% RH)	>100	>100	>100	—	—	—

[a] ASTM tests for mechanical and thermal properties.
Source: Brydson, J. A. 1982. *Plastics Materials*, Butterworth Scientific, London, UK.

Slowly cooled melts may form bigger spherulites, but rapidly cooled surface layers may be quite different from that of the more slowly cooled centers. The use of nucleating agents (e.g., about 0.1% of a fine silica) can give smaller spherulites and thus a more uniform structure in an injection molding. Such a product may have greater tensile strength, hardness, and abrasion resistance at the cost of some reduction in impact strength and elongation at break: the higher the degree of crystallinity the less the water absorption, and hence the less will be the effect of humidity on the properties of the polymer.

Nylon molding materials are available in a number of grades which many differ in molecular weight and/or in the nature of additives which may be present. The various types of additives used in nylon can be grouped as heat stabilizers, light stabilizers, lubricants, plasticizers, pigments, nucleating agents, flame retarders, and reinforcing fillers.

Heat stabilizers include copper salts, phosphoric acid esters, mercaptobenzothiazole, mercaptobenzi-midazole, and phenyl-β-naphthyl-amine. Among light stabilizers are carbon black and various phenolic materials. Self-lubricating grades of nylon which are of value in some gear and bearing applications incorporate lubricants such as molybdenum disulfide (0.2%) and graphite (1%).

Plasticizers may be added to nylon to lower the melting point and to improve toughness and flexibility particularly at low temperatures. A plasticizer used commercially is a blend of *o*- and *p*-toluene ethyl sulfonamide.

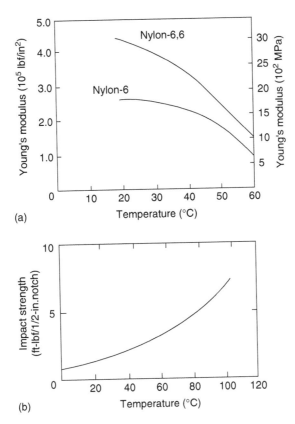

FIGURE 1.16 Effect of temperature on (a) Young's modulus of nylon-6,6 and nylon 6 and (b) impact strength of nylon-6,6.

Substances used as nucleating agents include silica and phosphorus compounds. Nucleating agents are used to control the size of morphological structures of the molding.

There have been substantial efforts to improve the flame resistance of nylons. Various halogen compounds (synergized by zinc oxide or zinc borate) and phosphorus compounds have been used (see the section on Flame Retardation in Chapter 1 of *Plastics Fundamentals, Properties, and Testing*). They are, however, dark in color.

Glass-reinforced nylons have become available in recent years. Two main types of glass fillers used are glass fibers and glass beads. From 20% to 40% glass in used. Compared to unfilled nylons, glass-fiber reinforcement leads to a substantial increase in tensile strength (160 vs. 80 MPa), flexural modulus (8000 vs. 3000 MPa), hardness, creep resistance (at least three times as great), and heat-distortion temperature under load (245 vs. 75°C under 264 psi), and to a significant reduction in coefficient of expansion (2.8×10^{-5} vs. 9.9×10^{-5} cm/cm-°C).

The glass-fiber-filled types can be obtained in two ways. One route involves passing continuous lengths of glass fiber (as rovings) through a polymer melt or solution to produce glass-reinforced nylon strand that is chopped into pellets. Another route involves blending a mixture of resin and glass fibers about 1/4 in. (0.6 cm) long in an extruder. Usually E-grade glass with a diameter of about 0.001 cm treated with a coupling agent, such as a silane, to improve the resin-glass bond is used.

Nylons filled with 4.0% glass spheres have a compressive strength about eightfold higher than unfilled grades, besides showing good improvement in tensile strength, modulus, and heat-distortion temperature. Having low melt viscosity, glass-bead-filled nylons are easier to process than the glass-fiber-filled varieties. They are also more isotropic in their mechanical properties and show minimum warpage. Glass fillers, both fibers and beads, tend to improve self-extinguishing characteristics of nylons.

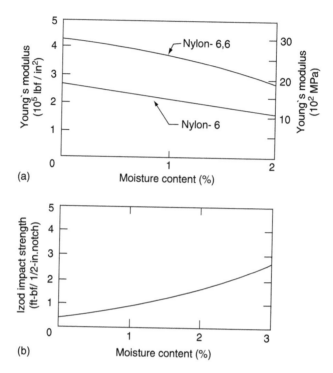

FIGURE 1.17 Effect of moisture content on (a) Young's modulus of nylon-6,6 and nylon-6 and (b) impact strength of nylon-6,6.

1.3.2.1.2 *Applications*

The most important application of nylons is as fibers, which account for nearly 90% of the world production of all nylons. Virtually all of the rest is used for plastic applications. Because of their high cost, they have not become general-purpose materials, such as polyethylene and polystyrene, which are available at about one-third the price of nylons. Nylons have nevertheless found steadily increasing application as plastics materials for specialty purposes where the combination of toughness, rigidity, abrasion resistance, reasonable heat resistance, and gasoline resistance is important.

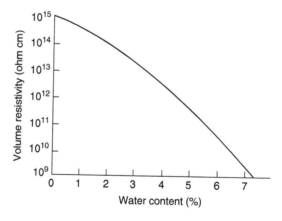

FIGURE 1.18 Effect of moisture content on the volume resistivity of nylon-6,6.

The largest plastics applications of nylons have been in mechanical engineering [14]—nylon-6, nylon-6,6, nylon-6,10, nylon-11, and nylon-12 being mainly used. These applications include gears, cams, bushes, bearings, and valve seats. Zippers made of nylon last longer than traditional ones of fabric or metal. Nylon moving parts have the advantage that they may often be operated without lubrication, and they may often be molded in one piece.

Among the aforesaid nylons, nylon-11 and nylon-12 have the lowest water absorption and are easy to process, but there is some loss in mechanical properties. For the best mechanical properties, the nylon-6,6 would be considered, but this material is also the most difficult to process and has high water absorption. Nylon-6 is easier to process but has even higher water absorption (see Figure 1.14).

Other applications include sterilizable nylon moldings in medicine and pharmacy, nylon hair combs, and nylon film for packaging foodstuffs (a typical example being milk pouches made of coextruded multilayered films of LDPE/LLDP/nylon-6, with nylon-6 as the barrier layer) and pharmaceutical products. The value of nylon in these latter applications is due to its low odor transmission and the boil-in-the-bag feature. Nylons have reasonable heat resistance. Spatula blades and spoons of nylon-6,6 withstand highest cooking temperatures.

Besides film, other extruded applications of nylons are as monofilaments, which have found applications in surgical sutures, brush tufting, wigs, sports equipment, braiding, outdoor upholstery, and angling.

Production of moldings by polymerization casting of caprolactam and the ability to produce large objects in this way have widened the use of nylon plastics in engineering and other applications. The process gives comparatively stress-free moldings having a reasonably consistent morphological structure with a 45–50% crystallinity, which is higher than melt-processed materials, and thus leads to higher tensile strength, modulus, hardness, and resistance to creep. Products made by polymerization casting include main drive gears for use in the textile and papermaking industries, conveyor buckets used in the mining industry, liners for coal-washing equipment, and propellers for small marine craft.

Glass-filled nylons form the most important group of glass-filled varieties of thermoplastics. Glass-reinforced nylon plastics have high rigidity, excellent creep resistance, low coefficient of friction, high heat-deflection temperature, good low-frequency electrical insulation properties, and they are nonmagnetic in nature. Therefore they have replaced metals in many applications.

Nylons reinforced with glass fibers are thus widely used in domestic appliances, in housings and casing, in car components, including radiator parts, and in the telecommunication field for relay coil formers and tag blocks. Glass-bad-filled nylons have found use in bobbins. Carbon-fiber reinforcement has been used with nylon-6 and nylon-6–nylon-12 mixtures. These materials have found use in the aerospace field and in tennis rackets.

A significant development is the appearance of supertough nylon plastics, which are blends in nylon-6,6 with other resins, such as an ionomer resin used in the initial grades or a modified ethylene–propylene–diene terpolymer rubber (EPDM rubber) used in later grades.

Nylon has been blended with PPO (Vydyne, Noryl GTX), PC, HDPE (Selar), PP, SAN, ABS (Triax, Elemid), PBT (Bexloy), and polyarylates (Bexlar). These blends have lower water absorption than nylon-6,6. Nylon-SAN, which has a high impact strength of 16 ft.-lb/in. of notch (854 J/m), has received a UL (Underwriters Laboratory) rating of 104°C. The nylon-arylate blend is transparent and has a heat-distortion temperature under load (264 lbf/in.^{2}) value of 154°C.

Interest has been aroused by the appearance of novel elastomeric polyamides. The products introduced by Huls under the designation XR3808 and X4006 may be considered as the polyether-amide analogue of the polyether-ester thermoplastic elastomers introduced in the 1970s by Du Pont as Hytrel (see Figure 1.10). The polyether-amide is a block copolymer prepared by the condensation of polytetramethylene ether glycol (i.e., polytetrahydrofuran) with laurin lactam and decane-1,10-dicarboxylic acid. The elastomeric polyamide XR3808 is reported to have a specific gravity of 1.02, yield stress of 24 MPa, a modulus of elasticity of 300 MPa, and an elongation at break of 360%.

1.3.2.2 Aromatic Polyamides

Aromatic polyamide fibers, better known as aramid fibers, have been defined as "a long chain synthetic polyamide in which at least 85% of the amide linkages are attached directly to two aromatic rings [42]." The first significant material of this type was introduced in 1961 by Du Pont as Nomex. It is poly (*m*-phenyleneisophthalamide), prepared from *m*-phenylenediamine and isophthaloyl chloride by interfacial polycondensation.

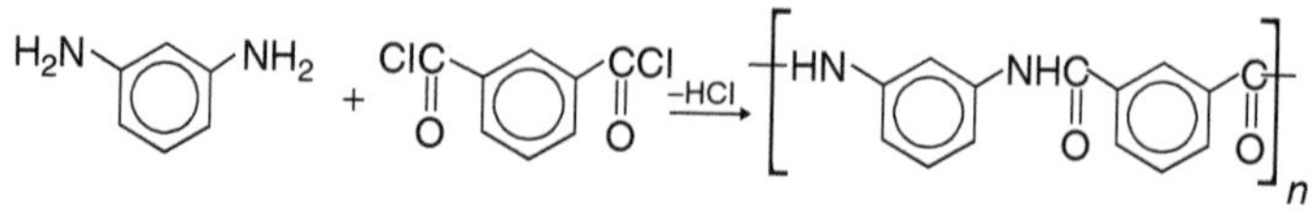

The fiber may be spun from a solution of the polymer in dimethylformamide containing lithium chloride. In 1973, Du Pont commenced production of another aromatic polyamide fiber, a poly(*p*-phenylene terephthalamide) marketed as Kevlar. It is produced by the reaction of *p*-phenylenediamine with terephthaloyl chloride in a mixture of hexamethylphosphoramide and *N*-methyl pyrrolidone (2:1) at −10°C.

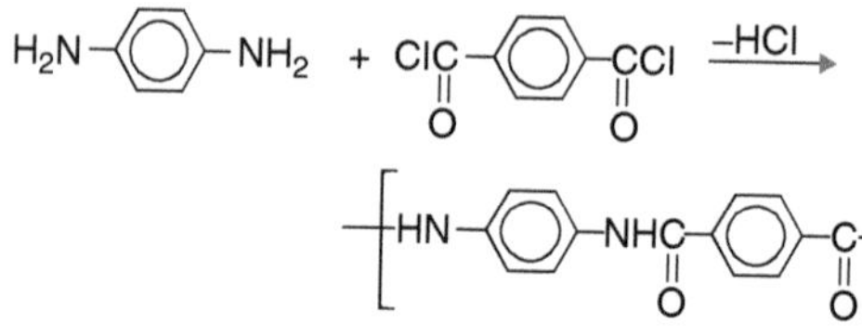

Kevlar fibers are as strong as steel but have one-fifth the weight. Kevlar is thus ideally suited as tire cord materials and for ballistic vests. The fibers have a high T_g ($>300°C$) and can be heated without decomposition to temperatures exceeding 500°C.

The dimensional stability of Kevlar is outstanding: It shows essentially no creep or shrinkage as high as 200°C. In view of the high melting temperatures of the aromatic polyamides and their poor solubility in conventional solvents, special techniques are required to produce the fibers. For example, Kevlar is wet spun from a solution in concentrated sulfuric acid.

Similar fiber-forming materials have been made available by Monsanto. Thus the product marketed as PABH-T X-500 is made by reacting *p*-aminobenzhydrazide with terephthaloyl chloride.

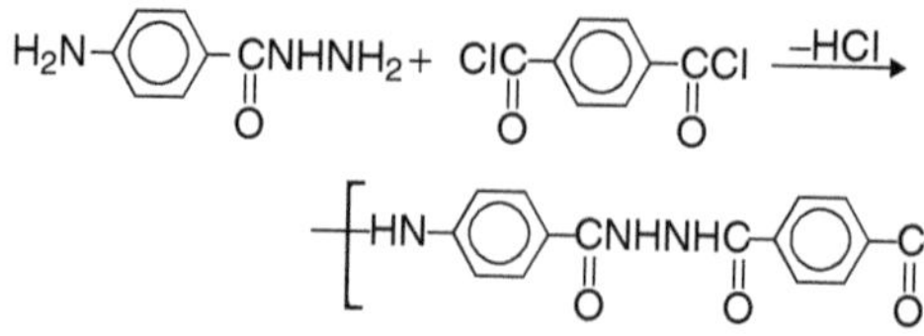

Polymers have also been prepared from cyclic amines such as piperazine and bis(*p*-aminocyclohexyl) methane. The latter amine is condensed with decanedioic acid to produce the silklike fiber Qiana (Du Pont).

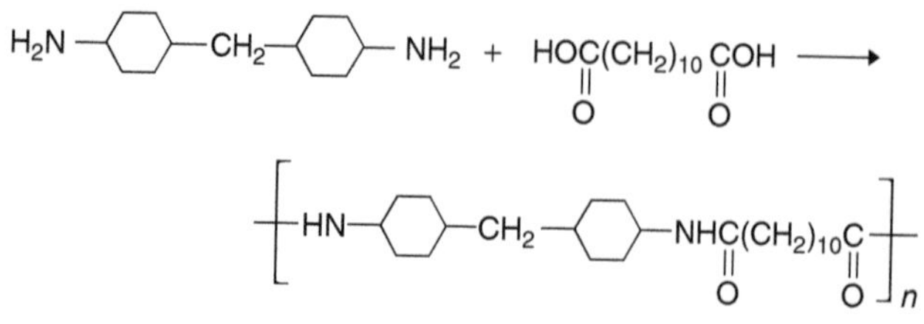

Qiana fibers have a high glass transition temperature (135°C, as compared to 90°C for nylon-6,6), which assures that the polymer will remain in the glassy state during fabric laundering and resist wrinkles and creases.

Synthetic fibers range in properties from low-modulus, high-elongation fibers like Lycra (see Section 1.3.4.1.3) to high-modulus high-tenacity fibers such as Kevlar. A breakthrough in fiber strength and stiffness has been achieved with Kevlar. Another high-performance fiber in commercial application is graphite. The use of these new fibers has resulted in the development of superior composite materials, generally referred to as fiber-reinforced plastics or FRPs (see Chapter 1 of *Plastics Fabrication and Recycling* and Chapter 2 of *Plastics Fundamentals, Properties, and Testing*), which have shown promise as metal-replacement materials by virtue of their low density, high specific strength (strength/density), and high specific modulus (modulus/density).

Today a host of these FRP products are commercially available as tennis rackets, golf clubs shaft, skis, ship masts, and fishing rods, which are filament wound with graphite and Kevlar fibers. Significant quantities of graphite composites and graphite/Kevlar hybrid composites are used in boeing 757 and 767 planes, which make possible dramatic weight saving. Boron, alumina, and silicon carbide fibers are also high-performance fibers but they are too expensive for large-scale commercial applications.

Partially aromatic, melt processable, polyamides are produced as random copolymers, which do not crystallize and are therefore transparent, but are still capable of high-temperature use because of their high T_g values. Several commercial polymers of this type that have glass-like clarity, high softening point, and oil and solvent resistance have been developed. For example, Trogamid T (Dynamit Nobel) contains repeat units of

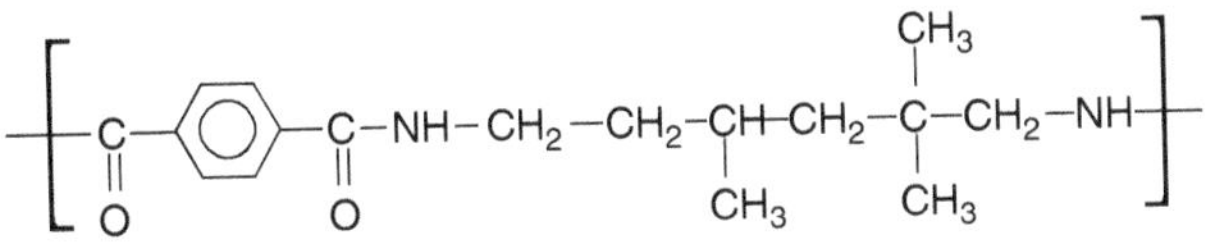

and of 2,4,4-trimethyl isomer, and has a T_g of about 150°C. Grilamid TR (Emser) with a T_g of about 160°C is a copolymer with units of

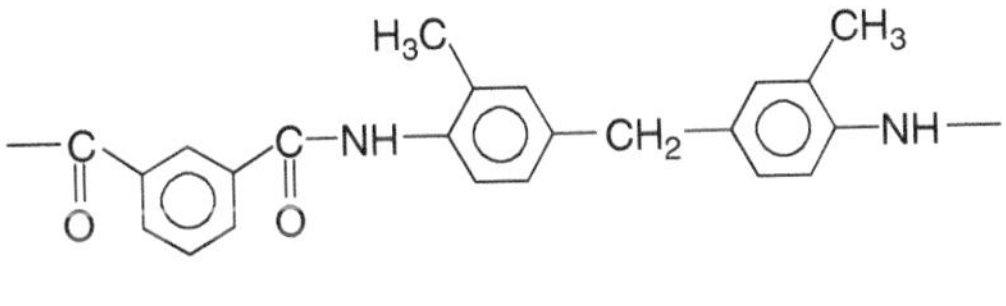

and of

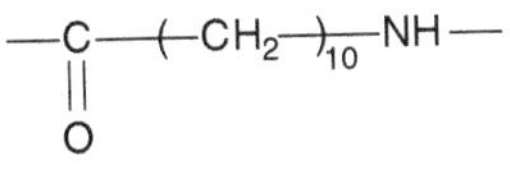

A crystalline, partially aromatic polyamide, poly-*m*-xylylene-adipamide, (also known as MXD-6) with repeat units of

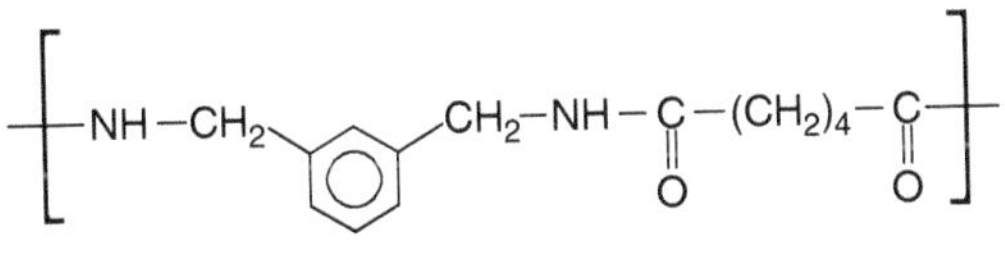

is available as a heat resistant engineering plastic (e.g., Ixef by Solvay) generally similar in properties to nylon-6,6, having a T_m of 243°C but with reduced water absorption, greater stiffness, and a T_g of about 90°C. Although its heat distortion temperature is only 96°C, with 30% glass filling this is increased to about 270°C. Typical properties of some of these polyamides, along with those of nylon-6,6 for comparison, are shown in Table 1.17.

Copolymers containing amide and imide units, the polyamideimides, are described in the following section on polyimides.

TABLE 1.17 Properties of Polyamides

Property	Nylon-6,6	Ixef(High-Impact Grade)	Trogamid T	Nomex (Fiber)	Kevlar
Tensile strength					
10^3 lbf/in.^{2}	9.4	25.7	8.7	97.2	435
MPa	65	177	60	670	3000
Tensile modulus					
10^5 lbf/in.^{2}	4.6	19.4	4.4	25.5	194
GPa	3.2	13.4	3.0	17.6	134
Elongation at break (%)	100	2.7	132	22	2.6
Flexural modulus					
10^5 lbf/in.^{2}	4.8	14.8	—	—	—
GPa	3.3	10.2	—	—	—
Impact strength, notched Izod					
ft.-lb/in.	1.3	3.0	—	—	—
J/m	69	159	—	—	—

1.3.2.3 Polyimides

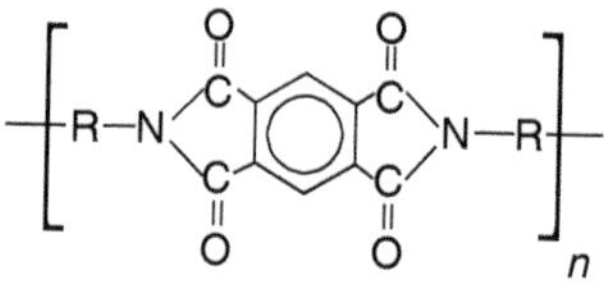

The polyimides have the characteristic functional group

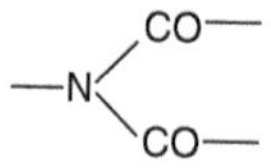

and are thus closely related to amides [43]. The branched nature of the imide functional group enables production of polymers having predominantly ring structures in the backbone and hence high softening points. Many of the structures exhibit such a high level of thermal stability that they have become important for application at much higher service temperatures than had been hitherto achieved with polymer.

The use of tetracarboxylic acid anhydride instead of the dicarboxylic acids used in the manufacture of polyamides yields polyimides. The general method of preparation of the original polyimides by the polymerization of pyromellitic dianhydride and aromatic diamine is shown in Figure 1.18a. A number of diamines have been investigated, and it has been found that certain aromatic amines, which include m-phenylendediamine, benzidine, and di-(4-aminophenyl)ether, give polymers with a high degree of oxidative and thermal stability.

The aromatic amine di(4-aminophenyl)ether is employed in the manufacture of polyimide film, designated as Kapton (Du Pont). Other commercial materials of this type introduced by Du Pont in the early 1960s included a coating resin (Pyre ML) and a machinable block form (Vespel). In spite of their high price these materials have found established uses because of their exceptional heat resistance and good retention of properties at high temperatures.

Since the polyimides are insoluble and infusible, they are manufactured in two stages. The first stage involves an amidation reaction carried out in a polar solvent (such as dimethylformamide and dimethylacetamide) to produce an intermediate poly(amic acid) which is still soluble and fusible. The poly(amic acid) is shaped into the desired physical form of the final product (e.g., film, fiber, coating, laminate) and then the second stage of the reaction is carried out.

In the second stage the poly(amic acid) is cyclized in the solid state to the polyimide by heating at moderately high temperatures above 150°C. A different approach, avoiding the intermediate poly(amic acid) step, was pioneered by Upjohn. The Upjohn process involves the self-condensation of the isocyanate of trimellitic acid, and the reaction by-product is carbon dioxide (Figure 1.19b).

Polypyromellitimides (Figure 1.19a) have many outstanding properties: flame resistance, excellent electrical properties, outstanding abrasion resistance, exceptional heat resistance, and excellent resistance to oxidative degradation, most chemicals (except strong bases), and high-energy radiation. After 1000 h of exposure to air at 300°C the polymers retained 90% of their tensile strength, and after 1500 h exposure to a radiation of about 10 rad at 175°C, they retained form stability, although they became brittle.

The first commercial applications of polypromellitimides were as wire enamels, as insulating varnishes, as coating for glass cloth (Pyre ML, Du Pont), and as film (Kapton, Du Pont). A fabricated solid grade was marketed as Vespel (Du Pont). Laminates were produced by impregnation of glass and carbon fiber, with the polyimide precursor followed by pressing and curing at about 200°C and further curing at temperatures of up to 350°C. Such laminates could be used continuously at temperatures up to 250°C and intermittently to 400°C. The laminates have thus found important application in the aircraft industry, particularly in connection with supersonic aircraft.

At the present time the applications of polyimides include compressor seals in jet engines, sleeves, bearings, pressure discs, sliding and guide rolls, and friction elements in data processing equipment, valve shafts in shutoff valves, and parts in soldering and welding equipment.

Polyimides have also found a number of specialist applications. Polyimide foams (Skybond by Monsanto) have been used for sound deadening of jet engines. Polyimides fibers have been produced by Upjohn and by Rhone-Poulenc (Kermel).

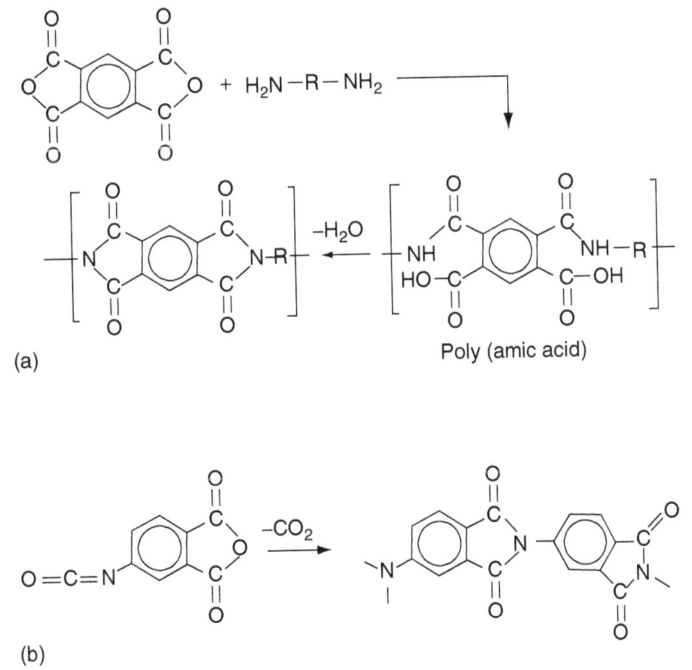

FIGURE 1.19 (a) Synthesis of polyimides by polycondensation. (b) Self-condensation of isocyanate of trimellitic acid.

A particular drawback of the polyimides is that they have limited resistance to hydrolysis and may crack in water or steam at temperatures above 100°C. Consequently, polyimides have encountered competition from polyetheretherketones (PEEK), which are not only superior in this regard but are also easier to mold.

1.3.2.3.1 Modified Polyimides

The application potential of polyimides is quite limited because, being infusible, they cannot be molded by conventional thermoplastics techniques [14]. In trying to overcome this limitation, scientists, in the early 1970s, developed commercially modified polyimides, which are more tractable materials than polyimides but still possessing significant heat resistance. The important groups of such modified polyimides are the polyamideimides (e.g., Torlon by Amoco Chemicals), the polybismaleinimides (e.g., Kinel by Rhone-Poulenc), the polyester–imides (e.g., Icdal Ti40 by Dynamit Nobel), and the polyether-imides (e.g., Ultem by General Electric).

If trimellitic anhydride is used instead of pyromellitic dianhydride in the reaction shown in Figure 1.19a, then polyamide-imide is formed (see Figure 1.20a). Other possible routes to this type of product involve the reaction of trimellitic anhydride with diisocyanates, (Figure 1.20b) or diurethanes (Figure 1.20c). Closely related is the Upjohn process for polyimide by self-condensation of the isocyanate of trimellitic acid, as illustrated in Figure 1.19b, although the product in this case is a true polyimide rather than a polyamide-imide.

Polyamide-imides may also be produced by reacting together pyromellitic dianhydride, a diamine, and a diacid chloride. Alternatively, it may be produced in a two-stage process in which a diacid chloride is reacted with an excess of diamine to produce a low-molecular-weight polyamide with amine end

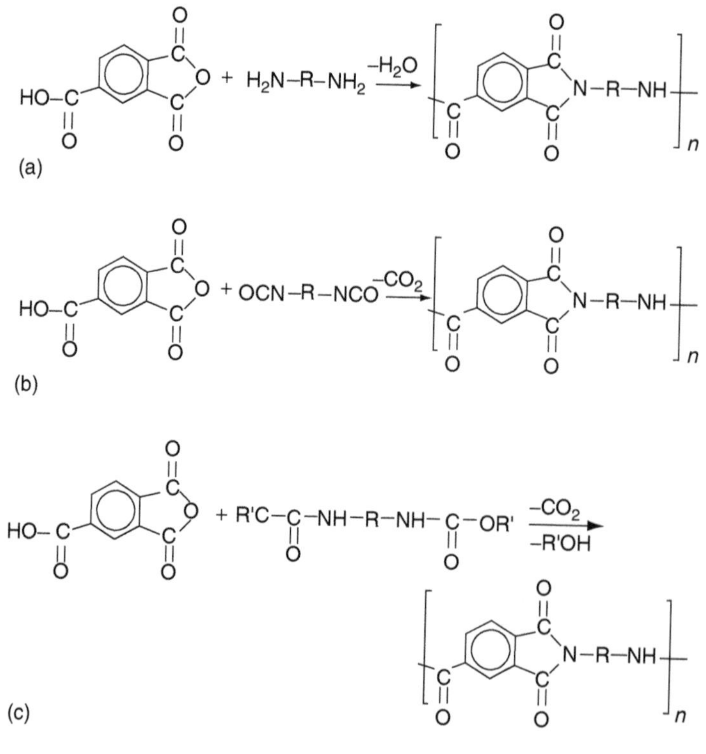

FIGURE 1.20 Synthesis of polyamide-imides from trimellitic anhydride and (a) diamine, (b) diisocyanate, and (c) diurethane.

groups which may then be chain extended by reaction with pyromellitic dianhydride to produce imide linkages.

The Torlon materials produced by Amoco Chemicals are polyamide-imides of the type shown in Figure 1.20a. Torlon has high strength, stiffness, and creep resistance, shows good performance at moderately high temperatures, and has excellent resistance to radiation. The polymers are unaffected by all types of hydrocarbons (including chlorinated and fluorinated products), aldehydes, ketones, ethers, esters, and dilute acids, but resistance to alkalis is poor.

Torlon has been marketed both as a compression-molding grade and as an injection-molding grade. The compression-molding grade, Torlon 2000, can accept high proportions of filler without seriously affecting many of its properties. For compression molding, the molding compound is preheated at 280°C before it is molded at 340°C at pressures of 4350 psi (30 MPa); the mold is cooled at 260°C before removal.

For injection molding, the melt at temperatures of about 355°C is injected into a mold kept at about 230°C. To obtain high-quality moldings, prolonged annealing cycles are recommended.

Uses of polyamide-imides include pumps, valves, refrigeration plant accessories, and electronic components. The polymers have low coefficient of friction, e.g., 0.2 (to steel), which is further reduced to as little as 0.02–0.08 by blending with graphite and Teflon. In solution form in *N*-methyl-2-pyrrolidone, Torlon has been used as a wire enamel, as a decorative finish for kitchen equipment, and as an adhesive and laminating resin in spacecraft.

The polyimides and polyamide-imides are produced by condensation reactions which give off volatile low-molecular-weight by-products. The polybismaleinimides may however be produced by rearrangement polymerization with no formation of by-products. The starting materials in this case are the bismaleimides, which are synthesized by the reaction of maleic anhydride with diamines (Figure 1.21).

The bismaleimides can be reacted with a variety of bifunctional compounds to form polymers by rearrangement reactions. These include amines, mercaptans, and aldoximes (Figure 1.22). If the reaction is carried out with a deficiency of the bifunctional compound, the polymer will have terminal double bonds to serve as a cure site for the formation of a cross-linked polymer via a double bond polymerization mechanism during molding. The cross-linked in this case occurs without the formation of any volatile by-products.

The Kinel materials produced by Rhone-Poulenc are polybismaleinimides of the type shown in Figure 1.22. These materials have chain-end double bonds, as explained previously, can be processed like conventional thermosetting plastics. The properties of the cured polymers are broadly similar to the polyimides and polyamide-imides. Molding temperatures are usually from 200°C to 260°C. Post-curing at 250°C for about 8 h is necessary to obtain the optimum mechanical properties.

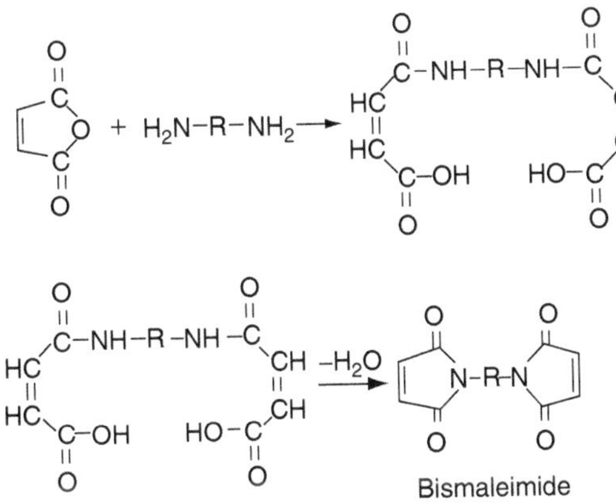

FIGURE 1.21 Synthesis of bismaleimides by the reaction of maleic anhydride with diamines.

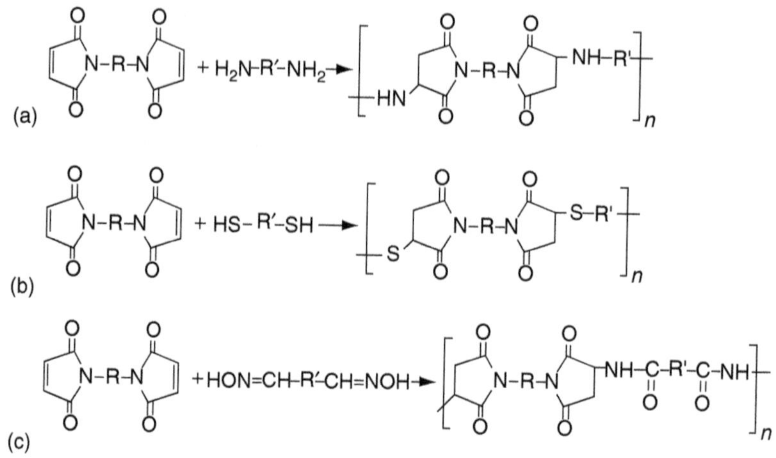

FIGURE 1.22 Formation of polymers by reaction of bismaleimides with (a) amines, (b) mercaptans, and (c) aldoximes.

Polybismaleinimides are used for making laminates with glass- and carbon-fiber fabrics, for making printed circuit boards, and for filament winding. Filled grades of polybismaleinimides are available with a variety of fillers such as asbestos, glass fiber, carbon fiber, graphite, Teflon, and molybdenum sulfide. They find application in aircraft, spacecraft, and rocket and weapons technology. Specific uses include fabrication of rings, gear wheels, friction bearings, cam discs, and brake equipment.

The polyester–imides constitute a class of modified polyimide. These are typified by the structure shown in Figure 1.23. Polyether-imides form yet another class of modified polyimide. These are high-performance amorphous thermoplastics based on regular repeating ether and imide linkages. The aromatic imide units provide stiffness, while the ether linkages allow for good melt-flow characteristics and processability.

One of several synthetic routes to polyetherimides of a general structure involves a cyclization reaction of form the imide rings and a displacement reaction to prepare the ether linkages and form the polymer

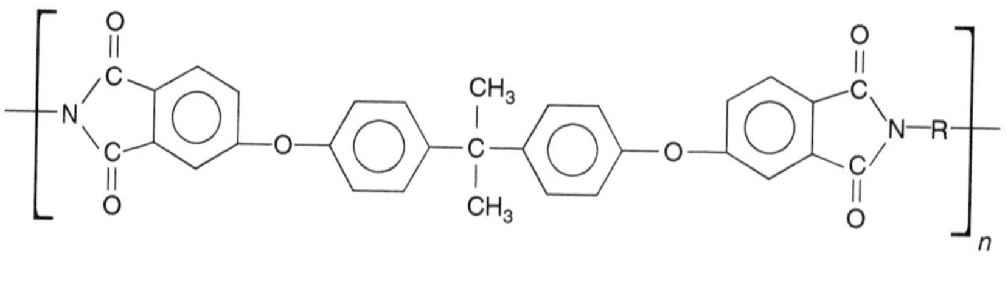

The first step of this synthesis is to form a bis-imide monomer formed by the reaction of nitrophthalic anhydride and a diamine (see Figure 1.21). The second step of polyetherimide synthesis involves the formation of a bisphenol dianion by treatment of a diphenol with two equivalents of base, followed by

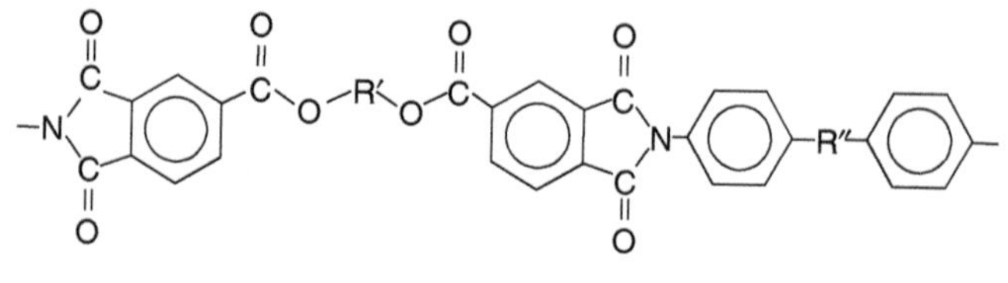

FIGURE 1.23 Typical structure of polyester–imides.

removal of water. The polymerization step involves displacement of the nitrogroups of the bis-imide by the bisphenol dianion to form the ether linkages of the polymer. A large number of polyetherimides can be prepared by the synthetic route.

Polyetherimides are suitable for applications that require high temperature stability, high mechanical strength, inherent flame resistance with extremely low smoke evolution, outstanding electrical properties over a wide frequency and temperature range, chemical resistance to aliphatic hydrocarbons, acids and dilute bases, UV stability, and ready processability on conventional equipment.

Ultem, introduced by General Electric in 1982, is a polyether-imide. It was designed to complete with heat- and flame-resistory, high-performance engineering polymers, polysulfones, and polyphenylene sulfide. Some typical properties of Ultem 1000 are specific gravity 1.27, tensile yield strength 105 MPa, flexural modules 3300 MPa, hardness Rockwell M109, Vicat softening point 219°C, heat-distortion temperature (1.82 MPa) 200°C, and limiting oxygen index 47. Specific applications include circuit breaker housings and microwave oven stirrer shafts.

Typical properties of unfilled polyimides are compared in Table 1.18.

1.3.3 Formaldehyde Resins

The phenol-formaldehyde and urea-formaldehyde resins are the most widely used thermoset polymers. The phenolic resins were the first truly synthetic polymers to be produced commercially. Both phenolic and urea resins are used in the highly cross-linked final form (C-stage), which is obtained by a stepwise polymerization process. Lower-molecular-weight prepolymers are used as precursors (A-stage resins), and the final form and shape are generated under heat and pressure. In this process water is generated in the form of steam because of the high processing temperatures. Fillers are usually added to reduced resin content and to improve physical properties. The preferred form of processing is compression molding.

The phenolics and urea resins are high-volume thermosets which owe their existence to the relatively low cost of the starting materials and their superior thermal and chemical resistance. Today these resins are widely used in molding applications, in surface coatings and adhesives, as laminating resins, casting resins, binders and impregnants, and in numerous other applications. However, as with all products based on formaldehyde, there is concern about the toxicity of these resins during processing and about the residual traces of formaldehyde in the finished product.

TABLE 1.18 Properties of Unfilled Polyimides

Property	Vespel (ICI)	Torlon (Amoco)	Kinel (Rhone-Poulenc)	Ultem (General Electric)
Tensile strength (MPa)				
25°C	90	186	~40	100
150°C	67	105	—	—
260°C	58	52	~25	—
Flexural modulus (GPa)				
25°C	3.5	4.6	3.8	3.3
150°C	2.7	3.6	—	2.5
260°C	2.3	3.0	2.8	—
Heat distortion temperature (°C)	357	282	—	200
Limiting oxygen index (%)	35	42	—	47

1.3.3.1   Phenol–Formaldehyde Resins

Monomers	Polymerization	Major Uses
Phenol, formaldehyde	Base- or acid-catalyzed stepwise polycondensation	Plywood adhesives (34%), glass-fiber insulation (19%), molding compound (8%)

Since the cross-linked polymer of phenol-formaldehyde reaction is insoluble and infusible, it is necessary for commercial applications to produce first a tractable and fusible low-molecular-weight prepolymer which may, when desired, be transformed into the cross-linked polymer [14,44,45]. The initial phenol-formaldehyde products (prepolymers) may be of two types: *resols* and *novolacs*.

1.3.3.1.1   Resols

Resols are produced by reacting a phenol with a molar excess of formaldehyde (commonly about 1:1.5–2) by using a basic catalyst (ammonia or sodium hydroxide). This procedure corresponds to Baekeland's original technique. Typically, reaction is carried out batchwise in a resin kettle equipped with stirrer and jacketed for heating and cooling. The resin kettle is also fitted with a condenser such that either reflux or distillation may take place as required.

A mixture of phenol, formalin, and ammonia (1–3% on the weight of phenol) is heated under reflux at about 100°C for 0.25–1 h, and then the water formed is removed by distillation, usually under reduced pressure to prevent heat hardening of the resin.

Two classes of resins are generally distinguished. Resols prepared with ammonia as catalysts are spirit-soluble resins having good electrical insulation properties. Water-soluble resols are prepared with caustic soda as catalyst. In aqueous solutions (with a solids content of about 70%) these are used mainly for mechanical grade paper and cloth laminates and in decorative laminates.

The reaction of phenol and formaldehyde in alkaline conditions results in the formation of *o*- and *p*-methylol phenols. These are more reactive towards formaldehyde than the original phenol and undergo rapid substitution with the formation of di- and trimethylol derivatives. The methylol phenols obtained are relatively stable in an alkaline medium but can undergo self-condensation to form dinuclear and polynuclear phenols (of low molecular weight) in which the phenolic nuclei are bridged by methylene groups. Thus in the base-catalyzed condensation of phenol and formaldehyde, there is a tendency for polynuclear phenols, as well as mono-, di-, and trimethylol phenols to be formed.

Liquid resols have an average of less than two phenolic nuclei per molecule, and a solid resol may have only three or four. Because of the presence of methylol groups, the resol has some degree of water tolerance. However, for the same reason, the shelf life of resols is limited.

Resols are generally neutralized or made slightly acidic before cure (cross-linking) is carried out. Network polymers are then obtained simply by heating, which results in cross-linking via the uncondensed methylol groups or by more complex mechanisms (see Figure 1.24). Above 160°C it is believed that quinone methide groups, as depicted on the bottom of Figure 1.24, are formed by condensation of the ether linkages with the phenolic hydroxyl groups. These quinone methide structures can be cross-linked by cycloaddition and can undergo other chemical reactions. It is likely that this formation of quinone methide and other related structures is responsible for the dark color of phenolic compression moldings made at higher temperatures. Note that cast phenol-formaldehyde resins, which are cured at much lower temperatures, are water white in color. If they are heated to about 180°C, they darken considerably.

1.3.3.1.2   Novolac

The resols we have described are sometimes referred to as *one-stage resins*, since cross-linked products may be made from the initial reaction mixture only by adjusting the pH. The resol process is also known as the *one-stage* process. On the other hand, the novolacs are sometimes referred to as *two-stage* resins because, in this case, it is necessary to add, as we will show, some agent to enable formation of cross-linked products.

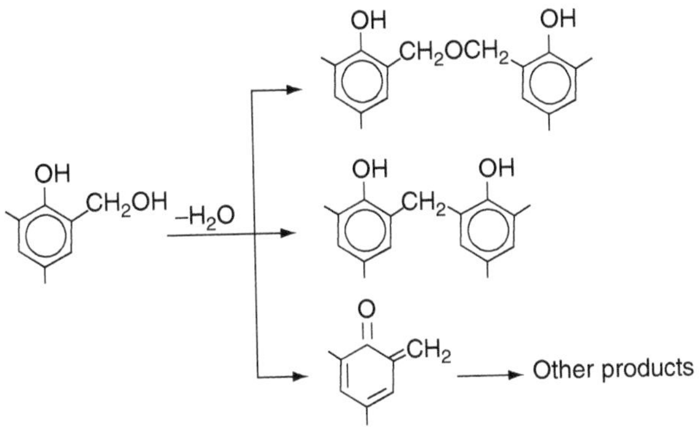

FIGURE 1.24 Curing mechanisms for resols.

Novolac resins are normally prepared by the reaction of a molar excess of phenol with formaldehyde (commonly about 1.25:1) under acidic conditions. The reaction is commonly carried out batchwise in a resin kettle of the type used for resol manufacture. Typically, a mixture of phenol, formalin, and acid is heated under reflux at about 100°C. The acid is usually either hydrochloric acid (0.1–0.3% on the weight of phenol) or oxalic acid (0.5–2%).

Under acidic conditions the formation of methylol phenols is rather slow, and the condensation reaction thus takes approximately 2–4 h. When the resin reaches the requisite degree of condensation, it become hydrophobic, and the mixture appears turbid. Water is then distilled off until a cooled sample of the residual resin shows a melting point of 65–75°C. The resin is then discharged and cooled to give a hard, brittle solid (novolac).

Unlike resols, the distillation of water for novolac is normally carried out without using a vacuum. Therefore the temperature of the resin increases as the water is removed and the reaction proceeds, the temperature reaching as high as 160°C at the end. At these temperatures the resin is less viscous and more easily stirred.

The mechanism and phenol-formaldehyde reaction under acidic conditions is different from that under basic conditions described previously. In the presence of acid the products *o*- and *p*-methylol phenols, which are formed initially, react rapidly with free phenol to form dihydroxy diphenyl methanes (Figure 1.25). The latter undergo slow reaction with formaldehyde and phenolic species, forming polynuclear phenols by further methylolation and methylol link formation. Reactions of this type continue until all the formaldehyde has been used up. The final product thus consists of a complex mixture of polynuclear phenols linked by *o*- and *p*-methylene groups.

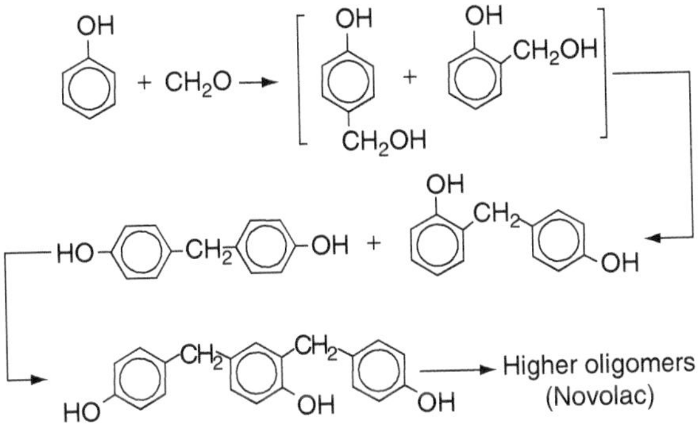

FIGURE 1.25 Formation of novolac in an acid-catalyzed reaction of phenol and formaldehyde.

The average molecular weight of the final product (novolac) is governed by the initial molar ratio of phenol and formaldehyde. A typical value of average molecular weight is 600, which corresponds to about six phenolic nuclei per chain. The number of nuclei in individual chains is usually 2–13.

A significant feature of novolacs is that they represent completed reactions and as such have no ability to continue increasing in average molecular weight. Thus there is no danger of gelation (cross-linking) during novolac production. Resols, however, contain reactive methylol groups and so are capable of cross-linking on heating.

To convert novolacs into network polymers, the addition of a cross-linking agent (hardener) is necessary. Hexamethylenetetramine (also known as hexa or hexamine) is invariably used as the hardener. The mechanism of the curing process is complex.

Because of the exothermic reaction on curing and the accompanying shrinkage, it is necessary to incorporate inert fillers to reduce resin content. Fillers also serve to reduce cost and may give additional benefits, such as improving the shock resistance. Commonly used fillers are wood flour, cotton flock, textile shreds, mica, and asbestos.

Wood flour, a fine sawdust preferably from soft woods, is the most commonly used filler. Good adhesion occurs between the resin and the wood flour, and some chemical bonding may also occur. Wood flour reduces exotherm and shrinkage, improves the impact strength of the moldings, and is cheap. For better impact strength cotton fabric or chopped fabric may be incorporated. Asbestos may be used for improved heat and chemical resistance, and iron-free mica powder may be used for superior electrical insulation resistance characteristics.

Other ingredients which may be incorporated into a phenolic molding powder include *accelerators* (e.g., lime or magnesium oxide) to promote the curing reaction, *lubricants* (e.g., stearic acid and metal stearates) to prevent sticking to molds, *plasticizers* (e.g., naphthalene, furfural, and dibutyl phthalate) to improve flow properties during cure, and *pigments* or *dyes* (e.g., nigrosine) to color the product. Some typical formulations of phenolic molding powders are given in Table 1.19.

Since the phenolic resins cure with evolution of volatiles, compression molding is performed using molding pressures of 1–2 tn/in.^{2} (15–30 MPa) at 155–170°C. Phenolic molding compositions may be preheated by high frequency or other methods. Preheating reduces cure time, shrinkage, and required molding pressures. It also enhances the ease of flow, with consequent reduction of mold wear and danger of damage to inserts. Molding shrinkage of general-purpose grades is about 0.005–0.08 in./in. Highly loaded mineral-filled grades exhibit lower shrinkage.

Phenol–formaldehyde molding compositions are traditionally processed on compression- and transfer-molding machines with a very small amount being extruded. However, the injection-molding process as modified for thermosetting plastics is being increasingly used and it is today the most common method used to process phenolic compounds. The shorter cycle times and low waste factor available with

TABLE 1.19 Typical Formulations[a] of Phenolic Molding Grades

Ingredient	General-Purpose Grade	Medium Shock-Resisting Grade	High Shock-Resisting Grade	Electrical Grade
Novolac resin	100	100	100	100
Hexa	12.5	12.5	17	14
Magnesium oxide	3	2	2	2
Magnesium stearate	2	2	3.3	2
Nigrosine dye	4	3	3	3
Wood Flour	100	—	—	—
Cotton flock	—	110	—	—
Textile shreds	—	—	150	—
Asbestos	—	—	—	40
Mica	—	—	—	120

[a] Parts by weight.

Source: Brydson, J. A. 1982. *Plastics Materials*, Butterworth Scientific, London, UK.

screw injection molding, which contribute to the lowest unit costs for extended runs, have induced phenolic molding compounders to develop products for this molding process.

1.3.3.1.3 Properties and Applications

Since the polymer in phenolic moldings is highly cross-linked and interlocked, the moldings are hard, infusible, and insoluble. The chemical resistance of the moldings depends on the type of resin and filler used. General-purpose PF grades are readily attacked by aqueous sodium hydroxide, but cresol- and xylenol-based resins are more resistant. Phenolic moldings are resistant to acids except formic acid, 50% sulfuric acid, and oxidizing acids. The resins are ordinarily stable up to 200°C.

The mechanical properties of phenolic moldings are strongly dependent on the type of filler used (Table 1.20). Being polar, the electrical insulation properties of phenolics are not outstanding but are generally adequate. A disadvantage of phenolics as compared to aminoplasts and alkyds is their poor tracking resistance under high humidity, but this problem is not serious, as will be evident from the wide use of phenolics for electrical insulation applications.

Perhaps the most well-known applications of PF molding compositions are in domestic plugs and switches. However, in these applications PF has now been largely replaced by urea-formaldehyde plastics because of their better antitracking property and wider range of color possibility. (Because of the dark color of the phenolic resins molded above 160°C, the range of pigments available is limited to relatively darker colors—blacks, browns, deep blues, greens, reds, and oranges.) Nevertheless, phenolics continue to be used as insulators in many applications because their properties have proved quite adequate.

There are also many applications of phenolics where high electrical insulation properties are not as important, and their heat resistance, adequate shock resistance, and low cost are important features: for example, knobs, handles, telephones, and instrument cases. In some of these applications phenolics have been replaced by ureaformaldehyde, melamine–formaldehyde, alkyd, or newer thermoplastics because of the need for brighter colors and tougher products.

In general, phenolics have better heat and moisture resistance than ureaformaldehyde moldings. Heat-resistant phenolics are used in handles and knobs of cookware, welding tongs, electric iron parts, and in the automobile industry for fuse box covers, distributor heads, and other applications where good electrical insulation together with good heat resistance is required.

Bottle caps and closures continue to be made in large quantities from phenolics. The development of machines for injection molding of thermosetting plastics and availability of fast-curing grades of phenolics have stimulated the use of PF for many small applications in spite of competition from other plastics.

Among the large range of laminated plastics available today, the phenolics were the first to achieve commercial significance, and they are still of considerable importance. In these applications one-stage resins (resols) are used, sine they have sufficient methylol groups to enable curing without the need of a curing agent.

Caustic soda is commonly used as the catalyst for the manufacture of resols for mechanical and decorative laminates. However, it is not used in electrical laminates because it adversely affects the electrical insulation properties. For electrical-grade resols ammonia is the usual catalyst, and the resins are usually dissolved in industrial methylated spirits. The use of cresylic acid (*m*-cresol content 50–55%) in place of phenol yields laminating resins of better electrical properties.

In the manufacture of laminates for electrical insulation, paper (which is the best dielectric) is normally used as the base reinforcement. Phenolic paper laminates are extensively used for high-voltage insulation applications.

Besides their good insulation properties, phenolic laminates also possess good strength, high rigidity, and machinability. Sheet, tubular, and molded laminates are employed. Phenolic laminates with cotton fabric reinforcement are used to manufacture gear wheels are quiet running but must be used at lower working stresses than steel. Phenolic-cotton or phenolic-asbestos laminates have been used as bearings for steel rolling mills to sustain bearing loads as high as 3000 psi (21 MPa). Because of the advent of

TABLE 1.20   Properties of Phenolic Moldings

Property	General-Purpose Grade	Medium Shock-Resisting Grade	High Shock-Resisting Grade	Electrical Grade
Specific gravity	1.35	1.37	40	1.85
Shrinkage (cm/cm)	0.006	0.005	0.002	0.002
Tensile strength				
lbf/in.^{2}	8000	7000	6500	8500
MPa	55	48	45	58
Impact strength				
ft.-lbf	0.16	0.29	0.8–1.4	0.14
J	0.22	0.39	1.08–1.9	0.18
Dielectric constant				
at 800 Hz	6.0–10.0	5.5–5.7	6.0–10.0	4.0–6.0
at 10^6 Hz	4.5–5.5	—	—	4.3–5.4
Dielectric strength (20°C)				
V/mil	150–300	200–275	150–250	275–350
kV/cm	58–116	78–106	58–97	106–135
Power factor				
at 800 Hz	0.1–0.4	0.1–0.35	0.1–0.5	0.03–0.05
at 10^6 Hz	0.03–0.05	—	—	0.01–0.02
Volume resistivity (ohm-m)	10^{12}–10^{14}	10^{12}–10^{14}	10^{11}–10^{13}	10^{13}–10^{16}
Water absorption (24 h, 23°C)				
mg	45–65	30–50	50–100	2–6

[a] Testing according to BS 2782.

Source: Brydson, J. A. 1982. *Plastics Materials.* Butterworth Scientific, London, UK.

cheaper thermoplastics, cast phenolic resins (resols) are no longer an important class of plastics materials.

1.3.3.2   Urea–Formaldehyde Resins

Monomers	Polymerization	Major Uses
Urea, formaldehyde	Stepwise polycondensation	Particle-board binder resin (60%), paper and textile treatment (10%), molding compound (9%) coatings (7%)

Aminoresins or *aminoplastics* cover a range of resinous polymers produced by reaction of amines or amides with aldehydes [14,46,47]. Two such polymers of commercial importance in the field of plastics are the urea-formaldehyde and melamine–formaldehyde resins. Formaldehyde reacts with the amino groups to form aminomethylol derivatives which undergo further condensation to form resinous products. In contras to phenolic resins, products derived from urea and melamine are colorless.

Urea and formaldehyde resins are usually prepared by a two-stage reaction. In the first stage, urea and formaldehyde (mole ratio in the range 1:1.3–1:1.5) are reacted under mildly alkaline (pH 8) conditions, leading to the production of monomethylol urea (Figure 1.26(I)) and dimethylol urea (Figure 1.26(II)). If the product of the first stage, which in practice usually also contains unreacted urea and formaldehyde, is subjected to acid conditions at elevated temperatures (stage 2), the solution increases in viscosity and sets to an insoluble and irreversible gel. The gel eventually converts with evolution of water and formaldehyde to a hard, colorless, transparent, insoluble, and infusible mass having a network molecular structure.

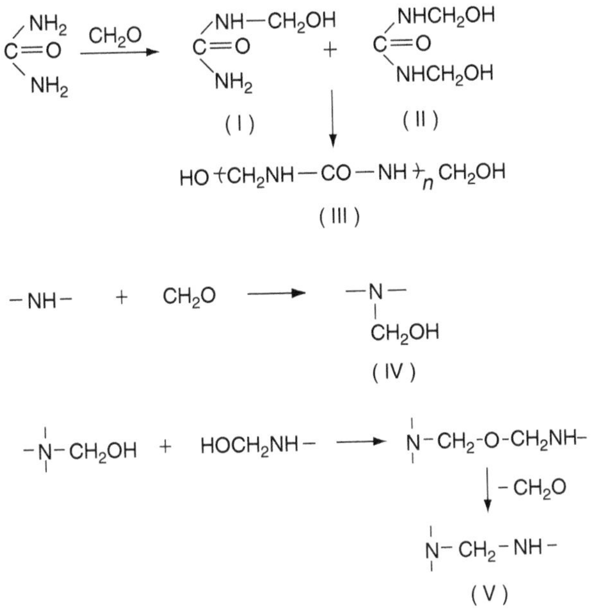

$$CO_2$$

HO $\text{+CH}_2\text{NH}-\text{CO}-\text{NH} +_n \text{CH}_2\text{OH}$

(III)

$-\text{NH}- \quad + \quad \text{CH}_2\text{O} \longrightarrow \quad \begin{matrix} -\text{N}- \\ | \\ \text{CH}_2\text{OH} \end{matrix}$

(IV)

$-\overset{|}{\text{N}}-\text{CH}_2\text{OH} \quad + \quad \text{HOCH}_2\text{NH}- \longrightarrow \overset{|}{\text{N}}-\text{CH}_2\text{-O-CH}_2\text{NH}-$

$\Big\downarrow -\text{CH}_2\text{O}$

$\overset{|}{\text{N}}-\text{CH}_2-\text{NH}-$

(V)

FIGURE 1.26 Reactions in the formation of urea-formaldehyde resins.

The precise mechanisms involved during the second stage are not fully understood. It does appear that in the initial period of the second stage methylol ureas condense with each other by reaction of a –CH$_2$OH group on one molecule with an –NH$_2$ of another molecule, leading to linear polymers of the form shown in Figure 1.26(III). These polymers are relatively less soluble in aqueous media and tend to form amorphous white precipitates on cooling to room temperature.

More soluble resins are formed on continuation of heating. This probably involves the formation of pendant methylol groups (Figure 1.26(IV)) by reactions of the –NH– groups with free formaldehyde. These methylol groups and the methylol groups on the chain ends of the initial reaction product can then react with each other to produce ether linkages, or with amine groups to give methylene linkages (Figure 1.26III). The ether linkages may also break down on heating to methylene linkages with the evolution of formaldehyde (Figure 1.26(V)). An idealized network structure of the final cross-linked product is shown in Figure 1.25 of *Plastics Fundamentals, Properties, and Testing*.

1.3.3.2.1 *Molding Powder*

The urea-formaldehyde (UF) Molding powder will contain a number of ingredients. Most commonly these include resin, filler, pigment, accelerator, stabilizer, lubricant, and plasticizer.

Bleached wood pulp is employed as a filler for the widest range of bright colors and in slightly translucent moldings. Wood flour, which is much cheaper, may also be used.

A wide variety of pigments is now used in UF molding compositions. Their principal requirements are that they should be stable to processing conditions and be unaffected by service conditions of the molding.

To obtain a sufficiently high rate of cure at molding temperatures, it is usual to add about 0.2–2.0% of an accelerator (hardener)—a latent acid catalyst which decomposes at molding temperatures to yield an acidic body that will accelerate the rate of cure. Many such materials have been described, the most prominent of them being ammonium sulfamate, ammonium phenoxyacetate, trimethyl phosphate, and ethylene sulfite. A stabilizer such as hexamine is often incorporated into the molding powder to improve its shelf life.

Metal stearates, such as zinc, magnesium, or aluminum stearates are commonly used as lubricants at about 1% concentration. Plasticizers (e.g., monocresyl glycidyl ether) are used in special grades of molding powders. They enable more highly condensed resins to be used in the molding powder and thus reduce curing shrinkage while maintaining good flow properties.

In a typical manufacturing process, the freshly prepared UF first-stage reaction product is mixed with the filler (usually with a filler-resin dry weight ratio of 1:2) and other ingredients except pigment in a trough mixer at about 60°C for about 2 h. Thorough impregnation of the filler with the resin solution and further condensation of the resin takes place during this process. Next, the wet mix is in a turbine or rotary drier for about 2 h at 100°C or about 1 h in a countercurrent of air at 120–130°C. The drying process reduces the water content from about 40% to about 6% and also causes further condensation of the resin.

After it is removed from the drier, the product is ground in a hammer mill and then in a ball mill for 6–9 h. The pigments are added during the ball-milling process, which ensures a good dispersion of the pigment and gives a fine powder that will produce moldings of excellent finish. The powder, however, has a high bulk factor and needs densification to avoid problems of air and gas trappings during molding.

There are several methods of densification. In one method, the heated powder is formed into strips by passing through the nip of a two-roll mill. The strips are then powdered into tiny flat flakes in a hammer mill. Other processes involve agglomeration of the powder by heating in an internal mixer at about 100°C or by treatment with water or steam and subsequent drying. Continuous compounders, such as the Buss Ko-Kneader, are also used.

1.3.3.2.2 Processing

Urea–formaldehyde molding powders have a limited storage life. They should therefore be stored in a cool place and should be used, wherever possible, within a few months of manufacture. Conventional compression and transfer molding are commonly used for UF materials, the former being by far the most important process in terms of tonnage handled. Compression molding pressures usually range from 1 to 4 tn/in.^{2} (15–60 MPa), the higher pressures being used for deep-draw articles. Molding temperatures from 125°C to 160°C are employed. The cure time necessary depends on the mold temperature and on the thickness of the molding. The cure time for a 1/8 in. thick molding is typically about 55 sec at 145°C. Bottle caps (less than 1/8 in. thick) and similar items, however, are molded industrially with much shorter cure times ($\sim$10–20 sec) at the higher end of the molding temperature range. For transfer molding of UF molding powders, pressures of 4–10 tn/in.^{2} (60–150 MPa), calculated on the area of the transfer pot, are generally recommended.

Special injection grades of UF molding powder have been developed for injection-molding applications which call for molding materials with good flow characteristics between 70°C and 100°C, unaffected by long residence time in the barrel but capable of almost instant cure in the mold cavity at a higher temperature.

Although the transition from compression molding to injection molding has been extensive for phenolics, the same cannot be said for UF Materials, because they are more difficult to mold, possibly because the UF are more brittle than a phenolic resin and so are less able to withstand the stress peaks caused by filler orientation during molding. A combination of compression and injection processes has therefore been developed in which a screw preplasticizing unit delivers preheated and softened material directly to a compression-mold cavity.

1.3.3.2.3 Properties and Applications

The wide color range possible with UF molding powders has been an important reason for the widespread use of the material. These moldings have a number of other desirable features: low cost, good electrical insulation properties, and resistance to continuous heat up to a temperature of 70°C. Some typical values of physical properties of UF molding compositions are given in Table 1.21. They do not impart taste and odor to food-stuffs and beverages with which they come in contact and are resistant to detergents and dry-cleaning solvents.

The foregoing properties account for major uses of UF in two applications, namely, bottle caps and electrical fittings. It is also used for colored toilet seats, vacuum flasks, cups and jugs, hair drier housings,

TABLE 1.21 Properties of Urea-Formaldehyde and Melamine-Formaldehyde Moldings.

Property	Urea-Formaldehyde (α-Cellulose Filled)	Melamine-Formaldehyde (Cellulose Filled)
Specific gravity	1.5–1.6	1.5–1.55
Tensile strength		
10^3 lbf/in.^{2}	7.5–11.5	8–12
MPa	52–80	55–83
Impact strength (ft.-lbf)	0.20–0.35	0.15–0.24
Dielectric strength (90°C)		
V/0.001 in.	120–200	160–240
Volume resistivity (ohm-m)	10^{13}–10^{15}	10^9–10^{10}
Water absorption (mg)		
24 h at 20°C	50–130	10–50
30 min at 100°C	180–460	40–110

[a] Testing according to BS 2782.
Source: Brydson, J. A. 1982. *Plastics Materials.* Butterworth Scientific, London, UK.

toys, knobs, meat trays, switches, lamp shades, and ceiling light bowls. In the latter applications it is important to ensure adequate ventilation to prevent overheating and consequent cracking of the molded articles.

However, only about 3% of UF resins are used for molding powders. The bulk (about 85%) of the resins are used as adhesives in the particleboard, plywood, and furniture industries. Resins for these applications are commonly available with U/F molar ratios ranging from 1:1.4 to 1:2.2.

To prepare a suitable resin for adhesive applications, urea is dissolved in formalin (initially neutralized to pH 7.5) to give the desired U/F molar ratio. After boiling under reflux for about 15 min to give demethylol urea and other low-molecular products, the resin is acidified, conveniently with formic acid, to pH 4, and reacted for a further period of 5–20 min. The resulting water-soluble resin with approximately 50% solids content is stabilized by neutralizing to a pH 7.5 with alkali. For use as an aqueous solution, as is normally the case, the resin is then partially dehydrated by vacuum distillation to give a 70% solids content.

Phosphoric acid, or more commonly ammonium chloride, is used as a hardener for UF resin adhesives. Ammonium chloride reacts with formaldehyde to produce hexamine and hydrochloric acid, and the latter catalyzes the curing of the resin. In the manufacture of plywood a resin (with U/F molar ratio typically 1:1.8) mixed with hardener is applied to wood veneers, which are then plied together and pressed at 95–110°C under a pressure of 200–800 psi (1.38–5.52 MPa). The UF resin-bonded plywood is suitable for indoor applications but is generally unsuitable for outdoor use. For outdoor applications phenol-formaldehyde, resorcinol–formaldehyde, or melamine–formaldehyde resins are more suitable.

Large quantities of UF resin are used in general wood assembly work. For joining pieces of wood the resin-hardener solution is usually applied to the surfaces to be joined and then clamped under pressure while hardening occurs. Alternatively, the resin may be applied to one surface and the hardener to the other, allowing them to come into contact in situ. This method serves to eliminate pot-life problems of the resin-hardener mixture.

Gap-filling resins are produced by incorporating into UF resins plasticizers, such as furfuryl alcohol, and fillers to minimize shrinkage and consequent cracking and crazing.

In the manufacture of wood chipboard, which represents one of the largest applications of UF resins, wood chips are mixed with about 10% of a resin-hardener solution and pressed in a multidaylight press at 150°C for about 8 min. Since some formaldehyde is released during the opening of the press, it is necessary to use a resin with a low formaldehyde content. Because it has no grain, a wood chipboard is nearly isotropic in its behavior and so does not warp or crack. However, the water resistance of chipboard is poor.

1.3.3.3   Melamine–Formaldehyde Resins

Monomers	Polymerization	Major Uses
Melamine (trimerization of cyanamide), formaldehyde	Stepwise polycondensation	Dinnerware, table tops, coatings

Reaction of melamine (2,4,6-triamino-1,3,5-triazine) with neutralized formalin at about 80–100°C leads to the production of a mixture of water-soluble methylol melamines. The methylol content of the mixture depends on the initial ratio of formaldehyde to melamine and on the reaction conditions. Methylol melamines possessing up to six methylol groups per molecule are formed (Figure 1.27).

On further heating, the methylol melamines undergo condensation reactions, and a point is reached where hydrophobic resin separates out. The rate of resinification depends on pH. The rate is minimum at about pH 10.0–10.5 and increases considerably both at lower and higher pH. The mechanism of resinification and cross-linking is similar to that observed for urea-formaldehyde (Figure 1.26) and involves methylol–amine and methylol–methylol condensations.

$$\text{~~NH·CH}_2\text{OH} + \text{H}_2\text{N~~} \longrightarrow \text{~~NH·CH}_2\text{·NH~~} + \text{H}_2\text{O}$$

$$\text{~~NH·CH}_2\text{OH} + \text{HO·CH}_2\text{NH~~} \longrightarrow \text{~~NH·CH}_2\text{·O·CH}_2\text{·NH~~} + \text{H}_2\text{O}$$

$$\text{~~NH·CH}_2\text{·O·CH}_2\text{·NH~~} \longrightarrow \text{~~NH·CH}_2\text{·NH~~} + \text{CH}_2\text{O}$$

In industrial practice, resinification is carried out to a point close to the hydrophobe point. This liquid resin is either applied to the substrate or dried and converted into molding powder before proceeding with the final cure.

In a typical process [14] a jacketed resin kettle fitted with stirrer and reflux condenser is charged with 240 parts of 40% w/v formalin (pH adjusted to 8.0–8.5 using a sodium carbonate solution) and 126 parts of melamine (to give a melamine–formaldehyde ratio of 1:3), and the temperature is raised to 85°C. The melamine forms methylol derivatives and goes into solution. This water-soluble A-stage resin may be used for treatment of paper, leather, and fabrics to impart crease resistance, stiffness, shrinkage control, water repellency, and fire retardance. It may be spray dried to give a more stable, water-soluble product.

For laminating and other purposes the initial product is subjected to further condensation reactions at about 85°C with continuous stirring for more than 30 min. The hydrophilicity of the resin, as shown by its water tolerance, decreases with increasing condensation. The reaction is usually continued until a stage is reached when addition of 3 cm³ of water will cause 1 cm³ of resin to become turbid. The condensation reactions may be carried out at higher temperatures and lower pH values to achieve this stage more rapidly.

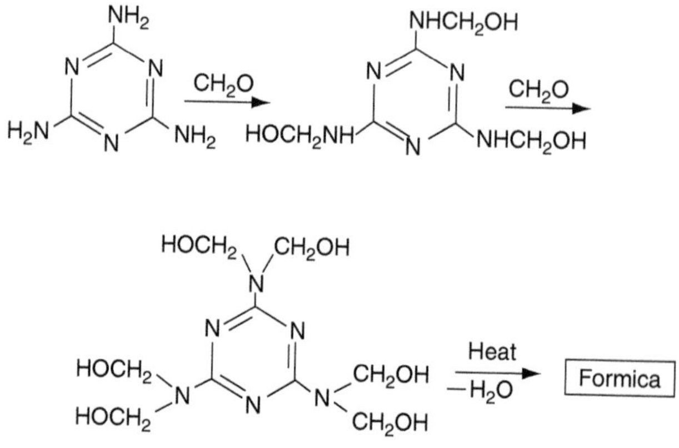

FIGURE 1.27   Reactions in the synthesis of formica.

In aqueous solutions the hydrophobic resins have a shelf life of just a few days. The resin may be diluted with methylated spirit to about 50% solids content and pH adjusted to 9.0–9.5 to achieve greater stability. The addition of about 0.1% borax (calculated on the weight of the solids content) as an aqueous solution is useful in obtaining this pH maintaining it for several months. The stabilized resin is stored preferably at 20–35°C, because too low a storage temperature will cause precipitation and too high at temperature will cause gelation.

Melamine–formaldehyde molding powders are generally prepared by methods similar to those used for UF molding powders. In a typical process an aqueous syrup of MF resin with melamine–formaldehyde ratio of 1:2 is compounded with fillers, pigments, lubricants, stabilizers, and accelerators in a dough-type mixer. The product is then dried and ball-milled by processes similar to those used for UF molding powders.

Alpha-cellulose is used as a filler for the more common decorative molding powders. Industrial-grade MF materials use fillers such as asbestos, silica, and glass fiber. These fillers are incorporated by dry blending methods. The use of glass fiber gives moldings of higher mechanical strength, improved dimensional stability, and higher heat resistance than other fillers. The mineralfilled MF moldings have superior electrical insulation and heat resistance and may be used when phenolics and UF compositions are unsuitable.

MF moldings are superior to UF products in lower water absorption (see Table 1.21), greater resistance to staining by aqueous solutions such as fruit juices and beverages, better retention of electrical properties in damp conditions, better heat resistance, and greater hardness. Compared with the phenolic resins, MF resins have better color range, track resistance, and scratch resistance. MF resins, are however, more expensive than general-purpose UF and PF resins.

MF compositions are easily molded in conventional compression- and transfer-molding equipment. Molding temperatures from 145°C to165°C and molding pressures 2–4 tonf/in.^{2} (30–60 MPa) are usually employed. In transfer molding, pressures of 5–10 tonf/in.^{2} (75–150 MPa) are used. The cure time for an 1/8-in.-thick molding is typically 2 1/2 min at 150°C.

Largely because of their wide color range, surface hardness, and stain resistance, MF resins are used as molding compositions for a variety of mechanical parts or household goods and as laminating resins for tops for counters, cabinets, and tables. The mineral-filled molding powders are used in electrical application and knobs and handles for kitchen utensils.

An interesting application of MF resins in compression molding involves decorative foils made by impregnating a printed or decorated grade of paper with resin and then drying. The foil may be applied to a compression molding shortly before the cure is complete, and the resin in the foil may be cured in that position to produce a bonding.

In a typical process of laminating paper layers to make materials useful in electrical applications as well as decorative laminates (best known as *formica*), kraft paper, about the weight used in shopping bags, is run through a solution of melamine–formaldehyde prepolymer. Drying out water or driving off the solvent leave an impregnated sheet that can be handled easily, since the brittle polymer does not leave the surface sticky.

As many as a dozen or more layers are piled up. For decorative purposes a printed rag or decorated cloth paper is put on top and covered with a translucent paper layer. The entire assembly is heated between smooth plates in a high-pressure press to carry out the thermosetting (curing) reaction that binds the sheets together into a strong, solvent-resistant, heat-resistant, and scratch-resistant surfacing material. The laminate, which is only about 1.5-mm thick, can be glued to a plywood base for use in tabletops, countertops, and the like.

1.3.4 Polyurethanes

A *urethane* linkage (–NHCOO–) is formed by the reaction of an isocyanate (–NCO) and an alcohol: $RNCO + R'OH \rightarrow RNHCOOR'$ [14,48,49]. By the same reaction, polyhydroxy materials will react with polyisocyanates to yield polyurethanes soft thermoplastic elastomers to hard thermoset rigid foams are readily produced from liquid monomers.

The basic building blocks for polyurethanes are polyisocyanates and macroglycols, also called *polyols*. The commonly used polyisocyanates are tolylene-diisocyanate (TDI), diphenylmethane diisocyanate or methylenediphenyl isocyanate (MDI), and polymeric methylenediphenyl isocyanate (PMDI) mixtures manufactured by phosgenating aromatic polyamines derived from the acid-catalyzed condensation of aniline and formaldehyde. MDI and PMDI are produced by the same reaction, and separation of MDI is achieved by distillation. The synthetic routes in the manufacture of commercial polyisocyanates are summarized in Figure 1.28.

A number of specialty aliphatic polyisocyanates have been introduced recently in attempts to produce a light-stable polyurethane coating. Triisocyanate made by reacting hexamethylene diisocyanate with water (Figure 1.28c) is reported to impart good light stability and weather resistance in polyurethane coatings and is probably the most widely used aliphatic polyisocyanate.

The macroglycols used in the manufacture of polyurethanes are either polyether or polyester based. Polyether diols are low-molecular-weight polymers prepared by ring-opening polymerization of olefin oxides (see also the section on polyethers), and commonly used polyester polyols are polyadipates. A polyol produced by ring-opening polymerization of caprolactone, initiated with low-molecular-weight glycols, is also used. The reactions are summarized in Figure 1.29.

Isocyanates are highly reactive materials and enter into a number of reactions with groups or molecules active hydrogen, such as water, amine, and also urethane. Isocyanates are also toxic and care should be exercised in their use. Their main effect is on the respiratory system.

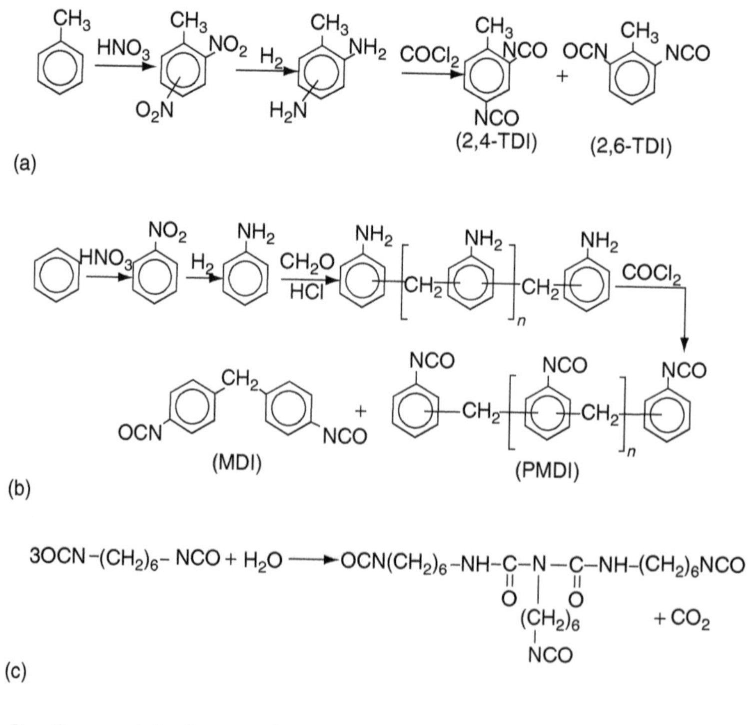

FIGURE 1.28 Reactions used in the manufacture of commercial isocyanates. (a) TDI. (b) PMDI and MDI. (c) Aliphatic triisocyanate.

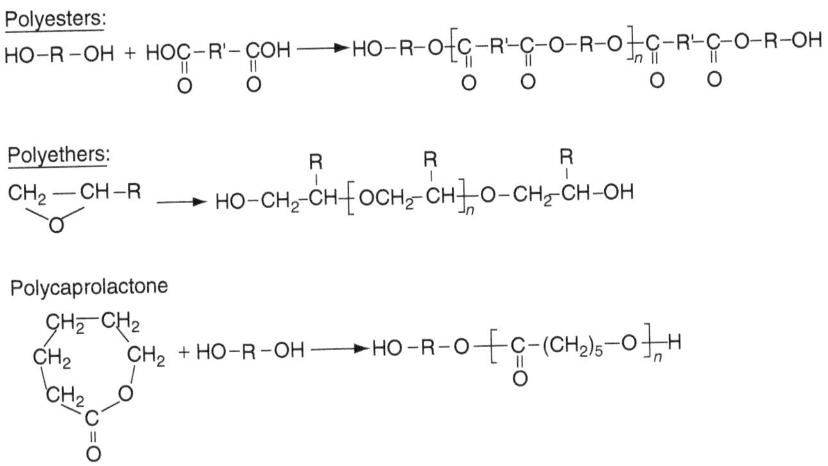

Polyesters:

$$HO-R-OH + HOC-R'-COH \longrightarrow HO-R-O \left[C-R'-C-O-R-O \right]_n C-R'-C-O-R-OH$$

Polyethers:

$$CH_2-CH-R \longrightarrow HO-CH_2-CH \left[OCH_2-CH \right]_n O-CH_2-CH-OH$$

Polycaprolactone

$$CH_2-CH_2 \quad CH_2 + HO-R-OH \longrightarrow HO-R-O \left[C-(CH_2)_5-O \right]_n H$$

FIGURE 1.29 Reactions used in the manufacture of macroglycols.

Major polyurethane products today include cellular materials such as water-blown flexible foams or fluorocarbon-blown rigid foams, elastomers, coatings, and elastic fibers, which are described subsequently. Closely related to polyurethanes is an isocyanate-based product called *isocyanurate foam.*

1.3.4.1 Polyurethane Rubbers and Spandex Fibers

By careful formulations it is possible to produce polyurethane rubbers with a number of desirable properties [50]. The rubbers can be thermoplastic (linear) or thermoset (slightly cross-linked) products.

1.3.4.1.1 Cross-Linked Polyurethane Rubbers

The starting point in the preparation of this type of rubber, typified by Vulkollan rubbers, is a polyester prepared by reacting a glycol such as ethylene or propylene glycol with adipic acid. The glycol is in excess so that the polyester formed has hydroxyl end groups. This polyester macroglycol is then reacted with an excess of a diisocyanate such as 1,5-naphthalene diisocyanate or MDI (Figure 1.28). The molar excess of diisocyanate is about 30%, so the number of polyesters joined together is only about 2–3, and the resulting prepolymer has isocyanate end groups (see Figure 1.30a).

The prepolymer can be chain extended with water, glycols, or amines which link up prepolymer chains by reacting with terminal isocyanate groups (see Figure 1.30b). (The water reaction liberates carbon dioxide, so it must be avoided in the production of elastomers, but it is important in the manufacture of foams.) The urea and urethane linkages formed in the chain extension reactions also provide sites for branching and cross-linking, since these groups can react with free isocyanate or terminal isocyanae groups to form *biuret* and *allophanate* linkages, respectively (see Figure 1.30c). Biuret links, however, predominate since the urea group reacts faster than the urethane groups. The degree of cross-linking can to some extent be controlled by adjusting the amount of excess isocyanate, whereas more highly cross-linked structures may be produced by the use of a triol in the initial polyester.

Vulkollan-type rubbers suffer from the disadvantage that the prepolymers are unstable and must be used within a day or two of their production. Moreover, these rubbers cannot be processed with conventional rubber machiners, so the products are usually made by a casting process. Attempts were then made to develop other polyurethane rubbers which could be processed by conventional techniques.

One approach was to react the diisocyanate with a slight excess of polyester so that the prepolymer produced has terminal hydroxyl groups. The prepolymers are rubberlike gums and can be compounded

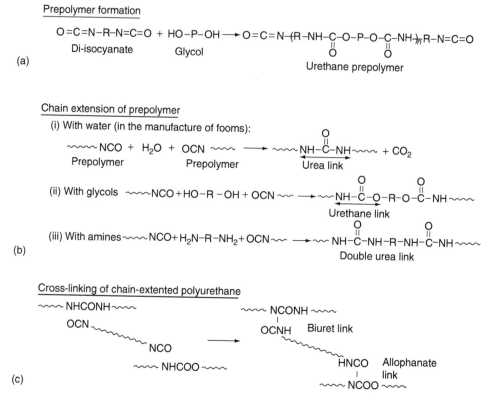

FIGURE 1.30   Equations for preparation, chain extension, and curing of polyurethanes.

with other ingredients on two-roll mills. Final curing can be done by the addition of a diisocyanate or, preferably, a latent diisocyanate, i.e., a substance which produces an active diisocyanate under the conditions of molding. Polyurethane rubbers of this class are exemplified by Chemigum SL (Goodyear), Desmophen A (Bayer), and Daltoflex 1 (ICI), which used polyester–amide for the manufacture of prepolymer.

Another approach was adopted by Du Pont with the product Adiprene C, a polyurethane rubber with unsaturated groups that allow vulcanization with sulfur.

Polyurethane rubbers, in general, and the Vulkollan-type rubbers, in particular, possess certain outstanding properties. They usually have higher tensile strengths than other rubbers and possess excellent tear and abrasion resistance. The urethane rubbers show excellent resistance to ozone and oxygen (in contrast to diene rubbers) and to aliphatic hydrocarbons. However, they swell in aromatic hydrocarbons and undergo hydrolytic decomposition with acids, alkalis, and prolonged action of water and steam.

Though urethane rubbers are more costly than most of other rubbers, they are utilized in applications requiring superior toughness and resistance to tear, abrasion, ozone, fungus, aliphatic hydrocarbons, and dilute acids and bases. In addition, they excel in low-temperature impact and flexibility. Urethane rubbers have found increasing use for forklift tires, shoes soles and heels, oil seals, diaphragms, chute linings, and a variety of mechanical applications in which high elasticity is not an important prerequisite.

Emphasis on reaction injection-molding (RIM) technology in the automotive industry to produce automotive exterior parts has created a large potential for thermoset polyurethane elastomers. *Reaction injection molding,* originally known as reaction casting, is a rapid, one-step process to produce thermoset polyurethane products from liquid monomers. In this process liquid monomers are mixed under high pressure prior to injection into the mold. The polymerization occurs in the mold. Commercial RIM polyurethane products are produced from MDI, macroglycols, and glycol or diamine extenders. The products have the rigidity of plastics and the resiliency of rubber.

Later advances include short, glass-fiber reinforced, high-modulus (flexural modulus greater than 300,000 psi, i.e., 2070 MPa) polyurethane elastomers produced by the reinforced RIM process. These reinforced high-modulus polyurethane elastomers are considered for automotive door panels, trunk lids, and fender applications.

Though originally developed for the automotive industry for the production of car bumpers, the RIM process has found its greatest success in the shoe industry, where semiflexible polyurethane foams have proved to be good soling materials.

1.3.4.1.2 *Thermoplastic Polyurethane Rubbers*

The reactions of polyols, diisocyanates, and glycols, as described, do tend to produce block copolymers in which hard blocks with glass transition temperatures well above normal ambient temperature are separated by soft rubbery blocks. These polymers thus resemble the SBS triblock elastomers and, more closely, the polyether–ester thermoplastic elastomers of the Hytrel-type described earlier.

In a typical process of manufacturing thermoplastic polyurethane elastomers, a prepolymer is first produced by reacting a polyol, such as linear polyester with terminal hydroxyl groups, or a hydroxyl-terminated polyether, of molecular weights in the range of 800–2500, with an excess of diisocyanate (usually of the MDI type) to give a mixture of isocyanate-terminated polyol prepolymer and free (unreacted) diisocyanate. This mixture is then reacted with a chain extender such as 1,4-butanediol to give a polymer with long polyurethane segments whose block length depends on the extent of excess isocyanate and the corresponding stoichiometric glycol.

The overall reaction is shown in Figure 1.31a. Provided that R (in free diisocyanate) and R′ (in glycol) are small and regular, the polyurethane segments will show high intersegment attraction (such as hydrogen bonding) and may be able to crystallize, thereby forming hard segments. In such polymers hard segments with T_g well above normal ambient temperature are separated by polyol soft segments, which in the mass are rubbery in nature. Hard and soft segments alternate along the polymer chain. This structure closely resembles that of polyester–polyether elastomers (Figure 1.11). Similar reactions occur when an amine is used instead of a glycol as a chain extender (see Figure 1.31b). The polymer in this case has polyurea hard segments separated by polyol soft segments.

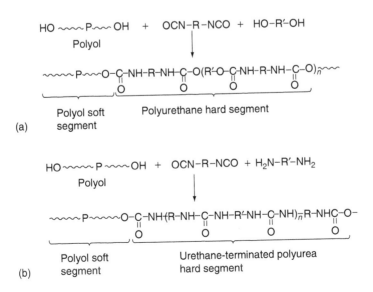

FIGURE 1.31 Reactions for the manufacture of polyurethane block copolymers.

The polymers produced by these reactions are mainly thermoplastic in nature. Though it is possible that an excess of isothiocyanate may react with urethane groups in the chain to produce allophanate cross-links (see Figure 1.30c), these cross-links do not destroy the thermoplastic nature of the polymer because of their thermal lability, i.e., breaking down on heating and reforming on cooling. However, where amines have been used as chain extenders, urea groups are produced (Figure 1.31b), which, on reaction with excess isocyanate, may give the more stable biuret cross-links (see Figure 1.30c).

Many of the commercial materials designated as thermoplastic polyurethanes are in reality slightly cross-linked. This cross-linking may be increased permanently by a post-curing reaction after shaping. The polyurethane product Estane (Goodrich) may, however, be regarded as truly thermoplastic. The thermoplastic rubbers have properties similar to those of Vulkollan-type cast polyurethane rubbers, but they have higher values for compression set.

Thermoplastic polyurethane elastomers can be molded and extruded to produce flexible articles. Applications include wire insulation, hose, tracks for all-terrain vehicles, solid tires, roller skate wheels, seals, bushing, convoluted bellows, bearings, and small gears for high-load applications. In the automobile industry thermoplastic polyurethanes are used primarily for exterior parts. Their ability to be painted with flexible polyurethane-based paints without pretreatment is valuable.

1.3.4.1.3 Spandex Fibers

One particular form of thermoplastic polyurethane elastomers is the elastic fiber known as Spandex. The first commercial material of this type was introduced by Du Pont in 1958 (Lycra). It is a relatively high-priced elastomeric fiber made on the principle of segmented copolymers. Here again the soft block is a polyol, and the hard block is formed from MDI (4,4'-diisocyanatodiphenyl methane) and hydrazine. The reactions are shown in Figure 1.32. Note that this fiber-forming polymer contains urethane and semicarbazide linkages in the chain. The product is soluble in amide solvent, and the fiber is produced by dry spinning from a solution. Major end uses of Lycra are in apparel (swimsuits and foundation garments).

Subsequently several other similar materials have been introduced, including Dorlastan (Bayer), Spanzelle (Courtaulds), and Vyrene (U.S. Rubber).

The polyol component with terminal hydroxyl groups used in the production of the foregoing materials may be either a polyether glycol or a polyester glycol (see Figure 1.29). For example, Du Pont uses polytetrahydrofuran (a polyether glycol) for Lycra. U.S. Rubber originally used a polyester of molecular weight of about 2000, obtained by condensation of adipic acid with a mixture of ethylene

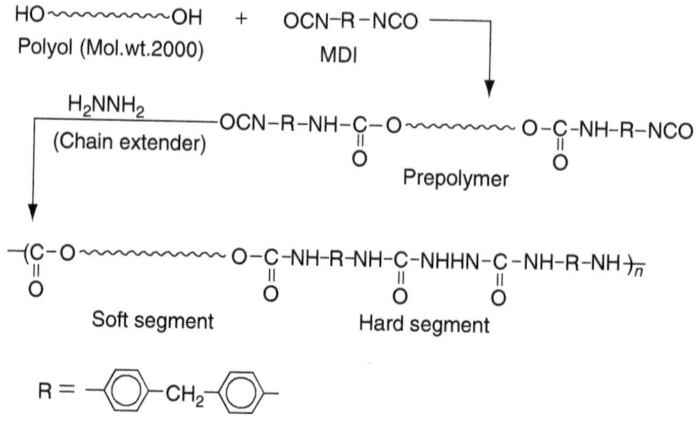

FIGURE 1.32 Reactions for the manufacture of segmented elastomeric fiber (Lycra).

glycol and propylene glycol, and a polyether-based mixture was used for Vyrene 2, introduced in 1967. These polyols are reacted with an excess of diisocyanate to yield an isocyanate-terminated prepolymer which is then chain extended by an amine such as hydrazine (NH_2NH_2) or ethylenediamine (see Figure 1.32). Fibers are usually spun from solution in dimethylformamide.

Possessing higher modulus, tensile strength, resistance to oxidation, and ability to be produced at finer deniers, spandex fibers have made severe inroads into the natural rubber latex thread market. Major end uses are in apparel. Staple fiber blends with nonelastic fibers have also been introduced.

1.3.4.2 Flexible Polyurethane Foam

Although polyurethane rubbers are specialty products, polyurethane foams are well known and widely used materials [51]. About half of the weight of plastics in modern cars is accounted for by such forms.

Flexible urethane foam is made in low densities of 1–1.2 lb/ft³ (0.016–0.019 g/cm³), intermediate densities of 1.2–3 lb/ft³ (0.019–0.048 g/cm³), high resilience (HR) foams of 1.8–3 lb/ft³ (0.029–0.048 g/cm³), and semiflexible foams of 6–12 lb/ft³ (0.096–0.192 g/cm³). Filed foams of densities as high as 45 lb/ft³ (0.72 g/cm³) have also been made.

In many respects the chemistry of flexible urethane foam manufacture is similar to that of the Vulkollan-type rubbers except that gas evolution reactions are allowed to occur concurrently with chain extension and cross linking (see Figure 1.30). Most flexible foams are made from 80/20 TDI, which refers to the ratio of the isomeric 2,4-tolylendiisocyanate to 2,6-tolylendiisocyanate. Isocyanates for HR foams are about 80% 80/20 TDI and 20% PMDI, and those for semiflexible foams are usually 100% PMDI.

Polyols usually are polyether based, since these are cheaper, give greater ease of foam processing, and provide more hydrolysis resistance than polyesters. Polyether polyols can be diols, such as polypropylene glycols with molecular weight of about 2000, or triols with molecular weight of 3000–6000. The most common type of the latter is triol adduct of ethylene oxide and propylene oxide with glycerol with a molecular weight of 3000. For HR foams polyether triols with molecular weight of 4500–6000 made by reacting ethylene oxide with polypropylene oxide based triols, are used.

The reaction of isocyanate and water that evolves carbon dioxide (see Figure 1.30b) is utilized for foaming in the production of flexible foams. The density of the product, which depends on the amount of gas evolved, can be reduced by increasing the isocyanate content of the reaction mixture and by correspondingly increasing the amount of water to react with the excess isocyanate. For greater softness and lower density some fluorocarbons (F-11) may be added in addition to water.

The processes for producing flexible polyurethane foams have been described in Chapter 1 of *Plastics Fabrication and Recycling*.

1.3.4.2.1 Applications

The largest-volume use of flexible foam is furniture and bedding, which account for nearly 47% of the flexible polyurethane foam market. Almost all furniture cushioning is polyurethane. Automobile seating is either made from flexible slabstock or poured directly into frames, so-called deep seating, using HR chemical formulations. Semiflexible polyurethane foams find use in crash pads, arm and head rests, and door panels.

Much flexible foam is used in carpet underlay. About 43% is virgin foam, and 57% is scrap rebounded with adhesives under heat and pressure.

Thermal interlining can be made by flame bonding thin-sliced, low-density (1–1.5 lb/ft³) polyester-based fabric. The flame melts about one-third of the foam thickness, and the molten surface adheres to the fabric.

Flexible foams also find use is packaging applications. Die-cut flexible foam is used to package costly goods such as delicate instruments, optical products, and pharmaceuticals. Semiflexible foam lining is used for cart interiors to protect auto and machine parts during transportation.

1.3.4.3 Rigid and Semirigid Polyurethane Foams

Most rigid foam is made from polymeric isocyanate (PMDI) and difunctional polyether polyols. PMDI of functionality 2.7 (average number of isocyanate groups per molecules) is used in insulation foam manufacturing. Functionalities greater than two contribute rigidity through cross-linking. Higher-functionality, low-molecular-weight polyols are sometimes added because they contribute rigidity by cross-linking and short chain length. Such polyols are made by reaction of propylene oxide with sucrose, pentaerythritol, or sorbitol.

Rigid insulation foams are usually hydrocarbon blown to produce a closed-cell foam with excellent insulation properties. High-density foams are water blown where structural and screw-holding strength is needed, and halocarbons are used as blowing agents when high-quality, decorative surfaces are required. A compromise between these two aims can be achieved by the addition of water to halocarbon-blown formulations.

Formulation, processing methods, properties, and applications of rigid and semirigid polyurethane foams have been described in have been described in Chapter 1 of *Plastics Fabrication and Recycling*. Major use of these foams are in building and construction (56%), transportation (12%), furniture, and packaging. Low-density, rigid foam is the most efficient thermal insulation commercially available and is extensively used in building construction. In transportation, urethane insulation is used in rail cars, containers, truck trailer bodies, and in ships for transporting liquefied natural gas. This requires thinner insulation and so yields more cargo space. Rigid foam is also used to give flotation in barge compartments.

High-density rigid foam, 5–15 lb/ft^3 (0.08–0.24 g/cm^3), is used for furniture items such as TV and stereo cabinets, chair shells, frames, arms, and legs, cabinet drawer fronts and doors, and mirror and picture frames. RIM-molded, integral-skin, high-density foams with core densities of 10–20 lb/ft^3 (0.16–0.32 g/cm^3) and skin densities of 55–65 lb/ft^3 (0.88–1.04 g/cm^3) are used in electronic, instrument, and computer cabinets.

Industrial uses of rigid foams include commercial refrigeration facilities as well as tanks and pipelines for cryogenic transport and storage.

1.3.4.3.1 Polyisocyanurates

Though closed-cell rigid polyurethane foams are excellent thermal insulators, they suffer form the drawback of unsatisfactory fire resistance even in the presence of phosphorus- and halogen-based fire retardants. In this context, polyisocyanurates, which are also based on isocyanates, have shown considerable promise. Isocyanurate has greater flame resistance then urethane. Although rigid polyurethane is specified for the temperatures up to 200°F (93°C), rigid polyisocyanurate foams, often called *trimer foams*, withstand use temperatures to 300°F (149°C). Physical properties and insulation efficiency are similar for both types.

The underlying reaction for polyisocyanurate formation is trimerization of an isocyanate under the influence of specific catalysts (Figure 1.33a). The most commonly used is a polymeric isocyanate (PMDI) prepared by reacting phosgene with formaldehyde–aniline condensates, as shown in Figure 1.28b.

PMDIs are less reactive than monomeric diisocyanate but are also less volatile. The polyisocyanurate produced from this material will be of type shown in Figure 1.33b. Some of the catalysts used for the polytrimerization reactions are alkali-metal phenolates, alcoholates and carboxylates, and compounds containing *o*-(dimethylaminomethyl)-phenol groups.

To produce foams, fluorocarbons such as trichlorofluoromethanes are use as the sole blowing agents. Polyisocyanurate foams may be prepared by using standard polyurethane foaming equipment and a two-component system, with isocyanate and fluorocarbon forming one component and the activator or activator mixture forming the second component.

Because of the high cross-link density of polyisocyanurates, the resultant foam tends to be brittle. Consequently, there has been a move toward making polyisocyanurate–polyurethance combinations. For example, the isocyanate trimerization reaction has been carried out with isocyanate end-capped TDI-based prepolymers to make isocyanurate-containing polyurethane foams. Isocyanate trimerization in the

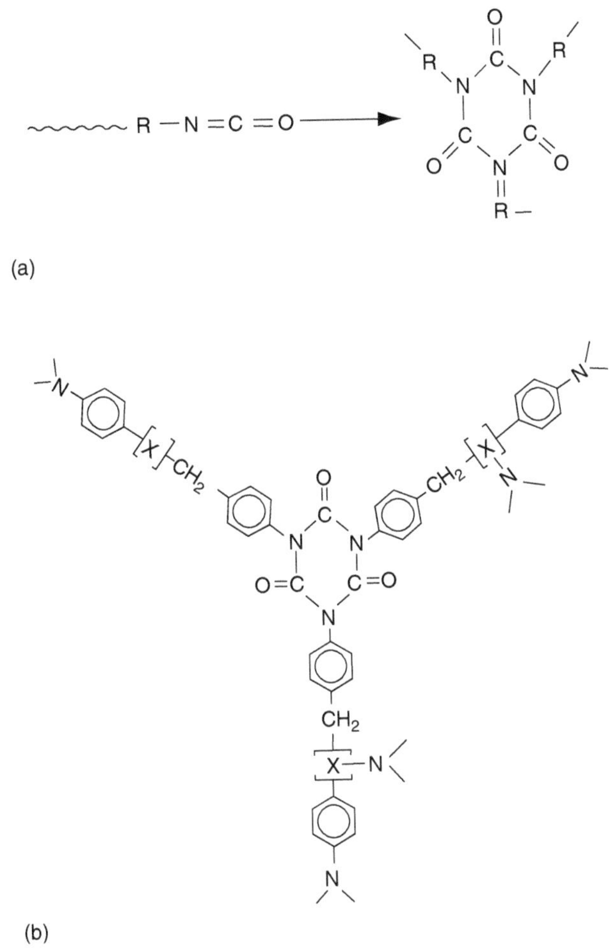

(a)

(b)

FIGURE 1.33 (a) Trimerization of an isocyanate. (b) Structure of polyisocyanurate produced from polymeric MDIs.

presence of polyols of molecular weights less than 300 has also been employed to produce foams by both one-shot and prepolymer methods.

1.3.4.4 Polyurethane Coatings

Polyurethane systems are also formulated for surface-coating applications. A wide range of such products has become available. These include simple simple solutions of finished polymer (linear polyurethanes), One component systems containing blocked isocyanates, two-component systems based on polyester and isocyanate or polyether and isocyanate, and a variety of prepolymer and adduct systems.

Coatings based reaon TDI and MDI gradually discolor upon exposure to light and oxygen. In contrast, aliphatic diisocyanates such as methylene dicyclohexyl isocyanate, hexamethylene diisocyanate derivatives, and isophorone diisocyanate all produce yellowing resistant, clear, or color-stable pigmented coatings. Of the polyols used, half are polyester type and half are polyehter types. The coatings can vary considerably in hardness and flexibility, depending on formulation.

Polyurethane coatings are used wherever applications require toughness, abrasion resistance, skin flexibility, fast curing, good adhesion, and chemical resistance. Use include metal finishes in chemical plants, wood finishes for boats and sports equipment, finishes for rubber goods, and rainerosion resistant coatings for aircraft.

TABLE 1.22 Blocked Isocyanates for One-Component System

Blocking Agent	Unblocking Temperature	
	°C	°F
Phenol	160	320
m-Nitrophenol	130	266
Acetone oxime	180	356
Diethyl malonate	130–140	266–284
Caprolactam	160	320
Hydrogen cyanide	120–130	248–266

The one-component coating systems require blocking of the isocyanate groups to prevent polymerization in the container. Typical blocking agents are listed in Table 1.22.

Generation of the blocking agent upon heating to cause polymerization is a disadvantage of blocked one-component systems. This problem can be overcome by using a masked aliphatic diisocyanate system, as shown in Figure 1.34. The cyclic bisurea derivative used in such a system is stable in the polyol or in water emulsion formulated with the polyol at ordinary temperatures. Upon heating, ring opening occurs, generating the diisocyanate, which reacts instantaneously with the macroglycol in form a polyurethane coating.

Polyurethane adhesives involving both polyols and isocyanates are used. These materials have found major uses in the boot and shoe industry.

1.3.5 Ether Polymers

For the purposes of this chapter ether polymers or polyethers are defined as polymers which contain recurring ether groupings in their backbone structure:

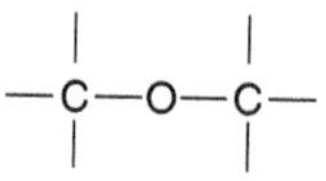

$$-\overset{|}{\underset{|}{C}}-O-\overset{|}{\underset{|}{C}}-$$

Polyethers are obtained from three different classes of monomers, namely, carbonyl compounds, cyclic ethers, and phenols. They are manufactured by a variety of polymerization processes, such as polymerization (polyacetal), ring-opening polymerization (polyethylene oxide, polyprophylene oxide, and epoxy resins), oxidative coupling (Polyphenylene oxide), and polycondensation (polysulfone).

Polyacetal polyphenylene oxide are widely used as engineering thermoplastics, and epoxy resins are used in adhesive and casting application. The main uses of poly(ethylene oxide) and poly(propylene oxide) are as macroglycols in the production of polyurethanes. Polysulfone is one of the high-temperature-resistant engineering plastics.

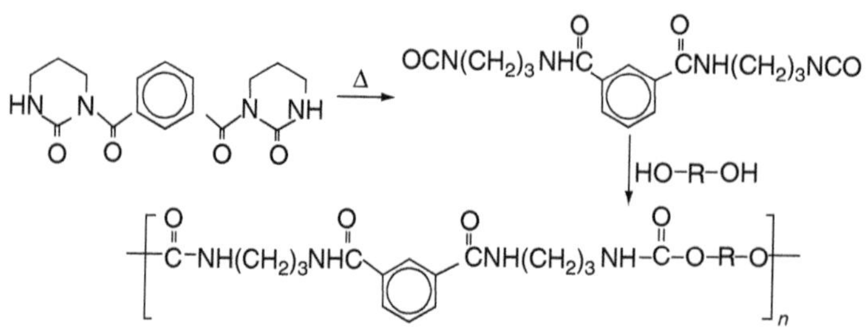

FIGURE 1.34 Reactions of macrocyclic ureas used as masked diisocyanates.

1.3.5.1 Polyacetal

$$-\!\!\!\left[-OCH_2-\right]_n\!\!\!-$$

Monomers	Polymerization	Major Uses
Formaldehyde, trioxane	Cationic or anionic chain polymerization	Appliances, plumbing and hardware, transportation

Polyoxymethylene (polyacetal) is the polymer of formaldehyde and is obtained by polymerization of aqueous formaldehyde or ring-opening polymerization of trioxane (cyclic trimer of formaldehyde, melting point 60–62°C), the latter being the preferred method [52]. This polymerization of trioxane is conducted in bulk with cationic initiators. In contrast, highly purified formaldehyde is polymerized in solution using using either cationic or anionic initiators.

Polyacetal strongly resembles polyethylene in structure, both polymers being liner with a flexible chain backbone. Since the structures of both the polymers are regular and the question of tacticity does not arise, both polymers are capable of high degree of crystallization. However, the acetal polymer molecules have a shorter backbone (–C–O–) bond and so pack more closely together than polyethylene molecules. The acetal polymer is thus harder and has a higher melting point (175°C for the homopolymer).

Being crystalline and incapable of specific interaction with liquids, acetal homopolymer resins have outstanding resistance to organic solvents. No effective solvent has yet been found for temperatures blow 70°C (126°F). Above this temperature, solution occurs in a few solvents such as chlorophenols. Swelling occurs with solvents of similar solubility parameter to that the polymer [$\delta = 11.1$ (cal/cm^3)$^{1/2}$ = 22.6 MPa$^{1/2}$]. The resistance of polyacetal to inorganic reagents in not outstanding, however. Strong acids, strong alkalis, and oxidizing agents cause a deterioration in mechanical properties.

The ceiling temperature for the acetal polymer is 127°C. Above this temperature the thermodynamics indicate that depolymerization will take place. Thus it is absolutely vital to stabilize the polyacetal resin sufficiently for melt processing at temperature above 200°C. Stabilization is accomplished by capping the thermolabile hydroxyl end groups of the macromolecule by etherification or esterification, or by copolymerizing with small concentrations of ethylene oxide. These expedients retard the initiation or propagation steps of chain reaction that could cause the polymer to unzip to monomer (formaldehyde). End-group capping is more conveniently achieved by esterification using acetic anhydride.

$$HOCH_2-\!\!\left[-OCH_2-\right]_n\!\!-O-CH_2OH \xrightarrow{\ Ac_2O\ }$$

$$CH_3\overset{\displaystyle O}{\underset{\displaystyle \|}{C}}-O-CH_2-\!\!\left[-OCH_2-\right]_n\!\!-O-CH_2-O-\overset{\displaystyle O}{\underset{\displaystyle \|}{C}}CH_3$$

If formaldehyde is copolymerized with a second monomer, which is a cyclic either such as ethylene oxide and 1,3-dioxolane, end-group capping is not necessary. The copolymerization results in occasional incorporation of molecules containing two successive methylene groups, whereby the tendency of the molecules to unzip is markedly reduced. This principle is made use of in the commercial products marketed as Celcon (Celanese), Hostaform (Farbwerke Hoechst), and Duracon (Polyplastic).

Degradation of polyacetals may also occur by oxidative attack at random along the chain leading to chain scission and subsequent depolymerization (unzipping). Oxidative chain scission is reduced by the use of antioxidants (see Chapter 1 of *Plastics Fundamentals, Properties, and Testing*), hindered phenols being preferred. For example, 2,2'-methylene-bis(4-methyl-6-*t*-butylphenol) is used in Celcon (Celanese) and 4,4'-butylidene bis(3-methyl-6-*t*-butylphenol) in Delrin (Du Pont).

Acid-catalyzed cleavage of the acetal linkage can also cause initial chain scission. To reduce this acid acceptors are believed to be used in commercial practice. Epoxides, nitrogen-containing compounds, and basic salts are all quoted in the patent literature.

Polyacetal is obtained as a linear polymer (about 80% crystalline) with an average molecular weight of 30,000–50,000. Comparative values for some properties of typical commercial products are given in Table 1.23. The principal features of acetal polymers which render them useful as engineering thermoplastics are high stiffness, mechanical strength over a wide temperature range, high fatigue endurance, resistance to creep, and good appearance. Although similar to nylons in many respects, acetal polymers are superior to them in fatigue resistance, creep resistance, stiffness, and water resistance (24-h water absorption at saturation being 0.22% for acetal copolymer vs. 8.9% for nylon-6,6). The nylons (except under dry conditions) are superior to acetal polymers in impact toughness. Various tests indicate that the acetal polymers are superior to most other plastics and die cast aluminum.

The electrical properties of the acetal polymers may be described as good but not outstanding. They would thus be considered in applications where impact toughness and rigidity are required in addition to good electrical insulation characteristics.

The end-group capped acetal homopolymer and the trioxane-based copolymers are generally similar in properties. The copolymer has greater thermal stability, easier moldability, better hydrolytic stability at elevated temperatures, and much better alkali resistance than the homopolymer. The homopolymer, on the other hand, has slightly better mechanical properties, e.g., higher flexural modulus, higher tensile strength, and greater surface hardness.

Acetal polymers and copolymers are engineering materials and are competitive with a number of plastics materials, nylon in particular, and with metals. Acetal resins are being used to replace metals because of such desirable properties as low weight (sp. gr. 1.41–1.42), corrosion resistance, resistance to fatigue, low coefficient of friction, and ease of fabrication.

The resins may be processed without difficulty on conventional injection-molding, blow-molding, and extrusion equipment. The acetal resins are used widely in the molding of telephone components, radios, small appliances, links in conveyor belts, molded sprockets and chains, pump housings, pump impellers, carburetor bodies, blower wheels, cams, fan blades, check values, and plumbing components such as valve stems and shower heads.

TABLE 1.23   Typical Values for Some Properties of Acetal Homopolymers and Copolymers

Property	Acetal Homopolymer	Acetal Copolymer
Specific gravity	1.425	1.410
Crystalline melting point (°C)	175	163
Tensile strength (23°C)		
lbf/in.^{2}	10,000	8500
MPa	70	58
Flexural modulus (23°C)		
lbf/in.^{2}	410,000	360,000
MPa	2800	2500
Deflection temperature °C)		
at 264 lbf/in.^{2}	100	110
at 66 lbf/in.^{2}	170	158
Elongation at break (23°C) (%)	15–75	23–35
Impact strength (23°C)		
ft.-lbf/in. notch	1.4–2.3	1.1
Hardness, Rockwell M	94	80
Coefficient of friction	0.1–0.3	0.2
Water absorption (%)		
24-h immersion	0.4	0.22
50% RH equilibrium	0.2	0.16
Continuous immersion equilibrium	0.9	0.8

Source: Brydson, J. A. 1982. *Plastics Materials.* Butterworth Scientific, London, UK.

Because of their light weight, low coefficient of friction, absence of slipstick behavior, and ability to be molded into intricate shapes in one piece, acetal resins find use as bearings. The lowest coefficient of friction and wear of acetal resin are obtained against steel.

Through counted as one of the engineering plastics, acetal resins with their comparatively high cost cannot, however, be considered as general-purpose thermoplastics in line with polyethylene, polypropylene, PVC, and polystyrene

1.3.5.2 Poly(Ethylene Oxide)

$$\left[-OCH_2CH_2-\right]$$

Monomer	Polymerization	Major Uses
Ethylene oxide	Ring-opening polymerization	Molecular weights 200 to 600— surfactants, humectants, lubricants Molecular weights >600— pharmaceutical and cosmetic bases, lubricants, mold release agents Molecular weights 10^5 to 5×10^6— water-soluble packaging films and capsules

Poly(ethylene oxide) of low molecular weight, i.e., below about 3000, are generally prepared by passing ethylene oxide into ethylene glycol at 120–150°C (248–302 F) and about 3 atm pressure (304 kPa) by using an alkaline catalyst such as sodium hydroxide [53]. Polymerization takes place by an anionic mechanism.

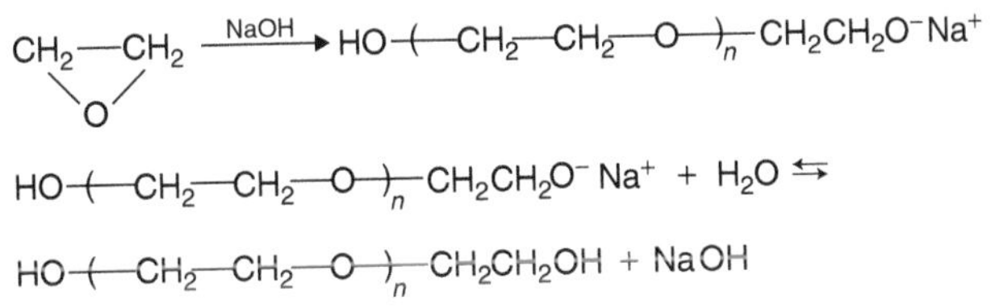

$$CH_2\!-\!CH_2 \xrightarrow{\ NaOH\ } HO\!-\!\left(CH_2\!-\!CH_2\!-\!O\right)_{\!n}\!-\!CH_2CH_2O^-\,Na^+$$

$$HO\!-\!\left(CH_2\!-\!CH_2\!-\!O\right)_{\!n}\!-\!CH_2CH_2O^-\,Na^+ + H_2O \rightleftarrows$$

$$HO\!-\!\left(CH_2\!-\!CH_2\!-\!O\right)_{\!n}\!-\!CH_2CH_2OH + NaOH$$

The polymers produced by these methods are thus terminated mainly by hydroxyl groups and are often referred to as *polyethylene glycols* (PEGs). Depending on the chain length, PEGs range in physical form at room temperature from water white viscous liquids (mol. wt. 200–700), through waxy semisolids (mol. wt. 1000–2000), to hard, waxlike solids (mol. wt. 3000–20,000 and above). All are completely soluble in water, bland, nonirritating, and very low in toxicity; they possess good stability, good lubricity, and wide compatibility.

Since PEGs form a homologous series of polymers, many of their properties vary in a continuous manner with molecular weight. The freezing temperature, which is less than −10°C for PEG of molecular weight 300, rises first rapidly with molecular weight through the low-molecular-weight grades, then increases more gradually through the solids while approaching 66°C, the true crystalline melting point for very high-molecular-weight poly(ethylene oxide) resins.

Other examples of such continuous variations are the increase in viscosity, flash points, and fire points with an increase in molecular weight and a slower increase in specific gravity. In a reverse relationship, hygroscopicity decreases as molecular weight increases, as does solubility in water.

In 1958, commercial poly(ethylene oxide)s of very high molecular weight became available from Union Carbide (trademark Polyox). Two Japanese companies, Meisei Chemical Works, Ltd., and Seitetsu Kagaku Company, Ltd., have begun producing poly(ethylene oxide) under the trademarks Alkox and PEO, respectively. Heterogeneous catalyst systems, which are mainly of two types, namely, alkaline earth compounds (e.g., oxides and carbonates of calcium, barium, and strontium) and organometallic compounds (e.g., aluminum and zinc alkyls and alkoxides, usually with cocatalysts) are used in their manufacture.

Commercial poly(ethylene oxide) resins supplied in the molecular weight range $1 \times 10^5 - 5 \times 10^6$ are dry, free-flowing, white powders soluble in an unusually broad range of solvents. The resins are soluble in water at temperatures up to 98°C and also in a number of organic solvents, which include chlorinated hydrocarbons such as carbon tetrachloride and methylene chloride, aromatic hydrocarbons such as benzene and toluene, ketones such as acetone and methyl ethyl ketone, and alcohols such as methanol and isopropanol.

1.3.5.3   Applications

1.3.5.3.1   *Polyethylene Glycol*

The unusual combination of properties of PEGs has enabled them to find a very wide range of commercial uses as cosmetic creams, lotions, and dressings; textile sizes; paper-coating lubricants; pharmaceutical sales, ointments, and suppositories, softeners and modifiers; metal-working lubricants; detergent modifiers; and wood impregnates. In addition, the chemical derivatives of PEGs, such as the mono- and diesters of fatty acids, are widely used as emulsifiers and lubricants.

The PEGs themselves show little surface activity, but when converted to mono- and diesters by reaction with fatty acids, they form a series of widely useful nonionic surfactants. The required balance of hydrophilic–hydrophobic character can be achieved by suitable combination of the molecular weight of the PEG and the nature of the fatty acid. For large-volume items a second production route of direct addition of ethylene oxide to the fatty acids is often preferred. End uses for the fatty esters are largely as textile lubricants and softeners and as emulsifiers in food products, cosmetics, and pharmaceuticals.

The PEGs have found a variety of uses in pharmaceutical products. Their water solubility, blandness, good solvent action for many medicaments, pleasant and nongreasy feel on the skin, and tolerance of body fluids are the reasons why they are frequently the products of choice. Blends of liquid and solid grades are often selected because of their desirable petrolatum like consistency.

An especially important example of pharmaceutical application of PEGs is as bases for suppositories, where the various molecular grades can be blended to provide any desired melting point, degree of stability, and rate of release of medication. The fatty acid esters of the PEGs are often used in pharmaceuticals as emulsifiers and suspending agents because of their nonionic, blandness, and desirable surface activity. The solid PEGs also find use a lubricants and binders in the manufacture of medicinal tablets.

The PEGs, providing they contain not over 0.2% ethylene and diethylene glycols, are permitted as food additives. The PEG fatty acid esters are especially useful emulsifying agents in food products.

For many of the same reasons which account for their use in pharmaceuticals, the PEGs and their fatty acid esters find many applications in cosmetics and toiletries. Their moisturizing, softening, and skin-smoothing characteristics are especially useful. Typical examples of applications are shaving creams, vanishing creams, toothpastes, powders, shampoos, hair rinses, suntan lotions, pomades and dressings, deodorants, stick perfumes, rouge, mascara, and so on. The nonionic surface-active PEG fatty acid esters find use in a variety of detergents and cleaning compositions.

Liquid PEGs, solid PEGs, or their solutions are often used in a variety of ink preparations, such as thixotropic inks for ballpoint pens, water-based stencil inks, steam-set printing inks, and stamp-pad inks.

There are numerous other industrial uses for the PEGs in which they serve primarily as processing aids and do not remain as integral components of the products. The PEGs add green strength and good formability to various ceramic components of the products. The PEGs add green strength and good formability to various ceramic compositions to be stamped, extruded, or molded. They can be burned out cleanly during subsequent firing operations.

In electroplating baths small amounts of PEGs improve smoothness and grain uniformity of the deposited coatings. Solid PEGs are effective lubricants in paper-coating compositions and promote better gloss and smoothness in calendaring operations. PEGs, and more particularly their fatty acid esters, are quite widely used as emulsifiers, lubricants, and softeners in textile processing.

PEGs and their esters find use as components of metal corrosion inhibitors in oil wells where corrosive brines are present. Fatty acid esters of PEGs are useful demulsifiers for crude oil–water separation.

The water solubility, nonvolatility, blandness, and good lubricating abilities are some of the reasons for the use of PEGs in metal-working operations. Metal-working lubricants for all but the most severe forming operations are made with the PEGs or with their esters or other derivatives.

1.3.5.3.2 Poly(Ethylene Oxide)

Since their commercialization in 1958, the reported and established applications for high-molecular-weight poly(ethylene oxide) resins have been numerous and diversified. Table 1.24 summarizes in alphabetical order the main applications of these resins in various industries.

In addition to their water solubility and blandness, the main functions and effects of poly(ethylene oxide) resins which lead to these diverse applications are lubrication, flocculation, thickening, adhesion, hydrodynamic drag reduction, and formation of association complexes.

The resins are relatively nontoxic and have a very low level of biodegradability (low BOD). Poly(ethylene oxide) resins with molecular weights from 1×10^5 to 1×10^7 have a very low level of oral toxicity and are not readily absorbed from the gastrointestinal tract. The resins are relatively nonirritating to the skin and have a low sensitizing potential.

Poly(ethylene oxide) resins can be formed into various shapes by using conventional thermoplastic processing techniques. Commercially, thermoplastic processing of these resins has been, however, limited almost exclusively to the manufacture of film and sheeting. Generally, the medium-molecular-weight resins (4×10^5–6×10^5) possess melt rheology best suited to thermoplastic processing. The films are produced by calendaring or blown-film extrusion techniques. Usually produced in thicknesses from 1 to 3 mils, the films have very good mechanical properties combined with complete water solubility.

TABLE 1.24 Applications of Poly(Ethylene Oxide) Resins

Industry	Applications
Agriculture	Water-soluble seed tapes
	Water-soluble packages for agricultural chemicals
	Hydrogels as soil amendments to increase water retention
	Soil stabilization using association complexes with poly(acrylic acid)
	Drift control agent for sprays
Ceramics and glass	Binders for ceramics
	Size for staple glass-fiber yarns
Chemical	Dispersant and stabilizer in aqueous suspension polymerization
Electrical	Water-soluble, fugitive binder for microporous battery and fuel-cell electrodes
Metals and mining	Flocculant for removal of silicas and clays in hydrometallurgical processes
	Flocculant for clarification of effluent streams from coal-washing plants
Paper	Filler retention and drainage aid in the manufacture of paper
	Flocculant for clarification of effluent water
Personal-care products	Lubricant and toothpaste
	Tickener in preparation of shaving stick
	Opthalmic solution for wetting, cleaning, cushioning, and lubricating contact lenses
	Adhesion and cushioning ingredient in denture fixatives
	Hydrogels as adsorptive pads for catamenial devices and disposable diapers
Petroleum	Thickener for bentonite drilling muds
	Thickener for secondary oil-recovery fluids in waterflooding process
Pharmaceutical	Water-soluble coating for tablets
	Suspending agent to inhibit settling of ceramic lotion
Printing	Microencapsulation of inks
Soap and detergent	Emollient and thickener for detergent bars and liquids
Textile	Additive to improve dyeability and antistatic properties of polyolefin, polyester, and polyamide fibers

Poly(ethylene oxide) films have been used to produce seed tapes, which consist of seeds sandwiched between two narrow strips of film scaled at the edges. When the seed tape is planted, water from the soil dissolves the water-soluble film within a day or two, releasing the seed for germination. Because the seeds are properly spaced along the tape, the process virtually eliminates the need for thinning of crops.

Films of poly(ethylene oxide) are also used to manufacture water-soluble packages for preweighed quantities of fertilizers, pesticides, insecticides, detergents, dyestuffs, and the like. The packages dissolve quickly in water, releasing the contents. They eliminate the need for weighing and offer protection to the user from toxic or hazardous substances.

Poly(ethylene oxide) forms a water-insoluble association complex with poly(acrylic acid). This is the basis of microencapsulation of nonaqueous printing inks. Dry, free-flowing powders obtained by this process can be used to produce "carbonless" carbon papers (see "Microencapsulation" in Chapter 2). When pressure is applied to the paper coated with the microencapsulated ink, the capsule wall ruptures and the ink is released.

The formation of a water-insoluble association complex of poly(ethylene oxide) and poly(acrylic acid) is also the basis for a soil stabilization process to prevent erosion of soil on hillsides and river banks.

A variety of different types of adhesives can be produced by forming association complexes of poly(ethylene oxide) with tannin or phenolic resins. Examples include wood glue, water-soluble quickset adhesive, and pressure-sensitive adhesives.

High-molecular-weight poly(ethylene oxide) resins are effective flocculants for many types of clays, coal suspensions, and colloidal silica, and so find application as process aids in mining and hydrometallurgy.

The turbulent flow of water through pipes and hoses or over surfaces causes the effect known as *hydrodynamic drag*. High-molecular-weight poly(ethylene oxide) resins are most effective in reducing the hydrodynamic drag and thus find use in fire fighting, where small concentrations of these resins (50–100 ppm) reduce the pressure loss in fire hoses and make it possible to deliver as much as 60% more water through a standard 2.5-in. (6.35-cm)-diameter fire hose. The Union Carbide product UCAR Rapid Water Additive, which contains high-molecular-weight poly(ethylene oxide) as the active ingredient, is a hydrodynamic-drag-reducing additive for this application.

The ability of poly(ethylene oxide) to reduce hydrodynamic drag has also led to its use in fluid-jet systems used for cutting soft goods, such as textiles, rubber, foam, cardboard, etc. In these systems specially designed nozzles produce a very-small-diameter water jet at a pressure of 30,000–60,000 psi (200–400 MPa). Although a plain water disperses significantly as it leaves the nozzle, with poly(ethylene oxide) addition the stream becomes more cohesive and maintains its very small diameter up to 4 in. (10 cm) from the nozzle.

Chemical or irradiation cross-linking of poly(ethylene oxide) resins yield *hydrogels*, which are not water soluble but water absorptive, capable of absorbing 25–100 times their own weight of water. These hydrogels are reportedly useful in the manufacture of absorptive pads for catamenial devices and disposable diapers.

The water absorbed by the hydrogels is also readily desorbed by drying the hydrogel. This characteristics is the basis for the use of these hydrogels as so-called *soil amendments*. When mixed with ordinary soil in a concentration of about 0.001–5.0 wt% of the soil, these hydrogels will reduce the rate of moisture loss due to evaporation but will still release water to the plants and thus eliminate the need for frequent watering of the soil.

1.3.5.4   Poly(Propylene Oxide)

$$\left[OCH_2-\underset{\underset{CH_3}{|}}{CH} \right]_n$$

Monomer	Polymerization	Major Uses
Propylene oxide	Base-catalyzed ring-opening polymerization	Polyols for polyurethane foams, surfactants, lubricants cosmetic bases

Propylene oxide is polymerized by methods similar to those described in the preceding section for poly (ethylene oxide). Like the latter, low- and high-molecular-weight polymers are of commercial interest.

Poly(propylene oxide)s of low molecular weight (i.e., from 500 to 3500), often referred to as *polypropylene glycols* (PPGs), are important commercial materials mainly because of their extensive use in the production of polyurethane foams (see Chapter 1 of *Plastics Fabrication and Recycling* and Chapter 1). PPGs are less hydrophilic and lower in cost and may be prepared by polymerizing propylene oxide in the presence of propylene glycol as an initiator and sodium hydroxide as a catalyst at about 160°C. The polymers have the general structure

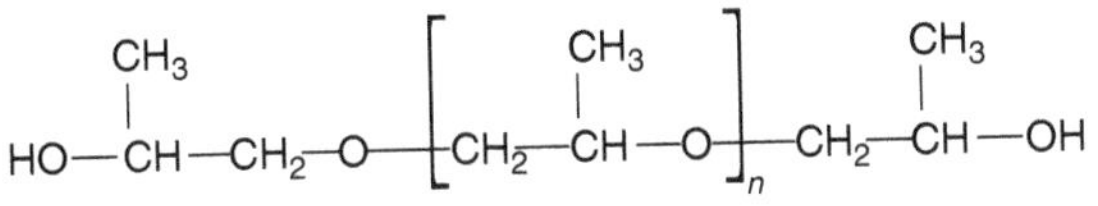

$$HO-\overset{\overset{\displaystyle CH_3}{|}}{CH}-CH_2-O\left[CH_2-\overset{\overset{\displaystyle CH_3}{|}}{CH}-O\right]_n CH_2-\overset{\overset{\displaystyle CH_3}{|}}{CH}-OH$$

The end hydroxy groups of the polymer are secondary groups and are ordinarily rather unreactive in the urethane reaction. Initially, this limitation was overcome by the preparation of isocyanate-terminated prepolymer and by the use of block copolymers with ethylene oxide. The latter products are known as *tipped polyols* and are terminated with primary hydroxy groups of enhanced activity.

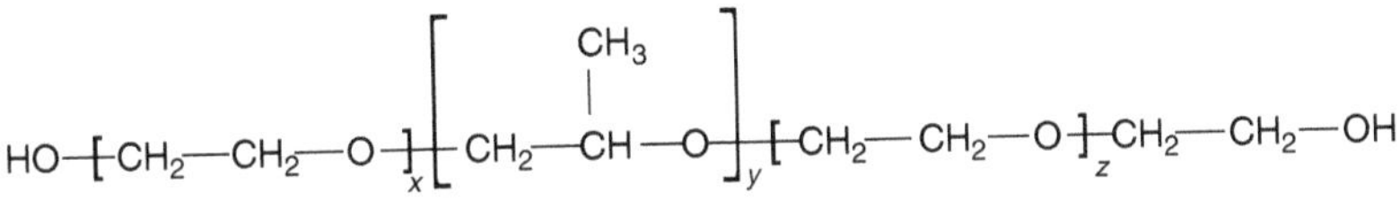

$$HO-[CH_2-CH_2-O]_x\left[CH_2-\overset{\overset{\displaystyle CH_3}{|}}{CH}-O\right]_y[CH_2-CH_2-O]_z CH_2-CH_2-OH$$

(Note that straight PEG is not satisfactory for polyurethane foam production due to its water sensitivity and tendency to crystallize.)

The later advent of more powerful catalysts, however, made it possible for straight PPG to be used in the preparation of flexible polyurethane foams (see one-shot processes in Chapter 1 of *Plastics Fabrication and Recycling*) without recourse to the foregoing procedures. Also, today the bulk of the polyethers used are triols rather than diols, since these lead to slightly cross-linked flexible foams with improved load-bearing characteristics.

The polyether triols are produced by polymerizing propylene oxide by using 1,1,1-trimethylolpropane, 1,2,6-hexane triol, or glycerol as the initiator. The use of, for example, trimethylolpropane leads to the following polyether triol.

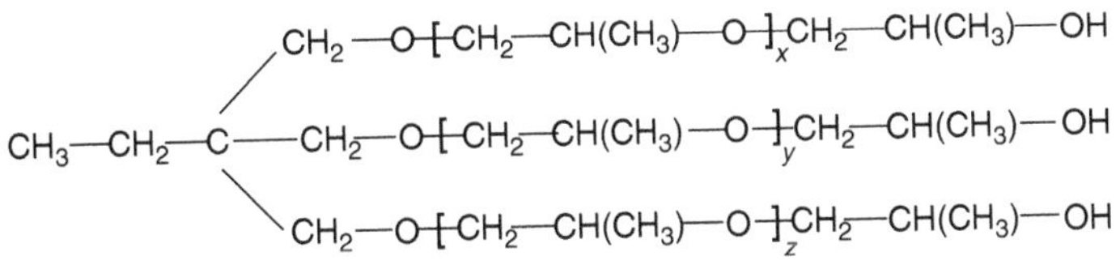

$$CH_3-CH_2-C \begin{cases} CH_2-O[CH_2-CH(CH_3)-O]_x CH_2-CH(CH_3)-OH \\ CH_2-O[CH_2-CH(CH_3)-O]_y CH_2-CH(CH_3)-OH \\ CH_2-O[CH_2-CH(CH_3)-O]_z CH_2-CH(CH_3)-OH \end{cases}$$

For flexible polyurethane foams, polyether triols of molecular weights from 3000 to 3500 are normally used because they give the best balance of properties. For the production of rigid foams, polyether triols of lower molecular weight (about 500) are used to increase the degree of cross-linking. Alternatively, polyether polyols of higher functionality, such as produced by polymerizing propylene oxide with pentaerythritol or sorbitol, may be used.

Copolymerization of ethylene oxide and propylene oxide yields quite valuable functional fluids of various sorts. The random copolymers of ethylene and propylene oxides of relatively low molecular weights are water soluble when the proportion of ethylene oxide is at least 40–50% by weight.

The block copolymers consist of sequences of "blocks" of all oxypropylene or all-oxyethylene groups, as shown.

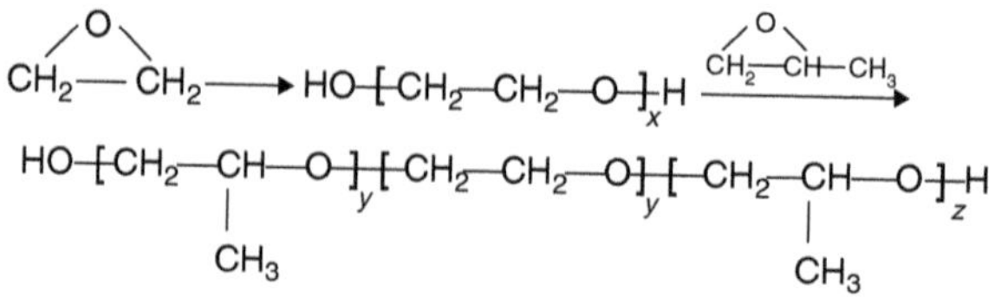

Properties vary considerably, depending on the lengths and arrangements of these blocks. The block copolymers comprise unique and valuable surface-active agents. They can act as breakers for water-in-oil emulsions, as defoamers, and as wetting and dispersing agents.

1.3.5.5 Epoxy Resins

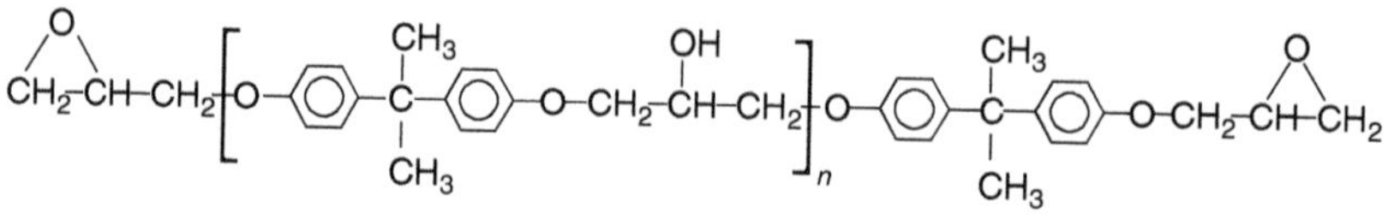

Monomers	Polymerization	Major Uses
Bisphenol A, epichlorohydrin	Condensation and ring-opening polymerization	Surface coating (44%), laminates and composites (18%, moldings (9%), flooring (6%), adhesives (5%)

Epoxide or epoxy resins [14,54,55] contain the epoxide group, also called the epoxy, oxirane, or ethoxyline group, which is a three-membered oxide ring:

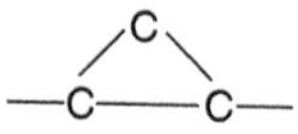

(The simplest compound in which the epoxy group is found is ethylene oxide.) In the uncured stage epoxies are polymers with a low degree of polymerization. They are most often used as thermosetting resins which cross-link to form a three-dimensional nonmelting matrix. A curing agent (hardener) is generally used to achieve the cross-linking. In room-temperature curing the hardener is generally an amine such as diethylene triamine or triethylenetetramine. For elevated temperature curing a number of different curing agents could be utilized, including aromatic amines and acid anhydrides.

Epoxy resins first developed commercially and still completely dominating the worldwide markets are those based on 2,2-bis(4′-hydroxyphenyl) propane, more commonly known as bisphenol A (as it is produced by condensation of phenol with acetone) and 1-chloro-2,3-epoxy-propane, also known as epichlorohydrin. It can be seen from the general formula of these resins that the molecular species concerned is a linear polyether with terminal glycidyl ether group

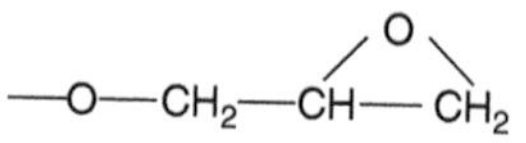

and secondary hydroxyl groups occurring at regular intervals along the length of the macromolecule. The number of repeating units (n) depends essentially on the molar ratio of bisphenol A and epichlorohydrin. When $n=0$, the product is diglycidyl ether and the molecular weight is 340. When $n=10$, the molecular weight is about 3000. Commercial liquid epoxy resins based on bisphenol A and epichlorohydrin have average molecular weights from 340 to 400, and it is therefore obvious that these materials are composed largely of diglycidyl ether. The liquid resin is thus often referred to as DGEBA (diclycidyl ether of bisphenol A). Similarly, the epoxy resin based on bis(4-hydroxyphenyl)methane, more commonly as bisphenol F (as it is produced by the condensation of phenol with formaldehyde) and epichlorohydrin is referred to as DGEBF.

DGEBA may be reacted with additional quantities of bisphenol A in an advancement reaction. This advancement produces higher-molecular-weight solid resins possessing a higher melting point ($>90°C$). Advancement generally increases flexibility, improves salt fog corrosion resistance, and increases hydroxyl content, which can be utilized later for cross-linking. Possessing generally low functionality (number of epoxy groups), their major use is in coatings. They provide outstanding adhesion and good salt fog corrosion resistance.

Commercial solid epoxy resins seldom have average molecular weights exceeding 4000, which corresponds to an average value of *n* of about 13. Resins with molecular weights above 4000 are of limited use since their high viscosity and low solubility make subsequent processing difficult.

1.3.5.5.1 *Resin Preparation*

The molecular weights of epoxy resins depend on the molar ratio of epichlorohydrin and bisphenol A used in their preparation (see Table 1.25). In a typical process for the production of liquid epoxy resins, epichlorohydrin and bisphenol A in the molar ratio of 10:1 are added to a stainless steel kettle fitted with a powerful anchor stirrer. The water content of the mixture is reduced to below 2% by heating the mixture until the epichlorohydrin–water distils off. After condensation, the epichlorohydrin layer is returned to the kettle, the water being discarded.

When the necessary water content of the reaction mixture is reached, the reaction is started by the slow addition of sodium hydroxide (2 mole per mole of bisphenol A) in the form of 40% aqueous solution, the temperature being maintained at about 100°C. The water content of the reaction mixture should be maintained between 0.31% and 2% by weight throughout the reaction if high yields (90–95%) are to be obtained. (No reaction occurs under anhydrous conditions, and undesired by-products are formed if the water content is greater than 2%). Besides distilling off the water as an azeotrope with the epichlorohydrin, the water content can also be partly controlled by the rate of addition of the caustic soda solution.

When all the alkali has been added, which may take 2–3 h, the excess epichlorohydrin is recovered by distillation at reduced pressure. A solvent such as toluene or methyl isobutyl ketone is then added to the cooled reaction product to dissolve the resin and leave the salt formed in the reaction.

The resin solution is washed with hot water, and after filtration and further washing, the solvent is removed by distillation and the resin is dried by heating under vacuum.

Though the pure diglycidyl ether of bisphenol A is a solid (m.p. 43°C), the commercial grades of the resin which contain a proportion of high-molecular-weight materials are supercooled liquid with viscosities of about 100–140 P at room temperature. The high-molecular-weight epoxy resins usually manufactured have values of *n* in the general formula ranging from 2 to 12. These resins are synthesized by allowing epichlorohydrin and bisphenol A to interact in the presence of excess sodium hydroxide.

TABLE 1.25 Effect of Reactant Ratio on Molecular Weight of Epoxy Resins

Molar Ratio Epichlorohydrin/ Bisphenol A	Molecular Weight	Epoxide Equivalent[a]	Softening Point (°C)
10:1	370	192	9
2:1	451	314	43
1.4:1	791	592	84
1.33:1	802	730	90
1.25:1	1133	862	100
1.2:1	1420	1176	112

[a] This is the weight of resin (in grams) containing one epoxide equivalent. For a pure diglycidyl ether (mol. wt. 340) with two epoxy groups per molecule, epoxide equivalent $= 340/2 = 170$.

In the *taffy process* usually employed, a mixture of bisphenol A and epichlorohydrin (the molar ratio of the reactants used depends on the resin molecular weight required; see Table 1.25) is heated to 100°C and aqueous sodium hydroxide (NaOH–epichlorohydrin molar ratio 1.3:1) is added slowly with vigorous stirring, the reaction being completed in 1–2 h. A white puttylike taffy (which is an emulsion of about 30% water in resin and also contains salt and sodium hydroxide) rises to the top of the reaction mixture. The lower layer of brine is removed; the resinous layer is coagulated and washed with hot water until free from alkali and salt. The resin is then dried by heating and stirring at 150°C under reduced pressure, poured into cooling pans, and subsequently crushed and bagged.

To avoid the difficulty of washing highly viscous materials, higher-molecular-weight epoxy resins may also be prepared by a two-stage process (advancement reaction, described earlier). This process consists of fusing a resin of lower molecular weight with more bisphenol A at about 190°C. The residual base content of the resin is usually sufficient to catalyze the second-stage reaction between the phenolic hydroxyl group and the epoxy group. Typical data of some commercial glycidyl ether resins are given in Table 1.26.

Solid resins have been prepared with narrow molecular-weight distributions. These resins melt sharply to give low-viscosity liquids, which enables incorporation of larger amounts of fillers with a consequent reduction in cost and coefficient of expansion. These resins are therefore useful in casting operations.

1.3.5.5.2 Curing

To convert the epoxy resins into cross-linked structures, it is necessary to add a curing agent. Most of the curing agents in common use can be classified into three groups: *tertiary amines*, *polyfunctional amines*, and *acid anhydrides*.

Examples of tertiary amines used as curing agents for epoxy resins include:

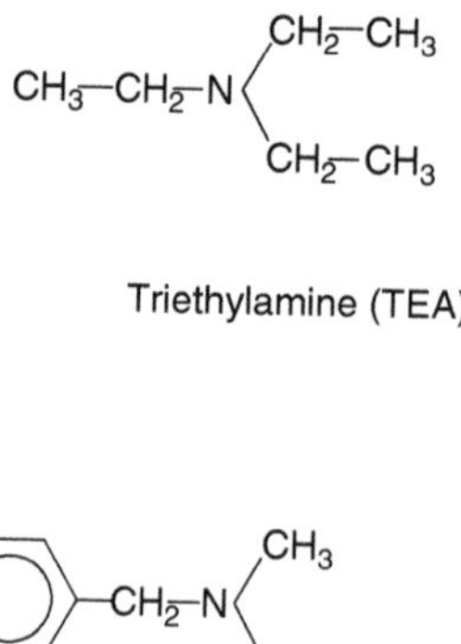

$$CH_3\text{--}CH_2\text{--}N \Big\langle\; \begin{matrix} CH_2\text{--}CH_3 \\ CH_2\text{--}CH_3 \end{matrix}$$

Triethylamine (TEA)

$$\bigcirc\text{--}CH_2\text{--}N \Big\langle\; \begin{matrix} CH_3 \\ CH_3 \end{matrix}$$

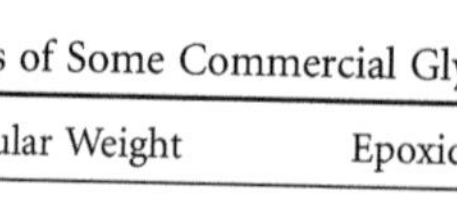

Bezyldimethylamine (BDA)

TABLE 1.26 Typical Properties of Some Commercial Glycidyl Ether Resins

Resin	Average Molecular Weight	Epoxide Equivalent	Melting Point (°C)
A	340–400	175–210	—
B	450	225–290	—
C	700	300–375	40–50
D	950	450–525	64–75
E	1400	870–1025	95–105
F	2900	1650–2050	125–132
G	3800	2400–4000	145–155

Source: Brydson, J. A. 1982. *Plastics Materials.* Butterworth Scientific, London, UK.

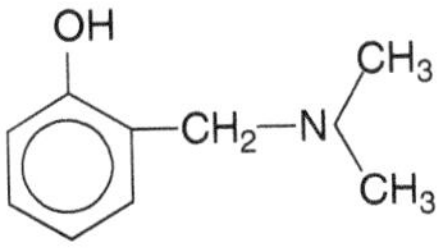

Dimethylaminomethylphenol (DMAMP)

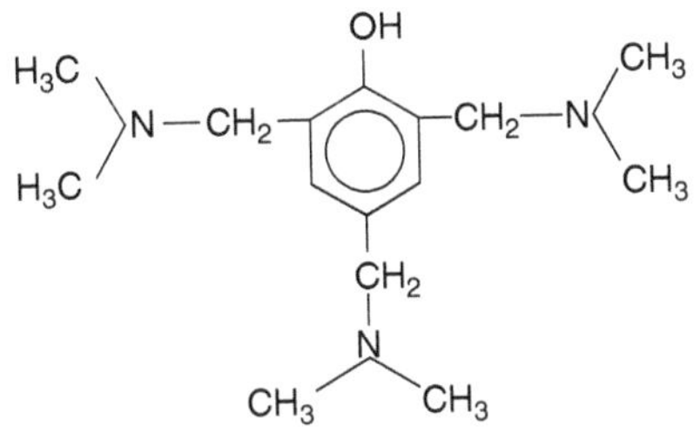

Tri(dimethylaminomethyl) Phenol (TDMAMP)

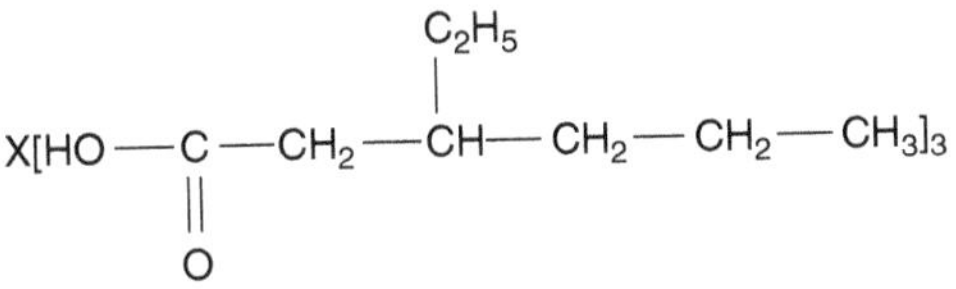

$$X[HO{-}\underset{\underset{O}{\|}}{C}{-}CH_2{-}\underset{\underset{C_2H_5}{|}}{CH}{-}CH_2{-}CH_2{-}CH_3]_3$$

Tri-2-ethylhexoate salt of TDMAMP (denoted as X)

Tertiary amines are commonly referred to as catalytic curing agents since they induce the direct linkage of epoxy groups to one another. The reaction mechanism is believed to be as follows:

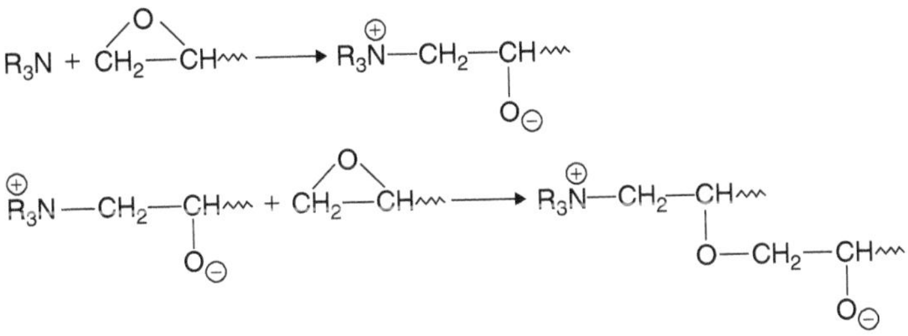

Since the reaction may occur at both ends of the diglycidyl ether molecule, a cross-linked structure will be built up. The overall reaction may, however, be more complicated because the epoxy group, particularly when catalyzed, also reacts with hydroxyl groups. Such groups may be present in the higher-molecular-weight homologues of the diglycidyl ether of bisphenol A, or they may be formed as epoxy rings are opened during cure.

In contrast to tertiary amine hardeners, which, as shown, cross-link epoxide resins by a catalytic mechanism, polyfunctional primary and secondary amines act as reactive hardeners and cross-link epoxy resins by bridging across epoxy molecules.

An amine molecule with two active hydrogen atoms can link across two epoxy molecules, as shown:

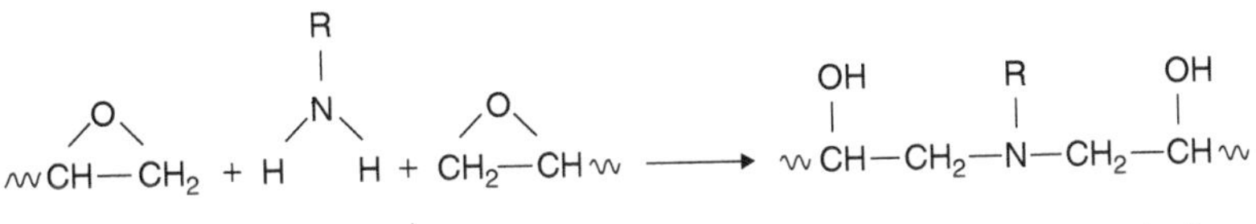

With polyfunctional aliphatic acid aromatic amines having three or more active hydrogen atoms in amine groups, this type of reaction results in a network polymer (see Figure 1.20 of *Plastics*

Fundamentals, Properties, and Testing). Generally speaking, aliphatic amines provide fast cures and are effective at room temperature, whereas aromatic amines are somewhat less reactive but give products with higher heat-distortion temperatures. Polyfunctional amines are widely used in adhesive, casting, and laminating applications.

Diethylenetriamine (DETA), $H_2N-CH_2-CH_2-NH-CH_2-CH_2-NH_2$, and triethylenetetramine (TETA), $H_2N-CH_2-CH_2-NH-CH_2-CH_2-NH-CH_2-CH_2-NH_2$, are highly reactive primary aliphatic amines with five and six active hydrogen atoms available for cross-linking, respectively. Both materials will cure (harden) liquid epoxy resins at room temperature and produce highly exothermic reactions.

With DETA the exothermic temperature may reach as high as 250°C in 200-g batches. With this amine used in the stoichiometric quantity of 9–10 pts phr (parts per hundred parts resin), the room temperature pot life is less then an hour. The actual time, however, depends on the ambient temperature and the size of the batch. With TETA, 12–13 pts phr are required.

Dimethylaminopropylamine (DMAPA), $(CH_3)_2N-CH_2-CH_2-CH_2-NH_2$, and diethylaminoropylamine (DEAPA), $(C_2H_5)_2N-CH_2-CH_2-CH_2-NH_2$, are slightly less reactive and give a pot life of about 140 min (for a 500-g batch) and are sometimes preferred. Note that both DMAPA and DEAPA have less than three hydrogen atoms necessary for cross-linking by reaction with epoxy groups.

Other examples of amine curing agents with less than three hydrogen atoms are diethanolamine (DEA), $NH(CH_2CH_2OH)_2$, and piperidine (see later). These curing agents operate by means of two-part reaction. Firstly, the active hydrogen atoms of the primary and secondary amine groups are utilized in the manner already described. Thereafter, the resulting tertiary amines, being sufficiently reactive to initiate polymerization of epoxy groups, function as catalytic curing agents, as described previously.

As a class, the amines usually suffer from the disadvantage that they are pungent, toxic, and skin sensitizers. To reduce toxicity, the polyfunctional amines are often used in the form of *adducts*. A number of such modified amines have been introduced commercially. For example, reaction of the amine with a mono- or polyfunctional glycidyl material will give a higher-molecular-weight product with less volatility.

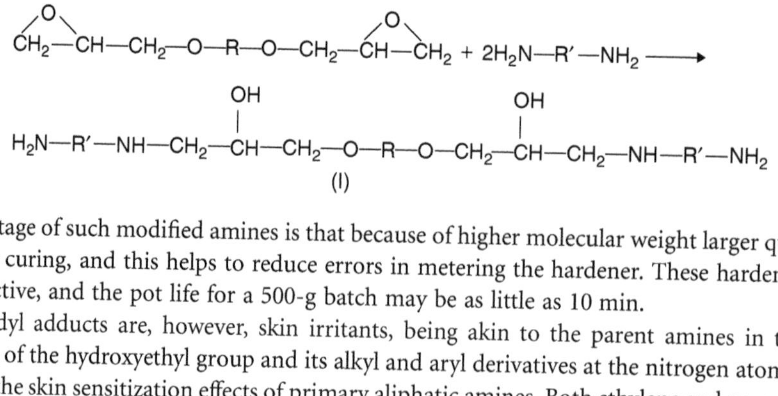

(I)

An advantage of such modified amines is that because of higher molecular weight larger quantities are required for curing, and this helps to reduce errors in metering the hardener. These hardeners are also extremely active, and the pot life for a 500-g batch may be as little as 10 min.

The glycidyl adducts are, however, skin irritants, being akin to the parent amines in this respect. Substitution of the hydroxyethyl group and its alkyl and aryl derivatives at the nitrogen atom is effective in reducing the skin sensitization effects of primary aliphatic amines. Both ethylene and propylene oxides have thus been used in the preparation of adducts from a variety of amines, including ethylene diamine and diethylenetriamine.

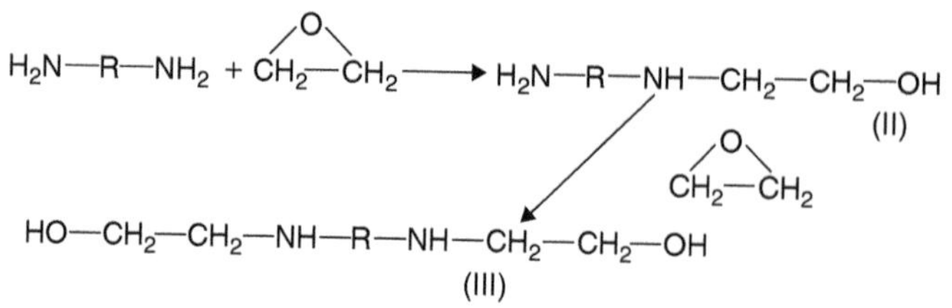

Such adducts from diethylenetriamine appear free of skin sensitizing effects. A hardener consisting of a blend of the reaction products II and III is a low-viscosity liquid giving a pot life (for a 500-g batch) of 16–18 min at room temperature.

Modification of primary amines with acrylonitrile results in hardeners with reduced activity.

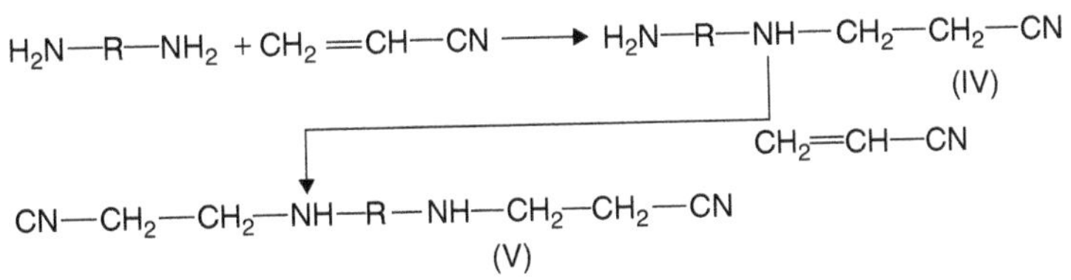

$$H_2N\!-\!R\!-\!NH_2 + CH_2\!=\!CH\!-\!CN \longrightarrow H_2N\!-\!R\!-\!NH\!-\!CH_2\!-\!CH_2\!-\!CN$$
(IV)

$$CH_2\!=\!CH\!-\!CN$$

$$CN\!-\!CH_2\!-\!CH_2\!-\!NH\!-\!R\!-\!NH\!-\!CH_2\!-\!CH_2\!-\!CN$$
(V)

Commercial hardeners are mixtures of the addition compounds IV and V. Since accelerating hydroxyl groups are not present in IV and V (in contrast to I, II, and III), these hardeners have reduced activity and so give longer pot lives.

A number of aromatic amines also function as epoxy hardeners. Since they introduce the rigid benzene ring structure into the cross-linked network, the resulting products have significantly higher heat-distortion temperatures than are obtainable with the aliphatic amines (see Table 1.27). For example, metaphenylenediamine (MPDA) (shown below) gives cure resins with a heat distortion temperature of 150°C and very good chemical resistance. The hardener finds use in the manufacture of chemical resistance laminates.

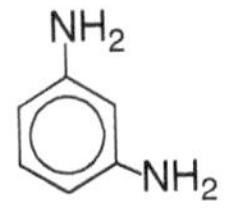

(MPDA)

Diaminodiphenylmethane (DAPPM) (below)

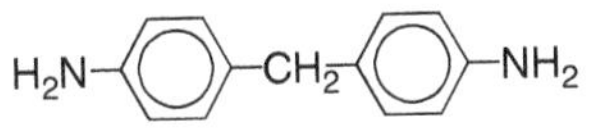

TABLE 1.27 Characteristics of Amine Hardeners Used in Low-Molecular-Weight Bisphenol A-Based Epoxy Resins

Hardener[a]	Parts used per 100 Parts Resin	Pot Life (500-g Batch)	Typical Cure Schedule	Max HDT of Cured Resin (°C)	Applications
DETA	10–11	20 min	Room temperature	110	General purpose
DEAPA	7	140 min	Room temperature	97	General purpose
DETA-glycidyl adduct	25	10 min	Room temperature	75	Adhesive laminating, fast cure
DETA-ethylene oxide adduct	20	16 min	Room temperature	92	—
MPDA	14–15	6 h	4–6 h at 150°C	150	Laminates, chemical resistance
DADPM	28.5	—	4–6 h at 165°C	160	Laminates
DADPS	30	—	8 h at 160°C	175	Laminates
Piperidine	5–7	8 h	3 h at 100°C	75	General purpose
Triethylamine	10	7 h	Room temperature	—	Adhesives
BDA	15	75 min	Room temperature	—	Adhesives
TDMAMP	6	30 min	Room temperature	64	Adhesives, coatings
2-Ethyl hexoate salt of TDMAMP	10–14	3–6 h	—	—	Encapsulation

[a] For chemical formulas see text.
Source: From Brydson, J. A. 1982. *Plastics Materials.* Butterworth Scientific, London, UK.

and diaminodiphenyl sulfone (DADPS) (below)

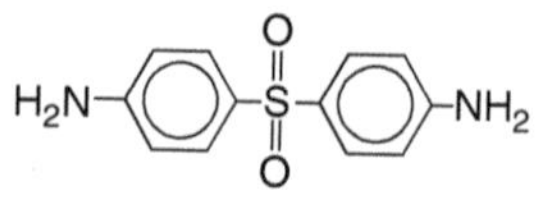

used in conjunction with an accelerator provide even higher heat-distortion temperatures but at some expense to chemical resistance.

Piperidine, a cyclic aliphatic amine

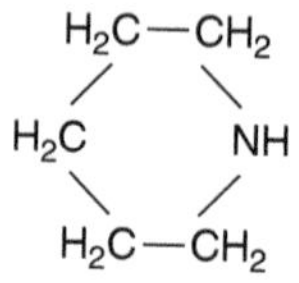

has been in use as an epoxy hardener since the early patents of Castan. Although a skin irritant, this hardener is still used for casting larger masses than are possible with primary aliphatic amines. Note that because it has only one active hydrogen atom in the amine group, piperidine operates by a two-part reaction in curing. Typical amine hardeners and their characteristics are summarized in Table 1.27.

Cyclic acid anhydrides are widely employed as curing agents for epoxy resins. Compared with amines they are less skin sensitive and generally give lower exotherms in curing reaction. Some acid curing agents provide cured resins with very high heat-distortion temperatures and with generally good mechanical, electrical, and chemical properties. The acid-cured resins, however, show less alkali resistance than amine-cured resins because of the susceptibility of ester groups to hydrolysis. Anhydride curing agents find use in most of the important applications of epoxy resins, particularly in casting and laminates.

In practice, acid anhydrides are preferred to acids since the latter are generally less soluble in the resin and also release more water on cure, leading to foaming of the product. Care must be taken, however, during storage since the anhydrides in general are somewhat hygroscopic. Examples of some anhydrides which are used are shown in Figure 1.35.

The mechanism of hardening by anhydride is complex. In general, however, two types of reactions occur: (1) opening of the anhydride ring with the formation of carboxy groups, and (2) opening of the epoxy ring. The most important reactions which may occur are shown in what follows, with phthalic anhydride as an example.

1. The first stage of the interaction between an acid anhydride and an epoxy resin is believed to be the opening of the anhydride ring by (a) water (traces of which may be present in the system) or (b) hydroxy groups (which may be present as pendant groups in the original resin or may result from reaction (2a).

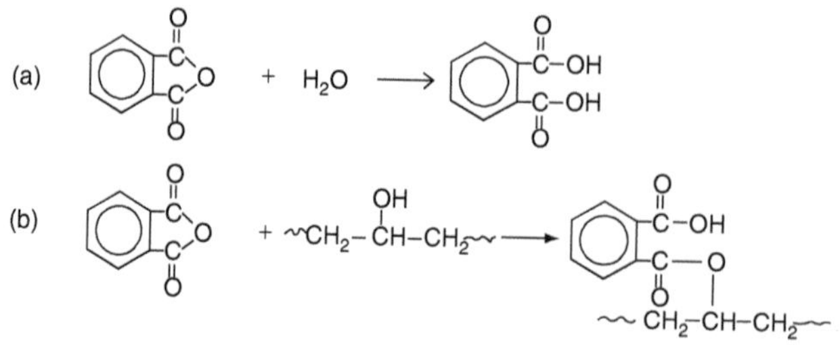

(1.1)

2. The epoxy ring may then be opened by reaction with (a) carboxylic groups formed by reactions (1a, b) or (b) hydroxyl groups [see (1.1)].

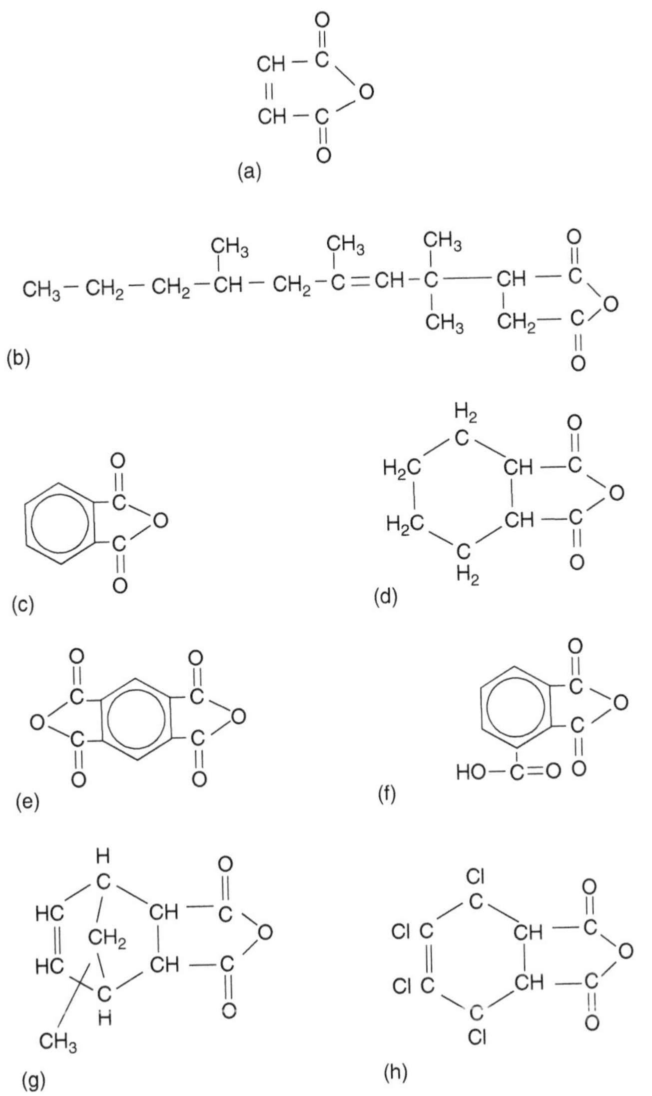

FIGURE 1.35 Anhydride curing agents. (a) Maleic anhydride. (b) Dodecenylsuccinic anhydride. (c) Phthalic anhydride. (d) Hexahydrophthalic anhydride. (e) Pyromellitic anhydride. (f) Trimellitic anhydride. (g) Nadic methyl anhydride. (h) Chlorendic anhydride.

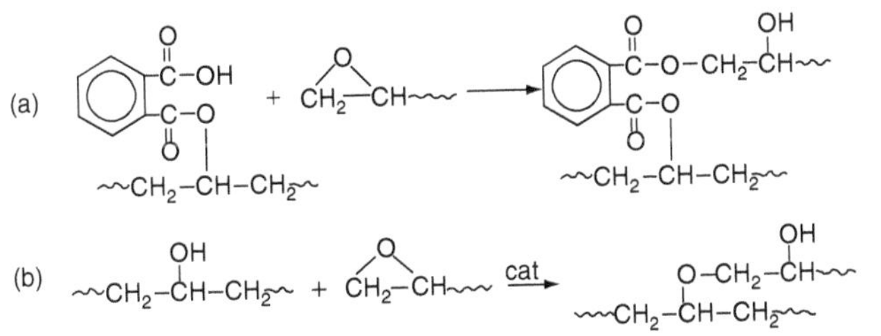

(1.2)

The reaction between an epoxy resin and an anhydride is rather sluggish. In commercial practice, the curing is accelerated by the use of organic bases to catalyze the reaction. These are usually tertiary amines such as α-methylbenzyl-dimethylamine and butylamine. The tertiary amine appears to react preferentially with the anhydride to generate a carboxy anion ($-COO^-$). This anion then opens an epoxy ring to give an alkoxide ion

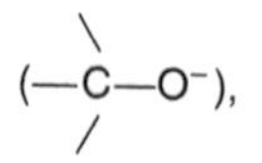

$$(-\overset{\displaystyle |}{\underset{\displaystyle |}{C}}-O^-),$$

which forms another carboxy anion from a second anhydride molecule, and so on.

Phthalic anhydride is the cheapest anhydride curing agent, but it has the disadvantage of being rather difficult to mix with the resin. Liquid anhydrides (e.g., dodecenylsuccinic anhydride and nadic methyl anhydride), low-melting anhydrides (e.g., hexahydrophthalic anhydride), and eutectic mixtures are more easily incorporated into the resin. Since maleic anhydride produces brittle products, it is seldom used by itself and is used as a secondary hardener in admixture with other anhydrides. Dodecenylsuccinic anhydride imparts flexibility into the casting, whereas chlorendic anhydride confers flame resistance.

Anhydride-cured resins generally have better thermal stability. Pyromellitic dianhydride with higher functionality produces tightly cross-linked products of high heat-distortion temperatures. Heat-distortion temperatures as high as 290°C have been quoted. Table 1.28 summarizes the characteristics of some of the anhydride hardeners.

In addition to the amine and anhydride hardeners, many other curing agents have been made available. Among them are the so-called fatty polyamides. These polymers are of low molecular weight (2000–5000) and are prepared by treating dimer acid (which is a complex mixture consisting of 60–75% dimerized fatty acids together with lesser amounts of trimerized acids and higher polymers) with stoichiometric excess of ethylenediamine or diethylenetriamine so that the resultant amides have free amine groups.

Fatty polyamides are used to cure epoxy resins where a more flexible product is required, particularly in adhesive and coating applications. An advantage of the system is that roughly similar quantities of hardener and resin are required and, because it is not critical, metering can be done visually without the need of measuring aids. They thus form the basis of some domestic adhesive systems.

Also used with epoxy resins for adhesives is dicyanodiamide, $H_2N-C(=NH)NH-CN$. It is insoluble in common resins at room temperature but is soluble at elevated temperatures and thus forms the basis of a one-pack system.

1.3.5.5.3 Other Epoxies

Resins containing the glycidyl ether group

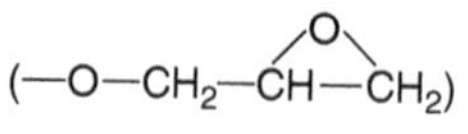

$$(-O-CH_2-\overset{\displaystyle O}{\overset{\displaystyle /\backslash}{CH-CH_2}})$$

TABLE 1.28 Properties of Some Anhydride Hardeners used in Low-Molecular-Weight Bisphenol A-Based Epoxy Resins

Hardener (Anhydride)	Parts used per 100 Parts Resin	Typical Cure Schedule	Max HDT of Cured Resin (°C)	Application
Phthalic	35–45	24 h at 120°C	110	Casting
Hexahydrophthalic (+accelerator)	80	24 h at 120°C	130	Casting
Pyromellitic	26	20 h at 220°C	290	High HDT
Nadic methyl	80	16 h at 120°C	202	High HDT
Dodecenylsuccinic (+accelerator)		2 h at 100°C+2 h at 150°C	38	Flexibilizing
Chlorendic	100	24 h at 180°C	180	Flame retarding

Source: Brydson, J. A. 1982. *Plastics Materials.* Butterworth Scientific, London, UK.

result from the reaction of epichlorohydrin and hydroxy compounds. Although bisphenol A is the most commonly used hydroxyl compound, a few glycidyl ether resins based on other hydroxy compound are also commercially available, including novolac epoxies, polyglycol epoxies, and halogenated epoxies.

A typical commercial novolac epoxy resin (VI) produced by epoxidation of phenolic hydroxyl groups of novolac (see the section on phenolformaldehyde resins) by treatment with epichlorohydrin has an

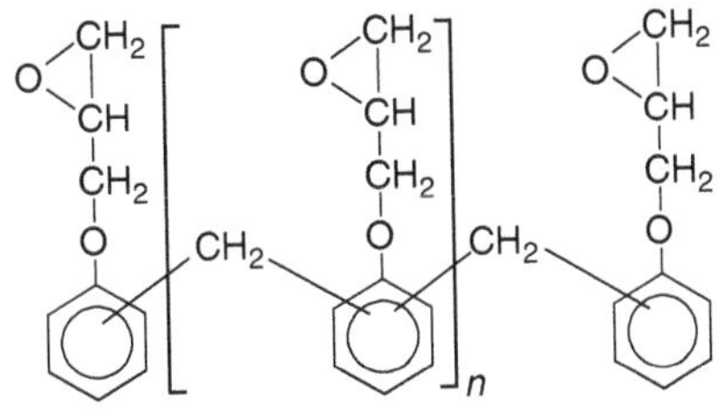

(VI)

average molecular weight of 650 and contains about 3.6 epoxy groups per molecule.

Because of higher functionality, the novolac epoxy resins give, on curing, more hightly cross-linked products than the bisphenol A-based resins. This results in greater thermal stability, higher heat-deflection temperatures, and improved chemical resistance. Their main applications have been in high-temperature adhesives, heat-resistant structural laminates, electrical laminates resistant to solder baths, and chemical-resistant filament-wound pipes. The use of novolac epoxies has been limited, however, by their high viscosity and consequent handling difficulties.

Polymeric glycols such as polypropylene glycol may be epoxidized through the terminal hydroxy groups to give diglycidyl ethers (VII).

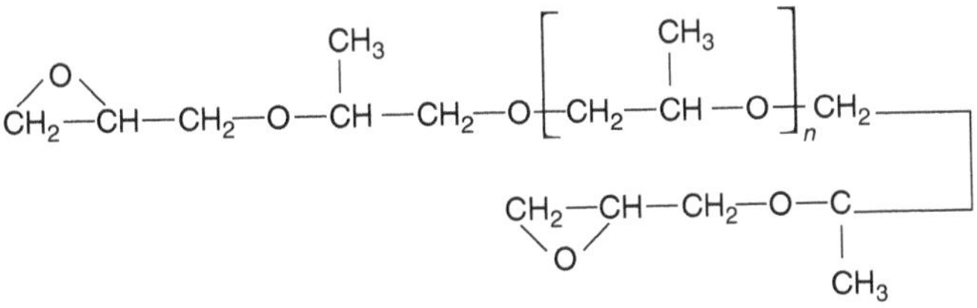

(VII)

In commercial products n usually varies from 1 to 6. Alone, these resins give soft products of low strength on curing. So they are normally used in blends with bisphenol A- or novolac-based resins. Added to the extent of 10–30%, they improve resilience without too large a loss in strength and are used in such applications as adhesives and encapsulations.

Epoxies containing halogens have flame-retardant properties and may be prepared from halogenated hydroxy compounds. Halogenated epoxies are available based on tetrabromobisphenol A and tetra-chlorobisphenol A (VIII).

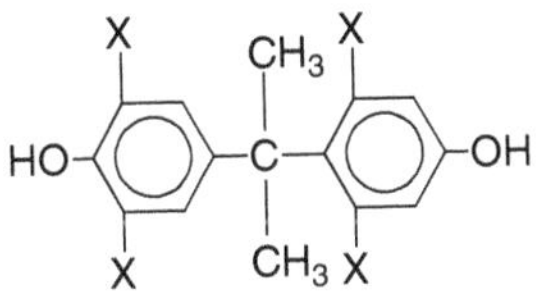

(VIII)

The brominated resins are more effective than the chlorinated resins and have become more predominant commercially. The ability of the resins to retard or extinguish burning is due to the evolution of hydrogen halide at elevated temperatures. Brominated epoxy resins are generally blended with other epoxy resins to impart flame retardance in such applications as laminates and adhesives.

Nonglycidyl ether epoxy resins are usually prepared by treating unsaturated compounds with peracetic acid.

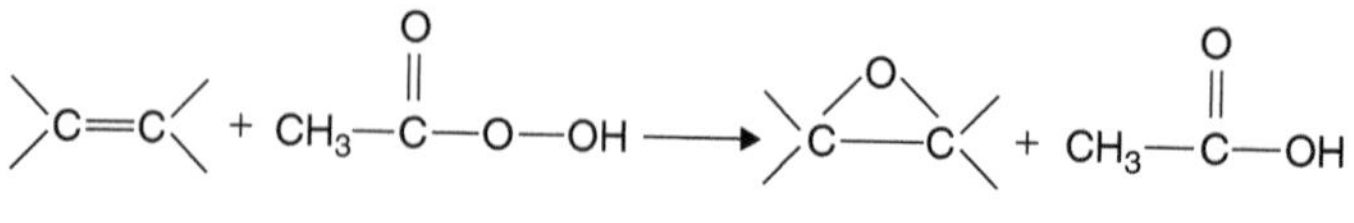

$$\text{C=C} + CH_3-\overset{\overset{\displaystyle O}{\|}}{C}-O-OH \longrightarrow \text{C}-\overset{O}{\text{C}} + CH_3-\overset{\overset{\displaystyle O}{\|}}{C}-OH$$

Two types of nonglycidyl ether epoxy resins are commercially available: cyclic aliphatic epoxies and acyclic aliphatic epoxies.

Cyclic aliphatic epoxy resins were first introduced in the United States. Some typical examples of commercial materials are 3,4-epoxy-6-methylcyclohexylmethyl-3,4-epoxy-6-methylcyclohexane carboxylate (Unox epoxide 201, liquid) (IX), vinylcyclohexene dioxide (Unox epoxide 206, liquid) (X), and dicyclopentadiene dioxide (Unox epoxide 207, solid) (XI).

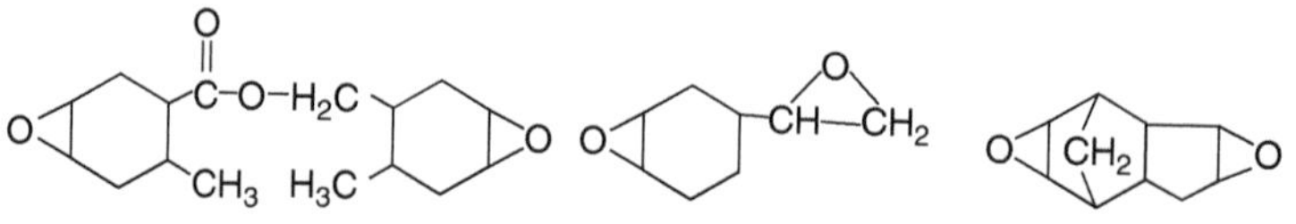

(IX)   (X)   (XI)

Generally, acid anhydrides are the preferred curing agents since amines are less effective. A hydroxy compound, such as ethylene glycol, is often added as initiator.

Because of their more compact structure, cycloaliphatic resins produce greater density of cross-links in the cured products than bisphenol A-based glycidyl resins. This generally leads to higher heat-distortion temperatures and to increased brittleness.

The products also are clearly superior in arc resistance and are track resistant. Thus although bisphenol A-based epoxies decompose in the presence of a high-temperature arc to produce carbon which leads to tracking and insulation failure, cycloaliphatic epoxies oxidize to volatile products which do not cause tracking. This has led to such applications as heavy-duty electrical castings and laminates, tension insulators, rocket motor cases, and transformer encapsulation.

Acyclic aliphatic resins differ from cyclic aliphatic resins in that the basic structure of the molecules in the former is a long chain, whereas the latter, as shown, contains ring structures. Two types of acyclic aliphatic epoxies are commercially available, namely, epoxidized diene polymers and epoxidized oils.

Typical of the epoxidized diene polymers are the products produced by treatment of polybutadiene with peracetic acid. Epoxidized diene polymers are not very reactive toward amines but may be cross-linked with acid hardeners. Cured resins have substantially higher heat-distortion temperatures (typically, 250°C) than do the conventional amine-cured diglycidylether resins.

Epoxidized oils are obtained by treatment of drying and semidrying oils (unsaturated), such as linseed and soybean oils, with peracetic acid. Epoxidized oils find use primarily as plasticizers and stabilizers for PVC.

1.3.5.5.4 Applications

Epoxy resins have found a wide range of applications and a steady rate of growth over the years mainly because of their versatility. Properties of the cured products can be tailored by proper selection of resin, modifier, cross-linking agent, and the curing schedule.

The main attributes of properly cured epoxy systems are outstanding adhesion to a wide variety of substrates, including metals and concrete; ability to cure over a wide temperature range; very low shrinkage on cure; excellent resistance to chemicals and corrosion; excellent electrical insulation properties; and high tensile, compressive, and flexural strengths.

In general, the toughness, adhesion, chemical resistance, and corrosion resistance of epoxies suit them for protective coating applications. It is not surprising that about 50% of epoxy resins are used in protective coating applications.

Two types of epoxy coatings are formulated: those cured at ambient temperature and those that are heat cured. The first type uses amine hardening systems, fatty acid polyamides, and polymercaptans as curing agents. Very high cure rates may be achieved by using mercaptans. Heat-cured types use acid anhydrides and polycarboxylic acids as well as formaldehyde resins as curing agents.

Typical coating applications for phenol-formaldehyde resin-modified epoxies include food and beverage can coatings, drum and tank liners, internal coatings for pipes, wire coating, and impregnation varnishes. Ureaformaldehyde resin-modified epoxies offer better color range and are used as appliance primers, can linings, and coatings for hospital and laboratory furniture.

Environmental concerns have prompted major developmental trends in epoxy coating systems. Thus epoxy coating systems have been developed to meet the high-temperature and hot-acid environments to which SO_2 scrubbers and related equipment are subjected. Ambient curable epoxy systems have been developed to provide resistance to concentrated inorganic acids.

In the development of maintenance and marine coatings, the emphasis has been on the development of low-solvent, or solvent-free, coatings to satisfy EPA volatile organic content standards. Thus liquid epoxy systems based on polyamidamines have been developed. Epoxies have also been formulated as powdered coatings, thus completely eliminating solvents.

Pipe coatings still represent a major market for epoxies. High-molecular-weight powdered formulations are used in this application.

The important maintenance coating area, particularly for pipe and tank coatings, is served by epoxy systems cured with polyamine or polyamidamines. Two-component, air-dried, solventless systems used in maintenance coatings provide tough, durable, nonporous surfaces with good resistance to water, acids, alkalies, organic solvents, and corrosion.

Emulsifiable epoxies of varying molecular weights, water-dilutable modified epoxies, and dispersions of standard resins represent promising developments for coating applications. Can coatings are an important waterborne resin application. Two-component waterborne systems are also finding use as architectural coatings.

For electrical and electronic applications epoxy formulations are available with low or high viscosity, unfilled or filled, slow or fast curing at low or high temperatures. Potting, encapsulation, and casting of transistors, integrated circuits, switches, coils, and insulators are a few electrical applications of epoxies. With their adhesion to glass, electrical properties, and flexural strength, epoxies provide high-quality printed circuit boards. Epoxies have been successfully used in Europe for outdoor insulators, switchgear, and transformers for many years. In these heavy electrical applications, the advantage of cycloaliphatic epoxies over porcelain has been demonstrated.

In a relatively new development, epoxy photopolymers have been used as solder masks and photoresists in printed circuit board fabrications.

Adhesion properties of epoxies, complete reactivity with no volatiles during cure, and minimal shrinkage make the materials outstanding for adhesives, particularly in structural applications. The most commonly known adhesive applications involve the two-component liquids or pastes, which cure at room or elevated temperatures.

A novel, latent curing system, which gives more than one year pot life at room temperature, has increased the use of epoxies for specialty adhesives and sealants, and for vinyl plastisols. The one-pack system provides fast cure when heated—for example 5 min at 100°C.

Epoxies are used in fiber-reinforced composites, providing high strength-to-weight ratios and good thermal and electrical properties. Filament-would epoxy composites are used for rocket motor casings, pressure vessels, and tanks. Glass-fiber reinforced epoxy pipes are used in the oil, gas, mining, and chemical industries.

Sand-filled epoxies are used in industrial flooring. Resistance to a wide variety of chemicals and solvents and adhesion to concrete are key properties responsible for this use. Epoxies are also used in patching concrete highways.

Decorative flooring and exposed aggregate systems make use of epoxies because of their low curing shrinkage, and the good bonding of glass, marble, and quartz chip by the epoxy matrix.

1.3.5.6   Poly(Phenylene Oxide)

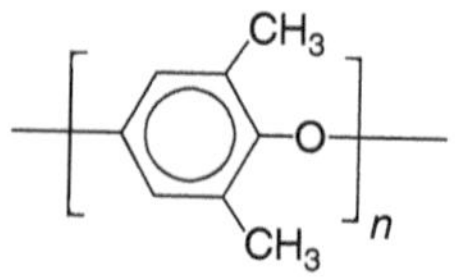

Monomer	Polymerization	Major Uses
2,6-Dimethylphenol	Condensation polymerization by oxidative coupling	Automotive, appliances, business machine cases, electrical components

Poly(2,6-dimethyl-1,4-phenylene oxide), commonly called poly(phenylene oxide) or PPO, was introduced commercially in 1964. PPO is manufactured by oxidation of 2,6-dimethyl phenol in solution using cuprous chloride and pyridine as catalyst. The monomer is obtained by the alkylation of phenol with methanol. End-group stabilization with acetic anhydride improves the oxidation resistance of PPO.

PPO is counted as one of the engineering plastics. The rigid structure of the polymer molecules leads to a material with a high T_g of 208°C. It is characterized by high tensile strength, stiffness, impact strength, and creep resistance, and low coefficient of thermal expansion. These properties are maintained over a broad temperature range (-45°C to 120°C). One particular feature of PPO is its exceptional dimensional stability among the so-called engineering plastics. The polymer is self-extinguishing.

PPO has excellent resistance to most aqueous reagents and is unaffected by acids, alkalis, and detergents. The polymer has outstanding hydrolytic stability and has one of the lowest water absorption rates among the engineering thermoplastics. PPO is soluble in aromatic hydrocarbons and chlorinate solvents. Several aliphatic hydrocarbons cause environmental stress cracking.

PPO has low molding shrinkage. The polymer is used for the injection molding of such items as pump components, domestic appliance and business machines, and electrical parts such as connectors and terminal blocks.

The high price of PPO has greatly restricted its application and has led to the introduction of the related and cheaper thermoplastic materials in 1966 under the trade name Noryl by General Electric. If PPO ($T_g=208$°C) is blended with polystyrene ($T_g\sim90$°C) in equal quantities, a transparent polymer is obtained with a single T_g of about 150°C, which apparently indicates a molecular level of mixing. Noryl thermoplastics may be considered as being derived from such polystyrene–PPO blends. Since the electrical properties of the two polymers are very similar, the blends also have similar electrical characteristics. In addition to Noryl blends produced by General Electric, grafts of styrene onto PPO are also available (Xyron by Asahi-Dow).

The styrenic component in polystyrene–PPO blends may not necessarily be straight polystyrene (PS) but instead high-impact polystyrene (HIPS) or some other related material. The most widely used blend is the blend of PPO and HIPS.

Like polystyrene, these blends have the following useful characteristics: (1) good dimensional stability and low molding shrinkage, thus allowing close dimensional tolerance in the production of moldings; (2) low water absorption; (3) excellent resistance to hydrolysis; and (4) very good dielectric properties over a wide range of temperature. In addition, unlike polystyrene, the blends have heat-distortion temperatures above the boiling point of water, and in some grades this is as high as 160°C.

The range of Noryl blends available comprises a broad spectrum of materials superior in many respects, particularly heat deformation resistance, to the general purpose thermoplastics but at a lower price than the more heat resistant materials such as polycarbonates, polyphenylene sulfides, and polysulfones (discussed later). The materials that come close to them in properties are the ABS/

polycarbonate blends. Noryl is also characterized by high dielectric strength (192 V/mil) and low dissipation factors (4.7×10^3 at 100 Hz and 3.9×10^3 at 10^6 Hz).

In common with other engineering thermoplastics, there are four main groups of modified PPOs available. They are: (1) non-self-extinguishing grades with a heat-distortion temperature in the range 110–106°C and with a notched Izod impact strength at 200–500 J/m; (2) self-extinguishing grades with slightly lower heat-distortion temperatures and impact strengths; (3) non-self-extinguishing glass-reinforced grades (10%, 20%, 30% glass fiber) with heat-distortion temperatures in the range of 120–140°C; and (4) self-extinguishing glass-reinforced grades. Among the special grades that should be mentioned are those containing blowing agents for use in the manufacture of structural foams (see Chapter 1 of *Plastics Fabrication and Recycling*).

Noryls maybe extruded, injection molded, and blow-molded without undue difficulty. Processing conditions depend on the grade used but in injection molding a typical melt temperature would be in the range 250–300°C.

The introduction of self-extinguishing, glass-reinforced, and structural foam grades has led to steady increase in the use of these materials in five main application areas. These are (1) the automotive industry; (2) the electrical industry; (3) radio and television; (4) business machines and computer housings; and (5) pumps and other plumbing applications.

Use in the automotive industries largely arises from the availability of high impact grades with heat-distortion temperatures above those of the general purpose thermoplastics. Specific uses include instrument panels, steering column cladding, central consoles, loudspeaker housings, ventilator grilles and nozzles, and parcel shelves. In cooling systems, glass-reinforced grades have been used for radiator and expansion tanks. Several components of car heating systems are also produced from modified PPOs. The materials have been increasingly used for car exterior trim such as air inlet and outlet grills and outer mirror housings.

In the electrical industry, well-known applications include fuse boxes, switch cabinets, housing for small motors, transformers, and protective circuits.

Uses in radio and television arise largely from the ability to produce components with a high level of dimensional accuracy coupled with good dielectric properties, high heat-distortion temperatures, and the availability of self-extinguishing grades. Specific uses include coil formers, picture tube deflection yokes, and insert card mountings.

In the manufacture of business machine and computer housings, structural foam grades have found use and moldings weighing as much as 50 kg have been reported. Glass-reinforced grades have widely replaced metals in pumps and other functional parts in washing equipment and central heating systems.

Another PPO, called PPE, is produced by the oxidative coupling of a mixture of 2,6-dimethylphenol and 2,3,6-trimethylphenol. This stiff polymer, like Noryl, is available from General Electric. It is usually modified by blending with PS or HIPS.

Blends of PPO and polyamide (PA, nylon) are incompatible but good properties can be obtained through the use of compatibilizing agents. The PPO is dispersed in a continuous nylon matrix in these blends. Because of the incompatibility of the two phases, the modulus decreases very little at the T_g of PA (71°C) and is maintained up to the T_g of the PPO phase (208°C).

The PPO–PA blends, which are sold by General Electric under the trade name Noryl GTX, can be baked and painted at 190°C without noticeable warpage or distortion, and have been used for producing automobile fenders.

The PPO–PA blends shown a mold shrinkage of 0.001 in./in. The molding pressure in 15×10^3 lbf/in.^{2} (103 MPa) and the processing temperature 260°C. The heat-deflection temperature of the molded specimen under flexural load of 264 lbf/in.^{2} is typically 190°C and the maximum resistance to continuous heat is 175°C. The coefficient of linear expansion is 10^{-5} cm/cm °C. The mechanical properties of the PPO–PA blends are shown in Table 1.29.

1.3.6 Cellulosic Polymers

Cellulose is a carbohydrate with molecular formula $(C_6H_{10}O_5)_n$, where n is a few thousand. Complete hydrolysis of cellulose by boiling with concentrated hydrochloric acid yields D-glucose, $C_6H_{12}C_6$, in 95–96% yield [56]. Cellulose can thus be considered chemically as a polyanhydroglucose. The structure of cellulose is

The regularity of the cellulose chain and extensive hydrogen bonding between hydroxyl groups of adjacent chains cause cellulose to be a tightly packed crystalline material which is insoluble and infusible. As a result, cellulose cannot be processed in the melt or in solution. However, cellulose derivatives in which there is less hydrogen bonding are processable.

The most common means of preparing processable cellulose derivatives are esterification and etherification of the hydroxyl groups. In the following, the more important commercial cellulosic polymers are described.

1.3.6.1 Regenerated Cellulose

As mentioned, many derivatives of cellulose are soluble, though cellulose itself is insoluble. A solution of a cellulose derivative can be processed (usually by extrusion) to produce the desired shape (commonly) fiber or film) and then treated to remove the modifying groups to reform or regenerated unmodified cellulose. Such material is known as *regenerated cellulose.*

Modern methods of producing regenerated cellulose can be traced to the discovery in 1892 by Cross, Bevan, and Beadle that cellulose can be rendered soluble by xanthate formation by treatment with sodium hydroxide and carbon disulfide and regenerated by acidification of the xanthate solution. This process is known as the *viscose process.* The reactions can be indicated schematically as

TABLE 1.29   Properties of Typical PPO–PA Blends (Noryl GTX 810)

Property	Value
Tensile strength	
10^3 lbf/in.^{2}	13
MPa	90
Tensile modulus	
10^5 lbf/in.^{2}	2.0
GPa	1.4
Elongation at break (%)	10
Flexural strength	
10^3 lbf/in.^{2}	19
MPa	131
Flexural modulus	
10^5 lbf/in.^{2}	2.25
GPa	1.6
Impact strength, notched Izod	
ft.-lb/in.	1.5
J/m	80

$$R-OH \xrightarrow{\text{NaOH}} R-ONa \xrightarrow{\text{CS}_2} R-O-\overset{\overset{\displaystyle S}{\|}}{C}-S-Na \xrightarrow{\text{H}^+} R-OH$$

| Cellulose | Alkali cellulose | Cellulose xanthate | Regenerated cellulose |

The viscose process is used for the production of textile fibers, known as *viscose rayon*, and transparent packaging film, known as *cellophane* (the name is coined from *cell*ulose and dia*phane*, which is French for transparent).

A suitably aged solution of cellulose xanthate, known as *viscose*, is fed through spinnerets with many small holes (in the production of fiber), through a slot die (in the production of film), or through a ring die (in the production of continuous tube used as sausage casing) into a bath containing 10–15% sulfuric acid and 10–20% sodium sulfate at 35–40°C, which coagulates and completely hydrolyzes the viscose. Cellulose is thus regenerated in the desired shape and form.

It is possible to carry out a drawing operation on the fiber as it passes through the coagulating bath. The stretching (50–150%) produces crystalline orientation in the fiber. The product, known as *high-tenacity rayon*, has high strength and low elongation and is used in such application as tire cord and conveyor belting.

For the production of cellophane, the regenerated cellulose film is washed, bleached, plasticized with ethylene glycol or glycerol, and then dried; sometimes a coating of pyroxylin (cellulose nitrate solution) containing dibutyl phthalate as plasticizer is applied to give heat sealability and lower moisture permeability.

Cellophane has been extensively and successfully used as a wrapping material, particularly in the food and tobacco industries. However, the advent of polypropylene in the early 1960s has produced a serious competitor to this material.

1.3.6.2 Cellulose Nitrate

Cellulose nitrate or nitrocellulose (as it is often erroneously called) is the doyen of cellulose ester polymers. It is prepared by direct nitration with nitric and sulfuric acid mixtures at about 30–40°C for 20–60 min. Complete substitution at all three hydroxyl groups on the repeating anhydroglucose unit will give cellulose trinitrate containing 14.14% nitrogen:

$$[C_6H_7O_2(OH)_3]_n \xrightarrow{\text{HNO}_3/\text{H}_2\text{SO}_4} [C_6H_7O_2(ONO_2)_3]_n + 3nH_2O$$

This material is explosive and is not made commercially, but products with lower degrees of nitration are of importance. The degree of nitration may be regulated by the choice of reaction conditions.

Industrial nitrocelluloses have a degree of substitution somewhere between 1.9 and 2.7 and are generally characterized for their various uses by their nitrogen content, usually about 11% for plastics, 12% for lacquer and cement base, and 13% for explosives.

The largest use of cellulose nitrate is as a base for lacquers and cements. Butyl acetate is used as a solvent. Plasticizers such as dibutyl phthalate and tritolyl phosphate are necessary to give films of acceptable flexibility and adhesion.

For use as plastic in bulk form, cellulose nitrate is plasticized with camphor. The product is known as *celluloid*. In a typical process alcohol-wet cellulose nitrate is kneaded at about 40°C with camphor (about 30%) to form a viscous plastic mass. Pigments or dyes may be added at this stage. The dough is then heated at about 80°C on milling rolls until the alcohol content is reduced to about 15%.

The milled product is calendered into sheets about 1/2-in. (1.25-cm) thick. A number of sheets are laid up in a press and consolidated into a block. The block is sliced into sheets of thickness 0.005–1 in. (0.012–2.5 cm), which are then allowed to season for several days at about 50°C so that the volatile content is reduced to about 2%. Celluloid sheet and block may be machined with little difficulty if care is taken to avoid overheating.

The high inflammability and relatively poor chemical resistance of celluloid severely restrict its use in industrial applications. Consequently the material is used because of its desirable characteristics, which include rigidity, dimensional stability, low water absorption, reasonable toughness, after-shrinkage around inserts, and ability of forming highly attractive colored sheeting. Today the principal outlets of celluloid are knife handles, table-tennis balls, and spectacle frames. Celluloid is marketed as Xylonite (BX Plastics Ltd.) in UK.

1.3.6.3   Cellulose Acetate

The acetylation of cellulose is usually carried out with acetic anhydride in the presence of sulfuric acid as catalyst. It is not practicable to stop acetylation short of the essentially completely esterified triacetate. Products of lower acetyl content are thus produced by partial hydrolysis of the triacetate to remove some of the acetyl groups:

$$[C_6H_7O_2(OH)_3]_n \xrightarrow{HOAc,\ Ac_2O} [C_6H_7O_2(OAc)_3]_n \xrightarrow[Heat]{H_2O}$$

Cellulose   Triacetate (44.8%acetyl)

$$[C_6H_7O_2(OAc)_2OH]_n$$

Diacetate (34.9%acetyl)

Cellulose triacetate is often known as *primary cellulose acetate*, and partially hydrolyzed material is called *secondary cellulose acetate.* Many physical and chemical properties of cellulose acetylation products are strongly dependent on the degree of esterification, which is measured by the acetyl content (i.e., the weight of acetyl radical (CH_3CO-) in the material) or acetic acid yield (i.e., the weight of acetic acid produced by complete hydrolysis of the ester).

The commercial products can be broadly distinguished as cellulose acetate (37–40% acetyl), high-acetyl cellulose acetate (40–42% acetyl), and cellulose triacetate (43.7–44.8% acetyl).

Cellulose acetate containing 37–40% acetyl is usually preferred for use in general-purpose injection-molding compounds. Cellulose acetate, however, decomposes below its softening point, and it is necessary to add a plasticizer (e.g., dimethyl phthalate or triphenyl phosphate), usually 25–35%, to obtain a moldable material. The use of cellulose acetate for molding and extrusion is now small owing largely to the competition of polystyrene and other polyolefins. At the present time the major outlets of cellulose acetate are in the fancy goods trade as toothbrushes, combs, hair slides, etc.

Cellulose acetate with a slightly higher degree of esterification (38.7–40.1% acetyl) is usually preferred for the preparation of fibers, films, and lacquers because of the greater water resistance. A significant application of cellulose acetate film has been found in sea-water desalination by reverse osmosis.

High-acetyl cellulose acetate (40–42% acetyl) has found occasional use in injection-molding compounds where greater dimensional stability is required. However, processing is more difficult.

Cellulose triacetate (43.7–44.8% acetyl) finds little use in molding compositions because its very high softening temperature is not greatly reduced by plasticizers. It is therefore processed is solution. A mixture of methylene chloride and methanol is the commonly used solvent.

The sheeting and fibers are made from cellulose triacetate by casting or by extruding a viscous solution and evaporating the solvent. The sheeting and film are grainless, have good gauge uniformity, and good optical clarity. The products have good dimensional stability and are highly resistant to water, grease, oils, and most common solvents such as alcohol and acetone. They also have good heat resistance and high dielectric constant.

Sheeting and films of cellulose triacetate are used in the production of visual aids, graphic arts, greeting cards, photographic albums, and protective folders. Cellulose triacetate is extensively used for photographic, x-ray, and cinematographic films. In these applications cellulose triacetate has displaced celluloid mainly because the triacetate does not have the great inflammability of celluloid.

1.3.6.4 Other Cellulose Esters

Homologues of acetic acid have been employed to make other cellulose esters. Of these, cellulose propionate, cellulose acetate-propionate, and cellulose acetate-butyrate are produced on a commercial scale. The are produced in a manner similar to that described previously for cellulose acetate. The propionate and butyrate esters are made by substituting propionic acid and propionic anhydride or butyric acid and butyric anhydride for some of the acetic acid and acetic anhydride.

Cellulose acetate-butyrate (CAB) has several advantages in properties over cellulose acetate: lower moisture absorption, greater solubility and compatibility with plasticizer, higher impact strength, and excellent dimensional stability. CAB used in plastics has about 13% acetyl and 37% butyryl content. It is an excellent injection-molding material (Tenite Butyrate by Kodak, Cellidor B by Bayer).

Principal end products of CAB have been for tabulator keys, automobile parts toys, pen and pencil barrels, steering wheels, and tool handles. In the United States CAB has been sued for telephone housings, and extruded CAB piping has been used for conveying water, oil, and natural gas. CAB sheet is readily vacuum formed and is especially useful for laminating with thin-gauge aluminum foil. It also serves particularly well for vacuum metallizing.

Cellulose propionate (Forticel by Celanese) is very similar in both coast and properties to CAB. It has been used for similar purposes as CAB. Cellulose acetate propionate (Tenite Propionate by Kodak) is similar to cellulose propionate. It find wide use in blister packages and formed containers, safety goggles, motor covers, metallized flash cubes, brush handles, steering wheels, face shields, displays, and lighting fixtures.

1.3.6.5 Cellulose Ethers

Of cellulose ethers only ethyl cellulose has found application as a molding material. Methyl cellulose, hydroxyethyl cellulose, and sodium carboxymethyl cellulose are useful water-soluble polymers. The first step in the manufacture of each of these materials is the preparation of alkali cellulose (soda cellulose) by treating cellulose with concentrated sodium hydroxide. Ethyl cellulose is made by reacting alkali cellulose with ethyl chloride.

$$[[R(OH)_3]_n + NaOH, H_2O] + Cl-CH_2CH_3 \xrightarrow[6-12\,h]{90-150°C}$$
$$\text{where } R = C_6H_7O_2$$

$$[R(OCH_2CH_3)_m(OH)_{3-m}]_n$$
$$m = 2.4 - 2.5$$

Ethyl cellulose is produced in pellet form for molding and extrusion and in sheet form for fabrication. It has good processability, is tough, and is moderately flexible; its outstanding feature is its toughness at low temperatures. The principal uses of ethyl cellulose moldings are thus in those applications where good impact strength at low temperatures is required, such as refrigerator bases and flip lids and ice-crusher parts. Ethyl cellulose is often employed in the form of hot melt for strippable coatings used for protection of metal parts against corrosion and marring during shipment and storage. A recent development is the use of ethyl cellulose gel lacquers for permanent coatings.

Methyl cellulose is prepared by reacting alkali cellulose with methyl chloride at 50–100°C (cf. Ethyl cellulose). With a degree of substitution of 1.6–1.8, the resultant either is soluble in cold water but not in hot water. It is used as a thickening agent and emulsifier in cosmetics, pharmaceuticals, ceramics, as a paper size, and in leather-tanning operations. Hydroxyethyl cellulose, produced by reacting alkali cellulose with ethylene oxide, is employed for similar purposes.

Reaction of alkali cellulose with sodium salt of chloroacetic acid yields sodium carboxymethyl cellulose (SCMC),

$$[[R(OH)_3]_n + NaOH, H_2O] + Cl—CH_2—CO—ONa \xrightarrow[\text{2–4 hr}]{\text{40–50°C}}$$

$$[R(OCH_2CO—ONa)_m(OH)_{3-m}]_n$$

where $R = C_6H_7O_2$ and $m = 0.65–1.4$. SCMC appears to be physiologically inert and is very widely used. Its principal application is as a soil-suspending agent in synthetic detergents, but it is also used as a sizing and finishing agent for textile, as a surface active agent, and as a viscosity modifier in emulsion and suspension. Purified grades of SCMC are used in ice cream to provide a smooth texture and in a number of cosmetic and pharmaceutical products. SCMC is also the basis of a well-known proprietary wallpaper adhesive.

1.3.7 Sulfide Polymers

1.3.7.1 Polysulfides

Monomers	Polymerization	Major Uses
Aliphatic dihalide, sodium sulfide	Polycondensation	Sealing, caulks, gaskets

Polysulfide elastomers are produced by the reaction of an aliphatic dihalide, usually bis(2-chloroethyl)formal, with sodium polysulfide under alkaline conditions:

$$ClCH_2CH_2OCH_2OCH_2CH_2Cl + Na_2S_x \longrightarrow$$
$$-\!\!\!\left[-CH_2CH_2OCH_2OCH_2CH_2S_x-\right]\!\!\!- + NaCl$$

The reaction is carried out with the dihalide suspended in an aqueous magnesium hydroxide phase. The value of x is slightly above 2. A typical polymerization system also includes up to 2% 1,2,3-trichloropropane. The polymerization occurs readily yielding a polymer with a very high molecular weight.

High molecular weight, however, is not desirable until its end-use application. The molecular weight is therefore lowered, and the *polysulfide rank* (value of x) is simultaneously brought close to 2, by reductive treatment with NaSH and Na_2SO_3 followed by acidification. The result is a liquid, branched polysulfide with terminal thiol end group and a molecular weight in the range of 1000–8000. Curing to a solid elastomer is accomplished by oxidation of thiol to disulfide links by oxidants such as lead dioxide and *p*-quinone dioxime:

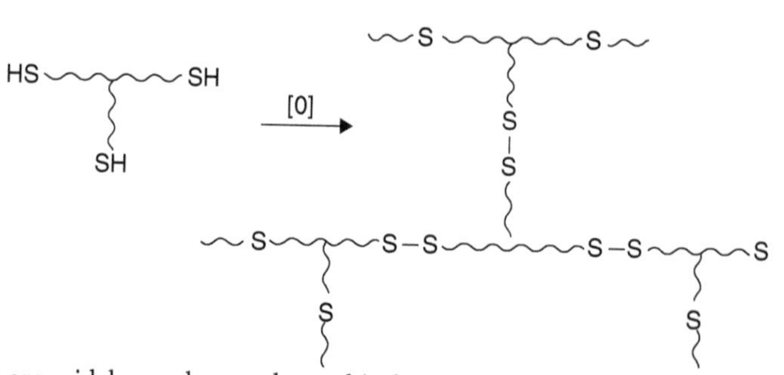

These materials are widely used as sealants, binders for solid propellants, caulking materials, and cements for insulating glass and fuel tanks.

Polysulfides, often referred to as *thiokols*, are produced at low volumes as specialty materials geared toward a narrow market. The advantages and disadvantages of polysulfides both reside in the disulfide linkage. Thus they possess low-temperature flexibility and very good resistance to ozone, oil, solvent (hydrocarbons as well as polar solvents such as alcohols, ketones, esters, etc.), and weathering. However, polysulfides have poor thermal stability and creep resistance, have low resilence, and are malodorous.

Thiokols are amorphous polymers which do not crystallize when stretched and hence reinforcing fillers, such as carbon black, must be added to obtain relatively high tensile strengths. Thiokol may be vulcanized in the presence of zinc oxide and thiuram accelerators, such as tetramethyl-thiuram disulfide (*Tuads*). The accelerators modify the sulfur links and serve as chemical plasticizers.

A typical thiokol with 60 parts carbon black per 100 of polymer has a tensile strength of 1200 lbf/in.^{2}, an elongation of 300%, a specific gravity of 1.25, and a Shore A hardness of 68. Thiokol has excellent resistance to ozone (O_3) and ultraviolet radiation. It has a low permeability to solvents, such as gasoline; esters, such as ethyl acetate; and ketones, such as acetone.

The principal application of solid Thiokol elastomers is as gaskets, O-rings, gasoline, and fuel hose lines, gas meter diaphragms, and as rollers, which are used for lacquering cans.

1.3.7.2 Poly(Phenylene Sulfide)

Monomers	Polymerization	Major Uses
p-Dichlorobenzene, sodium sulfide	Polycondensation	Electrical components, mechanical parts

Poly(phenylene sulfide) (PPS) is the thio analogue of poly(phenylene oxide) (PPO) [57]. The first commercial grades were introduced by Phillips Petroleum in 1968 under the trade name Ryton. Other manufacturers also have introduced PPS (e.g., Tedur by Bayer). The commercial process involves the reaction of *p*-dichlorobenzene with sodium sulfide in a polar solvent.

PPS is an engineering plastic. The thermoplastic grades of PPS are outstanding in heat resistance, flame resistance, chemical resistance, and electrical insulation characteristics. The linear polymers are highly crystalline with melting point in the range of 285–295°C and T_g of 193–204°C.

The material is soluble only above 200°C in aromatic and chlorinated aromatic solvents. It has the ability to cross-link by air-oxidation at elevated temperatures, thereby providing an irreversible cure. Thermogravimetric analysis shows no weight loss below 500°C in air but demonstrates complete decomposition by 700°C. It is found to retain its properties after four months at 233°C (450°F) in air.

Significant increases in mechanical properties can be achieved with glass-fiber reinforcement. In the unfilled form the tensile strength of the material is 64–77 MPa at 21°C, 33 MPa at 204°C, and the flexural modulus is 4200 MPa at 21°C. The corresponding values for PPS–glass fiber (60:40) composites are 150, 33, and 15,500 MPa.

Although rigidity and tensile strength are similar to those of other engineering plastics, PPS does not possess the toughness of amorphous materials, such as the polycarbonates and the polysulfones (described later), and are somewhat brittle. On the other hand, PPS does show a high level of resistance to environmental stress cracking.

Being one of the most expensive commercial moldable thermoplastics, the use of PPS is heavily dependent on its particular combination of properties. Good electrical insulation characteristics, including good arcing and arc-tracking resistance has led to PPS replacing some of the older thermosets in electrical parts. These include connectors, terminal blocks, relay components, brush holders, and switch components.

PPS is used in chemical process plants for gear pumps. It has found application in the automotive sector, in such specific uses as carburetor parts, ignition plates, flow control values for heating systems, and exhaust-gas return valves to control pollution. The material also finds uses in sterilizable medical, dental, and general laboratory equipment, cooking appliance, and hair dryer components.

Injection-molded products of PPS include high-temperature lamp holders and reflectors, pump parts, valve, and, especially when filled for example with PTFE or graphite, bearings. Processing temperatures are 300–350°C with mold temperatures of up to 200°C. PPS is also used for encapulation of electronic components and as a high temperature surface coating material.

PPS is resistant to neutron and gamma radiation. In nuclear installations, its flexural strength and modulus are essentially unchanged when it is exposed to gamma radiation of 5×10^9 rad and neutron radiation of 1×10^9 rad.

1.3.8 Polysulfones

Polysulfones are a family of engineering thermoplastics with excellent high-temperature properties. The simplest aromatic polysulfone, poly(*p*-phenylene sulfone)

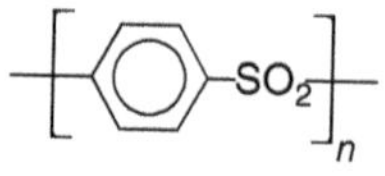

does not show thermoplastic behavior, melting with decomposition above 500°C. Hence, to obtain a material capable of injection molding in conventional machines, the polymer chain is made more flexible by incorporating ether links into the backbone.

The structures and glass transition temperatures of several commercial polysulfones are listed in Table 1.30. The polymers have different degrees of spacing between the *p*-phenylene groups and thus have a spectrum of glass transition temperatures which determine the heat-distortion temperature (or deflection temperature under load), since the materials are all amorphous.

The first commercial polysulfone (Table 1.30I) was introduced in 1965 by Union Carbide. This material, now known as Udel, has a continuous-use temperature of 150°C and a maximum-use temperature of 170°C, and it can be fabricated easily by injection molding in conventional machines.

In 1967, Minnesota Mining and Manufacturing (3M) introduced Astrel 360 (Table 1.30II), an especially high-performance thermoplastic, which requires specialized equipment with extra heating and pressure capabilities for processing. ICI's polyether sulfones, introduced in 1972—Victrex (Table 1.30III) and polyethersulfone 720P (Table 1.30IV)—are intermediate in performance and processing. In the late 1970s, Union Carbide introduced Radel (Table 1.30V), which has a higher level of toughness. Note that all of the commercial materials mentioned in Table 1.30 may be described as polysulfones, polyarylsulfones, polyether sulfones, or polyaryl ether sulfones.

In principle, there are two main routes to the preparation of polysulfones: (1) polysulfonylation and (2) polyetherification.

Polysulfonylation reactions are of the following general types:

$$\text{H—Ar—H} + \text{ClSO}_2\text{—Ar}'\text{—SO}_2\text{Cl} \xrightarrow[\text{Catalyst}]{\text{Friedel-Crafts}} \text{+Ar—SO}_2\text{—Ar}'\text{—SO}_2\text{+}_n + \text{HCl}$$

$$\text{H—Ar—SO}_2\text{Cl} \longrightarrow \text{+Ar—SO}_2\text{+}_n + \text{HCl}$$

The Ar and/or Ar′ group(s) contain an ether oxygen, and if Ar=Ar′, then basically identical products may be obtained by the two routes.

In the *polyetherification* route the condensation reaction proceeds by reactions of types

$$\text{HO—Ar—OH} + \text{Cl—Ar}'\text{—Cl} \xrightarrow{\text{NaOH}} \text{+O—Ar—O—Ar}'\text{+}_n + \text{NaCl}$$

$$\text{HO—Ar—Cl} \xrightarrow{\text{NaOH}} \text{+O—Ar+}_n + \text{NaCl}$$

The Ar and/or Ar′ group(s) contain sulfone groups, and if Ar=Ar′, then identical products may be obtained by the two routes.

Polyetherification processes form the basis of commercial polysulfone production methods. For example, the Udel-type polymer (Union Carbide) is prepared by reacting, 4,4′-dichlorodiphenylsulfone with an alkali salt of bisphenol A. The polycondensation is conducted in highly polar solvents, such as dimethylsulfoxide or sulfolane.

TABLE 1.30 Commercial Polysulfones

Type of Structure	T_g (°C)	Trade Name
I	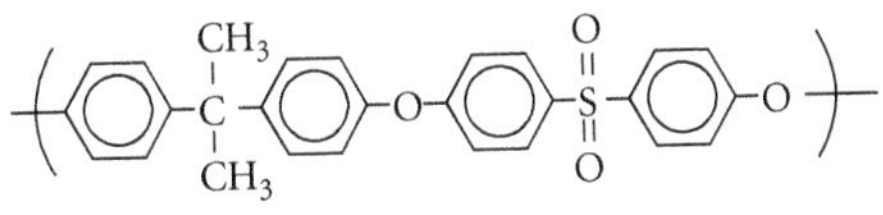 190	Udel (Union Carbide)
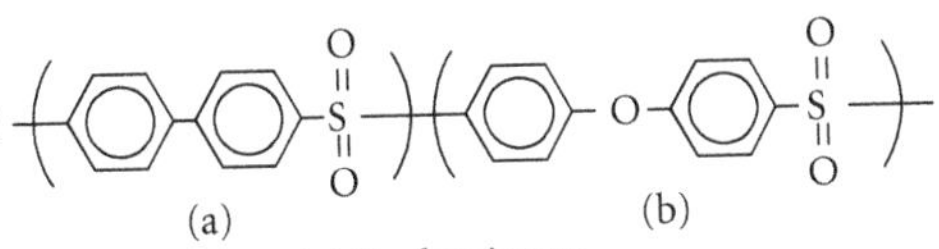 II (a) (b) (a) Predominates	285	Astrel (3M Corp.)
III	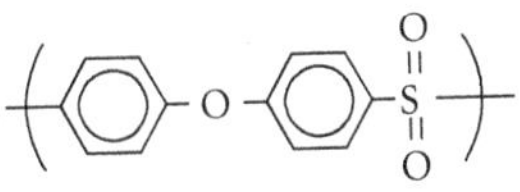 230	Victrex (ICI)
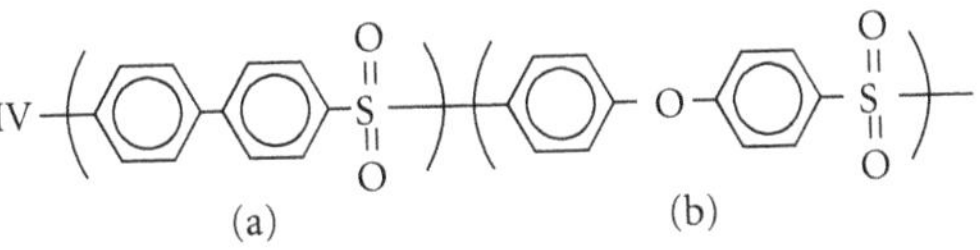 IV (a) (b) (b) Predominates	250	Polyether-Sulfone 720 P (ICI)
V	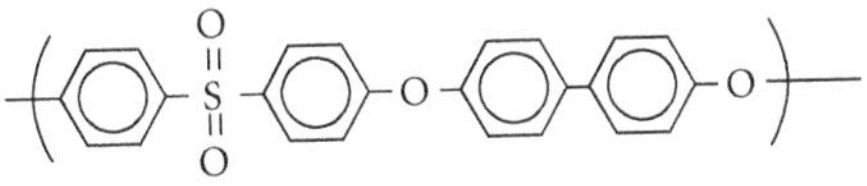 —	Radel (Union Carbide)

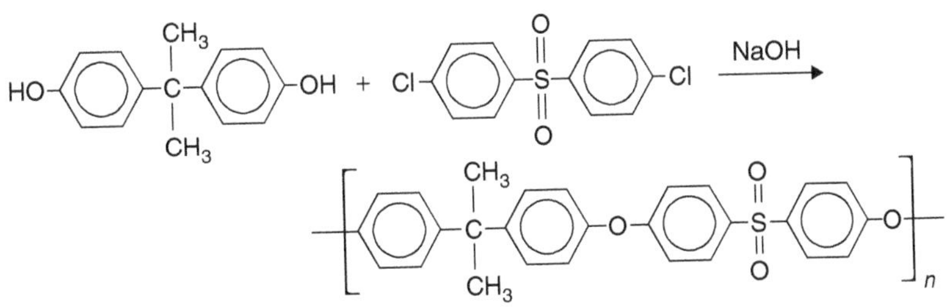

1.3.8.1 Properties

In spite of their linear and regular structure the commercial polysulfones are amorphous. This property might be attributed to the high degree of chain stiffness of polymer molecules which make crystallization difficult. Because of their high in-chain aromaticity and consequent high chain stiffness, the polymers have high values of T_g (see Table 1.30), which means that the processing temperatures must be above 300°C.

Commercial polymers generally resist aqueous acids and alkalis but are attacked by concentrated sulfuric acid. Being highly polar, the polymer is not dissolved by aliphatic hydrocarbons but dissolves in dimethyl formamide and dimethyl acetamide.

In addition to the high heat-deformation resistance, the polymers also exhibit a high degree of chemical stability. This has been ascribed to an enhanced bond strength arising from the high degree of resonance in the structure. The polymers are thus capable of absorbing a high degree of thermal and ionizing radiation without cross-linking.

The principal features of commercial polysulfones are their rigidity, transparency, self-extinguishing characteristics, exceptional resistance to creep, especially at somewhat elevated temperatures, and good high-temperature resistance.

The use temperatures of the major engineering thermoplastics are compared in Figure 1.36. Polysulfones are among the higher-priced engineering thermoplastics and so are only considered when polycarbonates or other cheaper polymers are unsuitable. In brief, polysulfones are more heat resistant and have greater resistance to creep, whereas polycarbonates have a somewhat higher Izod and tensile impact strength besides being less expensive.

In many fields of use polysulfones have replaced or are replacing metals, ceramics, and thermosetting plastics, rather than other thermoplastics. Since commercial polysulfones can be injection molded into complex shapes, they avoid costly machining and finishing operations. Polysulfones can also be extruded into film and foil. The latter is of interest for flexible printed circuitry because of its high-temperature performance.

Polysulfones have found widespread use where good dimensional stability at elevated temperatures is required and fabrication is done by injection molding. Some products made from polysulfones are electrical components, connectors, coil bobbins, relays, and appliances operating at high temperatures (e.g., hair driers, fan heaters, microwave ovens, lamp housings and bases).

Polysulfones are transparent (though often slightly yellow), have low flammability (limiting oxygen index typically 38), and burn with little smoke production. Typical properties of some of the commercial polysulfones are shown in Table 1.31.

1.3.9 Polyether Ketones

The chemistry and technology of aromatic polyether ketones may be considered as an extension to those of the polysulfones [58]. The two polymer classes have strong structural similarities, and there are strong parallels in preparative methods.

Preparations reported for aromatic polyether ketones are analogous to the polysulfonylation and polyetherification reactions for the polysulfones. Several aromatic polyether ketones have been prepared.

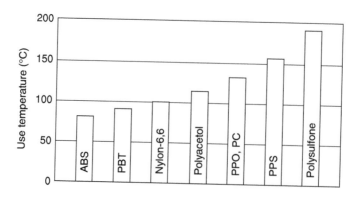

FIGURE 1.36 Use temperatures of major engineering thermoplastics.

TABLE 1.31 Properties of Aromatic Polysulfones

Property	Udel (Union Carbide)	Victrex (ICI)	Astrel (3M)
Tensile strength			
10^3 lbf/in.^{2}	9.3	12.2	13.0
MPa	64	84	90
Tensile modulus			
10^5 lbf/in.^{2}	3.6	3.5	3.8
GPa	2.5	2.4	2.6
Elongation at break (%)	50–100	40–80	10
Flexural modulus			
10^5 lbf/in.^{2}	3.0	3.8	4.0
GPa	2.1	2.6	2.8
Heat distortion temperature (°C)	174	203	274
Impact strength, notched Izod ft.-lb/in.	—	1.9	—
J/m	—	100	—
Limiting oxygen index (%)	—	38	—

The polyether ether ketone (PEEK) was test marketed in 1978 by ICI.

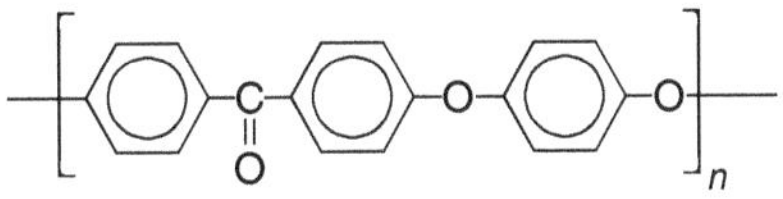

PEEK is made by the reaction of the potassium salt of hydroquione with difluorobenzophenone in a high boiling solvent, diphenylsulfone, at temperatures close to the melting point of the polymer:

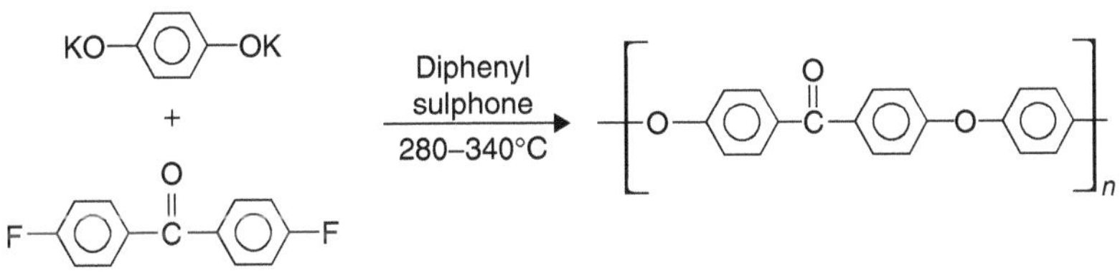

PEEK is semicrystalline with a melting temperature (T_m) of 335°C and a glass transition temperature (T_g) of 145°C. The degree of crystallinity can vary from 40% (slow cooling) to essentially amorphous (quenching), but is usually about 35%.

PEEK is a high-temperature-resistant thermoplastic suitable for wire coating, injection molding, film and advanced structural composite fabrication. The wholly aromatic structure of PEEK contributes to its high-temperature performance.

The polymer exhibits very low water absorption, very good resistance to water at 125°C (under which conditions other heat-resisting materials, such as aromatic polyamides, are liable to fail), and is resistant to attack over a wide pH range, from 60% sulfuric acid to 40% sodium hydroxide at elevated temperatures, although attack can occur with some concentrated acids.

PEEK has outstanding resistance to both abrasion and dynamic fatigue. Its tensile strength decreases less than 10% after 10^7 cycles at 23°C. It has low flammability with a limiting oxygen index of 35% and generates an exceptionally low level of smoke. Other specific features are excellent resistant to gamma radiation and good resistance to environmental stress cracking.

PEEK has greater heat resistance compared to poly(phenylene sulfide) and is also markedly tougher (and markedly more expensive). PEEK is melt processable and may be injection molded and extruded on conventional equipment.

Typical applications of PEEK include coating and insulation for high-performance wiring, particularly for the aerospace and computer industries, military equipment, nuclear plant applications, oil wells, and compressor parts.

Since it is a crystalline polymer, the strength and thermal resistance of PEEK are increased dramatically by incorporation of reinforcing agents. Composite prepegs with carbon fibers have been developed for structural aircraft components. Typical properties of PEEK are shown in Table 1.32.

1.3.10 Polybenzimidazole

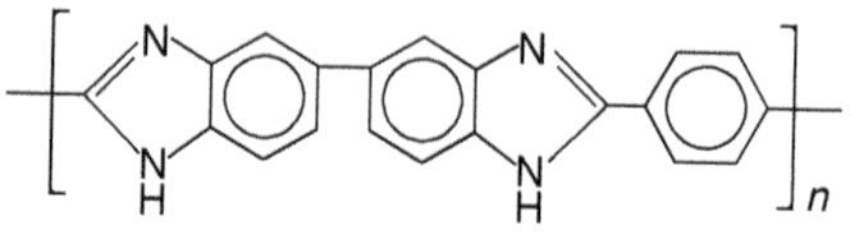

Monomers	Polymerization	Major Uses
Tetraaminobiphenyl, terephthalic acid	Polycondensation	Fiber

Polybenzimidazole (PBI) is the most well-known commercial example of aromatic heterocycles used as high-temperature polymers. The synthesis of PBI is carried out as follows (see also Figure 1.36 of *Plastics Fundamentals, Properties, and Testing*). The tetraaminobiphenyl required for the synthesis of PBI is obtained from 3,3'-dichloro-4,4'-diaminodiphenyl (a dye intermediate) and ammonia. Many other tetraamines and dicarboxylic acids have been condensed to PBI polymeric systems.

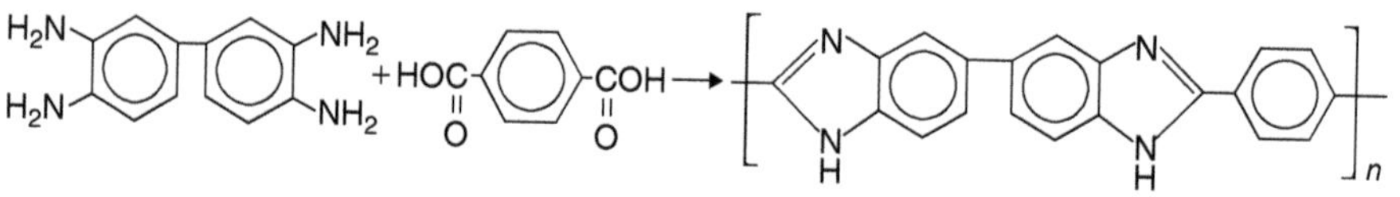

The high thermal stability of PBI (use temperature about 400°C compared to about 300°C for polyimides) combined with good stability makes it an outstanding candidate for high-temperature application despite its relatively high cost.

Fibers have been wet spun from dimethylacetamide solution, and a deep gold woven cloth has been made from this fiber by Celanese. The cloth is said to be more comfortable than cotton (due to high moisture retention) and has greater flame resistance than Nomex (oxygen index of 29% for PBI compared to 17% for Nomex). The U.S. Air Force has tested flight suits of PBI and found them superior to other materials.

TABLE 1.32   Properties of Polyetheretherketone

Property	Unfilled	30% Glass Fiber Filled
Tensile strength		
10^3 lbf/in.^{2}	13.3	22.8
MPa	92	157
Elongation at break (%)	4.9 (yield)	2.2
Flexural modulus		
10^5 lbf/in.^{2}	5.4	14.9
GPa	3.7	10.3
Impact strength, notched Izod		
ft.-lb/in.	1.6	1.8
J/m	83	96
Heat distortion temperature (°C)	140	315
Limiting oxygen index (%)	35	—

Other applications of PBI are in drogue parachutes and lines for military aircraft as well as ablative heat shields. The PBI fibers have also shown promise as reverse osmosis membranes and in graphitization to high-strength, high-modulus fibers for use in composites. The development of ultra-fine PBI fibers for use in battery separator and fuel cell applications has been undertaken by Celanese.

A new technique of simple precipitation has been used to process PBI polymers into films. High-strength molecular-composite films have been produced with tensile strength in the region of 20,000 psi (137 MPa). The PBI polymer has also been fabricated as a foam. The material provides a low-weight, high-strength, thermally stable, machinable insulation, much needed in the aerospace industry. PBIs exhibit good adhesion as films when cast from solution onto glass plates. This property leads to their use in glass composites, laminates, and filament-wound structures.

1.3.11 Silicones and Other Inorganic Polymers

The well-known thermal stability of minerals and glasses, many of which are themselves polymeric, has led to intensive research into synthetic inorganic and semi-inorganic polymers [14,59,60]. Numerous such polymers have been synthesized, but only a few have found industrial acceptance, due to the difficulties encountered in processing them.

1.3.11.1 Silicones

Monomer	Polymerization	Major Uses
Chlorosilanes	Polycondensation	Elastomer, sealants, and fluids

Silicones are by far the most important inorganic polymers and are based on silicon, an element abundantly available on our planet. The silicone polymers are available in a number of forms, such as fluids, greases, rubbers, and resins. Because of their general thermal stability, water repellency, antiadhesive characteristics, and constancy of properties over a wide temperature range, silicones have found many and diverse applications. [The structure used as the basis of the nomenclature of the silicon compounds is silane SiH_4, corresponding to methane CH_4. Alkyl-, aryl-, alkoxy-, and halogen-substituted silanes are referred to by prefixing silane by the specific group present. For example, $(CH_3)_2SiH_2$ is dimethyl silane, and CH_3SiCl_3 is trichloromethylsilane. Polymers in which the main chain consists of repeating $-Si-O-$ units together with predominantly organic side groups are referred to as *polyorganosiloxanes* or, more loosely, as *silicones*.]

The commercial production of a broad variety of products from a few basic monomers followed the development of an economically attractive direct process for chlorosilanes, discovered by E. G. Rochow in 1945 at the G. E. Research Laboratories. The process involves reaction of alkyl or aryl halides with elementary silicon in the presence of a catalyst, e.g., copper for methyl- and silver for phenylchlorosilanes. The basic chemistry can be described as

$$SiO_2 + C \rightarrow Si + 2CO$$

$$Si + RX \rightarrow R_nSiX_{4-n} \quad (n = 0\text{--}4)$$

In the alkylation of silicon with methyl chloride, mono-, di-, and trimethyl chlorosilanes are formed. The reaction products must then be fractionated. Because the dimethyl derivative is bifunctional, it produces linear methylsilicone polymers on hydrolysis.

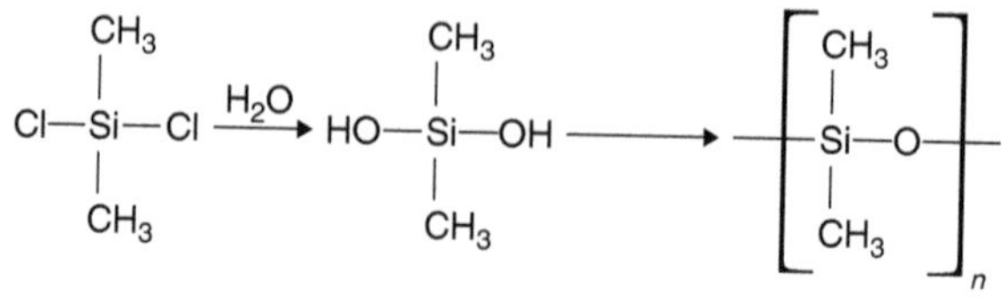

Since monomethyl trichlorosilane has a functionality of 3, the hydrolysis leads to the formation of a highly cross-linked gel.

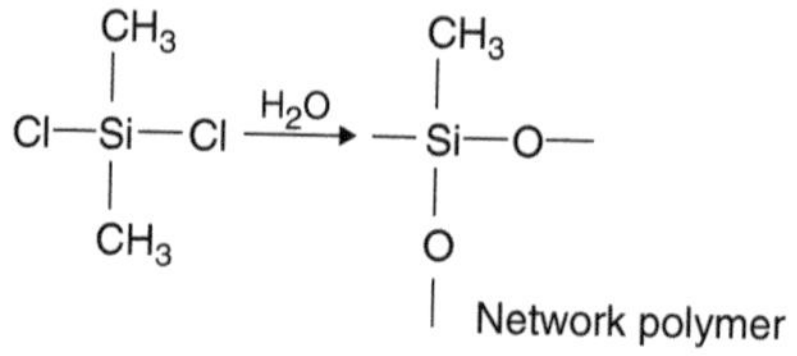

Since the trimethyl monochlorosilane is monofunctional, it forms only a disiloxane.

$$(CH_3)_3SiCl \xrightarrow{H_2O} (CH_3)_3Si\text{–}O\text{–}Si(CH_3)_3$$

Products of different molecular-weight ranges and degrees of cross-linking are obtained from mixtures of thee chlorosilanes in different ratios. In characterizing commercial branched and network structures, the CH_3/Si ratio (or, generally, R/Si ratio) is thus a useful parameter. For example, the preceding three idealized products have CH_3/Si ratios of 2:1, 1:1, and 3:1, respectively. A product with a CH_3/Si ratio of 1.5:1 will thus be expected to have a moderate degree of cross-linking.

Many different silicon products are available today. The major applications are listed in Table 1.33.

1.3.11.1.1 Silicone Fluids

The silicone fluids form a range of colorless liquids with viscosities from 1 to 1,000,000 centistokes (cs). The conversion of chlorosilane intermediates into polymer is accomplished by hydrolysis with water, which is followed by spontaneous condensation. In practice, the process involves three important stages: (1) hydrolysis, condensation, and neutralization (of the HCl evolved on hydrolysis); (2) catalytic equilibration; and (3) devolatilization.

The product after the first stage consists of an approximately equal mixture of cyclic compounds, mainly the tetramer, and linear polymer. To achieve a more linear polymer and also to stabilize the viscosity, it is common practice to equilibrate the products of hydrolysis by heating with a catalyst such as dilute sulfuric acid. For fluids of viscosities below 1000 cs, this equilibrium reaction is carried out for hours at 100–150°C.

After addition of water, the oil is separated from the aqueous acid layer and neutralized. To produce nonvolatile silicone fluids, volatile low-molecular products are removed by using a vacuum still. Commercial nonvolatile fluids have a weight loss of less than 0.5% after 24 h at 150°C.

Dimethylsilicone fluids find a wide variety of applications mainly because of their water repellency, lubrication and antistick properties, low surface tension, a high order of thermal stability, and a fair constancy of physical properties over a wide range of temperature ($-70°C$ to $200°C$).

As a class the silicone fluids have no color or odor, have low volatility, and are nontoxic. The fluids have reasonable chemical resistance but are attacked by concentrated mineral acids and alkalis. They are soluble in aliphatic, aromatic, and chlorinated hydrocarbons.

A well-known application of the dimethylsilicone fluids is as a polish additive. The value of the silicone fluids in this application is due to its ability to lubricate, without softening, the microcrystalline wax plates.

Dilute solutions or emulsions containing 0.5–1% of a silicone fluid are extensively used as a release agent for rubber molding. However, their use has been restricted with thermoplastics because of the tendency of the fluids to cause stress cracking in polymers.

Silicone fluids are used in shock absorbers, hydraulic fluids, dashpots, and in other damping systems in high-temperature operations.

TABLE 1.33 Major Applications of Silicones

Mold-release agents	Greases and waxes
Water repellants	Cosmetics
Antifoaming agents	Insulation
Glass-sizing agents	Dielectric encapsulation
Heat-exchange fluids	Caulking agents (RTV)
Hydraulic fluids	Gaskets and seals
Surfactants	Laminates
Coupling agents	Biomedical devices

The silicones have established their value as water-repellent finishes for a range of natural and synthetic textiles. Techniques have been developed which result in the pickup of 1–3% of silicone on the cloth. Leather also may be made water repellent by treatment with solutions or emulsions of silicone fluids. These solutions are also used for paper treatment.

Silicone fluids and greases are useful as lubricants for high-temperature operations for applications depending on rolling friction. Grease may be made by blending silicone with an inert filler such as fine silicas, carbon black, or a metallic soap. The silicone/silica greases are used as electrical greases for such applications as aircraft and car ignition systems. Silicone greases have also found uses in the laboratory for lubricating stopcocks and for high-vacuum work.

Silicone fluids are used extensively as antifoams, although concentration needed is normally only a few parts per million. The fluids have also found a number of uses in medicine. Barrier creams based on silicone fluids are particularly useful against cutting oils used in metal machinery processes.

High-molecular-weight dimethylsilicone fluids are used as stationary phase for columns in vapor-phase chromatographic apparatus.

Surfactants based on block copolymers of dimethylsilicone and poly(ethylene oxide) are unique in regulating the cell size in polyurethane foams. One route to such polymers used the reaction between a polysiloxane and an allyl ether of poly(ethylene oxide),

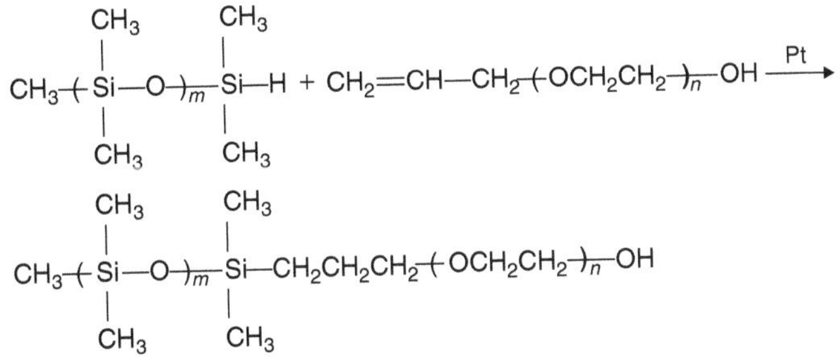

where $m = 2$–5 and $n = 3$–20. Increasing the silicone content makes the surfactant more lipophilic, whereas a higher poly(ethylene oxide) content makes it more hydrophilic.

1.3.11.1.2 *Silicone Resins*

Silicone resins are manufactured batchwise by hydrolysis of a blend of chlorosilanes. For the final product to be cross-linked, a certain amount of trichlorosilane must be incorporated into the blend. (In commercial practice, R/Si ratios are typically in the range of 1.2:1–1.6:1) The cross-linking of the resin is, of course, not carried out until it is in situ in the finished product. The cross-linking takes place by heating the resin at elevated temperatures with a catalyst, several of which are described in the literature (e.g., triethanolamine and metal octoates).

The resins have good heat resistance but are mechanically much weaker than cross-linked organic plastics. The resins are highly water repellent and are good electrical insulators particularly at elevated temperatures and under damp conditions. The properties are reasonably constant over a fair range of temperature and frequency.

Methyl phenyl silicone resins are used in the manufacture of heat-resistant glass-cloth laminates particularly for electrical applications. These are generally superior to PF and MF glass-cloth laminates. The dielectric strength of silicon-bonded glass-cloth laminates is 100–120 kV/cm compared to 60–80 kV/cm for both PF and MF laminates. The insulation resistance (dry) of the former (500,000 Ω) is significantly greater than those for the PF and MF laminates (10,000 and 20,000 Ω, respectively). The corresponding values after water immersion are 10,000, 10, and 10 Ω.

Silicone laminates are used principally in electrical applications such as printed circuit boards, transformers, and slot wedges in electric motors, particularly class H motors. Compression-molding powders based on silicone resins are available and have been used in the molding of switch parts, brush ring holders, and other electrical applications that need to withstand high temperatures.

1.3.11.1.3 Silicone Rubbers

Silicone elastomers are either *room-temperature vulcanization* (RTV) or *heat-cured silicone rubbers*, depending on whether cross-linking is accomplished at ambient or elevated temperature. [The term *vulcanization* (see Chapter 1 of *Plastics Fundamentals, Properties, and Testing* and Chapter 1 of *Plastics Fabrication and Recycling*) is a synonym for cross-linking. While *curing* is also a synonym for cross-linking, it often refers to a combination of additional polymerization plus cross-linking.] RTV and heat-cured silicone rubbers typically involve polysiloxanes with degrees of polymerizations of 200–1500 and 2500–11,000, respectively.

While the lower-molecular-weight polysiloxanes can be synthesized by the hydrolytic step polymerization process, the higher-molecular-weight polymers are synthesized by ring-opening polymerization using ionic initiators:

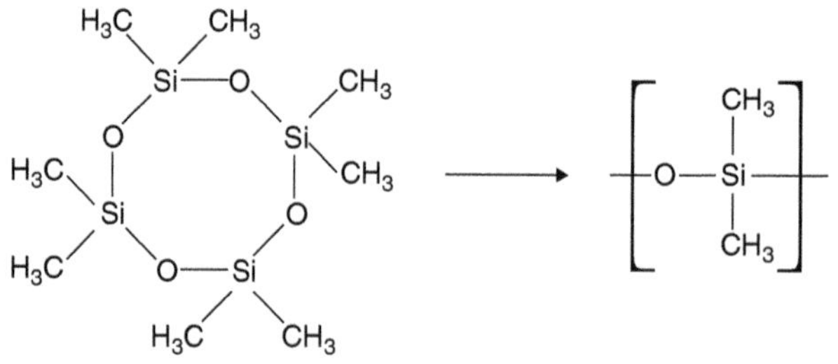

The cyclic tetramer (octamethylcyclotetrasiloxane) is equilibrated with a trace of alkaline catalyst for several hours at 150–200°C, the molecular weight being controlled by careful addition of monofunctional siloxane. The product is a viscous gum with no elastic properties.

Before fabrication it is necessary to compound the gum with fillers, a curing agent, and other special additives on a two-roll mill or in an internal mixer (see Rubber Compounding, Chapter 1 of *Plastics Fabrication and Recycling*). Unfilled polymers have negligible strength, whereas reinforced silicone rubbers may have strengths up to 2000 psi (14 MPa).

Silica fillers are generally used with silicone rubbers. These materials with particle sizes in the range 0.003–0.03 μm are prepared by combustion of silicon tetrachloride (fume silicas), by precipitation, or as an aerogel.

Heat-curing of silicone rubbers usually involve free-radical initiators such as benzoyl peroxide, 2,4-dichlorobenzoyl peroxide, and *t*-butyl per-benzoate, used in quantities of 0.5–3%. These materials are stable in the compounds at room temperature for several months but will start to cure at about 70°C. The curing (cross-linking) is believed to take place by the sequence of reactions shown in Figure 1.37. The process involves the formation of polymer radicals via hydrogen abstraction by the peroxy radicals formed from the thermal decomposition of the peroxide and subsequent cross-linking by coupling of the polymer radicals.

The cross-linking efficiency of the peroxide process can be increased by incorporating small amounts of a comonomer containing vinyl groups into the polymer, e.g., by copolymerization with small amounts of vinyl-methyl silanol:

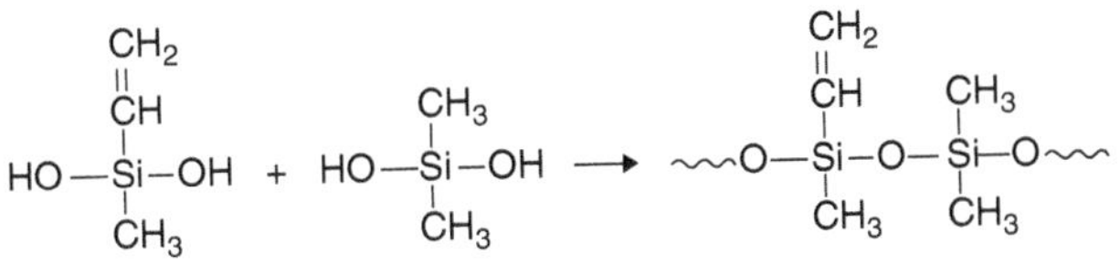

$$\underset{\substack{\text{HO}-\overset{\displaystyle\overset{\displaystyle CH_2}{\underset{\displaystyle CH}{\|}}}{\underset{\displaystyle CH_3}{\overset{\displaystyle |}{Si}}}-\text{OH}}{} \;+\; \underset{\substack{\text{HO}-\overset{\displaystyle CH_3}{\underset{\displaystyle CH_3}{\overset{\displaystyle |}{\underset{\displaystyle |}{Si}}}}-\text{OH}}{} \;\longrightarrow\; \sim\!\!\sim\text{O}-\overset{\displaystyle\overset{\displaystyle CH_2}{\underset{\displaystyle CH}{\|}}}{\underset{\displaystyle CH_3}{\overset{\displaystyle |}{Si}}}-\text{O}-\overset{\displaystyle CH_3}{\underset{\displaystyle CH_3}{\overset{\displaystyle |}{\underset{\displaystyle |}{Si}}}}-\text{O}\!\!\sim\!\!\sim}$$

Dimethyl silicone rubbers show a high compression set. (For example, normal cured compounds have a compression set of 20–50% after 24 h at 150°C.) Substantially reduced compression set values may be obtained by using a polymer containing small amounts of methylvinylsiloxane. Rubbers containing vinyl groups can be cross-linked by weaker peroxide catalysts. Where there is a high vinyl content (4–5% molar), it is also possible to vulcanize with sulfur.

Room-temperature vulcanizing silicone rubbers (RTV rubbers) are low-molecular-weight liquid silicones with reactive end groups and loaded with reinforcing fillers. Several types are available on the market.

"One-component" RTV rubbers consist of an air-tight package containing silanol-terminated polysiloxane, cross-linking agent (methylacetoxysilane), and catalyst (e.g., dibutyltin laurate). Moisture from the atmosphere converts the cross-linking agent to the corresponding silanol (acetic acid is a by-product), $CH_3Si(OH)_3$, which brings about further polymerization combined with cross-linking of the polysiloxane,

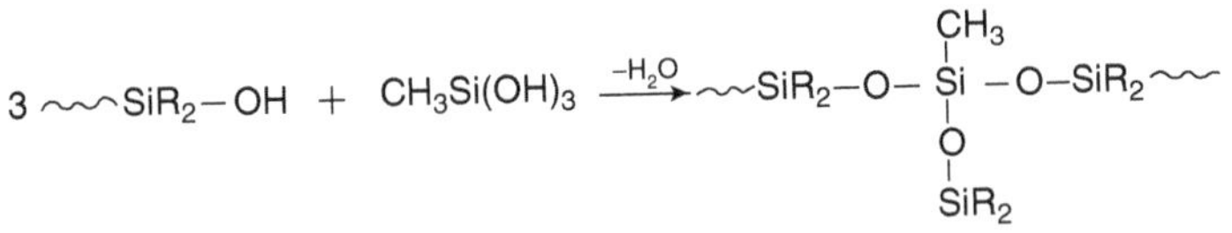

$$3 \sim\!\!\sim SiR_2-OH \;+\; CH_3Si(OH)_3 \;\xrightarrow{-H_2O}\; \sim\!\!\sim SiR_2-O-\overset{\displaystyle CH_3}{\underset{\displaystyle \underset{\displaystyle SiR_2}{\overset{\displaystyle |}{O}}}{\overset{\displaystyle |}{\underset{\displaystyle |}{Si}}}}-O-SiR_2\!\!\sim\!\!\sim$$

Two-component RTV formulations involve separate packages for the polysiloxane and cross-linking agent. A typical two-component RTV formulation cures by reaction of silanol end groups with silicate esters in the presence of a catalyst such as tin octoate or dibutyltin dilaurate (Figure 1.38).

Another two-pack RTV formulation cures by hydrosilation, which involves the addition reaction between a polysiloxane containing vinyl groups (obtained by including methylvinyldichlorosilane in the original reaction mixture for synthesis of polysiloxane) and a siloxane cross-linking agent that contains Si–H functional groups, such as $Si[OSi(CH_3)_2H]_4$. The reaction is catalyzed by chloroplatinic acid or other soluble platinum compounds.

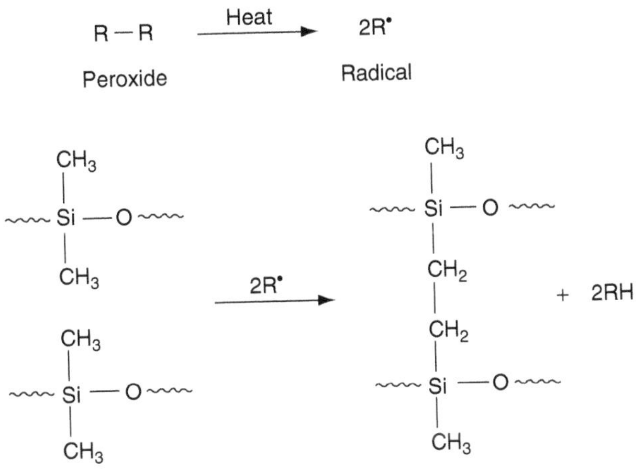

$$R-R \;\xrightarrow{\text{Heat}}\; 2R^\bullet$$
Peroxide $\qquad\qquad$ Radical

FIGURE 1.37 Peroxide curing of silicone rubbers.

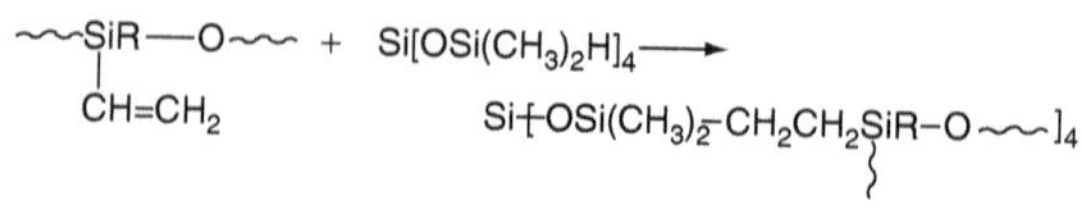

$$\sim\!\!\!\sim\!\overset{\displaystyle |}{\underset{\displaystyle CH=CH_2}{SiR}}\!-O\!\sim\!\!\!\sim\ +\ Si[OSi(CH_3)_2H]_4 \longrightarrow$$

$$Si\!\!\left[OSi(CH_3)_2\!-\!CH_2CH_2\!\underset{\displaystyle }{SiR}\!-\!O\!\sim\!\!\!\sim\right]_4$$

Hydride functional siloxanes can also cross-link silanol-terminated polysiloxanes. The reaction is catalyzed by tin salts and involves elimination of H_2 between Si–H and Si–O–H groups.

RTV rubbers have proved to be of considerable value as they provide a method for producing rubbery products with the simplest equipment. These rubbers find use in the building industry for caulking and in the electrical industry for encapsulation.

Nontacky self-adhesive rubbers (fusible rubbers) are obtained if small amounts of boron (~ 1 boron atom per 300 silicon atoms) are incorporated into the polymer chain. They may be obtained by condensing dialkylpolysiloxanes end-blocked with silanol groups with boric acid or by reacting ethoxyl end-blocked polymers with boron triacetate.

Bouncing putty is somewhat similar in that the Si–O–B bond occurs occasionally along the chain. It is based on a polydimethylsiloxane polymer modified with boric acid, additives, fillers, and plasticizers to give a material that shows a high elastic rebound when small pieces are dropped on a hard surface but flows like a viscous fluid on storage or slow application of pressure.

The applications of the rubbers stem from their important properties, which include thermal stability, good electrical insulation properties, nonstick properties, physiological inertness, and retention of elasticity at low temperatures. The temperature range of general-purpose material is approximately $-50°C$ to $+250°C$, and the range may be extended with special rubbers. Silicone rubbers are, however, used only as special-purpose materials because of their high cost and inferior mechanical properties at room temperature as compared to conventional rubbers (e.g., natural rubber and SBR).

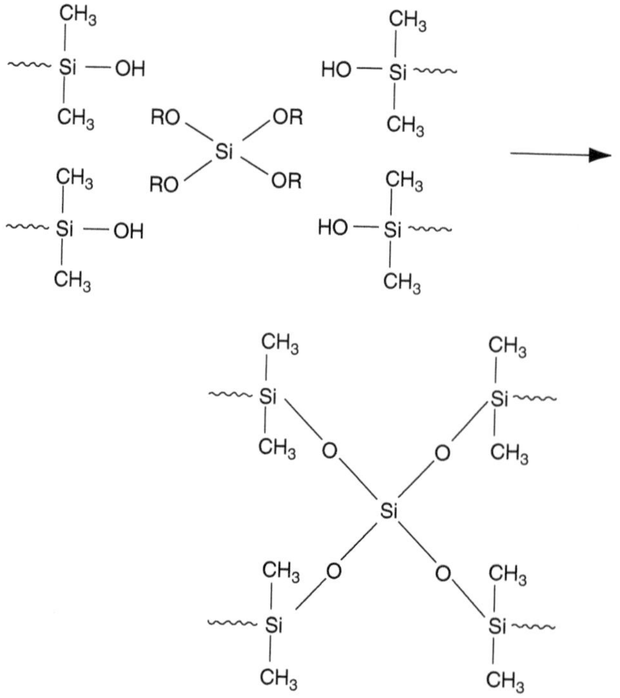

FIGURE 1.38 Curing of RTV rubbers by reaction of silanol end groups with silicate esters in the presence of a catalyst such as tin octoate or dibutyltin dilaurate.

Modern passenger and military aircraft each use about 500 kg of silicone rubber. This is to be found in gaskets and sealing rings for jet engines, vibration dampers, ducting, sealing strips, and electrical insulators. Silicone cable insulation is also used extensively in naval craft since the insulation is not destroyed in the event of fire but forms an insulating layer of silica.

The rubbers find use in diverse other applications which include electric iron gaskets, domestic refrigerators, antibiotic container closures, and for nonadhesive rubber-covered rollers for handling such materials as confectionary and adhesive tape.

Due to their relative inertness, new applications have emerged in the biomedical field. A silicone rubber ball is used in combination with a fluorocarbon seal to replace defective human heart valves. Silicone rubber has had many applications in reconstructive surgery on or near the surface of the body. Prosthetic devices are very successfully used in all parts of the body.

The cold-curing silicone rubbers are of value in potting and encapsulation.

Liquid silicone rubbers may be considered as a development from the RTV silicone rubbers but they have a better plot life and improved physical properties, including heat stability (in the cured state) similar to that of conventional silicone elastomers. Liquid silicone rubbers range from a flow consistency to a paste consistency and are usually supplied as a two-pack system, which requires simple blending before use. The materials cure rapidly above 110°C. In injection molding of small parts at high temperatures (200–250°C), cure times may be as small as a few seconds. One example of application is in baby bottle nipples, which, although more expensive, have a much longer working life.

Liquid silicone rubbers have also been used in some extruded applications. Vulcanization of the extruded material may be carried out by using infrared heaters or circulated hot air. The process has been applied to wire coating, ignition cables, optical fibers, various tapes, and braided glass-fiber sleeving, as well as for covering delicate products.

1.3.11.2 Polyphosphazenes

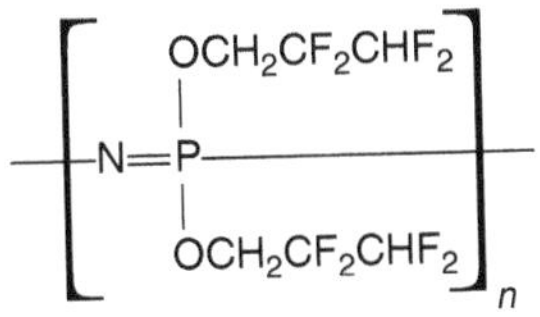

$$\left[-N{=}P \begin{array}{c} OCH_2CF_2CHF_2 \\ | \\ | \\ OCH_2CF_2CHF_2 \end{array} \right]_n$$

Monomers	Polymerization	Major Uses
Phosphorus pentachloride, ammonium chloride, fluorinated alcohols	Polycondensation followed by nucleophilic replacement of chloro-groups	Aerospace, military, oil exploration applications

Polyphosphazenes containing nitrogen and phosphorus have been synthesized by replacing the chlorine atoms on the backbone chain of polymeric phosphonitrilic chloride (dichlorophosphazene) by alkoxy or fluoroalkoxy groups. These derivative polymers do not exhibit the hydrolytic instability of the parent polymer. The general synthesis scheme is

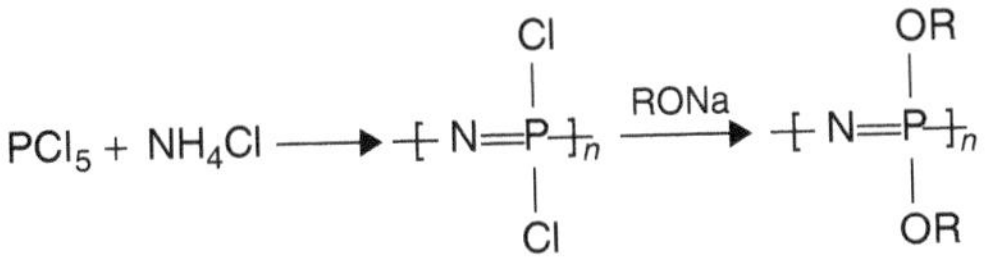

$$PCl_5 + NH_4Cl \longrightarrow \left[N{=}P \begin{array}{c} Cl \\ | \\ | \\ Cl \end{array} \right]_n \xrightarrow{RONa} \left[N{=}P \begin{array}{c} OR \\ | \\ | \\ OR \end{array} \right]_n$$

With mixtures of alkoxy substituents having longer alkyl chains, crystallization can be avoided to produce an amorphous rubber. The product is referred to as phosphonitrilic fluoroelastomer, a semiorganic rubber. The rubber can be cross-linked with free-radical initiators or by radiation. A commercial rubber (PNF by Firestone Tire and Rubber Co.) is based on alkoxides of trifluoroethyl alcohol or heptafluoroisobutyl alcohol.

The polyphosphazene rubbers have excellent resistance to oils and chemicals (except alcohols and ketones), good dynamic properties, good abrasion resistance, and a broad range of use temperatures ($-65°C$ to $+117°C$). The water resistance, however, is only fairly good.

1.3.11.3 Polythiazyl

Monomer	Polymerization	Major Uses
S_4N_4	Ring-opening polymerization	Semiconductive polymers

The solid four-member ring reaction product S_2N_2 obtained by pyrolysis of gaseous S_4N_4 under vacuum is polymerized at room temperature by ring-opening polymerization to give the linear chain polythiazyl.

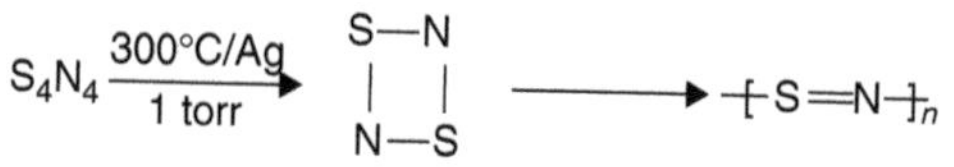

$$S_4N_4 \xrightarrow[\text{1 torr}]{\text{300°C/Ag}} \begin{matrix} S-N \\ | \quad | \\ N-S \end{matrix} \longrightarrow +S=N+_n$$

The product is a brasslike solid material that behaves like a metal or alloy but is lighter and more flexible. Polythiazyl has electrical conductivity (3700 reciprocal ohm-cm, or siemens/cm) at room temperature and super-conductivity at 0.3 K. Some doped polymers are photoconductive.

1.3.12 Polyblends

The concept of physically blending two or more existing polymers or copolymers to obtain new products or for problem solving is now attracting widespread interest and commercial utilization [61–65]. By definition, any physical mixture of two or more different polymers or copolymers that are not linked by covalent bonds is a *polymer blend* or *polyblend*.

Blends have been classified as miscible one-phase systems and partially miscible or immiscible two-phase systems. A negative change in free energy (ΔG) is required for miscibility. The change in entropy (ΔS) is essentially negligible, and hence the change in enthalpy (ΔH) must be negative or zero for the formation of miscible blends as shown by the Gibbs free energy equation:

$$\Delta G = \Delta H - T\Delta S$$

where T is the absolute temperature. The requirement for a negative ΔH value can be met when there is a physical attraction, such as hydrogen bonding, between the component polymers.

A polymer blend (PB) that is homogeneous down to the molecular level and is associated with the negative value of the free energy of mixing, $\Delta G = \Delta H \leq 0$, is termed a *miscible polymer blend*. An *immiscible polymer blend*, on the other hand, is any PB for which $\Delta G = \Delta H > 0$.

To make an analogy with metals polymer blends are sometimes referred to as *polymer alloys*. Thus the term alloy has been used to describe miscible or immiscible mixtures of polymers that are usually blended as melts. Another definition often used for a polymer alloy is that it is an immiscible PB having a modified interface and/or morphology. The general relationship between blends and alloys is shown in Figure 1.39. The term *compatibilization* in the figure refers to a process of modification of interfacial properties (discussed later) of an immiscible PB, leading to the creation of a polymer alloy.

The concept of blending polymers is not new; the rubber industry has used it for decades. In recent years, however, there has been a resurgence of interest arising primarily from the demand for engineering plastics and speciality polymers. There are sound economic reasons for this interest. Development of a new polymer to meet a specific need is a costly enterprise. If the desired properties can be realized simply by mixing two or more existing polymers, there is an obvious economic advantage. Since the time normally required to develop a new blend is commercially 3–5 years, compared to 8–10 years required for new plastics materials. Moreover, the cost is reduced if an expensive polymer can be diluted by a less expensive one.

Blending provides a convenient way of combining the mechanical, physical, or thermal properties of more than one material. One example is the blend of ABS and PVC. The ABS contributes high heat-distortion temperature, toughness, and easy moldability, while the PVC imparts weatherability and flame retardance, as well as reducing the cost of the blend. Applications of the blend include automotive interior trim, luggage shells, and canoes.

Important disadvantages of a basic polymer (for example, difficulties of fabrication) may be overcome by blending with one or more other polymers. In some cases, synergistic improvement in properties, exceeding the value for either polymer alone, is achieved by blending, though such cases are relatively rare. A good example is the improved Izod notched impact strength shown when ABS and polycarbonate (PC) are blended. At subzero temperatures, the blend has better notched impact strength than either of its component polymers.

The first commercial blend of two dissimilar polymers was Noryl, a miscible polyblend of poly (phenylene oxide) and polystyrene, introduced by General Electric in the 1960s. Since that time a large number of different blends have been introduced. A number of technologies have been devised to prepare polyblends; these are summarized in Table 1.34. For economic reasons, however, mechanical blending predominates.

It so happens that most polymers are not miscible; rather they separate into discrete phases on being mixed. Differences between miscible and immiscible polyblends are manifested in appearance (miscible blends are usually clear, immiscible blends are opaque) and in such properties as glass transition temperature (miscible blends exhibit a single T_g intermediate between those of the individual components, whereas immiscible blends exhibit separate T_gs characteristic of each component).

Though miscibility is by no means a prerequisite to commercial utility, homogeneous polymer blends are more convenient from the standpoint of being able to predict properties or processing characteristics.

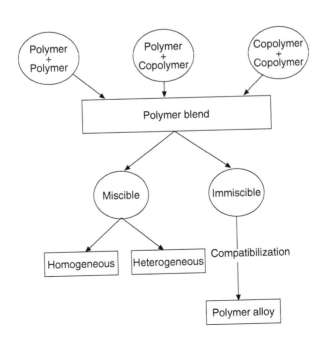

FIGURE 1.39 Schematic representation of the general relationship between polymer blends and alloys.

For example, if additives are used, there are no problems of migration from one phase to another. Physical or mechanical properties usually reflect, to a degree, the weighted average of the properties of each component. For a binary homogeneous blend, this can be expressed quantitatively for a particular property (P) by means of a semiempirical relationship:

$$P = \phi_1 P_1 + \phi_2 P_2 + \phi_1 \phi_2 I \qquad (1.3)$$

where ϕ is the volume fraction in the mix and I is an interaction term that may be negative, zero, or positive. For $I=0$, the properties are strictly additive. If I is positive, the property in question is better than the weighted average and the blend is said to be *synergistic* for that property. However, if I is negative, the property is worse than the weighted average (*incompatible, nonsynergistic*). This is illustrated schematically in Figure 1.40 by a plot of property vs. composition.

In the case of complete miscibility, we would expect properties to follow a simple monotonic function, more or less proportional to the contents of the two polymer components in the blend (Figure 1.40a). This is particularly useful for processors because it permits them to inventory a few commodity polymers and simply blend them in different proportions to meet the specific requirements of each product they manufacture.

Noryl, for example, is composed of polystyrene, an inexpensive polymer, and poly(phenylene oxide) or PPO, a relatively expensive polyether. For the most, the properties of Noryl are additive. For example, Noryl has poorer thermal stability than the polyether alone, but is easier to process. Its single glass transition temperature increases with increasing polyether content. In terms of tensile strength, however, the polyblend is synergistic.

For two polymers to be completely miscible, the optimum requirements are similar polarity, low molecule weight, and hydrogen bonding or other strong intermolecular attraction. Most polymer pairs do not meet these requirements for complete theoretical miscibility. The free energy of mixing is positive, and they tend to separate into two phases.

If they are slightly immiscible, each phase will be a solid solution of minor polymer in major polymer, and the phases will separate into submicroscopic domains with the polymer present in major amount forming the continuous matrix phase and contributing most toward its properties. Plots of properties vs.

TABLE 1.34   Types and Methods of Producing of Polyblends

Type	Method of Blending
Mechanical blends	Polymers are mixed at temperatures above T_g or T_m for amorphous and semicrystalline polymers, respectively
Mechanochemical blends	Polymers are mixed at shear rates high enough to cause degradation. Resultant free radicals combine to form complex mixtures including block and graft components
Solution-cast blends	Polymers are dissolved in common solvent and solvent is removed
Latex blends	Fine dispersions of polymers in water (latexes) are mixed, and the mixed polymers are coagulated
Chemical blends	
Interpenetrating polymer networks (IPN)	Cross-linked polymer is swollen with a different monomer; the monomer is then polymerized and cross-linked
Semi-interpenetrating polymer networks (semi-IPN, also called pseudo-IPN)	Polyfunctional monomer is mixed with thermoplastic polymer, then monomer is polymerized to network polymer
Simultaneous interpenetrating polymer networks (SIN)	Different monomers are mixed, then homopolymerized and cross-linked simultaneously
Interpenetrating elastomeric networks (IEN)	Latex polyblend is cross-linked after coagulation

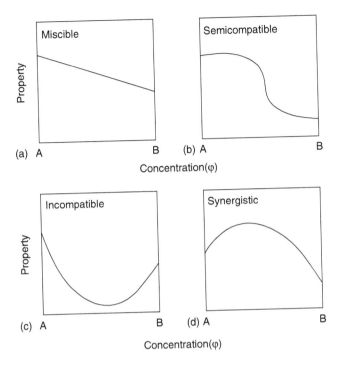

FIGURE 1.40   Properties vs. concentration of polymer components in a polyblend.

composition will be S-shaped showing an intermediate transition region where there is a phase inversion from one continuous phase to the other (Figure 1.40b). Most commercial polyblends are of this type, with the major polymer forming the continuous phase and retaining most of its useful properties, while the minor polymer forms small discrete domains, which contribute synergistically to certain specific properties.

When the polymer components in a blend are less miscible, phase separation will form larger domains with weaker interfacial bonding between them. The interfaces will therefore fail under stress and properties of polyblends are thus likely to be poorer than for either of the polymers in the blend. U-shaped property curves (Figure 1.40c) thus provide a strong indication of immiscibility. In most cases they also signify practical incompatibility, and hence lack of practical utility.

A fourth type of curve for properties vs. polyblend composition representing synergistic behavior (Figure 1.40d) has been obtained in a few cases of polymer blending. This shows improvement of properties, beyond what would be expected from simple monotonic proportionality, and sometimes far exceeding the value for either polymer alone. Synergism may result from a very favorable dipole-dipole attraction between the polymer components.

A blend of low-density polyethylene (LDPE) polyethylene (LDPE) with the terpolymer ethylene–propylene–diene monomer rubber (EPDM) exhibits a synergistic effect on tensile strength if EPDM is partially crystalline, but a nonsynergistic effect if the EPDM is amorphous [65]. This example shows the dramatic effect that morphology can have on properties of polymer blends. The synergism apparently arises from a tendency for crystallites in the LDPE to nucleate crystallization of ethylene segments in the EPDM.

Stereochemistry is also important in determining properties of polyblands. For example, syndiotactic poly(methyl methacrylate) is miscible with poly(vinyl chloride) at certain concentrations, whereas the isotactic form is immiscible over the entire composition range [66].

1.3.12.1   Prediction of Polyblend Properties

A major problem in polyblend development is trying to predict polymer miscibility. The incompatibility of various pairs of polymers has been correlated with the mutual effects on intrinsic viscosities and dipole moment differences of the component polymers [67,68]. These results can give a guide for finding compatible polymer or polymer pairs or with very low incompatibility.

The viscometric study of the ternary system, that is, polymer A—polymer B—solvent, has been used in order to determine the interaction parameter χ_{AB} which characterizes the incompatibility of the couple polymer A—polymer B [60].

The viscometric study has also led to a method of estimation of the incompatibility of the polymer without the need to evaluate χ_{AB}. This method evaluates the quantity $\Delta[\eta]_{AB}$ defined by

$$\Delta[\eta]_{AB} = (1/2)(\Delta[\eta]_A + \Delta[\eta]_B)\% \tag{1.4}$$

where $\Delta[\eta]_A$ is the decrease of the intrinsic viscosity of the polymer A by the presence of the polymer B in the solvent, given by

$$\Delta[\eta]_A = ([\eta]_A - [\eta]_{AB})/[\eta]_A \tag{1.5}$$

The decrease is due to the incompatibility existing between polymer A and polymer B. The quantity $\Delta[\eta]_B = ([\eta]_B - [\eta]_{BA})/[\eta]_B$ similarly gives the decrease of the intrinsic viscosity of polymer B due to the presence of polymer A in the solvent. The constant quantity of the polymer A or B in the mixture, solvent plus polymer, is always the same (typically 0.125×10^2 g/cm^3). Then the quantity $\Delta[\eta]_{AB}$ given by Equation (1.4) expresses a measure of the incompatibility existing between polymer A and polymer B. A high value of $\Delta[\eta]_{AB}$ for a polymer pair indicates a high value of incompatibility and vice versa.

The incompatibility of polymer pairs has been correlated with the dipole moment μ of two polymers. As it is known, the value of μ of each polymer [69] is given per monomer repeating unit and is given by the relation $\mu = (\bar{\mu}^2/\overline{DP}_n)^{1/2}$, where $\overline{DP}_n$ is the number average degree of polymerization and $\bar{\mu}^2$ is the mean square dipole moment of the long-chain molecules.

The values of $\Delta[\eta]_{AB}$ and $\Delta\mu$ for several pairs of vinyl polymers drawn from polystyrene, poly(methyl methacrylate), poly(vinyl chloride) and poly(vinyl acetate) are given in Table 1.35. It is seen that when the dipole moment of polymer A is very close to the dipole moment of polymer B, the incompatibility of the mixture, polymer A—polymer B, is very low. In the contrary, a mixture of polymer A and polymer B presents a high incompatibility when the dipole moment of polymer A is too different from that of polymer B.

In the case of two polymers presenting a high value of the difference $|\mu_A - \mu_B|$, the dissimilarity is high and the monomers of each polymer give interactions of the type $\delta^- - \delta^+$ with other monomers that belong to the same polymer and this leads to the phase separation. In the contrary, in the case of two polymers presenting a very low value of the difference $|\mu_A - \mu_B|$, the dissimilarity is low and the monomers of each polymer give interactions with other monomers that belong to the same or to the other polymer indifferently; hence the polymers are compatible.

Attempts to predict miscibility using simple solubility parameters of the type described in Chapter 1 of *Plastics Fundamentals, Properties, and Testing* have been largely unsuccessful because strong dipolar interactions are not taken into account. The importance of these interactions has been demonstrated, for example, with miscible mixtures of poly(acrylic esters) and poly(vinyl fluoride), where the compatibility has been attributed to dipolar interactions of the type

TABLE 1.35 Values of $\Delta[\eta]_{AB}$ and $\Delta\mu$ for Six Polymer Couples

Couples of Polymers	$\Delta[\eta]_{AB}$	$\Delta\mu$
PS–PMMA	7.00	0.99
PS–PVC	10.55	1.19
PS–PVA	11.20	1.34
PMMA–PVA	3.6	0.35
PMMA–PVC	10.55	1.19
PMMA–PVA	3.30	0.15

Predictions of properties for immiscible polyblends is much more complicated. This is partly due to the effects of varying morphologies that might arise as a result of processing variables. Frequently, one polymer will form a continuous phase with the second being dispersed as a noncontinuous phase in the form of spheres, lamellae, fibrils, and so on. It is, however, the polymer in the continuous phase that largely determines the properties of the polyblend.

For example, a 50:50 blend of polystyrene (a hard, glassy polymer at ordinary temperature) and polybutadiene (an elastomer) will be hard if polystyrene is the continuous phase, but soft if polystyrene is the dispersed phase. In some cases, however, an immiscible polyblend may have both components dispersed as continuous phases. Evidently, a proper control of phase morphology is of utmost importance with immiscible blends. The size of the dispersed phase should be optimized considering the final performance of the blend.

The major problem with immiscible blends is the poor physical attraction at phase boundaries that can lead to phase separation under stress resulting in poor mechanical properties. A number of ingenious approaches have been adopted to overcome this problem by improving compatibility between immiscible phases. One is through formation of interpenetrating polymer networks (IPN) as described in Table 1.34.

In IPNs (discussed later) the polymers are physically "locked" together by the interdispersed three-dimensional network, a phenomena referred to as *topological bonding* [70]. Such mixtures, however, still undergo phase separation into microdomains that vary in size according to the degree of immiscibility. Kinetic control of phase separation during the formation of IPN [or the semi-interpenetrating polymer networks (SINs)] provides the method of generation of desired properties.

Another approach is to incorporate *compatibilizers* or interfacial agents into the blend to improve adhesion between phases. The concept of compatibilization of polymers is described in a later section.

1.3.12.2 Selection of Blend Components

The properties of engineering polymer blends and alloys that are sought to be improved by blending are impact strength, processability, tensile strength, rigidity/modulus, heat-deflection temperature, flame retardancy, thermal stability, dimensional stability, and chemical/solvent resistance. Among them, toughening and processability are of major concern. The second group of importance includes strength, modulus, and heat-deflection temperature, while the third group concerns flame retardancy, solvent resistance, as well as thermal and dimensional stability.

The patent literature also provides information on means of achieving these goals. As Table 1.36 indicates, the blending effect in most case is nonspecific. Therefore, the data in Table 1.36 should only be considered as a general guide to blending. The fact that two polymers with desired properties are immiscible should, however, not be a deterrent since the modern compatibilization and reactive processing methods (discussed later) can overcome such problems.

The main reason for polymer blending is economy. If a material can be produced at a lower cost with properties meeting specifications, the manufacturer must use it to remain competitive. The main and most difficult task in polyblend production is the development of materials with a full set of desired properties. This is usually achieved by selecting blend components in such a way that the principal advantages of one polymer will compensate for deficiencies of the second and vice versa. For example, the disadvantages of PPO with regard to processability and impact strength (see Table 1.37) are compensated by advantageous properties of either PA or HIPS. Due to PPO/PS miscibility, the original Noryls were formulated as PPO/HIPS polyblend.

The reactive methods of compatibilization developed subsequently allowed the second generation Noryl (a blend of PPO with PA) to be developed. The compositions claimed usually cover 30–70% of each of the main ingredients, PPO and PA, with additionally up to four parts of such modifier as polycarboxylic acid, trimellitic anhydride acid chloride, quinine, oxidized polyolefin wax, and so on. In most cases, PA forms a matrix with spherical inclusions of PPO acting as compatibilized low-density filler.

Similarly the disadvantages of PC are the stress cracking and chemical sensitivity. Stress cracking can be treated as a part of impact properties and a simple solution may thus be addition of ABS or ASA. On the other hand, to improve the solvent resistance—a property that is particularly important in automobile applications—a semicrystalline polymer may be added. From Table 1.37, it is apparent that TPEs (e.g., PBT, PET) could provide that property, but they also lack warp resistance and impact strength. Hence an ideal blend for automobile application based on PC and TPEs should be impact modified with, for example, an acrylic latex copolymer. A schematic of preparation of this type of toughened blend introduced by GEC-Europe in 1979 under the tradename Xenoy is shown in Figure 1.41.

As exemplified by PC/TPE/latex system, modern blends are increasingly required to play multiple roles so as to provide a balance of such diverse properties as mechanical behavior, chemical/solvent resistance, dimensional stability, paintability, weatherability, and, of course, economy. Such a complex balance is usually achieved by means of mulicomponent blending often with unavoidable compromises.

1.3.12.3 Compatibilization of Polymers

Blends made from incompatible polymers are usually weak. These poor blends are a result of high interfacial tension and poor adhesion between the two phases. For many years, blending of polymers was unsuccessful due to the fact that many polymers were incompatible. Lately new polymer blends have been successfully made by the use of compatibilizing agents. This has yielded polymers of unique properties not attainable from either of the polymer components of the blends. Sometimes these blends

TABLE 1.36 How to Modify Properties by Blending

Property	Matrix Resin	Modifying Polymer
Impact strength	PVC, PP, PE, PC, PA, PPE, TPE	ABS, ASA, SBS, EPR, EPDM, PBR, SAN, SMA, MBA, polyolefin, HIPS
HDT, stiffness	PC, PA	TPEs, PEI, PPE,
	ABS, SAN	PC, PSO
Flame retardancy	ABS, acrylics	PVC, CPE
	PA, PC	Aromatic-PA, PSO, copolysiloxanes or phosphazanes
Chemical/solvent resistance	PC, PA, PPO	TPEs
Barrier properties	Polyolefins	PA, EVOH, PVCl₂
Processability	PPO	Styrenics
	PET, PA, PC	PE, PBR, MBS, EVOH
	PVC	CPE, acrylics
	PSO	PA
	PO	PTFE, SI

For abbreviations see Appendix A2.
Source: Utracki, L. A. 1989. *Polymer Alloys and Blends.* Hanser Publishers, Munich, Germany.

Industrial Polymers

TABLE 1.37 Advantages and Disadvantages of Some Engineering Polymers and Modifiers

Polymer	Advantages in	Disadvantages in
Polyamide (PA)	Processability, impact strength, crystallinity	Water absorption, HDT
Polycarbonate (PC)	Low-temperature toughness, HDT	Stress-crack sensitivity, solvent and chemical resistance
Polyoxymethylene (POM)	Tensile strength, modulus	Stress-crack sensitivity, impact strength
Polyphenylene oxide (PPO)	HDT, rigidity, flame retardancy	Processability, impact strength
Thermoplastic polyesters (PET, PBT)	Chemical and solvent resistance	Shrinkage, low-temperature toughness, processability
High-impact polystyrene (HIPS)	Processability, impact strength	HDT
Acrylonitrile–butadiene–styrene copolymer (ABS or ASA)	Impact strength, processability, weatherability	HDT

Source: Utracki, L. A. 1989. *Polymer Alloys and Blends.* Hanser Publishers, Munich, Germany.

are preferred over completely miscible blends since they may combine important characteristics of both polymers. Their performance greatly depends on the size and morphology of the dispersed phase.

Polymers can be compatibilized in a number of ways. Polymers that act as mutual solvents can be used as compatibilizing agent, e.g., polycaprolactone in a blend of polycarbonate and poly(styrene-*co*-acrylonitrile).

Polymers can also be compatibilized by the use of functionalized polymers. Thus polymers can be post-reacted with a reactive monomer to form functional groups that react with a second polymer to form grafts in a blend, e.g., EPDM-*g*-maleic anhydride/nylon-6,6 blends are produced by grafting maleic anhydride onto EPDM, followed by blending with nylon-6,6. The grafted maleic anhydride reacts with the terminal—NH_2 groups of the nylon to form the following copolymer:

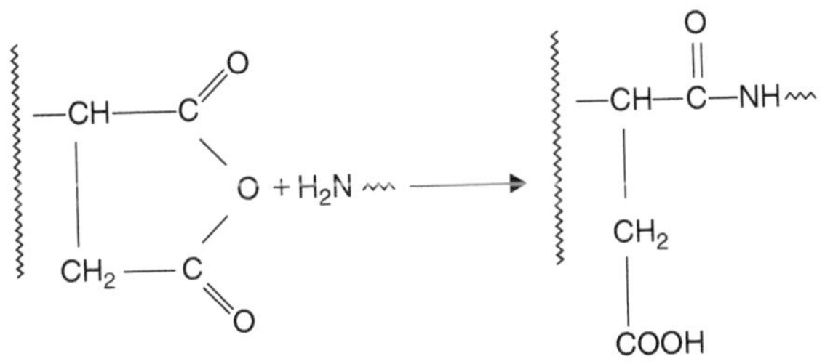

This method allows us to generate the compatibilizer in situ during blending (e.g., by adding benzoyl peroxide and maleic anhydride to EPDM followed by nylon in a twin screw extruder) and is often called *ractive belending* or reactive *compatibilization*. Block or graft copolymers can also be added separately to produce blends with controlled particle sizes. This type of compatibilizing agent is called an interfacial agent since it acts at the boundary between two immiscible polymers.

An interfacial compatibilizing agent functions as surfactant, somewhat the same as low molecular weight surfactant used in emulsion polymerization. It is usually a block or graft copolymer and is located at the interface between two polymers (Figure 1.42). It reduces interfacial tension, alleviates gross segregation, and promotes adhesion. The effective concentrations are from 0.1% to 5% depending on the desired particle size and the effectiveness of the compatibilizing agent.

The block or graft copolymer used as a compatibilizing agent must have the ability to separate into their respective phases and not be miscible as a whole molecule in one phase. The best choice of block or graft copolymers would be to choose polymers identical to the blended polymers, e.g., poly(A-*b*-B) or poly(A-*g*-B) copolymers for blending poly(A) and poly(B). Another alternative is to use poly(C-*b*-D) or poly(C-*g*-D) where C and D are compatible with A and B, respectively or where C and D adhere to A and B, respectively, but are not miscible with them.

Examples of such compatibilized systems that have been studied include EPDM/PMMA blends compatibilized with EPDM-*g*-MMA, polypropylene/polyethylene blends with EPM or EPDM, polystyrene/nylon-6 blends with polystyrene/nylon-6 block copolymer, and poly(styrene-*co*-acrylonitrile)/poly(styrene-*co*-butadiene) blends with butadiene rubber/PMMA block copolymer.

Reactive compatibilizing agents of the type A–C can also compatibilize an A/B blend as long as C can chemically react with B. Such studied systems include polyethylene (PE)/nylon-6 blends compatibilized with carboxyl functional PE, polypropylene (PP)/poly(ethylene–terephthalate) with PP-*g*-acrylic acid, nylon-6,6/EPDM with poly(styrene-*co*-maleic anhydride), and nylon-6/PP with PP-*g*-maleic anhydride.

In several commercial polymer blends, "modifiers" that also act as compatibilizing agents are used. The modifier is usually an interactive copolymer containing a rubbery component. Acrylic-based copolymers, chlorinated polyolefins, ethylene–propylene-diene, poly(ethylene-*co*-vinyl acetate), etc., are frequently used. These play a dual role, compatibilizing and toughening the blend. For this reason, they are used at much higher loading than pure compatibilizers. Thus, while 0.1–5 wt% of the latter may be sufficient, 10–40 wt% of a modifier may be needed (see Figure 1.41).

Co-reaction of blends to improve the performance has for decades been a practice in the rubber industry. In high-shear mixers some of the chains in rubbers are broken and are reformed by the free-radical mechanism. A similar phenomenon occurs during intensive mixing of polyolefins (see Table 1.34). To enhance this process, sometimes a free-radical source, e.g., peroxides, can be added.

1.3.12.4   Industrial Polyblends

The automotive sector is an important area for polymer blends. Alloys of polycarbonate (PC) and poly(butylenes terephthalate) (PBT), which combine the high impact strength of PC with the resistance of gasoline and oil and processing ease of PBT, are being increasingly used in car bumpers. General Electrics PC/PBT alloy is used for bumpers in several cars made in the United States and Europe. Bayer's PC/PBT blends, called Makrolon PR, are used for several automobile impact-resistant parts. Makrolon EC900 is also used for crash helmets.

An important requirement for plastics materials in the automotive industry is paintability alongside metal components in high temperature areas. Noryl GTX series of General Electric has been developed to meet this requirement. The blend consisting of polyphenylene oxide (PPO) in a nylon matrix combines the heat and dimensional stability and the low water absorption of PPO with the flow and semicrystalline properties of nylons. Its impact strength, however, is not as high as PC/PBT blends.

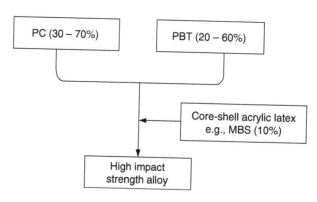

FIGURE 1.41   Schematic of preparation of polycarbonate/poly(butylenes terephthalate) blend. (After Utracki, L. A. 1989. *Polymer Alloys and Blends.* Hanser Publishers, Munich, Germany.)

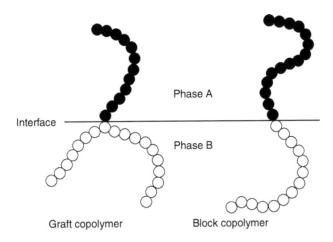

FIGURE 1.42 Ideal location of block and graft copolymers at the interface between polymer phases A and B in a polyblend.

Several grades of Noryl GTX are available, some of these being glass-filled. The unfilled grades are used for automobile exterior components such as fenders, spoilers and wheeltrims, whereas the glass-reinforced grades are used in under-the-bonnet applications, such as cooling fans, radiator end caps, and impeller housings, where in addition to high heat resistance extra stiffness is required.

Bayer's Bayblend PC/ABS alloys are used in dashboard housings, interior mirror surrounds, and steering column covers. The high stability of the alloy allows pigmentation in pastel colors.

Some alloys of PPO have been developed specifically for office equipment and housings. Noryl AS alloys, which are blends of PPO/PS and special fillers, are expected to replace metals in such applications as structural chassis for typewriters and printers.

Monsanto Chemical Co. produce injection molded Nylon/ABS alloys (Triax) that are characterized by good toughness, high chemical resistance, and easy moldability. Applications of these alloys include industrial power tools, lawn and garden equipment housing and handles, gears, pump impellers, and car fascia.

Bayer and BASF have developed a PC/ASA blend where ASA is the rubber variant of ABS in which the polybutadiene phase is completely replaced by a polyacrylate rubber. Compared to PC/ABS alloys, the PC/ASA blend has greater resistance to weathering, which permits outdoor applications and an extra heat stability that reduces the risk of yellowing in processing. BASFs alloy, Terblend S, is recommended for outdoor applications as it has six times the UV resistance of PC/ABS and an even higher resistance when pigmented.

A number of polyphenylene ether (PPE) blends have been introduced by Borg-Warner, Hulls, and BASF. These blends have the advantage that they can be made flame retardant without halogens. Borg-Warner offers many grades of PPE/high impact polystyrene (HIPS) blend under the trade name Prevex. The materials are characterized by high impact resistance and high heat distortion temperature. Applications are for office equipment, business machines, and telecommunication equipment. Luranyl PPE/HIPS blends of BASF include general purpose types, reinforced grades, impact modified types, and flame-proof blends. BASF also has a PPE/nylon blend tradenamed Ultranyl, which is aimed at a similar market.

Blends have also been used to a limited extent in packaging. Du Ponts Solar is a nylon/polyethylene blend that can be blow-molded into bottles and can contain volatile hydrocarbons found in household cleaners and agricultural chemicals. Further developments in such barrier blends are expected.

A number of commercial blends are listed in Appendix A5 of *Plastics Fabrication and Recycling* indicating their compositions, properties, and typical uses.

1.3.12.5   Nanoblends

For most polymer blends processed under typical extrusion conditions, the particle size of the dispersed phase is rarely below 0.1 μm (or 100 nm), whatever the compatibilization method employed [71]. Recently, however, a novel extrusion process involving an in situ polymerizing and compatibilizing system has been reported [72] for the preparation of nanostructured blends (*nanoblends*). [By nanoblends it is meant that the scale of dispersion of one polymer phase in the other is below 100 nm.] The method essentially consists of polymerizing a monomer of polymer A in the presence of polymer B, while a fraction of polymer B chains bears initiating sites at the chain ends or along the chain backbones from which polymer A chains can grow. In the process, polymer A and a graft or block copolymer of A and B are formed simultaneously, leading to in situ polymerized and in situ compatibilized A/B polymer blends [73]. The feasibility of such a process has been demonstrated by synthesizing nanoblends of polypropylene (PP)/polyamide-6 (PA-6) [72].

In a typical method to synthesize PP/PA-6 nanoblends, the monomer ε-caprolactam (ε-CL) is polymerized in the matrix of PP, a fraction of the latter having 3-isopropenyl-α,α-dimethylbenzene isocyanate (TMI) grafted on it to act as centers for initiating PA-6 chain growth. As such, the formation of PA-6 homopolymer and a graft copolymer of PP and PA-6 takes place simultaneously in the matrix of PP, leading to compatibilized PP/PA-6 blends, with the dispersed phase (PA-6) particle size between 10 and 100 nm that cannot be achieved otherwise by melt-blending pre-made PP and PA-6. The underlying chemistry of the process is shown in Figure 1.43.

Anionic polymerization of ε-CL in the presence of a catalyst (sodium caprolactamate, NaCL) and an activator (such as an isocyanate) yields PA-6. In the reaction scheme shown in Figure 1.43, the TMI-grafted PP (denoted by PP-*g*-TMI) acts as the macroactivator as it has isocyanate groups hanging from the polymer chain. (It should be noted that, unlike classical isocyanates, TMI with an isocyanate group is sufficiently stable under extrusion conditions involving high temperature and exposure to moisture.) Basically, polymerization starts with the formation of an acyl caprolactam as a result of the reaction between $-N=C=O$ and ε-CL (reaction 1). This acyl caprolactam then reacts readily with NaCL forming a new reactive sodium salt (reaction 2). As the latter initiates the polymerization of ε-CL, the catalyst (NaCL) is regenerated (reaction 3). Repetition of reactions 2 and 3 finally leads to the formation of a PA-6 chain grafted to PP (propagation reactions 4 and 5). Similar steps involving a microactivator, such as $R-N=C=O$, also result in the formation of PA-6 homopolymer. This simultaneous formation of PA-6 and graft copolymer of PP and PA-6 in the PP matrix, when the system PP-*g*-TMI/ε-CL/NaCL/$R-N=C=O$ is used, results in a well stabilized PP/PA-6 nanoblend. As compared to this, if ε-CL is polymerized only in the presence of PP, as in the system PP/ε-CL/NaCL/$R-N=C=O$, a highly immiscible mixture of PP and PA-6 will be obtained with morphology quite similar to that obtained by simply blending PP and PA-6 polymers. PP/PA-6 nanoblends cannot also be obtained by reactive blending in a system such as PP-*g*-MA (maleic anhydride grafted PP)/PA-6.

The mechanism of copolymer formation in the PP-*g*-TMI/ε-CL/NaCL/$R-N=C=O$ system is quite different from that in the PP-*g*-MA/PA-6 system. In the latter, the interfacial region surrounding a PA-6 particle is composed of the PP-*g*-MA and PA-6 bearing a terminal amine group (see Figure 1.44a). Their reaction at the interfaces leads to the formation of a graft copolymer PP and PA-6. The amount of copolymer formed by this method is limited primarily by the total interfacial volume [74], which is very small compared to the bulk. The situation, however, is very different for the PP-*g*-TMI/ε-CL/NaCL/$R-N=C=O$ system. Being immiscible with the PP-*g*-TMI, the mixture of ε-CL/NaCL/$R-NC=O$ is in the form of fine droplets in the PP-*g*-TMI. The interfacial region surrounding a droplet of ε-CL/NaCL/$R-N=C=O$ is PP-*g*-TMI/ε-CL/NaCL (see Figure 1.44b). The polymerization of ε-CL occurs in these droplets leading to PA-6 particles, which are stabilized by a layer of the graft copolymer resulting from the polymerization of PP-*g*-TMI/ε-CL/NaCL in the interfacial region. In this case, the problem of interfacial volume is no longer important for copolymer formation which depends largely on the capacities of the PP-*g*-TMI (macroactivator) and $R-N=C=O$ (microactivator) to initiate ε-CL. If the amount of the microactivator is zero, one would obtain a pure PP-*g*-PA-6 graft copolymer. It is the capacity of the in situ

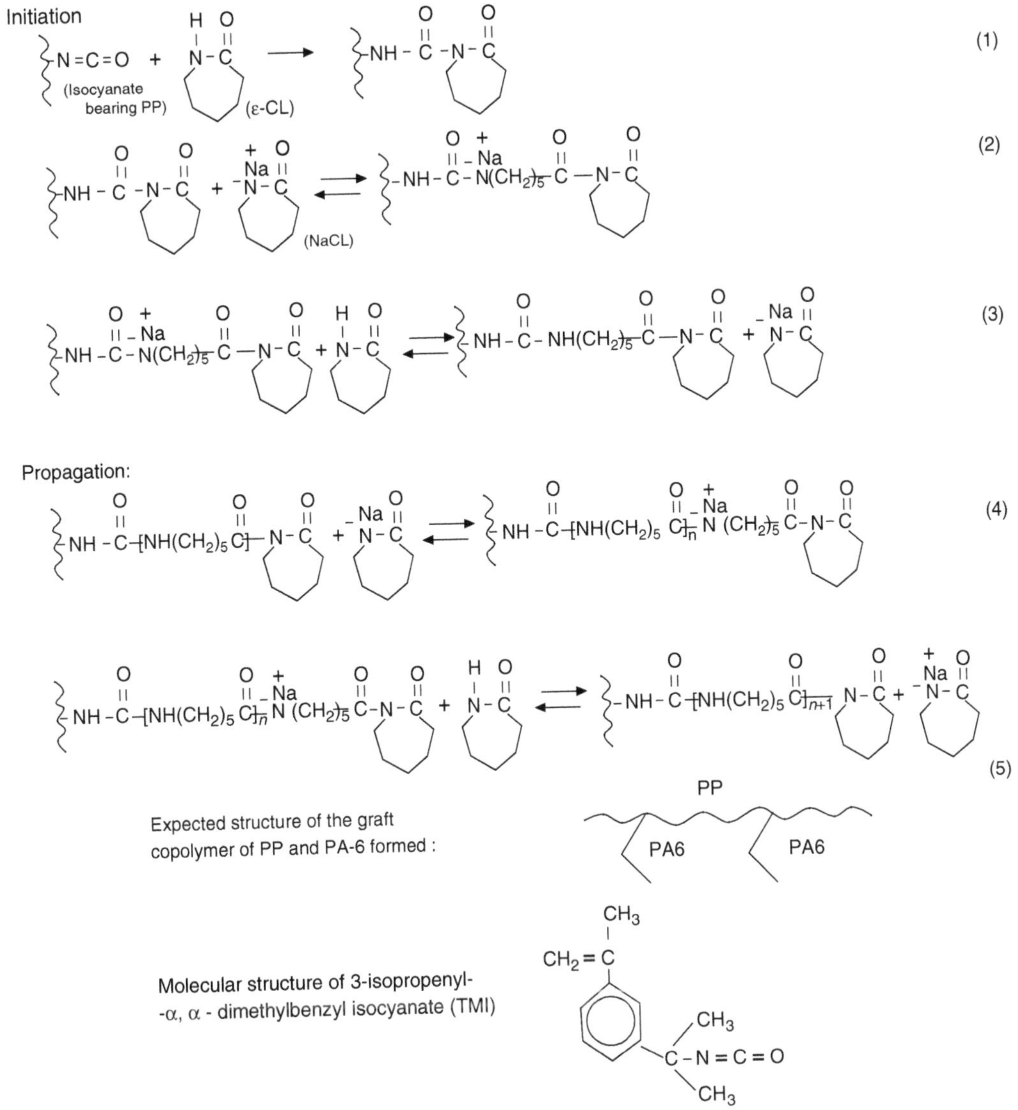

FIGURE 1.43 Schematic description of the mechanism of formation of a graft copolymer of PP and PA-6 using an isocyanate bearing PP, such as TMI-grafted PP (PP-*g*-TMI) as macroactivator. The mechanism of polymerization of ε-CL can be described by the same sequence of steps 1–5 with the macroactivator replaced by a microactivator R–N=C=O. (After Hu, G. H., Cartier, H., and Plummer, C. 1999. *Macromolecules*, 32, 4713. With permission.)

polymerization and in situ compatibilization process to generate very high amounts of copolymer that allows the formation and stabilization of nanodispersion [72].

1.3.13 Interpenetrating Polymer Networks

An interpenetrating polymer network (IPN) has been defined [75,76] as "an intimate combination of two polymers both in network form, at least one of which is synthesized or cross-linked in the presence of the other." There are no induced covalent bonds between the two polymers, i.e., monomer A reacts only with other molecules of monomer A.

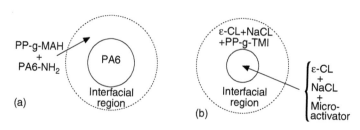

FIGURE 1.44 Comparison of the mechanism of copolymer formation in (a) conventional reactive blending of PP/ PA-6 system with PP-*g*-MAH as compatibilizer which reacts with –NH$_2$ end groups of PA-6 and (b) *in situ* polymerizing and compatibilizing system (see text). (After Hu, G. H., Cartier, H., and Plummer, C. 1999. *Macromolecules*, 32, 4713. With permission.)

A schematic representation of an ideal IPN (*full IPN*) in which both the polymers are cross-linked is shown in Figure 1.45a. If one of the two polymers is in network form (cross-linked) and the other a linear polymer, i.e., not cross-linked, the product is called a *semi-IPN* (Figure 1.45b).

An IPN is distinguished from simple polymer blends and block or graft copolymers in two ways: it swells but does not dissolve in solvents, and creep and flow are suppressed [75].

IPNs represent a third mechanism, in addition to mechanical blending and copolymerization, by which different polymers can be intimately combined. IPNs can be synthesized either by sequential polymerization or simultaneous polymerization of two monomers. In a sequential synthesis, monomer I is polymerized using an initiator and a cross-linking agent to form network I. Network I is then swollen in monomer II containing cross-linking agent and initiator to form network II by polymerization in situ. Such polymers are called *sequential interpenetrating networks* (SIPN).

Polymer networks synthesized by mixing monomers or linear polymers (prepolymers) of the monomers together with their respective cross-linking agents and catalysts in melt, solution, or dispersion, followed, usually immediately, by simultaneous polymerization by noninterfering modes, are called *simultaneous interpenetrating networks* (SIN). In the latter process, the individual monomers are polymerized by chain or stepwise polymerization, while reaction between the polymers is usually prevented due to different modes of polymerization.

Interpenetrating polymerization is the only way of combining cross-linked polymers. This technique can be used to combine two or more polymers into a mixture in which phase separation is not as extensive as it would be otherwise. Normal blending or mixing, it may be noted, result in a multiphase morphology unless the polymers are completely compatible. However, complete compatibility, which is almost impossible, is not necessary for complete phase mixing by interpenetrating polymerization, since permanent entanglements produced by interpenetration prevent phase separation.

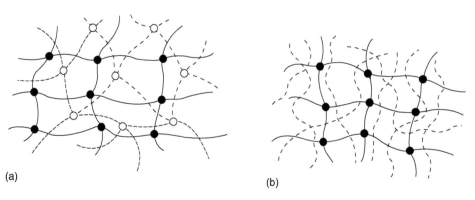

FIGURE 1.45 Schematic representation of (a) a full IPN and (b) a semi-IPN.

With moderate compatibility, intermediate and complex phase behavior results. Thus IPNs with dispersed phase domains have been reported ranging from a few micrometers (the largest) to a few hundred nanometers (intermediate) to those without a resolvable domain structure (complete mixing). With highly incompatible polymers, on the other hand, the thermodynamics of phase separation is so powerful that it occurs before it can be prevented by cross-linking.

Combining polymeric networks in different compositions often results in IPNs with controlled different morphologies showing synergistic behavior. For example, a glassy polymer (T_g above room temperature) combined with an elastomeric polymer (T_g below room temperature) gives a reinforced rubber, if the elastomeric phase is predominant and continuous, or a high impact plastic, if the glassy phase is continuous. More complete phase mixing enhances mechanical properties due to increased physical cross-link density. IPNs have thus been synthesized in various network compositions with optimum bulk properties such as tensile strength, impact strength, and thermal resistance.

1.3.13.1 Industrial IPNs

Commercial IPNs have been developed to combine useful properties of two or more polymer systems. For example, high levels of silicone have been combined with the thermoplastic elastomer (TPE) based on Shells Kraton styrene–ethylene/butadiene–styrene TPE and Monsanto's Santoprene olefin TPE. These IPN TPEs are said to provide the biocompatibility and release properties of silicone with tear and tensile strength up to five times greater than medical-grade silicone. Thermal and electronic properties and elastic recovery are also improved.

IPN TPEs offer physical and thermal properties of thermoset rubber, the processability of a thermoplastic, and a wider hardness range than available to other TPEs.

Rimplast materials from Petrarch are silicone/thermoplastic IPNs that combine the warpage and wear resistance properties of silicone with nylon, thermoplastic polyurethane, or styrene/butadiene block copolymers. The combination of properties these IPNs possess make them suitable for high quality, high tolerance gear and bearing applications that can benefit from the addition of internal lubrication and isotropic shrinkage of silicone to the fatigue endurance and chemical resistance of crystalline resin, while processability still remains the same.

References

1. Renfrew, A. and Morgan, P. eds. 1960. *Polythene—The Technology and Uses of Ethylene Polymers, 5th Ed.*, p. 255. Iliffe, London, UK.
2. Kresser, T. O. J. 1960. *Propylene*, Van Nostrand, New York.
3. Frank, H. P. 1966. *Polypropylene*, Macdonald, London, UK.
4. Van der Ven, S. 1990. *Polypropylene and Other Olefins*, Elsevier, Amsterdam.
5. Tusch, R. 1966. *Polym. Eng. Sci.*, 6, 3, 255.
6. Forsman, J. 1972. *Hydrocarbon Process.*, 51, 130.
7. Vandenburg, E. Repka, B. 1977. Ziegler type polymerizations. In *Encyclopedia of Polymer Science and Technology, Vol. 29*, Wiley, New York.
8. Huang, J. and Rempel, G. L. 1995. *Prog. Polym. Sci.*, 20, 459.
9. Kaminsky, W., Kulper, K., Britzinger, H. H., and Wild, F. R. W. P. 1985. *Angew. Chem. Int. Ed. Engl.*, 24, 507.
10. Schwank, D. 1993. *Modern Plastics*, p. 49. August 1993.
11. Sarvetnick, H. A. 1969. *Polyvinyl Chloride*, Van Nostrand Reinhold, New York.
12. Whelan, A. and Craft, J. L. eds. 1977. *Development in PVC Production and Processing*, p. 2375. Applied Science Publishers, London, UK.
13. Nass, L. I. 1976. *Encyclopedia of PVC, Vol 1 and 2*, Marcel Dekker, 1977.
14. Brydson, J. A. 1982. *Plastics Materials*, Butterworth Scientific, London, UK.

15. Ritchie, P. D., ed., 1972. Vinyl chloride and vinyl acetate polymers. In *Vinyl and Allied Polymers, Vol. 2,* Iliffe, London, UK.

16. Ritchie, P. D. ed. 1968. Aliphatic polyolefins and polydienes, In *Vinyl and Allied Polymers, Vol. 1,* p. 2375. Iliffe, London, UK.

17. Rudner, M. A. 1958. *Fluorocarbons,* Van Nostrand Reinhold, New York.

18. Wall, L. A. ed. 1972. *Fluoropolymers, High Polymer Series, Vol. 25,* p. 2375. Wiley Interscience, New York.

19. Boundy, R. H. and Boyer, R. F. 1952. *Styrene, Its Polymers, Copolymers and Derivatives,* Van Nostrand Reinhold, New York.

20. Brighton, C. A., Pritchard, G., and Skinner, G. A. 1979. *Styrene Polymers: Technology and Environmental Aspects,* Applied Science, London, UK.

21. Brydson, J. A. 1978. *Rubber Chemistry,* Applied Science, London, UK.

22. Morton, M. ed. 1973. *Rubber Technology,* Van Nostrand Reinhold, New York; Kaminsky, W. and Steiger, R. 1973. *Polyhedra,* 7, 22, p. 2375.

23. Basdekis, C. H. 1964. *ABS Plastics,* Van Nostrand Reinhold, New York.

24. Kinsey, R. H. 1969. Ionomers, chemistry and new developments. *Appl. Polym. Symp.,* 11, 77.

25. Beevers, R. B. 1968. *Macromol. Rev.,* 3, 113.

26. Kine, B. B. and Novak, R. W. 1985. Acrylic and methacrylic ester copolymers, In *Encyclopedia of Polymer Science and Engineering, Vol. 1,* J. I. Kroschwitz, ed., Interscience, New York; Rehberg, C. E. and Fisher, C. H. 1948. *Ind. Eng. Chem.,* 40, 143.

27. Wu, J., Lin, J., Zhou, M., and Wei, C. 2000. *Macromol. Rapid Commun.,* 21, 1032.

28. Coover, H. W. and McIntyre, J. M. 1985. 2-Cyanoacrylic ester polymers, In *Encyclopedia of Polymer Science and Engineering, Vol. 1,* J. I. Kroschwitz, ed., p. 161. Interscience, New York.

29. Finch, C. A. ed. 1973. *Polyvinyl Alcohol: Properties and Applications,* p. 161. Wiley, New York.

30. Goodman, I. and Rhys, J. A. 1965. *Polyesters, Vol. 1: Saturated Polymers,* Iliffe, London, UK.

31. Al-Ghatta, H., Cobror, S., and Severini, T. 1997. *Polym. Adv. Technol.,* 8, 161.

32. Karayannidis, G. P. and Psalida, E. A. 2000. *J. Appl. Polym. Sci.,* 77, 2206.

33. Awaja, F. and Pavel, D. 2005. *Eur. Polym. J.,* 41, 1453.

34. Parkyn, B., Lamb, F., and Clifton, B. V. 1967. *Polyesters, Vol. 2: Unsaturated Polyesters and Polyester Plasticizers,* Iliffe, London, UK.

35. Seymour, R. B. and Krishenbaum, G. S. eds. 1986. *High Performance Polymers: Their Origins and Development,* p. 981. Elsevier, New York, Chapters 8–10.

36. Schnell, H. 1964. *Chemistry and Physics of Poycarbonates,* Wiley Interscience, New York.

37. Carhart, R. O. 1985. *Polycarbonates, Engineering Thermoplastics,* Marcel Dekker, New York, Chapter 3.

38. Liu, C. F. and Itoi, H., (to GE). 2000. US Patent 6 043 310.

39. Levechik, S. V. and Weil, E. D. 2005. *Polym. Int.,* 54, 981.

40. Kohan, M. I. ed. 1973. *Nylon Plastics,* p. 201. Wiley, New York.

41. Nelson, W. E. 1976. *Nylon Plastics Technology,* Newnes-Butterworths, London, UK.

42. Black, W.B. and Preston, J. eds 1973. *High Modulus Wholly Aromatic Fibers,* Marcel Dekker, Newyork.

43. Mittal, K. L. 1984. *Poyimides: Synthesis, Characterization and Applications,* Plenum Press, New York.

44. Gould, D. F. 1959. *Phenolic Resins,* Van Nostrand Reinhold, New York.

45. Whitehouse, A. A. K., Pritchett, E. G. K., and Barnett, G. 1967. *Phenolic Resins,* Iliffe, London, UK.

46. Blais, J. F. 1959. *Amino Resins,* Van Nostrand Reinhold, New York.

47. Vale, C. P. and Taylor, W. G. K. 1964. *Aminoplastics,* Iliffe, London, UK.

48. Phillips, L. N. and Parker, D. B. V. 1964. *Polyurethanes: Chemistry, Technology and Properties,* Iliffe, London, UK.

49. Frisch, K. C. and Reegan, S. L. *Advances in Urethane Science and Technology*, Vol. 1, (1972), Vol. 2 (1973), Vol. 3 (1974), Vol. 4 (1976), Vol. 5 (1976), Vol. 6 (1978), Vol. 7 (1979), Technomics, Westport, Conn.

50. Oertal, G. 1985. *Polyurethane Handbook*, Hanser Publishers, Munich, Germany.

51. Dunnols, J. 1979. *Basic Urethane Foam Manufacturing Technology*, Technomics, Westport, CT.

52. Barker, S. J. and Price, M. B. 1970. *Polyacetals*, Iliffe, London, UK.

53. Bailey, F. E. and Koleske, J. V. 1976. *Poly(ethylene Oxide)*. Academic Press, New York.

54. Potter, W. G. 1970. *Epoxide Resins*, Iliffe, London, UK.

55. May, C. A. and Tanaka, Y. eds. 1973. *Epoxy Resins: Chemistry and Technology*, p. 201. Marcel Dekker, New York.

56. Yarsley, V. E., Flavett, W., Adamson, P. S., and Perkins, N. G. 1964. *Cellulosic Plastics*, Iliffe, London, UK.

57. Margolis, J. M. 1985. *Engineering Thermoplastics*, Marcel Dekker, New York.

58. May, R. 1988. Polyetheretherketones, In *Encyclopedia of Polymer Science and Engineering*, J I. Kroschwitz,, ed., p. 201. Wiley Interscience, New York.

59. Rochow, E. G. 1951. *Introduction to the Chemistry of Silicones*, Wiley, New York.

60. Noll, W. 1968. *Chemistry and Technology of Silicones*, Academic Press, London, UK.

61. Mason, J. A. and Sperling, L. H. 1976. *Polymer Blends and Composites*, Plenum Press, New York.

62. Olabisi, O., Robeson, L. M., and Snow, M. T. 1979. *Polymer–Polymer Miscibility*, Academic Press, New York.

63. Paul, D. R. and Neuman, S. eds. 1978. *Polymer Blends, Vols. 1 and 2*, p. 201. Academic Press, New York.

64. Utracki, L. A. 1989. *Polymer Alloys and Blends,* Hanser Publishers, Munich, Germany.

65. Cooper, S. L. and Estes, G. M., eds. 1977. Multiphase Polymers, *Adv. Chem. Ser.*, 176, 367. American Chemical Society, Washington, DC.

66. Schurer, J. W., de Boer, A., and Challa, G. 1975. *Polymer*, 16, 201.

67. Dondos, A. and Benoit, H. 1968. *Macromol. Chem.*, 118, 165.

68. Dondos, A. and Pierri, E. 1986. *Polym. Bull.*, 16, 567.

69. Brandrup, J. and Immergut, E. eds. 1975. *Polymer Handbook*, p. 149. Wiley, New York.

70. Frisch, H. I. 1985. *Br. Polym. J.*, 17, 149.

71. Nishio, T., Suzuki, Y., Kojima, K., and Kakugo, M. 1991. *J. Polym. Eng.*, 10, 123.

72. Hu, G.-H., Cartier, H., and Plummer, C. 1999. *Macromolecules*, 32, 4713.

73. Cartier, H. and Hu, G.-H. 2001. *Polymer*, 42, 8807.

74. Hu, G.-H. and Kadri, H. J. 1998. *J. Polym. Sci., Part B: Phys. Ed.*, 36, 2153.

75. Sperling, L. H. 1981. *Interpenetrating Polymer Networks and Related Materials*. Plenum Press, New York.

76. Klemner, D. and Berkowski, L. 1987. Interpenetrating polymer networks, In *Encyclopedia of Polymer Science and Engineering*, Vol. 8, H. F. Mark, N. M. Bikales, C. G. Overberger, and G. Menges, eds., P. Wiley Interscience, New York.

2

Polymers in Special Uses

2.1 Introduction

There are a number of polymeric materials that distinguish themselves from others by virtue of their limited use, high prices, or very specific application or properties. The expression specialty polymer is, however, slightly ambiguous to use for such materials as the definition covers any polymeric material that does not have high volume use. There are thus some materials that were originally developed as specialties have now become high volume commodities, while a number of materials developed some years ago still fall into the specialty category. Some examples of the latter are polytetrafluoroethylene, polydimethyl siloxane, poly(vinylidene fluoride), and engineering materials such as poly(phenylene oxide), poly(phenylene sulfide) (PPS), polyether sulfone, polyether ether ketone, and polyetherimide.

In this chapter, attention is focused on a number of polymers that are either themselves characterized by special properties or are modified for special uses. These include high-temperature and fire-resistant polymers, electroactive polymers, polymer electrolytes, liquid crystal polymers (LCPs), polymers in photoresist applications, ionic polymers, and polymers as reagent carriers and catalyst supports.

2.2 High-Temperature and Fire-Resistant Polymers

Compared to traditional materials, especially metals, organic polymers show high sensitivity to temperature [1–3]. Most importantly, they exhibit very low softening points, which is attributed to the intrinsic flexibility of their molecular chains. Thus, whereas most metals do not soften appreciably below their melting points, which may be 1000°C or higher, many polymers commonly used as plastics such as polyethylene, polystyrene, and poly(vinyl chloride), soften sufficiently by about 100°C to be of no use in any load-bearing applications.

The poor thermal resistance of common polymers has greatly restricted some of their application potential. In two particular application areas, namely electrical and transport applications, this restriction has long been particularly evident.

Owing to their unique electric insulation properties, polymers are widely used in electrical products. However, many electrical components are required to operate at high temperatures, for example, electric motors and some domestic appliances.

Another characteristic property of polymers, namely their high specific stiffness and strength (which are due to their low density, especially when used in fiber-reinforced composite materials), has led to the use of polymers in transport applications, especially in aerospace industries, where weight saving is of vital importance and materials cost is secondary. However, here again many applications also demand high temperature resistance.

Road vehicle manufacturers in their attempt to save weight (and hence reduce fuel consumption) have been replacing heavy metal components with light plastic ones. The ease of molding plastics into intricately shaped part has also been used to advantage for fabricating many under-the-bonnet products. Here again, resistance to elevated temperatures is often also needed, necessitating use of high-temperature resistant polymers.

Although electrical and transport applications have perhaps provided the biggest demand for thermally resistant specialty polymers, such polymers are also sought for use in more mundane consumer goods, especially appliances where exposure to elevated temperature can occur, such as hair dryers, toasters, and microwave ovens.

2.2.1   Temperature-Resistant Polymers

To measure the thermal stability of polymers, one must define the thermal stress in terms of both time and temperature. An increase in either of these factors shortens the expected lifetime. In general terms, for a polymer to be considered thermally stable, it must retain its physical properties at 250°C for extended periods, or up to 1000°C for a very short time (seconds). As compared to this, some of the more common engineering thermoplastics such as ABS, polyacetal, polycarbonates, and the molding grade nylons have their upper limit of use temperatures (stable physical properties) at only 80°C–120°C.

The principal ways to improve the thermal stability of a polymer are to increase crystallinity, introduce cross-linking, increase inherent stiffness of the polymer chain, and remove thermoxidative weak links. Although cross-linking of oligomers is certainly useful and does make a real change in properties (see Chapter 1 of *Plastics Fundamentals, Properties, and Testing*), crystallinity development has limited application for very high temperatures, since higher crystallinity results in lower solubility and more rigorous processing conditions. Chain stiffening or elimination of weak links is a more fruitful approach.

The weakest bond in a polymer chain determines the overall thermal stability of the polymer molecule. The aliphatic carbon–carbon bond has a relatively low bond energy (see Table 2.1). Oxidation of alkylene groups is also observed during prolonged heating in air. Thus the weak links to be avoided are mostly those present in alkylene, alicyclic, unsaturated, and nonaromatic hydrocarbons. On the other hand, the functions proven to be desirable are aromatic (benzenoid or heterocyclic) ether, sulfone, and some carboxylic acid derivatives (amide, imide, etc.). Aromatic rings in the polymer chain also give intrinsically stiff backbone.

It follows from this reasoning that aromatic polymers will have greater thermal stability. For example, poly(*p*-phenylene) synthesized by stereospecific 1,4-cyclopolymerization of cyclohexadiene, followed by dehydrogenation:

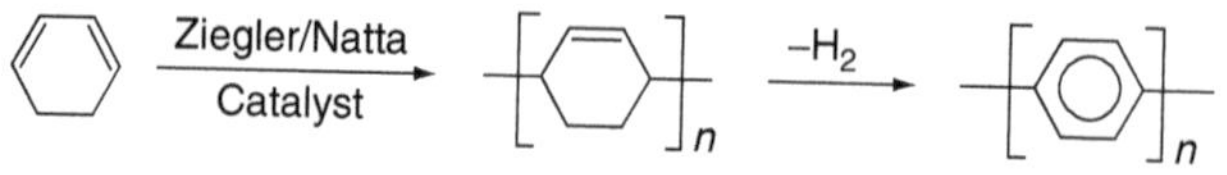

is infusible and insoluble.

TABLE 2.1   Bond Energies of Common Organic and Inorganic Polymers

Bond	Bond Energy	
	(kcal/mol)	(kJ/mol)
C_{al}–C_{al}	83	347
C_{ar}–C_{ar}	98	410
C_{al}–H	97	405
C_{ar}–H	102	426
C–F	116	485
B–N	105	439
Si–O	106	443

Another example of aromatic polymer is poly (*p*-xylene), which is usually vacuum deposited as thin films on substrates. In one application the monomer is di-*p*-xylylene formed by the pyrolysis of *p*-xylylene at 950°C in the presence of steam. When this monomer is heated at about 550°C at a reduced pressure, a diradical results in the vapor phase, which, when deposited on a surface below 70°C, polymerizes instantaneously and forms a thin, adherent coating.

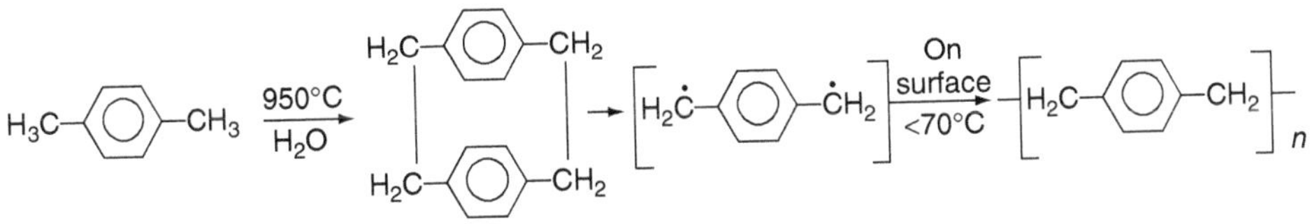

Metals or other substrates can be coated in this way. A major application has been the production of miniature capacitors that have the polymer as a dielectric.

Commercial polymers based on the principle of synthesis of polyaromatic compounds include the previously discussed commercial polymers—aromatic polyamides, polyimides, poly(phenylene oxide), polysulfone, and polybenzimidazole (see Chapter 1).

Another approach to achieve thermal stability is to synthesize ladder polymers, so called because of their ladderlike structure ( ). For example, pyrolysis of polyacrylonitrile gives a ladder polymer of high thermal stability:

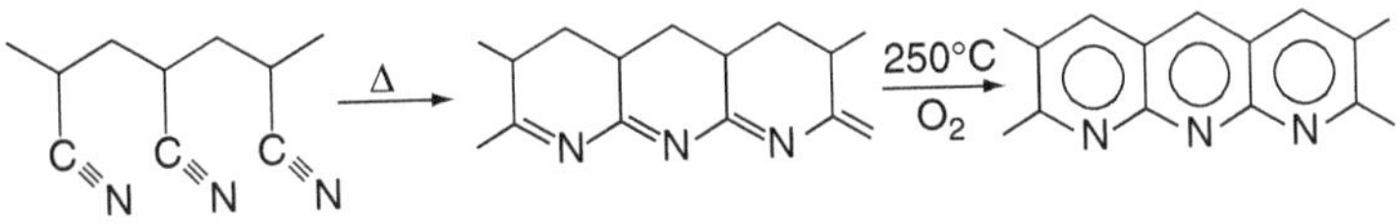

The product (black orlon) is so stable that it can be held directly in a flame in the form of woven cloth and not be changed physically or chemically. Further heating of black orlon to 1400°C–1800°C and simultaneous stretching produces graphite HT. If the heating and stretching is conducted at 2400–2500°C, the high modulus graphite HM is obtained. Other carbonizable polymers that produce carbon fibers on heating include poly(vinyl acetate), poly(vinyl chloride), poly(vinylidene chloride), and cellulose. Thermosets, such as phenolic resins, generally produce nongraphitizing or glassy carbon.

Ladder polymers are also produced by polycondensation reactions of tetrafunctional monomers. If a tetrafunctional monomer is reacted with a bifunctional monomer, as in the formation of polyimides, the derived polymer is referred to as a partial ladder or stepladder polymer.

If two tetrafunctional monomers are used, as in the formation of polyquinoxaline (PQ) from an aromatic tetramine and an aromatic tetraketone, the resulting polymer is a ladder polymer:

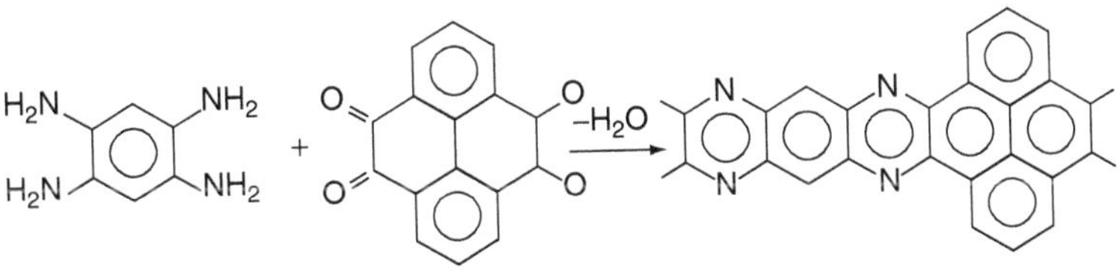

Other tetraketones have also been used in the preparation of PQs. The PQs have proven to be one of the better high temperature polymers with respect to both stability and potential application. The PQs are also one of the three most highly developed systems—the others being benzimidazoles and oxadiazoles. The interest in the PQs increased considerably after the development in 1967 of the soluble phenylated polyquinoxaline (PPQ). PQs are stable to 550°C and are used for high-temperature composites and adhesives.

2.2.2 Fire-Resistant Polymers

To understand the burning of polymers better, the combustion region may be divided into five zones, as shown in Figure 2.1. The first zone is defined by the polymer layer where pyrolysis takes place due to heat produced in the flame, but very little oxidation takes place. Thermal oxidation takes place in the second superficial zone. In the third zone, which is a gaseous one, low-molecular-weight products formed in these two zones are mixed with heated air and are decomposed or oxidized by oxygen or by radicals coming out of the flame.

The fourth zone begins where the concentration of degradation products is sufficient for the flame to grow, emitting light radiation. This is where most of the thermal energy is released, and the temperature reaches it maximum value. The fifth zone, also called the post-combustion zone, is where oxidation reactions terminate and combustion products are found in maximum concentration. The thermal energy produced in this zone and in the fourth zone, together with the light radiation, are responsible for the degradation that occurs in the inner part of the polymer not touched by the flame.

It is evident from the above discussion that the burning of polymers may be considered to consist of three principal stages, namely, (1) thermal (nonoxidative) decomposition of the polymer, (2) evolution and transport of volatiles from the burning surface to the flame, and (3) high temperature (600°C–900°C) oxidation of the volatiles in the flame itself. Interruption of the processes occurring at any of the above three stages will reduce polymer flammability. Thus, polymers with good thermal stability that are, therefore, resistant to thermal decomposition also offer good fire resistance.

Although, because many factors are involved, no single test can provide an adequate picture of the fire hazards of polymers, the limiting oxygen index (LOI) is the single most widely used test of polymer flammability. The test (ASTM D-2863) is made in a very simple way. A small rod of a plastic material is burned in a Pyrex tube in the presence of oxygen and nitrogen (see Figure 2.83 of *Plastics Fundamentals, Properties, and Testing*). The tubing is connected to flowmeters and the minimum amount of oxygen, in an oxygen–nitrogen mixture, required to just sustain burning is determined. The LOI is defined by

$$LOI = \frac{[O_2]}{[O_2] + [N_2]} \times 100$$

Thus a smaller LOI value indicates a greater flammability. LOI values for some common and not so common polymers are shown in Table 2.2.

The polymers listed in Table 2.2 have been classified into non-self-extinguishing and self-extinguishing types depending on their LOI values.

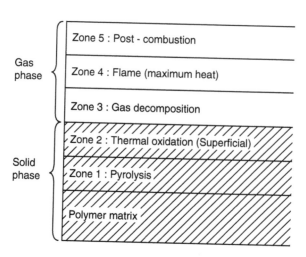

FIGURE 2.1   Schematic representation of the combustion region of a burning polymer.

TABLE 2.2 Limiting Oxygen Indices (LOI) of Polymers

Non-Self-Extinguishing Polymer	LOI	Self-Extinguishing Polymer	LOI
Polyoxymethylene	16	Polycarbonate	27
Polyethylene	17	Polyarylate	34
Polypropylene	18	Polyethersulfone	38
Polystyrene	18	Polyether ether ketone	40
Poly(methyl methacrylate	18	Poly(vinyl chloride), rigid	42
Polyisoprene (natural rubber)	18	Polyamide-imide	43
ABS	18	Poly(phenylene sulfide)	44
Nylon-6,6	24	Polyether-imide	47
Poly(ethylene terephthalate)	25	Polypyromellitimide	50
Polychloroprene	26	Poly(vinylidene chloride)	60
		Polytetrafluoroethylene	95

The LOI value for self-extinguishing behavior is often taken as 27, not 21 (which is the volume % concentration of oxygen in air), to correct for a lack of convective heating in the LOI test.

It is seen that aromatic polymers have low flammability and are self-extinguishing, which reflects their much greater thermal stability compared with aliphatic polymers. It is because of this property that aromatic polymers are often specified in fire-hazardous electrical insulation applications or for fire-resistant textiles. In contrast, most of the common polymers, with the exception of PVC, are non-self-extinguishing and flammable.

The presence of halogen in a polymer produces the most dramatic increase in LOI and decrease in flammability. Thus, rigid (unplasticized) PVC, with a LOI value of 42, has the lowest flammability of any common and many not so common polymers. This advantage is, however, offset by the large amount of smoke and toxic gas (hydrochloric acid) produced when PVC and other chloropolymers burn. Moreover, plasticization of PVC markedly reduces the LOI; for example, the value decreases to 25 when 50 phr dioctylphthalate is used as a plasticizer.

Fluoropolymers also low flammability. The fully fluorinated polymer polytetrafluoroethylene, for example, burns only in 95% oxygen under the test conditions of LOI. The burning, however, produces a highly toxic and corrosive gas hydrogen fluoride.

2.3 Liquid Crystal Polymers

Ordinarily a crystalline solid melts sharply at a single, well-defined temperature to produce a liquid phase that is amorphous and isotropic. A different behavior is exhibited by a class of organic compounds known as liquid crystals. The oldest examples are cholesterol derivatives, e.g., cholesteryl benzoate. This substance, for instance, does not have a sharp transition to amorphous liquid at 145.5°C, but changes to a cloudy liquid, which becomes clear and isotropic only at 178.5°C. This cloudy intermediate state that possesses an ordered structure with some resemblance to a crystalline solid, while still in the liquid state, is called a mesophase or mesomorphic phase from the Greek mesos, meaning in between or intermediate.

Liquid crystal materials can thus be defined as those which are characterized by the appearance of mesophases between the crystalline solid and isotropic liquid phases. Many organic compounds with this property have been synthesized. The temperature range of stability of a mesophase may be anything up to 150°C.

Liquid crystals can be divided into two main classes; those forming liquid crystalline phases in the melt are called thermotropic and those forming liquid crystalline phase in solution are referred to as lyotropic. Depending on the type of molecular order in the mesophase these classification can be broken down further into three main categories: smectic, nematic, and cholesteric [4–7].

Smectic liquid crystals consist of elongated molecules that line up with the long axes of the molecules aligned in one direction and the ends of the molecules lying on parallel planes to produce a type of layered structure (Figure 2.2a) like that in a layered box of cigars. A layer of smectic molecules is just one molecule (longitudinally) thick. The molecules in nematic liquids are similar in shape to smectic molecules and also point in one direction, but unlike the latter they do not line up with the ends lying on parallel planes (Figure 2.2b).

Cholesteric liquid crystals consist of long flat molecules that line up in the same manner as nematic molecules with molecular long axes parallel to each other in a plane. However, the molecules in one plane are slightly displaced (due to side groups of molecules) with respect to those in the neighboring planes to form a helical pattern, as indicated Figure 2.2c. Since the cholesteric structure is basically a derivative of the nematic, a transition from the cholesteric structure to the nematic structure can be effected by using a magnetic or electric field to unwind the helical configuration of the former.

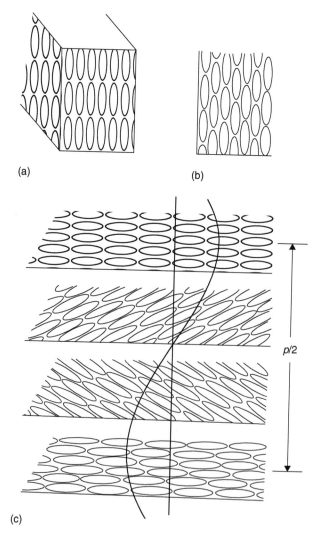

FIGURE 2.2 Liquid crystals. (a) Smectic crystals: the ends of the molecules are on a plane. (b) Nematic crystals: the ends of the molecules do not match. (c) Cholesteric crystals: the molecules in each layer are arranged in a manner similar to nematic crystals, but the angle changes from plane to plane of the molecules, forming a helix of pitch length p.

Liquid crystals have interesting electro-optical properties. When subjected to small electric fields, reorientation and alignment of the liquid crystal molecules takes place, which produces striking optical effects because light travels more slowly along the axes of the molecules than across them. This has led to their use in optical display devices for electronic instruments such as digital voltmeters, desk calculators, clocks, and watches. Nematic liquid crystals are most commonly used in these applications. Cholesteric materials are added to provide memory effects.

Molecules that have a tendency to form liquid crystalline phase usually have either rigid, long rod-like shapes with a high length to breadth (aspect) ratio, or disc-shaped molecular structures. These rigid groups, referred to as mesogens, may be chemically composed of a central core comprising aromatic or cycloaliphatic units joined by rigid links, and having either polar or flexible alkyl and alkoxy terminal groups.

Polymers exhibiting liquid crystalline properties can be constructed from these mesogens in three different ways: (1) incorporation into chain-like structures by linking them together through both terminal units to form main-chain LCPs; (2) attachment through one terminal unit to a polymer backbone to produce a side-chain comb-branch structure; and (3) a combination of both main- and side-chain structures. LCPs typically show either smectic or nematic liquid crystal behavior. A schematic diagram of the two main phase types is shown in Figure 2.3.

The liquid crystalline phases in polymeric materials are sometimes difficult to identify unequivocally. However, several techniques can be used that provide information on the nature of the molecular organization within the phase, and if used in a complimentary fashion these can provide reliable information on the state of order of the mesogenic groups.

While differential scanning calorimetry is widely used as a means of detecting the temperatures of thermotropic mesophase transitions, the phases can be identified by observing the characteristic textures developed in thin layers of the polymer when viewed through a microscope using a linearly polarized light source. X-ray diffraction can be used to characterize the mesophases and to provide reliable information on the number of phases present.

The type of phase formed in a polymer liquid crystal can also be identified often by examining the manner in which it mixes with a small molecule mesogen of known mesophase type. If these textures are the same, then a mixed liquid crystal phase is formed with no observable transition between the two types of molecule.

In some polymer crystals, moreover, several mesophases can be identified. In main-chain LCPs there is usually a transition from the crystal to a mesophase, while in more amorphous systems where a glass

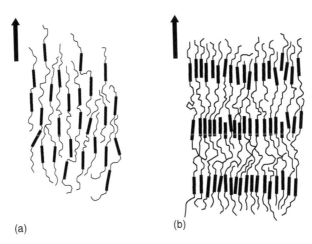

(a)   (b)

FIGURE 2.3 Schematic representation of (a) nematic phase and (b) smectic phase for main-chain liquid crystalline polymers, showing the director as the arrow. The relative ordering is the same for side-chain-polymer liquid crystals.

transition (T_g) is present, the mesophase may appear after this transition has occurred. In thermotropic system having multiphase transitions, the increase in temperature leads to changes from the most ordered to the least ordered states, i.e., crystal → smectic → nematic → isotropic (see Figure 2.4).

From the standpoint of polymer applications, two properties of LCPs are of major interest—the effect of order on polymer melt viscosity, and the ability of the polymer to retain its ordered configuration when cooled down to the solid state.

Among the first polymers observed to exhibit the aforesaid properties of LCPs were copolyesters (I) prepared from terephthalic acid, ethylene glycol, and *p*-hydroxybenzoic acid

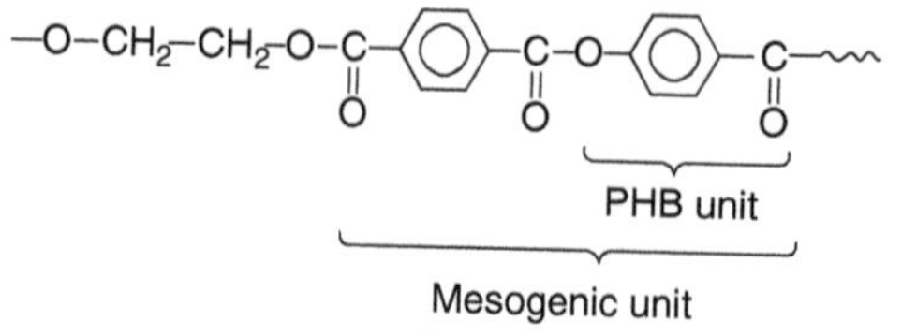

(I)

As the amount of *p*-hydroxybenzoate (PHB) units is increased, the polymer melt viscosity initially increases, which is expected because of the decreased flexibility caused by incorporation of the "rigid" PHB unit. At levels of 30 mol% PHB, however, the melt viscosity begins to decrease, reaching a minimum at about 60–70 mol%. This is shown in Figure 2.5a at three different shearing rates. Significantly, as the melt viscosity begins to decrease, the melt's appearance also changes from clear to opaque.

Both the decrease in viscosity and appearance of opaqueness arise from the onset of liquid crystalline morphology, which in turn is due to increased backbone rigidity. The rigid polymeric mesophases become aligned in the direction of flow, thus minimizing the frictional drag. Liquid crystal melts or solutions have, in consequence, lower viscosities than melts or solutions of random-coil polymers. The significance of this effect from the standpoint of polymer processing is obvious: the lower the viscosity, the more readily the polymer can be fabricated into a useful plastic or fiber.

An equally important observation for the above copolyester LCPs is that the ordered arrangement of polymeric mesophases in the melt is retained upon cooling, which is manifested in greatly improved mechanical properties (see Figure 2.5b). The liquid crystalline behavior is therefore advantageous from the standpoint of both processing and properties. Thermotropic liquid crystal copolyesters of structures similar to (I) are now available commercially.

Ordered behavior is also observed in solutions of some liquid crystal polymers (lyotropic LCPs). Unlike flexible polymers that assume a random coil conformation in solution, the rigid polymers being

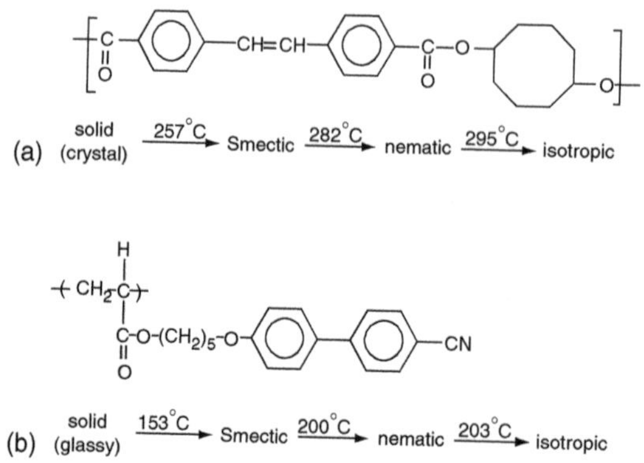

FIGURE 2.4 Examples of thermotropic liquid crystal polymers showing multiphase transitions: (a) main-chain polymer; (b) side-chain polymer.

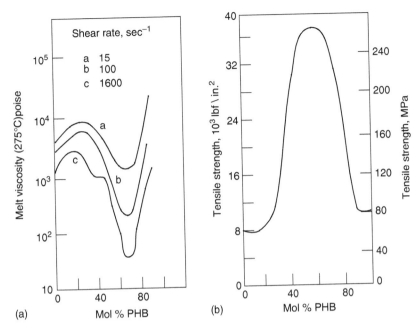

FIGURE 2.5 Effect of *p*-hydroxybenzoic acid concentration on (a) melt viscosity and (b) tensile strength of a terephthalic acid/*p*-hydroxybenzoic acid/ethylene glycol copolyester. (After Lenze, R. W. and Lin, J. I. 1986. *Polym. News*, 11, 200.)

rod-like tend to cluster together in bundles of quasi-parallel rods as their concentration in solution in increased. These form domains that are anisotropic and within which there is nematic order of the chains. There is, however, no directional correlation between the domains themselves unless the solutions are shared. When shearing takes place the domains tend to become aligned parallels to the direction of flow, thereby reducing viscosity of the system.

Lyotropic LCPs exhibit quite characteristic viscosity behavior in solution as the polymer concentration in solution is changed. Typically the viscosity follows the trend shown in Figure 2.6. As more and more polymer is added to the solvent, the viscosity increases while the solution remains isotropic and clear. At a critical concentration (which depends on the polymer and the solvent) the solution becomes opaque and anisotropic and there occurs a sharp fall in viscosity with further increase in the polymer concentration. This results from the formation of oriented nematic domains in which the chains are now aligned in the direction of flow, thereby reducing the frictional drag on the molecules.

The additional chain orientation in the direction of the fiber long axis, obtained from the nematic self-ordering in the system, leads to a dramatic enhancement of the mechanical properties of the polymer. A number of aromatic polyamides have thus achieved commercial importance because of the very high tensile strengths and moduli of the fibers that can be spun from the nematic solutions. These have consequently become attractive alternatives to metal or carbon fiber for use in composites as reinforcing material. The most significant of these aramid fibers are:

1. Poly(*m*-phenylene isophthalamide), trade name Nomex

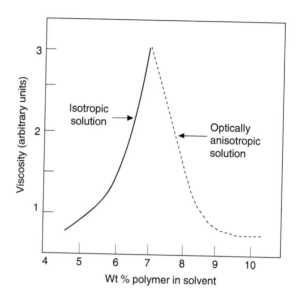

FIGURE 2.6 Variation of viscosity of solutions of partially chlorinated poly(1,4-phenylene-2,6-naphthalamide) dissolved in solvent mixture of hexamethylene phosphoramide and *N*-methylpyrrolidone containing 2.9% LiCI, as a function of solution concentration showing transition from isotropic to anisotropic behavior. (After Morgan, P. W. 1979. *Chem. Tech.*, 316.)

2. Poly(*p*-benzamide) or Fiber B

3. Poly(*p*-phenylene terephthalamide), trade name Kevlar

The last-named polymer exhibits liquid crystalline phase in sulfuric acid solution. The solution is extruded to form a fiber, resulting in further alignment of the molecules. The product, once the sulfuric acid is removed, is a fiber with a more uniform alignment than could be obtained simply by drawing and thus has much better mechanical properties. Tensile strength of Kevlar (see Table 1.17), for example, is considerable higher than that of steel, whereas its density is much lower. Although most Kevlar produced is used in tie cord, the polymer also finds use in specialty clothing. Light-weight bulletproof vests are made containing up to 18 layers of woven Kevlar cloth.

LCPs have significantly increased crystalline melting temperatures as a result of the extended chain morphology. In fact, major drawbacks to the type of rigid polymers that exhibit liquid crystalline behavior are that they have a very high melting point—e.g., poly(*p*-hydroxybenzoic acid) melts at about 500°C—and are difficult, it not impossible, to dissolve in the usual organic solvents. This makes them difficult to process, so alternative structures with much lower melting points are more useful.

The melting points of LCPs can be reduced in a number of different ways (schematically represented in Figure 2.7), namely,

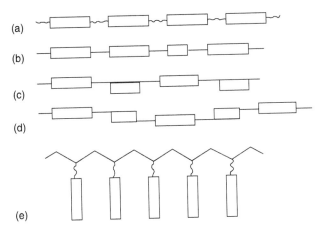

FIGURE 2.7 Schematic representation of several arrangements of mesogens (□) and flexible spacers (∼) in main chain and side chain of liquid crystalline polymers.

1. Incorporation of flexible spacer units, e.g., $-(CH_2)_{\overline{n}}$, $-(CH_2-CH_2-O)_{\overline{n}}$, $-S-R-S-$, and $-(SiR_2-O)_{\overline{n}}$ in chain to separate the rigid backbone groups (mesogens), which are responsible for the mesophases (Figure 2.7a).
2. Copolymerization of several mesogenic monomers of different sizes to give a random and more irregular structure (Figure 2.7b).
3. Introduction of kinks in the main chain, such as by using meta substituted monomers (Figure 2.7c) or a crankshaft monomer (e.g., 6-hydroxy-2-naphthoic acid) (Figure 2.7d).
4. Attachment of mesogens to the polymer backbone via flexible spacers (Figure 2.7e).

2.3.1 Thermotropic Main-Chain Liquid Crystal Polymers

The use of flexible spacers is a popular approach for producing thermotropic main-chain polymer liquid crystals. The mesogenic moiety consists of two cyclic units, normally joined by a short rigid bridging group. These are then linked through functional groups to flexible spacers of varying length that separate the mesogens along the chain and thus reduce the overall rigidity. The schematic chemical constitution of these main chain LCPs, together with example of the various types of groups that have been used are shown in Table 2.3.

The bridging groups are usually multiple bond units, because they must be rigid to maintain the overall stiffness of the mesogens. Ester groups also serve this purpose, particularly when the cyclic units are aryl rings where the conjugation leads to a stiffening of the overall structure (II):

$$-O-\underset{\underset{O}{\parallel}}{C}-\under{\bigcirc}-\underset{\underset{O}{\parallel}}{C}-O-$$

(II)

Many of the examples of thermotropic main-chain LCPs are polyesters that are synthesized by condensation reactions including interfacial polymerizations, or by high-temperature solution polymerizations using diols and diacid chlorides. However, the preferred method is often an ester interchange reaction in the melt. Among the commonly used monomer units are *p*-hydroxybenzoic acid, terephthalic acid, 2,6-naphthalene dicarboxylic acid, 2-hydroxy-6-naphthoic acid, and 4,4′-biphenol. The main-chain LCPs prepared in this way, however, tend to be very insoluble polymers with high melting points and mesophase ranges that make them difficult to process. A common approach is thus to introduce flexible

TABLE 2.3 Chemical Constitution Typical of Thermotropic Main-Chain-Polymer Liquid Crystals [4]

Cyclic Unit	Bridging Group	Functional Group	Spacer
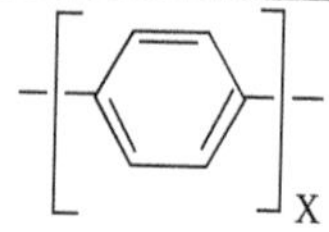 X = 1-3	$-C(=O)-O-$	$-C(=O)-O-$	$-(CH_2)_n-$
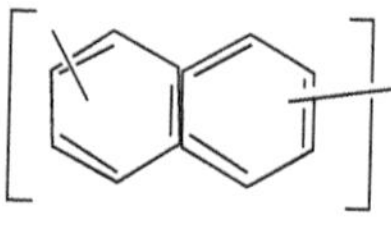 1,4 1,5 2,6	$-C(R)=N-N=C(R)-$	$-O-C(=O)-$	$-(CH_2-CH(R)O)_n-$
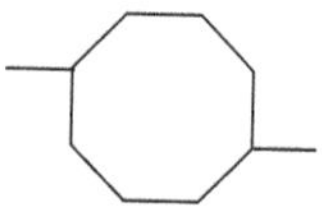	$-CH=N-$ $-N=N-$ $-N(\rightarrow O)=N-$	$-(CH_2)_n-$	$-Si(R)_2-O-$
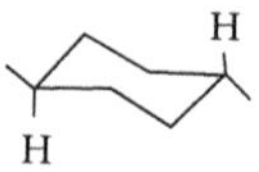	$-CH=C(R)-$	$-O-$	$-S-R-S-$

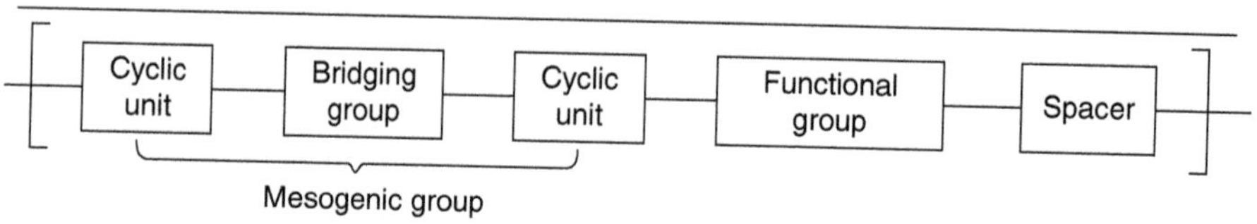

spacers in order to obtain more tractable LCPs (Table 2.3). The spacer units are usually introduced by a copolymerization reaction and the proportion of the spacer units relative to the mesogens can be varied resulting in alteration of the melting point and the temperature range of mesophase.

The majority of the main chain LCPs having group arrangements as shown in Table 2.3 exhibit a nematic phase after melting, but in some cases small variations in structure can lead to formation of a smectic mesophase. Thus for polyesters with the following structures:

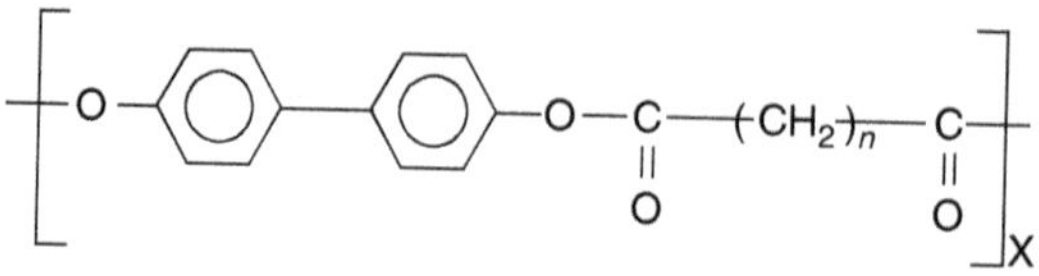

A nematic phase is observed when the number of methylene units (n) in the spacer is odd, but a smectic mesophase results when n is even. Similarly, for polyesters with multiple rings, the phases can be nematic or smectic depending on the orientation of the ester units, e.g., (III) and (IV):

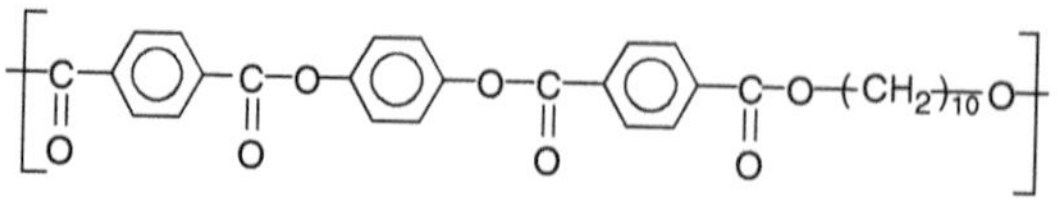

(III)

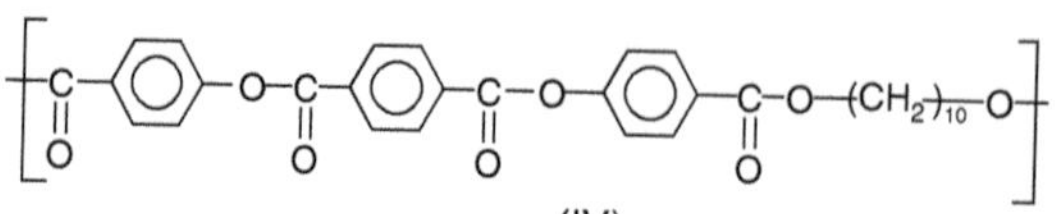

(IV)

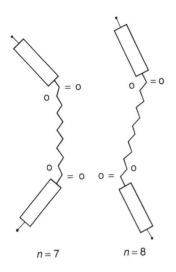

$n = 7$ $n = 8$

FIGURE 2.8 Schematic diagram showing the effect of odd and even number of—CH$_2$—spacer units on the relative orientation of the mesogenic units in a main-chain liquid crystal polymer. (After Krigbaum, W. R., Watanabe, J., and Ishikawa, T. 1983. *Macromolecules*, 16, 1271.)

The length of the spacer unit has significant effect on the melting point (T_m) and isotropic transition temperature (T_i), and hence on the temperature range (T_i–T_m) in which the mesophase is stable.

Polymers with spacers having an even number of CH$_2$ units usually have higher melting (T_m) and clearing (T_i) temperatures than those with an odd number. This suggests that the spacer length influences the ordering in the liquid crystal phase. As shown schematically in Figure 2.8, it is easier with even-numbered methylene unit spacers to maintain the orientation of the mesogen parallel to the director axis in the all *trans* zig–zag conformation.

2.3.2 Side-Chain Liquid Crystal Polymers

It has been demonstrated that polymers with mesogens attached as side chains can exhibit liquid crystalline properties. The extent to which mesophases can develop in these system is influenced by the flexibility of the backbone chain and whether the mesogen is attached directly to the chain or is separated from the chain by a flexible spacer unit.

The degree of flexibility of the polymer chain to which the mesogens are bonded can affect both the glass transition temperature (T_g) and the mesophase to isotropic phase transition temperature (T_i). This is illustrated in Table 2.4 for a number of side-chain LCPs having the same mesogen. The temperatures of the transitions are seen to decrease with the increase in chain flexibility, the latter being in the order methacrylate < acrylate < siloxane. The thermal range of the mesophase (ΔT) is thus the greatest when the chain is most flexible and its conformational changes largely do not interfere with, or disrupt, the anisotropic alignment of the mesogens in the liquid crystalline phase.

The influence of the polymer backbone on the alignment of the side-chain mesogens can be minimized by decoupling the motions of the main chain from those of the mesogens. This can be achieved by introducing long flexible spacer units between the backbone and the mesogen. A typical side-chain LCP structure would thus be that shown at the top of Table 2.5. Structures of this type can be synthesized in a number of ways. One such scheme is shown in Figure 2.9.

It is generally observed that as longer spacer units are introduced, the T_g of the polymer is lowered by internal plasticization and the tendency for the more ordered smectic phase to develop is increased. A similar ordering effect is also encouraged by lengthening the alkyl tail unit (see Table 2.7). Both these phenomena reflect the known tendency for long side chain to order and eventually to crystallize, when sufficiently long, and this condition persists also in the liquid crystal state.

The ordered state of the mesophase in the aforesaid crystalline polymers is readily frozen and locked into a glassy state if the temperature is rapidly brought down below the T_g and remain stable until heated above T_g again. The phenomenon offers several interesting application possibilities in optoelectronics and information storage (discussed latter). These applications often depend on the ability of the mesogenic groups to align under the influence of an external magnetic or electric field, as discussed below.

As the dielectric constant and diamagnetic susceptibility of many mesogens are anisotropic, the orientation of side-chain LCPs in the nematic state can also be changed by the application of a magnetic

TABLE 2.4 Effect of Chain Flexibility on the Transition Temperatures of Side-Chain Liquid Crystal Polymers Having a Common Mesogen [4]

Polymer	Transitions (°C)	ΔT (°C)
$\begin{array}{c} CH_3 \\ | \\ \{CH_2{-}C\}_n \\ | \\ COOR \end{array}$	Glassy $\xrightarrow{187}$ Nematic $\xrightarrow{201}$ Isotropic	14
$\begin{array}{c} \{CH_2{-}CH\}_n \\ | \\ COOR \end{array}$	Glassy $\xrightarrow{160}$ Nematic $\xrightarrow{177}$ Isotropic	17
$\begin{array}{c} CH_3 \\ | \\ \{O{-}Si\}_n \\ | \\ CH_2R \end{array}$	Glassy $\xrightarrow{142}$ Nematic $\xrightarrow{168}$ Isotropic	26

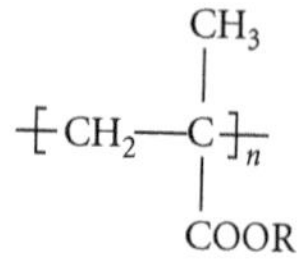

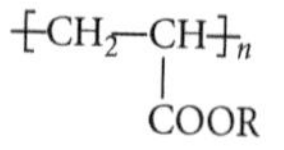

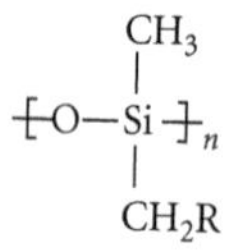

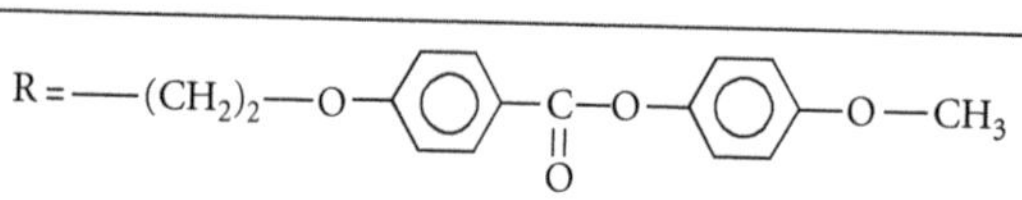

$$R = -(CH_2)_2 - O - \bigcirc - \overset{O}{\underset{\|}{C}} - O - \bigcirc - O - CH_3$$

of electric field. The parameter of interest is the magnitude of the critical field, which is required to affect transition (Fredericks transition) from the homogeneous to the homeotropic aligned state (see Figure 2.10).

Whereas the relaxation time for this transition in low-molecular-weight mesogens is of the order of seconds, this may be several orders of magnitude larger in polymer systems due to viscosity effects. Though this makes the use of polymeric liquid crystals less attractive in rapid-response display devices, the additional stability that can be gained in polymeric systems can be distinctly advantageous for some applications as thermorecording in optical storage systems.

The principles of using side-chain LCPs as optical storage systems have been demonstrated using a polymer film prepared from a side-chain polymer with the structure as shown in Figure 2.4b. The mesogenic side groups are first oriented by application of an electric field to the polymer above its T_g, such that homeotropic alignment is obtained. On cooling below the T_g, the alignment is locked into the glassy phase, and a transparent film that will remain stable on removal of the electric field is produced. If a laser beam is now used to address the film, localized heating occurs at the point where the beam impinges on the film and the material passes into the isotropic, disordered, melt state. This results in a local loss of the homeotropic orientation at the place of laser exposure and, on cooling, an unoriented region with a polydomain texture, which scatters light and thus produces a nontransparent spot, forms in the film (see Figure 2.11). Information can thus be "written" onto the film, and can subsequently be erased by simply raising the temperature of the whole film to regain the isotropic, disordered, melt state.

2.3.3   Chiral Nematic Liquid Crystal Polymers

Cholesteric mesophase (Figure 2.2c) is a special form of nematic phase, first observed in low-molecular-weight esters of cholesterol. This structure can often be detected in mesogenic systems containing a chiral center and hence is also called the chiral nematic state. The structure, as shown in Figure 2.2c, is a helically disturbed nematic phase, where each alternate layer is displaced by an angle θ relative to its immediate neighbors, thus imparting a helical twist (with a pitch p) to the phase. A system with this type of ordering has very high optical activity and an ability to selectively reflect circularly polarized light of a

TABLE 2.5 Schematic Representation of the Organization of a Side-Chain Liquid Crystal Polymer (SCLCP) [4]

SCLCP = [Flexible backbone]—[Functional unit]—[Spacer]—[Functional unit]—[Cyclic unit]—[Bridging group]—[Cyclic unit]—[Flexible tail]

Flexible Backbone	Functional Unit	Spacer	Cyclic Unit	Bridging Group	Flexible Tail
—CH—CHR—	None	None	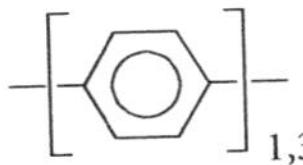 benzene ring (1,3; 1,4)	None	None
—SiR—O—	—O—	—(CH$_2$)$_n$—	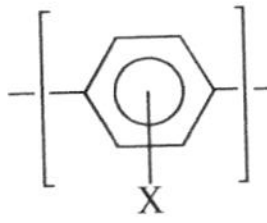 benzene ring with X (X = Me, Ph, Cl)	—CO—O—	R
—SiR—O—SiR$_2$—O—	—CO—O—	—(CH$_2$—CHR)$_n$—	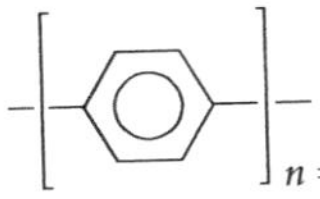 benzene ring (n = 1,2,3)	—CR=CR—	OR
—P=N—	—O—CO—	—NR'—R—NR'—	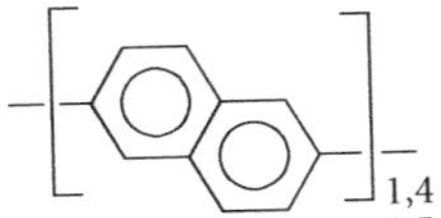 naphthalene (1,4; 1,5; 2,6)	—C≡C—	CN
		—S—R—S—	 (Cholesteryl)	—CR=NO—	
		—SiR$_2$—O—		—NO=N—	
				—CR=N—N=CR—	

Source: Cowie, J. M. G. 1991. *Polymers: Chemistry and Physics of Modern Materials.* Blackie, Glasgow and London.

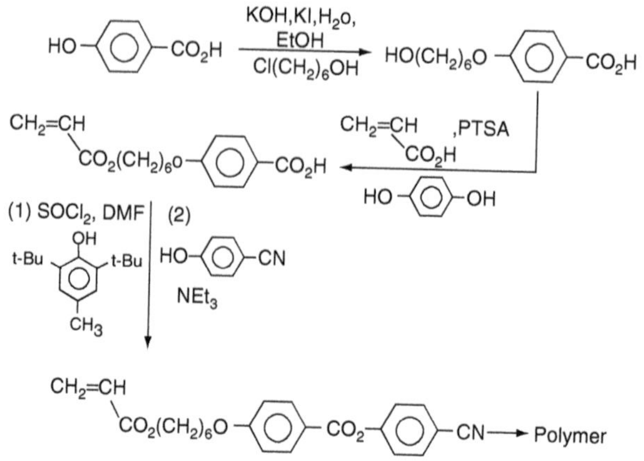

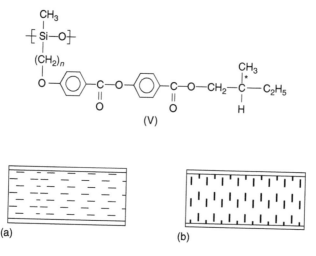

FIGURE 2.9 A representative scheme of synthesis of side-chain liquid crystal polymer of the type shown in Table 2.5. The monomer containing the mesogenic unit is first synthesized and then polymerized to obtain the liquid crystal polymer.

specific wavelength, when irradiated by normal light. The wavelength (λ_R) of this reflected light is related to the pitch p of the helical structure by

$$\lambda_R = np \tag{2.1}$$

where n is the average refractive index of the liquid crystalline phase.

   The chiral nematic state, first observed in cholesteryl derivatives, was later detected in other chiral mesogens, and can also be induced by adding small chiral molecules to a host nematic LCP. Chiral nematic LCPs have thus been synthesized as side-chain polymers by introducing a chiral unit in the tail moiety of mesogens or by copolymerizing cholesterol-containing monomers with another potential mesogenic monomer. Examples [4] of these types are shown as structures (V) and (VI):

FIGURE 2.10 Schematic representation of (a) homogeneous and (b) homeotropic alignment of mesogens in a measuring cell. (After Folmer, B. J. B., Sijbesma, R. P., Versteegen, R. M., Van der Rijt, J. A. J., and Meijer, E. W., 2000. *Adv. Mater.*, 12, 874. With permission.)

$$\text{---}(\text{CH}_2\text{--CH})_{\overline{n}}\text{\scriptsize \hspace{1mm}}(\text{CH}_2\text{--CH})_{\overline{m}}$$

$$\begin{array}{cc}
\text{C=O} & \text{C=O} \\
| & | \\
\text{O} & \text{O} \\
| & | \\
(\text{CH}_2)_5 & (\text{CH}_2)_5 \\
| & | \\
\text{C=O} & \text{O---\hspace{-1mm} benzene ---\hspace{-1mm} benzene ---CN} \\
| & \\
\text{O} & \\
| & \\
(\text{Chol}) &
\end{array}$$

Chol =

(VI)

The data in Table 2.6 indicate that as the content of the chiral monomer (cholesterol) is varied in the copolymer, the wavelength of the reflected light (in the average temperature region between T_g and T_i) changes. This suggests that the composition changes alter the pitch (p) of the helical structure in the chiral nematic phase, with consequent changes in the reflected light wavelength [see Equation 2.1]. For (VI), the increase in cholesterol content apparently causes the helix twist to become tighter, i.e., p becomes smaller and so λ_R moves to shorter wavelengths. The pitch is also sensitive to temperature. When the temperature is raised, the helix tends to unwind (i.e., p increases) with a consequential increase in λ_R.

Like the other side-chain LCPs described in the previous section, these materials also offer the possibility of locking the chiral nematic phase into the glassy state by rapid cooling to temperatures below T_g. This leads to a preservation of the structure and its reflected color. With suitable systems, the process can thus be used to produce stable and light-fast monochromatic films.

2.3.4 Properties of Commercial LCPs

The first LCP to be launched commercially was Dartco's Xydar, introduced in 1984 [11]. Xydar injection-molding resins are aromatic polyesters based on terephthalic acid, *p*-hydroxybenzoic acid, and *p,p'*-dihydroxybiphenyl. Xydar has a high melting point (close to 400°C), which necessitates certain modifications to processing equipment. It also has a high melt viscosity making it difficult to mold in

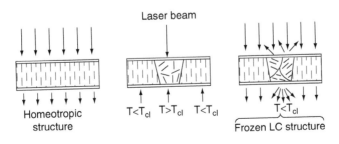

FIGURE 2.11 Thermal recording using a homeotropically aligned side-chain liquid crystal polymer as a transparent film. A laser beam is used to address the film by producing local heating and disorder, which is subsequently frozen in by cooling. (After Cowie, J. M. G., 1991. *Polymers: Chemistry and Physics of Modern Materials,* Blackie, Glasgow and London.)

TABLE 2.6   Variation of λ_R with Copolymer Composition for Structure (VI) [4]

% Chol	T_g (°C)	T_i (°C)	λ_R (nm)
34	10	32	850
40	10	39	660
55	13	40	555
65	13	65	500

Source: Cowie, J. M. G. 1991. *Polymers: Chemistry and Physics of Modern Materials.* Blackie, Glasgow and London.

small sections. An advantage of Xydar, however, is that it offers high retention of mechanical properties even at temperatures in excess of 300°C.

A second range of LCPs were introduced by Hoechst Celanese under the trade name Vectra. Vectra is an aromatic polyester condensate derived from 2,6-dihydroxynaphthalene. Its main advantages are its fast cycling and ease of processing, the melt temperature for both injection and extrusion grades being typically about 285°C.

Cycle time of Vectra can be less than half that of other engineering plastics, i.e., on the order of 530 sec depending on part design. This is due to a number of factors. Since the melt is already ordered, parts require little or no crystallization time. LCPs moreover have an exceptionally low heat of transition; so little heat of crystallization must be removed, thereby allowing for fast cooling. And because compounds have excellent stiffness at high temperatures, parts are stiff enough to eject without cooling.

The material is noncorrosive to molds and clean running, while its exceptionally low melt viscosity ensures easy filling of complex, small-part geometries. No modifications are required to processing equipment. However, as with filled or reinforced polymers, tools should be hardened to minimize the effects of abrasion.

Vectra LCPs consist of parallel, closely packed, fibrous chains in injection molded parts which give the polymer a wood-like appearance at fracture surfaces and also give it self-reinforcing properties that are at least as good as those of conventional fiber-reinforced thermoplastics. ICI supplies LCPs (self-reinforcing polymers) under the trade name Victrex SRP. Excellent physical properties, low combustibility and smoke generation, resistance to chemicals and solvents, and stability towards radiation are some of the important properties of LCPs.

LCPs being anisotropic, their properties are enhanced along with the direction of flow and decreased across the flow direction. This anisotropic ratio is reduced by the presence of fillers, which is in contrast to conventional polymer composites. Thus the properties of fiber-filled LCPs, such as shrinkage, modulus, and creep, are less anisotropic than those of the unfilled LCP. These LCPs also retain their excellent mechanical properties when used for long periods of time at elevated temperatures.

Because of the low melt viscosity of LCPs, thin sections can be molded with highly filled (70%) LCPs. The addition of bulk lubricants reduces the melt viscosity. Injection molded parts are produced with a minimum of molded-in stress, little or no flash, and high dimensional stability. Fast cycle times, reduced mold wear, elimination of secondary operations, and other processing and performance gains of LCPs often lead to lower overall finished part costs even below those of less-expensive resins.

During injection mold filling, the fountain flow effect of LCPs causes surface molecules to stretch in the flow direction. These molecules ultimately form a highly oriented skin (15%–30% of a part's thickness) that is self-reinforcing. This results in exceptional flexural strength and modulus, as well as good tensile performance. For example, LCPs have a modulus from 1.4×10^6 to 3.5×10^6 lbf/in.^{2} (10–24 GPa), a tensile strength from 18×10^3 to 37×10^3 lbf/in.^{2} (124–255 MPa), and notched Izod impact values up to 7 ft-lb/in (374 J/m).

The LCPs retain 70% of their notched Izod impact values down to a temperature of about -270°C (-450°F). They are also resistant to bleaches, chlorinated organic solvents and alcohols. They have extremely low gas-permeation rates compared to commercial packaging films. Their ionic extractables are well below those needed for corrosion-free environments for integrated circuit chip applications. The

LCPs are also inherently flame retardant and possess a high LOI ($>35\%$). Their combustion produces a low smoke density and relatively nontoxic products.

Because LCPs are thermoplastic and have excellent thermal stability, high levels of regrind can be used in injection molding of components. As a guide, 50% regrind is usually acceptable.

2.3.5 Applications

One of the first applications of LCPs was a range of dual-ovenable cookware of Xydar made by Tupperware, a subsidiary of Dartco Manufacturing. The resistance of LCPs to chemicals and solvents and good performance in hostile environments have led to their several specialized applications. Vectra materials have been used commercially for the molding of formic acid separation tower packings as they have proved to be more efficient and longer lasting than conventional ceramic packing materials.

LCPs are also used in such demanding areas as surgical instruments, aircraft and automotive engine systems, fiber optic devices, chemical equipment, and photocopier components.

LCPs are finding use as replacement for epoxy and phenolic resins in electrical and electronic components, printed circuit boards (PCBs), and fiber optics. In these applications, the high mechanical properties, low coefficient of thermal expansion, inherent inflammability, good barrier properties, and ease of processing of LCPs (Vectra in particular) are important.

In fiber optic applications, LCPs can be extruded into a variety of shapes and sizes using conventional extrusion equipment found in typical fiber optic wire and cable production plants.

Having one of the highest flow rates of any polymer and virtually no deformation or shrinkage on molding, LCPs are typically used for precision parts with thin walls and complex shapes. In the electronics industry, the benefits mentioned above—plus the resin's resistance to soldering temperatures of 200°C–250°C—give it an edge over other high-performance plastics in many surface-mounted devices. For example, finely dimensioned electronic components such as SIMM socket with 0.050-in. spacing between pins are typical uses of LCPs.

In several electronic applications, such as connectors and capacitors, LCPs are also being used in preference to high-performance materials, such as PPS, because of their ease of processing, greater impact resistance, and overall lower part cost.

2.4 Electroactive Polymers

As computers and sophisticated electronics devices began to move out of their shielded rooms into offices, stores, and homes, it became highly desirable to take advantage of the light weight, low cost, and aesthetics that could be gained by substituting plastics for metal housings for the instruments [12–16]. Conductive plastics have therefore been increasingly used to provide flexible, lightweight, and moldable parts having good static bleed-off and electromagnetic interference (EMI) shielding properties. A variety of uses of such materials are encountered, ranging from compliant gasketing to rigid housing for business machines.

Conductive polymers fall into two distinct categories: filled polymers, which are used for a wide range of anti-static and static-discharging applications, and intrinsically conductive polymers, which contain no metals but conduct electricity when chemically modified with dopants.

A number of other polymeric solids have also been the subject of much interest because of their special properties, such as polymers with high photoconductive efficiencies, polymers having nonlinear optical properties, and polymers with piezoelectric, pyroelectric and ferroelectric properties. Many of these polymeric materials offer significant potential advantages over the traditional materials used for the same application, and in some cases applications not possible by other means have been achieved.

2.4.1 Filled Polymers

Polymers can be made to conduct electricity relatively easily by compounding them with high loadings of conductive materials as fillers. Apart from the inherent properties of the fillers, parameters such as concentration, particle form (sphere, flake, fiber), size, distribution, and orientation are deciding factors that influence the properties of filled polymers.

When choosing a filler, the following requirements merit consideration: (1) the filler has high conductivity to avoid excess weight; (2) it does not impair the physical/mechanical properties of the plastic; (3) it is easily dispersed in the plastic; (4) it does not cause wear on forming tools (injection molding, extrusion); (5) it has favorable cost picture; and (6) it produces good surface structure of the finished product. The commonly used electrically conducting fillers are carbon, aluminum, and steel. The most common metallic conductor, copper, is not used because it oxidizes within the plastic and impairs its physical properties.

Typical dimensions for common fillers are:

Aluminum flakes: length 1–1.4 mm, thickness 25–40 μm
Carbon black: spherical, diameter 25–50 μm
Graphite fiber: length mm-cm, thickness 8 μm
Steel fiber: length 3–5 mm, diameter 2–22 μm

All of them are to be found in a number of qualities with varying prices and conduction properties.

Depending on particle form and orientation, there is a certain critical volume concentration at which the resistance decreases, i.e., the conductivity increases, drastically. This is shown in Figure 2.12. At the critical concentration, the filler can form a continuous phase through the matrix in the form of microscopic conductive channels. As shown in Figure 2.12, the specific resistance of filled polymers also depends on the inherent conductivity and particle form of the filler besides its concentration. The critical concentration can be reduced to low levels by using conductive particles that are fibrous in shape. The reduction in critical volume loading is proportional to the magnitude of the fiber's aspect ratio (length/diameter).

It has been shown that even extremely small concentrations of additives can make plastics conductive if they are in the form of conductive fibers with length to diameter (L/D) ratio of 100 or more. The striking difference between the use of chunky fragments and fibrous materials in their effect on conductivity can be seen from the diagram in Figure 2.13.

If one loads a plastic with 5 percent by volume of 6-mil size in a random distribution (Figure 2.13a), there will be, on the average, a 6-mil gap to the nearest sphere. Heat or electrons flowing through such a matrix would thus cross alternate paths of about equal lengths in the two media. The picture, however

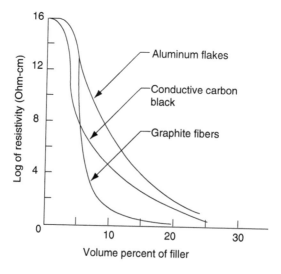

FIGURE 2.12   Composite resistivity as a function of filler volume loading and filler type.

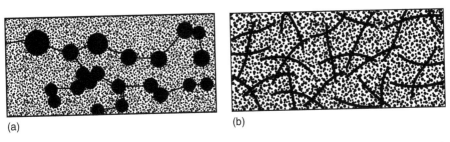

(a)   (b)

FIGURE 2.13 Comparison of typical flow path through composites using the same volume percentage of material as spheres and fibers ($L/D = 100$). (a) Chunky particles—long path through plastic; (b) fibers—short gaps in plastic.

becomes quite different if the same 5 percent of material is dispersed as 1-mil diameter and 100-mil long fibers. It is inevitable that at random orientations they will touch one or more of their nearest neighbors, as shown in Figure 2.13b, thereby providing an almost continuous path through the composite along the conductive fibers. Fibers will therefore be more effective in lowering the electrical resistivity of plastic and increasing the thermal conductivity than chunky fragments.

The above concept led to the creation of metalloplastics—a family of conductive plastics in which the conductivity is great enough to make the material resemble metal both electrically and thermally. To serve as a practical engineering material, metalloplastics should have electrical resistivities of less than 1 ohm-cm and thermal conductivities of at least a factor of ten higher than those of normal plastics.

The importance of high thermal conductivity in plastics is being recognized as the automobile business tries hard to use more and more plastic parts to reduce the weight of the automobile. With conventional plastics having poor thermal conductivity, the problem of getting heat into the part to form it and then getting the heat out again results in cycle times of the order of minutes, not seconds. This implies an

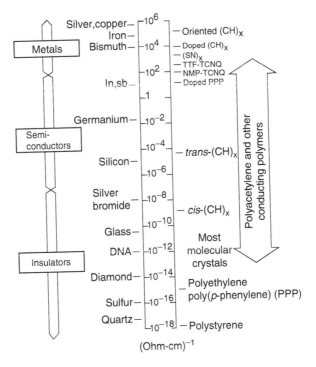

FIGURE 2.14 Chart of conductivities for selected materials. [Note: 1 (ohm-cm)$^{-1}$ = 1 S cm^{-1}.]

increase of one to two orders of magnitude in the number of tools necessary to make the same number of parts. So, for the auto industry, an order of magnitude increase in thermal conductivity could be a very significant advance.

Metalloplastics with electrical resistivities as low as 0.01 ohm-cm and up to 100-fold higher thermal conductivity than ordinary plastics have been developed by the addition of a few percent of metal and/or metallized glass fibers (*L/D* of the order of 100/1) to plastics. Use of such materials can have very significant effects on molding cycle times, uniform heating and cooling rates, and heat transfer rates in the final product.

Whereas, thermal conductivity is proportional to the concentration of conductive fillers and is increased even by such low concentrations of the fibers that one fiber does not touch its neighbors, the electrical resistivity is not significantly modified until an almost continuous path is available through the conductive fibers. Plastics can thus be developed that are improved in thermal conductivity but can be used for electrical insulation or resistive heating. Suitably filled polymers are thus used to drain off heat in pressure switches as well as in polymeric tapes intended for self-regulating, resistive heating of water pipes, railroad switches, etc.

Filled conductive polymers are 10–12 orders of magnitude more conductive than unfilled polymers but are still several orders of magnitude lower than copper (Figure 2.14). Carbon-black filled polymers are the most common. Fillers other than carbon blank include finely divided metal flakes and fibers, metallized glass fibers, and metallized inorganics such as mica.

Filled conductive polymers used for packaging include polycarbonate, polyolefin, and styrenics incorporating fillers such as carbon, aluminum, and steel flakes and fibers. A polycarbonate/ABS blend (Bayblend ME) introduced by Bayer is 4% aluminum filled and suited to many screening functions.

The electrically conductive polymers have their greatest use in EMI-shielded casings and protective housings of electronic and telecommunications equipment, where they have rapidly substituted metal because of their low weight and easy workability. Steel fiber is mostly used in filled conductive polymers intended for EMI-shielding applications, which will be discussed below.

Conductive rubber has been used in a number of applications. In most cases, silicone rubber is used because of its greater resistance to temperature, UV light, oxygen, corrosive gases, chemicals, and solvents as compared to organic rubber. Conductive rubber is used in tires to leak off static electricity. Ethylene–propylene conductive rubber is used as a covering in cables to reduce high voltage problems. In all these cases, carbon black is generally used as the filler, one reason being that it also acts as a reinforcing filler, increasing the strength and tear resistance of the product. In the electronics industry, conductive rubber finds use as a connector for liquid crystal displays in electronic digital watches.

2.4.1.1 EMI Shielding

EMI is the random, uncontrolled, broad-range frequency radiation emitted from many natural and man-made sources. The recent rapid growth in man-made sources such as computers, telecommunication, and other business machinery has led to the legislation in the United States and Europe covering the levels of electromagnetic radiation emitted by electronic devices. The EMI-shield on most computers and electronic digital equipment on the American market has to conform to Federal Communications Commission (FCC) standards within the frequency range 30–1000 MHz. Similar regulations exist in Germany and are being implemented in other countries.

Metal or carbon particles in plastic reflect and absorb electromagnetic radiation. When dimensioning a shielding material, the absorption of non-reflected radiation must be made sufficient. An electromagnetic wave is absorbed according to the following correlation [13]:

$$A = 3.34 \ t\sigma\mu\sqrt{F} \tag{2.2}$$

where

A absorption expressed in decibels (dB)
t material thickness in one thousandths of an inch (mils)

σ conductivity relative to copper
μ relative magnetic permeability
F frequency in MHz

Experimentally, EMI shielding is determined by the attenuation of a high frequency signal transmitted through the test sample. The attenuation or shielding effect α expressed in decibels is calculated from the equation

$$\alpha(dB) = 20 \log_{10}(E_b/E_a) \tag{2.3}$$

where E_b is the field intensity (V/m) without shielding and E_a is the field intensity (V/m) with shielding.

Data for some homogeneous metallic materials are given in Table 2.7. Aluminum and steel are the commonly used metallic fillers, and of these, steel provides the most effective EMI-shielding because of its high relative magnetic permeability.

Steel fibers give very good shielding properties at considerably low percentages. At only 1 percent of steel fiber in plastic, a shielding of 45–60 dB at 1 GHz is obtained. The conductivity of the filled plastic then is approximately 1 (ohm-cm)$^{-1}$. One great advantage of such low percentages permitted by steel fibers is that the other properties of the plastics are left almost unaffected.

In carbon, the conductivity varies from 10^{-2} (ohm-cm)$^{-1}$ for amorphous carbon to approximately 300 (ohm-cm)$^{-21}$ in the longitudinal direction for PAN-based high modulus carbon fibers. Apart from relatively low conductivity, carbon has the same magnetic permeability as aluminum, i.e., approximately 1. In order to obtain a given damping, carbon-based fillers have to be added in higher concentrations in comparison with metallic fillers such as steel. However, special carbon black grades with microporous structure and increased conductivity can now be found that allow the construction of a conductive network at relatively low concentrations.

Although a plastic with conductive fillers does not provide the same degree of shielding as metal, a great number of filled plastic materials have been introduced commercially, which in 1–3 mm thicknesses give shielding effects of 30–40 dB. As can be seen from the relationship given in Table 2.8 between shielding effect expressed in dB and in percent, respectively, these dB numbers correspond to 99.9%– 99.99% damping of the interference signal, which is considered satisfactory.

There are a range of conductive polymers on the market that are based on metal fillers such as aluminum flake, brass fibers, stainless steel fibers, graphite-coated fibers, and metal-coated graphite fibers. However, the most cost effective conductive filler is carbon black. Mention should also be made of

TABLE 2.7   Conductivity, Permeability, and Absorption Characteristics of Major Metals

			Absorption Loss (dB/mil)		
Metal	Relative[a] Conductivity	Relative[a] Permeability (100 kHz)	100 Hz	10 kHz	1 MHz
Copper	1.00	1	0.03	0.33	3.33
Silver	1.05	1	0.03	0.34	3.40
Aluminum	0.61	1	0.03	0.26	2.60
Zinc	0.29	1	0.02	0.17	1.70
Brass	0.26	1	0.02	0.17	1.70
Nickel	0.20	1	0.01	0.15	1.49
Iron	0.17	1,000	0.44	4.36	43.60
Steel (SAE 1045)	0.10	1,000	0.33	3.32	33.20
Stainless steel	0.02	1,000	0.15	1.47	14.70
Mu-metal	0.03	80,000	1.63	16.30	163.00
Permalloy	0.03	80,000	1.63	16.30	163.00

[a] Relative to copper.

TABLE 2.8 Shielding Effect Expressed in Decibels, Percentage of Leakage, and Classification of Shielding Effect

Shielding Effect $(\alpha)^a$, dB	EMI Leakage (%)	Value Judgement (technical)
0–10	10^2–10^1	Insignificant
10–30	10^1–10^{-1}	Minimal
30–60	10^{-1}–10^{-4}	Average
60–90	10^{-4}–10^{-7}	Above average
90–120	10^{-7}–10^{-10}	Maximum (surpassing present technology)

^{a} Defined by Equation 2.3.

other more exclusive fillers, such as fiber glass with a metallized surface (aluminum) and silver-coated glass beads.

Where color is not critical, carbon black filled conductive polymers provide the most cost effective way of producing EMI shielding. A range of carbon black filled conductive polymers are produced by Cabot Plastics under the trade name Cabelec. One application is in telephone microcomponents.

In addition to being invariably black, another disadvantage of carbon-black-filled polymers is that they do not have the impact resistance of commonly used materials. One solution to these problems has been found by the use of sandwich molding. The process, originally developed by ICI in the 1960s, involves the production of a housing with an inner core of conductive plastics surrounded by an outer skin of conventional engineering plastic. The process thus gives the designer the full aesthetic freedom of design and also gives a good in-molded surface finish.

A system for sandwich molding developed by Aron Kasei uses fiber-reinforced ABS as the outer skin and a PLS conductor (see below) as the inner core. Applications include keyboard housings, printer housings, CRT enclosures, and medical equipment housings.

Conductor PLS, offered by Aron Kasei, is a range of brass or aluminum-filled thermoplastics such as ABS, polypropylene, PBT, and polycarbonate. Housing for personal computers made of PLS is claimed to cost less than housing made of plastics coated with conductive paint.

Ube Industries offer a brass fiber-filled nylon-6 that is reported to have excellent EMI-shielding properties, high resistance to abrasion, high conductivity, and easy malleability.

Mobay has introduced a 40% aluminum flake-filled, flame-retardant grade polycarbonate/ABS blend that is molded into internal cover configurations for EMI shielding.

LNP Corporation offers conductive plastics with fillers such as aluminum flakes, nickel-coated fibers, stainless steel fibers, and carbon fibers in a range of engineering polymers including nylon-6,6, polycarbonate, ABS, PPS, and PEEK. The company's 40% flake-filled polycarbonate is used for microprocessor covers in mainframe computers and nickel coated fiber-filled PEEK for avionic enclosures.

A 10% stainless steel fiber-filled ABS, produced by Mitech Incorporated under the tradename Magnex DC is used to shield electronic components in Xerox's Model 2100 laser printer. Mitech also produces a flame retardant stainless steel fiber-filled ABS for large interior covers.

Lacana Mining Company has developed nickel-coated mica fillers (Suzerite E. Micon) that can withstand compounding, extrusion, and injection molding with no special treatment. The coating is done by a hydrometallurgical process patented by Sheritt Gordon Company. Wilson-Fiberfil also offers nickel-coated mica that at 45% loadings in polyolefins provides 40 dB shielding and can be used for small-part moldings.

Wilson-Fiberfil offers 1% steel fiber-filled polycarbonate (Electrapil R-5147) for EMI-shielding, which has the advantage that the other properties of the plastic are left mostly unaffected during both tooling and use.

2.4.1.2 Conductive Coating

Some of the more commonly used methods to produce EMI shielding consist of coating the plastic with a conductive layer by post-molding or in-molding processes. The coating methods mostly used are conductive paints, zinc arc spraying, vacuum metallizing, electroplating, and foil application.

The simplest method of coating is the use of conventional coating systems such as brush coating or spray gun. Nickel, copper, silver, or graphite can be coated in this way onto vinyl polymers, acrylics, and polyurethane. This method is the least expensive but it suffers from the disadvantage that the resin has to be tailored to the substrate to avoid cracking.

Nickel coating is the most popular. A film thickness of 50–70 microns can achieve attenuation levels of 30–60 dB. An acrylic-based nickel coating, MDT1001, from Mitsubishi Rayon Company is claimed to be 100 times more effective as an EMI shield than normal carbon-based coating: it dries at room temperature and does not require multiple coatings.

For solvent-sensitive plastics where paint systems are unsuitable, zinc arc spraying is sometimes used. In this method, the metal is melted by an electric arc and sprayed as droplets by compressed air onto the part to be coated. Consequently, only materials with very high melting points can be used. The method gives a hard, dense coating of good conductivity, but it requires expensive equipment.

In vacuum metallizing, which also requires a major capital investment in equipment, a metallic film, usually aluminum, is deposited onto a plastic by condensation from vapor under high vacuum. By rotating parts during application, a controlled film thickness can be obtained producing high levels of attenuation with layers as thin as 4–6 microns.

Electroplating with copper, nickel, or chromium using conventional techniques provides good conductivity and durability. It, however, requires paint finishing. Electroless plating, which differs from electroplating in that no external source is required to initiate and control the process, has proved to be an effective EMI-shielding process for components that may be conductive in some areas and nonconductive in others, such as computer housing.

In a selective electroless plating, a conductive film of copper or nickel is deposited onto a specific part of the molding as opposed to overall plating by conventional techniques that require paint finishing. Attenuation levels of 50 dB can be obtained at layers of 1–3 microns.

In one electroless plating technique, which reportedly achieves attenuation levels of 70 dB in layers of 1–2 microns thick, the molded part is immersed in an aqueous solution of copper salt, reducing agent, and initiator to obtain copper deposition as the primary shielding layer. Nickel is then added to form an outer layer that provides corrosion resistance and improved impact strength.

In the process of metal coating by ion plating or cathode sputtering, which is claimed to produce better adhesion of metal to the substrate compared to vacuum metallizing, positively charged gas ions are discharged between two electrodes and bombarded onto a metal target. As a result, the metal vaporizes and condenses onto the molding. A major drawback of the process is the cost of ion-plating equipment.

A relatively new approach to EMI shielding is in-mold coating. In a process developed by Dai Nippon Toryo and Tokai Kogyo, tradenamed NTS, the mold surface is coated with a metallic material tradenamed Metrafilm by electrostatic spraying. The coating and compound then bond during injection producing excellent adhesion. Since the metallic powder is encapsulated in a resin and does not contain any solvents, the process can be used for resins that are sensitive to solvents such as polystyrene.

2.4.1.3 Signature Materials

Operative conditions of weapons carriers can be drastically influenced (stealth technology) by the so-called signature adaptation, which provides reduced probability of detection by radar, infrared, visual, or acoustical reconnaissance. The active materials are applied mostly as top layers, such as camouflage painting, radar layers, IR-layers, and acoustical damping layers, and are often used in combination. Thickness and material properties are adjusted to give minimal reflection against incident electromagnetic radiation. During radar reconnaissance, this radiation falls in the frequency range of 3–30 GHz. In other cases, it may be a question of damping the natural radiation from the weapons carrier in the IR range of 2–20 μm.

In order to obtain low radar reflection with very thin layers, the following correlations should be fulfilled:

$$\varepsilon' = \mu'$$

$$\varepsilon'' = \mu'' \tag{2.4}$$

where ε' and ε'' are, respectively, the real part and the imaginary part of the permittivity ε, which in alternating fields is generalized to a complex quantity:

$$\varepsilon = \varepsilon' - j\varepsilon'' \tag{2.5}$$

The imaginary part ε'' is a measure for the dielectric losses of the material. Similarly, μ' and μ'' are defined in relation to the permeability μ' and μ'' are defined in relation to the permeability μ:

$$\mu = \mu' - j\mu'' \tag{2.6}$$

The correlations (2.4) above, result in the layer having the same wave impedance as air ($120\,\pi$), i.e., there is no reflection from the outer surface. The radiation passing through the layer is reflected back by the metallic base. If the dielectric and magnetic losses, represented by ε'' and μ'' respectively, are large, the greater part of the radiation can be absorbed by the layer even if it is thin.

For several decades, efforts have been made to produce materials with these properties. Since the end of the sixties, great interest has been shown towards magnetic absorption materials based on ferrite powder dispersed in organic binder agents. Ferrite represents a group of ceramic materials with the principal formula $(M^{II}O)(F_2O_3)$, where M^{II} is a divalent metal, e.g., Fe, Co, Ni, Mn, Zn, or Cd.

The magnetic losses of the ferrite produce a high resonance peak at a certain frequency, above which they diminish linearly to the wavelength. The phenomenon can be used to obtain thin, absorbent, wide-band layers. No reflection occurs at the outer surface when the impedance at the surface air/ferrite becomes equal to in-impedance of the layer. The in-impedance is, in effect, a sum impedance because the greatest part of the incident radiation is reflected several times between the interfaces of the layer.

The development in the ferrite are is intensive and is aimed at increasing the medium-frequency in the GHz-range with preserved band-width. Much of the development in the area is covered in secrecy.

Mention should also be made of signature adaptation that makes use of negative interference to reduce the probability of detection. In this case, the signature protective layer is adjusted so that negative interference occurs between the radar radiation that is reflected respectively from the outer surface and from the underlying metal, thus causing a black-out of the net reflection from the surface. The reflections in this case need to be adjusted so that 50% is reflected at the outer layer and 50% at the metal surface and so that the phase shift between these reflected radiations becomes half a wavelength at the outer surface.

Interference layers in the form of filled polymers have been used on warships. A simple system is polyester filled with TiO_2. It, however, provides protection only within a very narrow frequency range.

2.4.2   Inherently Conductive Polymers

Interest in electrically conducting polymers began in the mid-1960s with the suggestion of Little [17] in his theoretical studies that some organic materials could become superconductors. The first example that has been extensively studied is tetrathiofulvalene-tetracyanoquinodimethane (TTF–TCNQ) and its derivatives. The first covalent polymer exhibiting the electronic properties of a metal was polymeric sulfur nitride, $(SN)_x$ which attracted much attention in the mid- to late 1970s.

In 1977, the first covalent organic polymer, polyacetylene $(CH)_x$ that could be doped through the semiconducting to the metallic range, was reported. Another significant breakthrough occurred in 1980 with the discovery that PPP could be doped to conductivity levels quite comparable to those in polyacetylene. This polymer is the first example of non-acetylene hydrocarbon polymer that can be doped to give polymers with metallic properties.

Polyacetylene has been investigated much more extensively than any other conducting polymer and has served as a prototype for the synthesis and study of a large number of other conjugated, dopable organic polymers. Of these, the greatest research efforts are devoted to polypyrrole, PPS, PPP, polythiophene, and polyaniline.

In a nondoped state, the basic polymers have low conductivity (Figure 2.14). The two polyacetylene conformations *cis* and *trans* are in the semiconductor range; PPP is a good insulator.

The synthesis of the polyacetylene powder has been known since the late 1950s, when Natta used transition metal derivatives that have since become known as Ziegler–Natta catalysts. The characterization of this powder was difficult until Shirakawa and coworkers [18] succeeded in synthesizing lustrous, silvery, polycrystalline films of polyacetylene (which has become known as "Shirakawa" polyacetylene) and in developing techniques for controlling the content of *cis* and *trans* isomers:

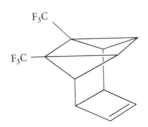

Coordinative polymerization of acetylene using the Ziegler-Natta catalytic system, $Ti(OBu)_4$-$AlEt_3$, is the most powerful way of forming polyacetylene powders, gels, or films. Polyacetylene films may be prepared by simply wetting the inside walls of a glass vessel with a toluene solution of $AlEt_3$ and $Ti(OBu)_4$ and then immediately admitting acetylene gas at any pressure up to 1 atm. A cohesive film grows over a period of time with thickness varying from 10^{-5} to 0.5 cm, depending on the pressure of acetylene and the temperature used. The film is formed completely as a *cis* isomer if a polymerization temperature of $-78°C$ is used and as the *trans* isomer if a temperature of 150°C is used (with decane solvent), while with room temperature polymerization, the film is approximately 60% *cis* and 40% *trans* isomer.

The *cis* isomer may be converted to the *trans* isomer, which is the thermodynamically stable form, by heating at 150°C for 30 min to 1 hr. The *cis*–*trans* content is conveniently determined from the relative intensities of bands in the IR spectra.

The morphology of all polyacetylenes made with various Ziegler-Natta catalysts is fibrillar. The conductivity of films of polyacetylene depends on the *cis*–*trans* content, varying from 10^{-5} (ohm-cm)$^{-1}$ for the *trans* material to 10^{-9} (ohm-cm)$^{-1}$ for the *cis* isomer. Doping increases the conductivity of both *cis*- and *trans*-polyacetylene drastically.

The polyacetylene made with the Ziegler-Natta catalytic system is infusible, insoluble, usually contaminated by catalyst residues, and tends to become brittle and dull when exposed to air due to slow oxidation. These features make it difficult to process or handle, and attempts have been made to either improve the polymer or make derivatives or precursors that are soluble in organic solvents.

Many of these problems have been solved by Feast and coworkers [19], who at the University of Durham developed a very elegant synthetic method, now commonly known as the Durham route. This is a two-stage process in which soluble precursor polymers are prepared by a metathesis ring-opening polymerization reaction and these are subsequently heated to produce polyacetylene by a thermal elimination reaction. An example of the method, given in Figure 2.15, involves the tricyclic monomer (A) which undergoes metathesis polymerization to precursor polymer (B). Thermal degradation of the latter yields *trans*-polyacetylene.

A refinement of the process in Figure 2.15 involves photochemical conversion of (A) with R=CF_3 into

F_3C

F_3C

which on polymerization produces a precursor that is stable at room temperature and is converted to *trans*-polyacetylene on heating at 57°C–67°C.

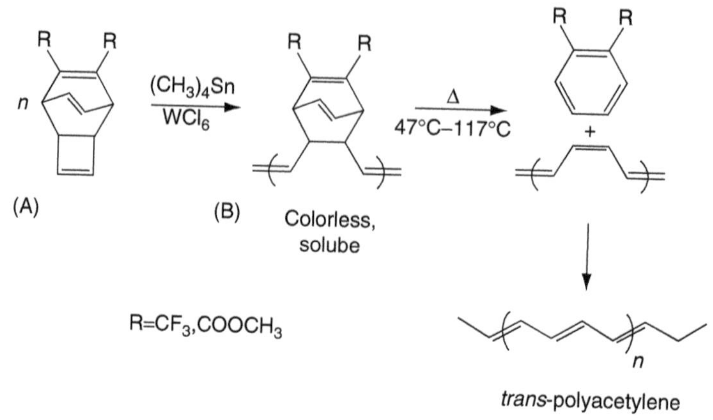

FIGURE 2.15   Durham route for the synthesis of polyacetylene.

The advantages of the Durham route are: (1) contaminating catalyst residues can be removed because the precursor polymers are soluble and can be purified by dissolution and reprecipitation; and (2) the precursors can be cast as films or drawn and oriented prior to conversion to the all-trans form of polyacetylene. This allows a degree of control over the morphology of the final product which in the pristine state appears to be fibrous and disordered. Because conductivity increases by alignment of the polymer chains, stretching the film or fiber assists this process and this can be performed using the prepolymer.

The PPP structure has all the characteristics required of a potential polymeric conductor, but it has proved difficult to synthesize a high-molecular-weight material. One method is the polycondensation

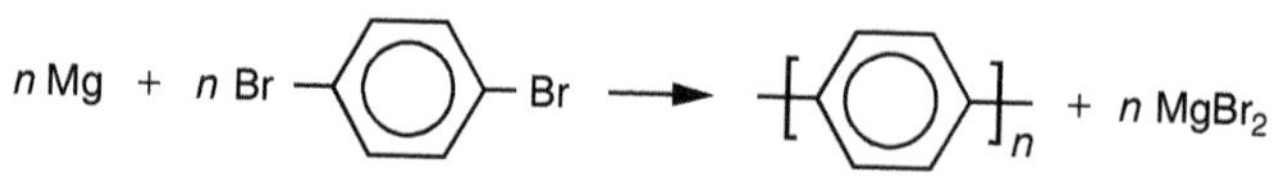

but this only yields oligomeric material, and even this is insoluble.

A novel route developed by workers at ICI overcomes these problems by again making use of a tractable intermediate polymer. Thus radical polymerization of an ester of 5,6-dihydroxycyclohexa-1,3-diene leads to a soluble precursor polymer that can be processed prior to the final thermal conversion into PPP:

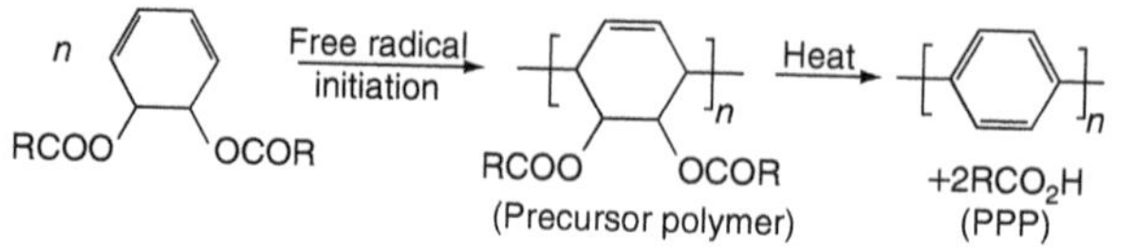

The polymer is very stable and can withstand temperatures up to 450°C in air without degrading. It is an insulator in the pure state but can be both *n*- and *p*-doped using methods similar to those for polyacetylene. However, as PPP has a higher ionization potential it is more stable to oxidation and requires strong *p*-dopants. It responds well to AsF_5, with which it can achieve conductivity levels of 10^2 $(ohm\text{-}cm)^{-1}$.

2.4.2.1   Doping

Because electron mobility is almost nonexistent in most organic polymers, a mechanism for the creation of that mobility is needed and is generally accomplished by the doping process (oxidizing or reducing). In practice, doping occurs by exposing the polymer, most often in the form of thin film or powder, to the doping agent in gas phase or in liquid phase. The doping can also be done electrochemically by using the polymer as an electrode. The principal methods of doping are as follows [16,20].

1. Chemical doping. The doping is accomplished by exposing the polymer to the vapor of a dopant such as I_2, AsF_3, or H_2SO_4, or by immersing the polymer films in a dopant solution, such as

sodium naphthalide in tetrahydrofuran, I_2 in hexane, and so on. The amount of dopant incorporated in the material (doping level) depends on the vapor pressure of the dopant or its concentration, the doping time, and the temperature.

2. Electrochemical doping. Polymers can be doped electronically in an electrochemical cell, such as by immersing the material as an electrode in an organic electrolyte solution ($LiClO_4$ in tetrahydrofuran or propylene carbonate, with lithium as a counter electrode) or in aqueous electrolytes ($PbSO_4/PbO_2$ in sulfuric acid solutions, with lead as a counter electrode).

  Figure 2.16 shows, schematically, electrochemical doping in comparison with chemical doping. During electrochemical doping, the electrons are driven by an outside source of current. The lack of electrons and then the surplus of electrons which occurs, is balanced electrostatically by the dissociated ions in the electrolyte, which diffuse into the polymer. The end result is a polymer that is *p*-doped and *n*-doped, respectively.

3. *Ion implantation.* Polymers, when exposed to proper ion beams, can become conducting due to the implantation of ions in the polymer's lattice, which form covalent bonds with the material. The doping level depends on the energy of the ion beam.

4. *Photochemical doping.* This is accomplished by treating the polymer with a dopant species that is initially inert towards the materials, followed by irradiation. For example, diphenyliodonium hexafluoride arsenate in methylene chloride or triarylsulfonium salts, followed by ultraviolet irradiation.

The academic research in the area of conductive polymers was mainly directed towards polyacetylene, while industry, gave priority to more stable basic materials such as PPP, polypyrrole, PPS, and

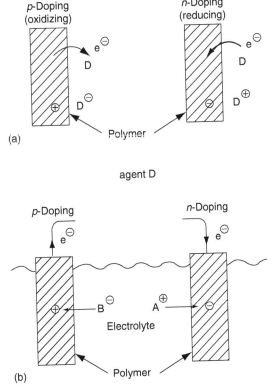

FIGURE 2.16 Schematic comparison of chemical doping and electrochemical doping. For electrochemical doping an outside power source is required (not drawn). (a) Direct charge transfer with doping agent D in chemical doping. (b) Charge transfer in electrochemical doping.

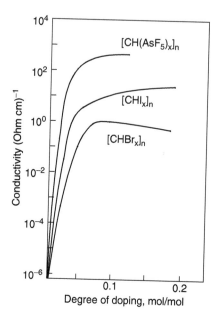

FIGURE 2.17 Variation of conductivity of polyacetylene on degree of doping. (After Chiang, C. K., Gau, S. C., Fincher, Jr., C. R., Park, Y. W., Macdiarmid, A. G., and Heeger, A. J. 1978. *Appl. Phys. Lett.*, 33(1), 18.)

polythiophene. Today, the academic research is extensive even in those areas, as well as a number of other conjugating polymers. A large number of dopants have been used to dope polyacetylene and other conjugated polymers.

Electron-attracting dopants, such as Br_2, I_2, AsF_5, H_2SO_4 and $HClO_4$ may oxidize polyacetylene to produce p-type conductivity. As shown in Figure 2.17, the conductivity increases continuously with the degree of doping to a saturation level. Some of the powerful dopants have been used to oxidize other conjugated polymers. Some examples of p-type dopants and their measured room temperature conductivities are given in Table 2.9.

Electron-donating dopants may reduce polyacetylene or other conjugated polymers giving rise to n-type conductivity. Doping occurs when the polymer is immersed in a tetrahydrofuran solution of radical anion/alkalide where the alkalide components can be Li, Na, K, Cs, or Rb, and the radical anion can be naphthalene, anthracene, or benzophenone. A doping reaction such as the following occurs (Nph = naphthalene):

$$(CH)_x + 0.10x(Na^+Nph^-) \rightarrow [Na^{0.1}(CH)]_x + 0.10xNph \qquad (2.7)$$

When the polymer is allowed to contact a K/Na alloy at room temperature, n-doping also occurs. In n-doped polymers, the polymer chain is considered as a polycarbanion associated with the corresponding M^+ metal ion. A few examples are given in Table 2.10.

Polyacetylene, because it represents the simplest of the conjugated organic polymers, has received much attention as a fundamental material, although other polymers such as polyphenylene, polypyrrole,

TABLE 2.9 Examples of Oxidation/p-Type Doping [16]

Doped Polymer	Conductivity, $(ohm\text{-}cm)^{-1}$
$[CH(I_3^-)_{0.10}]_n$ [a]	5.5×10^2
$[CH(AsF_5)_{0.10}]_n$ [a]	1.2×10^3
$[CH(SbF_5)_{0.006}]_n$ [a]	5.0×10^1
$[CH(H_3O^+HSO_4^-)_{0.10}]_n$ [a]	1.2×10^3
$[CH(FeCl_3)_{0.09}]_n$ [a]	8.5×10^2
$[CH(MoCl_5)_{0.10}]_n$ [a]	2.0×10^2
$[CH(ClO_4)_{0.065}]_n$ [b]	5.0×10^2
$[C_6H_4(AsF_5)_{0.26}]_n$ [a]	1.5×10^2
$[C_6H_4S(AsF_5)_x]_n$ [a]	2.5×10^1
Polypyrrole/$(C_6H_5)_2IAsF_6$ [c]	3.0×10^{-3}
Polyaniline/60% protonation	5.0×10^0

Note: 1 $(ohm\text{-}cm)^{-1} = 1$ S cm^{-1}.
[a] Chemical doping.
[b] Electrochemical doping.
[c] Photochemical doping.

TABLE 2.10 Examples of Reduction/n-Type Doping [16]

Doped Polymer	Conductivity, $(ohm\text{-}cm)^{-1}$
$[Li_{0.30}(CH)]_n{}^a$	2.0×10^2
$Na_{0.21}(CH)]_n{}^a$	2.5×10^1
$[K_{0.16}(CH)]_n{}^a$	5.0×10^1
$[(n\text{-}Bu_4N)_{0.03}(CH)]_n{}^b$	2.5×10^1
Polyquinoline/Li^b	2.0×10^1
Polyquinoline/Na^a	5.6×10^{-1}
$(CH)_n$/LiAlH$_4{}^a$	6.0×10^0

^{a} Chemical doping.
^{b} Electrochemical doping.

polyquinoline, polyquinoxyline, polyphenylene sulfide, and polythiophene are also being increasingly studied. PPP was the first example of a nonacetylene hydrocarbon polymer having an intrinsic conductivity of less than 10^{-14} $(ohm\text{-}cm)^{-1}$, but, as with polyacetylene, PPP can be doped with AsF$_5$ to give conductivities of around 10^2 $(ohm\text{-}cm)^{-1}$. It may also be doped with alkali metals to provide highly conductive n-type materials.

The main advantage of poly-p-phenylene is that, due to its nonacetylenic composition, it has a much higher thermal stability (450°C in air and 550°C in inert atmosphere). Potential applications of poly-p-phenylene are similar to those envisaged for polyacetylene, such as Schottky barriers in photocells. (A Schottky barrier is a metal semiconductor contact that has rectifying characteristics similar to a p-n junction.)

One of the most scientifically and technologically studied organic polymers is polypyrrole. Polypyrroles (PPys) have prompted considerable research because they are the only conducting polymers that can be produced directly in the doped state. The highly stable, flexible films of polypyrrole produced by one-step electro-oxidation have room temperature p-type conductivities ranging from 10 to 100 $(ohm\text{-}cm)^{-1}$. No additional dopants are required to produce electrical conductivity.

Typically, when a solution of pyrrole (0.06 M) and tetraethylammonium tetrafluoroborate $Et_4N^+BF_4^-$ (0.1 M) in acetonitrile containing 1% water is electrolyzed, an insoluble blue-black film of conducting polymer is produced at the anode. The film contains BF_4^-, has a conductivity of about 100 $(ohm\text{-}cm)^{-1}$, and the composition is:

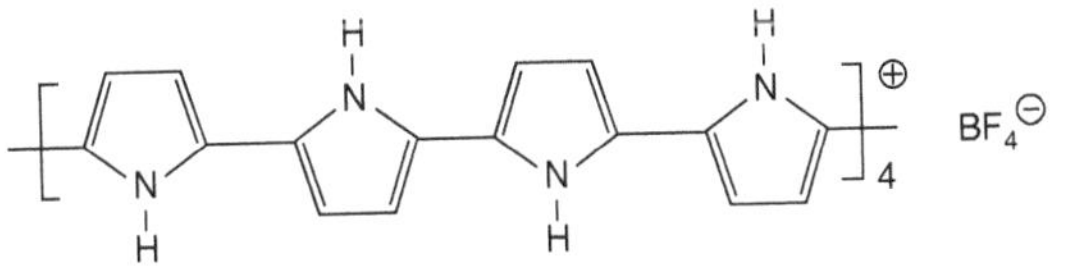

In addition to showing high electrical conductivity, polypyrrole films can be repeatedly electro-chemically driven or switched between the blue-black conducting (100 ohm^{-1} cm^{-1}) oxidized form and a yellow nonconducting neutral. The switching rate for thin films is approximately 1 per second.

One of the principal advantages of polypyrrole over other doped polymers is its excellent thermal stability in air. It is thermally stable up to 250°C, showing little degradation of its conducting properties below this temperature. Above 250°C, the rate of weight loss increases with temperature.

Polyaniline results from the reaction of aniline with ammonium persulfate in aqueous HCl. The reaction produces polyaniline as a dark blue powder with a conductivity of 5 $(ohm\text{-}cm)^{-1}$. The structure of the conducting form of the polymer is believed to be the di-iminium salt:

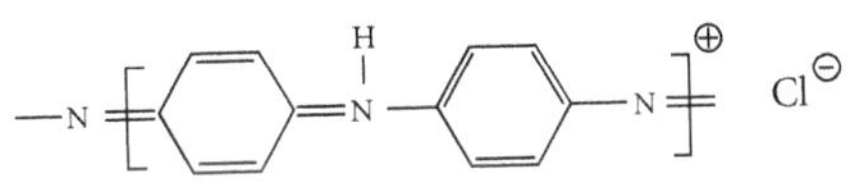

Polyanilines, from an industrial point of view, are in many applications the preferred conducting polymer system, as they offer a number of advantages over other conducting polymers. They are generally soluble and environmentally stable polymers that are made by a one-step synthesis involving inexpensive raw materials. They offer extensive chemical versatility, which allows the properties of the polymer to be tuned to the needs of different applications. Thus, many polyaniline derivatives exist today as a result of chemical modification of polymer backbone, dopant, and oxidation state.

A number of water-soluble polyaniline derivatives have been developed in recent years. Incorporation of sulfonate groups onto the polymer backbone imparts water solubility to the polymer. In one process, this is accomplished by treating the polymer with fuming sulfuric acid which results in a sulfonic acid ring-substituted derivative that is alkali soluble but only upon conversion to the nonconducting sulfonate salt form.

A second method of introducing sulfonate groups is accomplished by deprotonating polyaniline base and reacting with a sultone, i.e., 1,3-propanesultone [22]. This gives rise to an *N*-substituted polyaniline derivative that is water soluble. Another route involves the polymerization of sulfonated aniline monomer such as sodium salt of diphenylaminesulfonic acid [23].

IBM has introduced a family of water-soluble polyanilines under the trademark PanAquas [24]. This is a series of polymers that are highly soluble in neutral water in the conducting form. They are made in a one-step straightforward synthesis involving a template-guided polymerization (Figure 2.18). In this process, the aniline monomer is first complexed to a polymeric acid and then polymerized in a controlled fashion to allow the polyaniline chain, as it grows, to wrap around the polyacid chain. The resulting polyaniline/polyacid blend is water soluble and is conducting. Different polyaniline derivatives can be made by this method by variations in the nature of the aniline monomer (i.e., variation in R in Figure 2.18) and in the nature of the polyacid. The conductivity of these polymers is on the order of 10^{-2}–10^{-3} (ohm-cm)$^{-1}$.

PPS is unique amongst polymers utilized for electrical conduction in that it is available commercially as an engineering thermoplastic (Ryton by Philips Petroleum) and is finding use as a molding component for the encapsulation of electronic components.

Unfilled PPS is a white polymer with a melting point of 288°C and is intrinsically insulating, having a conductivity of less than 10^{-16} (ohm-cm)$^{-1}$. Doping with AsF_5 leads to an increase in conductivity to around 1–10 (ohm-cm)$^{-1}$, although there is also a deterioration in mechanical properties and a color change to dark blue.

Liquid-phase doping has been used for PPS. Thus, PPS has been doped and dissolved simultaneously by exposing PPS particles suspended in AsF_3 to the doping agent AsF_5. Films of doped PPS with good mechanical properties and conductivities up to 200 (ohm-cm)$^{-1}$ can be cast from such solutions of

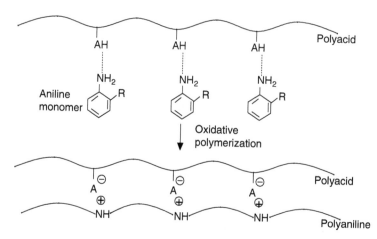

FIGURE 2.18 Template-guided polymerization to make water-soluble polyaniline (PanAquas). Aniline monomer is complexed to a polyacid template and then oxidatively polymerized inacontrolled fashion. (After Nguyen, M. T., Kasai, P., Miller, J. L., and Diaz, A. F. 1994. *Macromolecules*, 27, 3625.)

AsF$_5$-doped PPS in AsF$_3$. The resulting films show improved strength and flexibility compared to those obtained by standard doping of solid PPS with AsF$_5$.

2.4.2.2 Conducting Mechanisms

The nature of the processes that induce high conductivity in polymers is different from those in the inorganic semiconductors. In the doping of inorganic semiconductors such as silicon, the dopant species occupies positions within the lattice of the host material, resulting in the formation of either electron-rich or electron-deficient sites with no charge transfer occurring between the sites. The doping reaction in polymers, on the other hand, is a charge-transfer reaction, resulting in the partial oxidation or reduction of the polymer, rather than the creation of holes.

It is now well established [25] that the exposure of polyacetylene to an oxidizing agent X (or reducing agent M) leads to the formation of a positively (or negatively) charged polymeric complex and of a counterion which is the reduced X$^-$ (or the oxidized M$^+$) form of the oxidant (or the reductant). Thus studies based on Raman spectroscopy and Mössbauer spectroscopy show that after exposure to iodine, the polyacetylene chain acts as a positively charged polycation with I_3^- or I_5^- acting as counterions.

Similar studies based on optical absorption and ESR spectra indicate that the reduction of polyacetylene by exposure to sodium naphthalide in tetrahydrofuran results in negatively charged $(CH)_x^-$ carbanions and N$^+$ counterions.

The doping process in the case of conducting polymers may therefore be more correctly classified as redox processes of the following general scheme:

$$\text{Polymer} + X = (\text{Polymer})^{n+} + X^{n-} \tag{2.8}$$

where X = I$_2$, Br$_2$, AsF$_5$,..., in the case of an oxidation (*p*-doping) process and

$$\text{Polymer} + M = (\text{Polymer})^{n-} + M^{n+} \tag{2.9}$$

where M = Na, Li,.... for a reduction (*n*-doping) process.

The above reactions are most likely to occur in the case of unsaturated polymers with π electrons as they can be easily removed from the polymeric chains to form polycations or added to the chains to form polyanions and, therefore, these are the types of polymers which assume high conductivity on doping.

Contrary to inorganic crystalline semiconductors, where charge is transported in general by electrons in the conduction band and holes in the valence band, in doped conjugated polymers charged solitons, polarons, and bipolarons act as charge carriers. These quasi-particles are the direct consequence of the strong electron-phonon interaction present in these quasi-one-dimensional polymers.

In polyacetylene, the *trans* structure is unique in that it has a twofold degenerate ground state as shown by sections (a) and (b) in Figure 2.19, which are mirror images, and the single and double bonds can be interchanged without changing the energy. Where an A sequence meets a B sequence, a free radical is produced (Figure 2.19c). This is a relatively stable entity and the resulting defect in the chain is called a neutral soliton (from solitary wave solutions, a mathematical treatment of conductivity), which corresponds, in simple terms, to a break in the pattern of bond alternation, i.e., it separates the

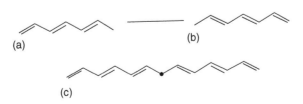

(a)

(b)

(c)

FIGURE 2.19 (a) and (b) Degenerate ground state of *trans*-polyacetylene. (c) Polyacetylene chain with a defect (neutral soliton).

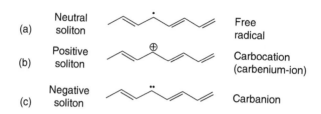

FIGURE 2.20   Formation from (a) a neutral soliton, (b) a positive soliton and (c) a negative soliton by controlled addition of *p*- and *n*-doping agents, respectively.

degenerate ground state structures. The electron has an unpaired spin and is located in a nonbonding state in the energy gap, midway between the two bands. It is the presence of these neutral solitons that gives *trans* polyacetylene the characteristics of a semiconductor with an intrinsic conductivity of 10^{-7}–10^{-8} $(ohm\text{-}cm)^{-1}$.

This conductivity can be magnified by doping. Controlled addition of an acceptor or *p*-doping agent such as AsF_5, Br_2, I_2, or $HClO_4$ removes an electron and creates a positive soliton. In chemical terms this is the same as forming a carbenium ion (Figure 2.20b) that is stabilized by having the charge spread over several monomer units. Similarly, a negative soliton (Figure 2.20c) can be formed by treating the polymer with a donor or *n*-doping agent that adds an electron to the mid-gap energy level.

At high doping levels, the soliton regions tend to overlap and create new mid-gap energy bands that may merge with the valence and conduction bands, allowing freedom for extensive electron flow and thus making the polymer a conductor.

In the case of defect-free *trans*-polyacetylene chain, a charge transfer directly between doping agents and valence and conduction band, respectively, will produce an ion radical in the chain, i.e., a defect pair instead of an isolated defect. Figure 2.21 shows the mechanism of *p*-doping of *trans*-polyacetylene.

It can be shown theoretically that a separation of ion-radical components along the chain into one charged and one neutral soliton corresponds to a higher energy state compared to when the defect pair is held together. This more stable ion-radical pair is called polaron, and its formation is accompanied by two energy levels obtained symmetrically above and below the middle level of the band gap. When the doping degree is increased further, i.e., when polaron concentration is increased, the polarons begin to interact with each other. Two polarons interact to give two charged solitons (i.e., two unpaired electrons are united leaving behind two charged solitons), which produce a band halfway between the valence band and the conduction band.

A lattice of charged solitons (Figure 2.21) is obtained at about 4 mol% doping. At metallic conductivity, the soliton band overlaps the entire band gap between the valence band and the conduction band so that unpaired electrons with spin can contribute to the conductivity. This occurs at 11% doping of *trans*-polyacetylene.

In contrast to *trans*-polyacetylene, all other conjugated polymers, including *cis*-polyacetylene, possess nondegenerate ground states and this affects the nature of charge carriers they can support. In such

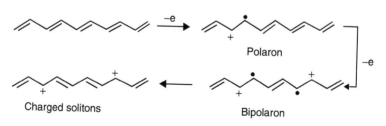

FIGURE 2.21   *p*-Doping of defect-free *trans*-polyacetylene.

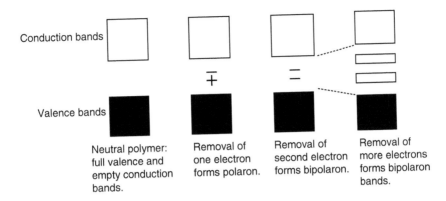

Conduction bands

Valence bands

Neutral polymer: full valence and empty conduction bands.

Removal of one electron forms polaron.

Removal of second electron forms bipolaron.

Removal of more electrons forms bipolaron bands.

FIGURE 2.22 Band structure of the *p*-doped poly(*p*-phenylene).

polymers, it has been found that the formation of single solitons, whether as a result of doping or from inherent defects, is energetically unfavorable and the energetically preferred configurations involve paired sites. This can be illustrated by considering the example of PPP.

In PPP and most other polyconjugated conducting polymers, the conduction occurs via the polaron or bipolaron. The band structure for the *p*-doped PPP is shown schematically in Figure 2.22.

2.4.2.3 Applications

The interest on the part of industry in inherently conductive polymers is great because of the possibilities that lie in the workable qualities and light weights of polymer materials. The principal interest is, however, in their potential use for rapid, low-cost processing using film-forming polymer solutions.

2.4.2.3.1 *Rechargeable Batteries*

One of the most promising areas of application of inherently conducting polymers is as electrode materials in rechargeable batteries. Conducting polymers are promising charge storage materials because of their high charge carrier concentrations.

A polymeric electrode is doped electrochemically when the battery is charged, and is undoped at discharge (see Figure 2.16). The fundamentally appealing aspect of polymer electrodes is that charge and discharge are based on purely physical and reversible processes. There are no chemical processes to consume material. Whereas metal electrodes are continuously subjected to dissolution and redeposition during the charge-discharge cycles, resulting in mechanical wear, with polymer electrodes the ions can enter or leave without significant disturbance of the polymer structure. Polymer electrodes thus have a longer life than metal electrodes.

A battery cell where both the electrodes consist of dopable polymer is shown in Figure 2.23. The electrolyte in this case consists of $Li^+ClO_4^-$ dissolved in an inert organic solvent, usually tetrahydrofuran or propylene carbonate. When two sheets of polyacetylene or PPP are separated by an insulating film of polycarbonate saturated in an electrolyte (lithium perchlorate), and completely encapsulated in a plastic casing, a plastic battery can be made. The two sheets of polyacetylene or PPP act as both anode and cathode for the battery. A schematic is shown in Figure 2.24. Although doped polyacetylene and polyaniline electrodes have been developed, polypyrrole-salt films are the most promising for practical application.

It is also possible for only one electrode to consist of dopable polymer and for the other be of metal, usually lithium (Figure 2.25), which is preferred because of its low weight and high thermodynamic electrode potential.

The comparable data for two conventional and two polymeric batteries, which are given in Table 2.11, show that, in energy absorption per unit weight, the polymeric batteries are not superior to the lead-acid battery. What is most significant, however, is the fact that the power density is calculated here to be higher

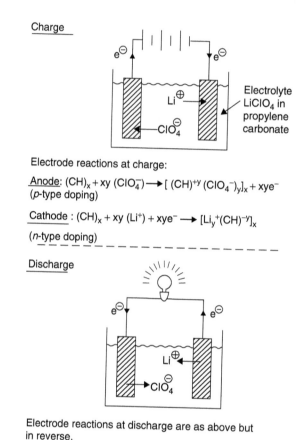

FIGURE 2.23   Secondary battery with PAc electrodes. Here PAc is given the chemical symbol $(CH)_x$.

by an order of magnitude. This reflects, apart from the higher voltage, the considerably higher current density in polymer batteries that is attributed to the act that the ions in the electrolyte are able to wander quickly in and out in the electrode as it is doped and undoped. This is facilitated by the porous nature of polyacetylene film consisting of very thin fibers with a diameter of about 200 A $(2 \times 10^{-8}$ m), which only

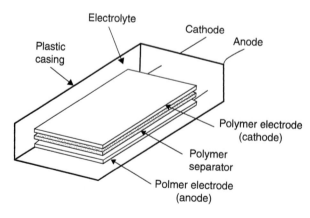

FIGURE 2.24   Schematic of plastic battery.

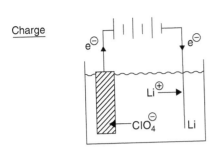

Charge

Electrode reactions at charge:

Anode: $(CH)_x + xy\ (ClO_4^-) \longrightarrow [\ (CH)^{+y}\ (ClO_4^-)_y]_x + xye^-$

Cathode : $xyLi^+ + xy\ e^- \longrightarrow xyLi$

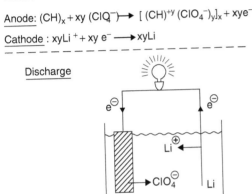

Discharge

Electrode reactions at discharge are as above but in reverse.

FIGURE 2.25 Electrochemical oxidation and reduction reactions in polyacetylene/lithium batteries.

occupies one-third of the volume of the film and gives the material an exceptionally large effective surface ($40–50\ m^2/g$).

Though the rate of diffusion for a doping agent in polyacetylene itself is very low ($D = 10^{-15}\ cm^2/s$ for Li in polyacetylene), it is compensated for by the fiber structure, which reduces the diffusion length when the material is impregnated with a liquid electrolyte.

Though the charging and discharging of polymer electrodes are based on purely physical and reversible processes, in practice, however, disturbing side reactions occur that give reduced efficiency. This happens especially at complete discharge-recharge. Therefore, only about one-third of the total charge should be utilized. (However, this also applies to lead-acid batteries where the normal operating range is between 40% and 80% of full charge.) Better cell construction and more stable electrolytes need to be developed in order to increase efficiency and useful life of polymeric batteries.

In order for electrically operated vehicles to become generally accepted, batteries with higher energy- and power-density in combination with lower weight and volume are required. Polymeric batteries show promise of making this a reality. Indeed, polymer batteries have been heralded as the invention that would make electric cars a reality, because the lightweight polymer batteries would not weigh a car down to the extent as the heavy lead acid batteries would do.

An all-plastic battery may have many advantages [26]. For example, a car battery made of polyacetylene could weigh only one-tenth of that of a conventional lead-acid battery. Moreover, batteries with plastic electrodes could be fabricated into odd shapes, such as a flat disc that could be slotted into a car door. Prototype batteries have been made using polyacetylene and poly-*p*-phenylene electrodes, but a number of technical problems, such as long-term mechanical integrity need to be solved.

A derivative of polyacetylene that is more stable in air and soluble in organic solvents when doped with iodine, is polytrimethyl-silylacetylene (PTMSA). Though the derivative has a conductivity only one ten-millionth that of doped polyacetylene, it can be used as precursor to a wide range of other conductive derivatives. One such derivative is polyfluoroacetylene (polyacetylene in which all of he hydrogen atoms are replaced by fluorine), which according to calculations, would be electrically conductive without the need of dopants.

TABLE 2.11 Comparison of Conventional and Polymeric Batteries [13]

Battery Type	Initial Voltage (V)	Energy Density (Wh/kg)	Power Density (kW/kg)
Lead-acid battery	2	30	0.4
Zinc–bromine	1.8	65	0.14
$(CH)_x/(CH)_x^a$	3.7	30	5
$(C_6H_4)_x/(C_6H_4)_x^b$	4.4	40	5

[a] Polyacetylene electrodes.
[b] Poly(*p*-phenylene) electrodes.

The Bridgestone Corporation of Japan has developed coin-type rechargeable polymer lithium batteries with a conducting polymer polyaniline and the high capacity lithium aluminum alloy as the two electrodes. The characteristics of these batteries are low self-discharge, high voltage, and long-term reliability. One of the unique features of this rechargeable polymer lithium battery is that it can be used as a power source in combination with solar cells. This feature opens up significant possibilities for application as a power source. The operation of electronic clocks, watches, calculators, remote control devices of audiovisual units, etc., are some of the potential application areas of this technology.

2.4.2.3.2 *Electrochromic Devices*

An electrochromic display (ECD) is a thin solid state device that changes color reversibly when subjected to a small electrical potential. Since the doping processes of certain conducting polymers are accompanied by changes in the color, this effect has been conveniently exploited in the realization of ECD devices. Thin films of a conducting polymer polythiophene, for example, are red in the doped state and deep blue in the undoped state.

The fact that a material changes color is, however, not sufficient to permit its use in a display system; its switching time between these two states must be very short and it should maintain these properties over many doping/undoping cycles. In the case of polythiophene, switching time of around 30 millisec and a stability exceeding one million doping/undoping cycles have been measured. These values correspond to about 2 years of operation for a frequency of one cycle per minute.

Some of the applications of ECDs include time tables in airports and train stations, calculators, computers, clocks, and any other piece of equipment that utilizes a liquid crystal display.

ECD devices have several advantages over liquid crystal devices. ECDs have low power consumption, good optical contrast, a wide viewing angle, and an all solid-state construction. ECD devices may be constructed in large dimensions and have optical memory since the color acquired remains also after the driving voltage has been removed.

The disadvantage of an ECD is mainly associated with its comparatively slow response time since the driving mechanism deals with ions which generally have low diffusion coefficients in polymers.

The principle of electrochromic devices can be exploited in tinting ordinary window glass. Very thin polymer layers embedded in a colorless Solid electrolyte and sandwiched between two layers of glass may tint a window when an electric potential is applied. The degree of tinting can be controlled by the size of the electric potential.

2.4.2.3.3 *Sensors*

Because their chemical and physical properties may be tailored over a wide range of characteristics, the use of polymers is finding a permanent place in sophisticated electronic measuring devices such as chemical sensors, pH sensors, ion-selective sensors, humidity sensors, biosensor devices, etc [27]. Chemical sensors are used for detecting and measuring the concentration of various chemical species in liquid or gas phase. The conductivity of conducting polymers with conjugated π bonds depends on the electronic structure, which can undergo changes under the influence of chemical species adsorbed on the surface. For example, such changes may result from redox or acid-base type interactions between polymer and the chemical species.

Conducting polymer sensor arrays for gas and odor sensing based on substituted polymers of pyrrole, thiophene, aniline, indole and others were reported in 1984 at the European Chemoreception Congress (ECRO), Lyon, followed by a detailed paper in 1985 [28]. Nucleophilic gases (ammonia and alcohol, such as methanol and ethanol, vapors) cause a decrease in conductivity, while electrophilic gases (NO_x, PCl_3, SO_2) have the opposite effect. Most of the widely studied conducting polymers in gas sensing applications are polythiophene and its derivatives [29], polypyrroles [30], polyaniline and their composites [31,32].

Electrically conducting polystyrene/polythiophene, polystyrene/polypyrrole, polyacrylonitrile/poly-pyrrole, polycarbonate/polythiophene, polycarbonate/polypyrrole composites are prepared by electropolymerization of the conducting polymers into the matrix of the insulating polymers polystyrene, polyacrylonitrile, and polycarbonate, respectively.

Acrylic acid doped polyaniline exhibits sensor response in terms of the dc electric resistance on exposure to ammonia. The change in resistance is found to decrease linearly with NH_3 concentration up to 58 ppm and saturate thereafter [33]. The decrease in resistance is explained on the basis of removal of a proton from the acrylic acid dopant by the ammonia molecules, thereby rendering free conduction sites in the polymer matrix.

Electroactive nanocomposite ultrathin films of polyaniline (PANI) and isopolymolybdic acid (IPMA) for detection of NH_3 and NO_2 gases have been fabricated by alternate deposition of PANI and IPMA following Langmuir–Blodgett (see later) and self-assembly techniques [34]. While the NH_3-sensing involves de-doping of PANI by basic ammonia, in NO_2 sensing NO_2 plays the role of an oxidative dopant, causing an increase in the conductivity.

For optical pH sensors based on PANI, the effect of pH on the change in electronic spectrum of the polymer is explained by the different degree of protonation of the imine nitrogen atoms in the polymer chain [35]. For the quantitative analysis of ions in solutions by ion selective sensors (ISS), polymers have been utilized to entrap the sensing elements, the ion selectivity being conveyed by ionophore (e.g., ion-exchange agents, charged carriers, and neutral carriers) doped in polymeric membranes. Ion sensors find wide application in medical, environmental and industrial analysis. They are also used for measuring the hardness of water. Polyacetylene can be used for determining the concentration of nitrate ions in acid solutions because as a result of the intercalation oxidation the conductivity of the polymer changes. The use of conducting polymers for preparing enzymatic electrochemical microsensors sensitive to glucose content has been reported.

Polyaniline and its substituted derivatives, such as poly(*o*-toluidine), poly(*o*-anisidine), poly(*N*-methylaniline), poly(*N*-ethylaniline), poly(2,3-dimethylaniline), poly(2,5-dimethylaniline) and poly (diphenylamine) have been reported [36] to show measurable responses (sensitivity ∼60%) for short chain alcohols (viz., methanol, ethanol and propanol) at concentrations up to 3000 ppm. The change (decrease) in resistance of the polymers on exposure to alcohol vapors has been explained based on the vapor-induced change in the crystallinity of the polymer. Polypyrrole (PPy) incorporated with dodecyl benzene sulfonic acid and ammonium persulfate has been reported to show a linear change in resistance when exposed to methanol vapor in the range 87–5000 ppm [37]. The response is rapid and reversible at room temperature.

Conducting polymers such as polyfuran and polythiophene may have valuable uses in humidity sensors and radiation detectors as the conductivity of these doped polymers varies considerably when exposed to humidity or radiation.

PPy has been widely used for gas sensors and biosensors [38,39]. The physical and chemical properties of PPy depend strongly on the nature of the dopant anions and their interactions with other chemical species. Though thin films of PPy can be very easily made by electrochemical deposition, a challenge in their applications in thin film-based devices is that the active sensing components remain embedded in the bulk, which limits both the efficiency and the sensitivity. This can be improved by making PPy in a nanofiber form to generate high surface area for a given mass or volume, thus providing a large number of surface functional groups to which sensing chromophores or enzymes can be anchored [40]. Moreover, the nanofiber texture can enhance the transport of ions and chemicals from the solution to the interior of the sensor component.

The production of PPy in a nanofiber form has been traditionally accomplished through templated synthesis methods which employ mesoporous silica, anodized aluminum oxide membrane, and particle track-etched membranes [41–43], while a bulk growth approach using V_2O_5 seed as a template has been reported more recently [44]. However, nanofibers produced by these methods are typically very short and not easy to handle for device fabrication. An electrospinning technique has thus been widely used for producing polymeric nanofibers in a nonwoven mat that is amenable to handling macroscopically [45]. (Note: Electrospinning is a process that produces continuous polymer fibers through the action of an external electric field imposd on a polymer solution. For a review of the process see previous publications [45–47].)

Nanofibers can also be deposited directly on device substrates. Polyaniline being soluble, the electrospinning process has been successfully utilized for the production of polyaniline nanofibers [48]. However, this process cannot be directly employed for producing PPy nanofiber due to the intractability of PPy.

To overcome the problem of intractability of PPy, a simple two-step process has been reported recently [49] for the synthesis of electrically conducting PPy nanofibers. The method consists of electrospinning of poly(ethylene oxide) (PEO) nanofibers that contain Fe(III) oxidants, followed by vapor phase polymerization of pyrrole. PEO is chosen as it forms a complex with $FeCl_3$, which is known to be one of the most efficient oxidants for pyrrole polymerization and leaves chloride ions in the produced PPy making it electrically conducting. Since the Fe^{3+} ions are bound by the coordinating oxygen atoms of the PEO chains, it suppresses crystallization of $FeCl_3$ and ensures homogeneous distribution of $FeCl_3$, and hence of the produced PPy, along the PEO nanofibers as PPy is produced by the vapor phase polymerization [50] through diffusion of pyrrole monomer into the $FeCl_3$-containing PEO nanofibers.

In a typical process [49], a homogeneous solution of PEO-$FeCl_3$ (10.7 wt.%) prepared by dissolving a mixture of PEO and $FeCl_3$ (1:2.5 by weight) in a solvent consisting of ethanol and water (1:2.3 by weight) is loaded into a glass pipette (Figure 2.26) and a positive bias of 8 kV is applied to the solution. The electrospun fibers are collected on a ground electrode (aluminum foil) placed ~ 10 cm from the glass pipette tip. The electrospun PEO-$FeCl_3$ nanofibers are then exposed to pyrrole vapor at ambient conditions (8.3 torr at 25°C), the exposure time varying from 1 to 14 d for vapor phase polymerization to take place. The sheet conductivity of the as produced PPy-PEO composite nanofiber (average diameter 96 nm) mat at room temperature is in the order of 10^{-3} S.cm^{-1} [49].

2.4.2.3.4 *Microelectronics*

Inherently conducting polymers offer a unique combination of properties that make them attractive for use in microelectronics [51]. The conductivity of these materials can be tuned by chemical manipulation of the polymer backbone, by variation of the dopant and degree of doping, and by blending with other polymers. In addition, they offer the advantages of light weight, processability, and flexibility. The conducting polymers have potential for use in an array of microelectronic applications ranging from the device level to the final electronic product.

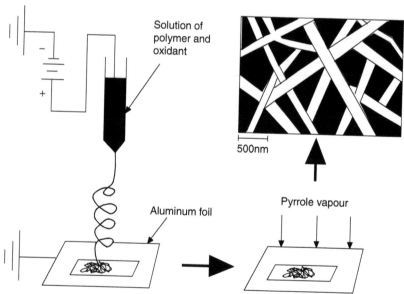

FIGURE 2.26 Schematic of electrospinning of PEO nanofiber templates containing $FeCl_3$ followed by vapor phase polymerization of pyrrole. An SEM image of the resulting PPy-PEO composite nanofibers is shown at top right corner. The scale bar for the image is about 500 nm. (After Nair, S., Natarajan, S., and Kim, S. H. 2005. *Macromol. Rapid Commun.*, 26, 1599. With permission.)

During the e-beam patterning process, charging of the resist is a significant problem. Thick insulating resist materials can trap charge and the trapped charge can deflect the path of the e-beam, resulting in image distortion as well as level-to-level registration errors. To eliminate this problem, conducting materials are incorporated into resist systems (see later) as coatings above or below the imaging resist to function as discharge layers. Intrinsically conducting, in particular the soluble, derivatives are attractive charge dissipators for e-beam lithography. These materials combine high conductivity with ease of processability. The first conducting polymer to be used in this type of application is polyaniline.

IBM introduced a family of polyanilines, referred to as PanAquas, which are highly soluble in neutral water in the conducting form. The PanAqua provides a simple and effective discharge layer for e-beam lithography and can be cleanly removed during the alkaline development of the resist. A water-soluble polythiophene derivative referred to as ESPACER (a trade mark and product of Showa Denko K. K.) is also very effective at eliminating resist charging and can be removed by water wash.

A method used to prevent charging of the sample during scanning electron microscopic inspection is to coat the sample with a conducting metal such as gold, which is a destructive process as the metal cannot be removed from the surface of the substrate. Conducting polymers that can be spin-applied onto the sample and subsequently removed cleanly are ideal. Polyaniline has been demonstrated to provide such a solution.

In microelectronics, metallization generally refers to the deposition of a patterned film of conducting material on a substrate to form interconnections between electronic components. Conducting polymers have been demonstrated to provide a new route to metallization, in particular in PCB technology [52]. The conducting polymers that have been of interest in this area include polyaniline, polypyrrole, and polythiophene.

The fact that many conducting polymers are semiconductors in the undoped or slightly doped state has prompted attempts by both industrial and academic groups to use them in microelectronics such as in semiconductor devices and especially field effect transistors (FETs) [53]. For a FET's active layer, i.e., semiconducting layer between the source and drain electrodes, micrometer-thick films are required. Film as thin as this has been prepared by spin coating a solution of a precursor polymer on a substrate having the required electrode pattern and subsequent heat treatment in a stream of gaseous dopant to convert it into a conducting polymer. Improvement of technique has led to carrier mobilities as high as 10^{-1} cm^2/V-s in polymeric FETs. This has opened the way to uses such as flat-panel color displays for computers and possibly flat color televisions when tied to a liquid crystal matrix.

Electroluminescence from semiconducting conjugated polymers was first reported in 1990 [54] using poly(*p*-phenylenevinylene) (PPV) as the single semiconductor layer between metallic electrodes, as illustrated in Figure 2.27. PPV has energy gap between π and π^* states of about 2.5 eV and produces

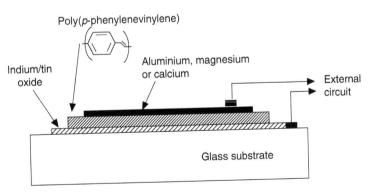

FIGURE 2.27 Schematic structure of a single-layer polymer LED device formed with poly(*p*-phenylene vinylene). (After Burroughes, J. H., Bradley, D. D. C., Brown, A. R., Marks, R. N, Mackay, K., Friend, R. H., Burn, P. L., and Holmes, A. B. 1990. *Nature*, 347, 539.)

luminescence in a band below this energy. Light emitting diodes (LEDs) have been fabricated from PPV as the emissive layer. Operation of an LED is achieved when the diode is biased sufficiently to achieve injection of positive and negative charge carriers from opposite electrodes. Capture of these oppositely charged carriers in the polymer layer can result in the formation of singlet exciton, which is generated by photoexitation across the $\pi-\pi^*$ gap and this can then decay radiatively to produce the same emission spectrum as that produced by photoexcitation.

Further advances in the chemistry of processible conjugated polymers and focused work on the physics of electroluminescence in these materials have led to the development of flexible, almost entirely metal-free LEDs [55]. These polymer-based LEDs could be competitive in display applications because of the potential ease, low cost of fabrication, and large surface area of devices based upon processible polymers.

2.4.2.3.5 *Electrostatic Discharge Protection*

Eletrostatic charge (ESC) and electrostatic discharge (ESD) are a serious and expensive problem for many industries, in particular for microelectronics. Electrostatic charge can accumulate to thousands of volts and discharge in the form of a lightning bolt, which can destroy devices on ICs. To protect devices against ESC and ESD, conducting materials are extensively used. The materials currently used include ionic conductors, carbon- or metal-filled resins and in certain cases metal coatings. They do not offer the ideal solution to ESD protections. Ionic conductors, though inexpensive, have significant drawbacks, such as very low surface conductivity ($10^{-9}-10^{-11}$ ohm^{-1}/square) and so are not dissipative. Moreover, the conductivity is humidity dependent, as water is needed for ionic conduction. Carbon-filled systems, on the other hand, pose contamination concerns due to sloughing of carbon particles.

Intrinsically conducting polymers offer a new alternative to these materials for ESD protection with numerous advantages. The conductivity can be tuned and can easily meet the high end of the dissipative range. In addition, conducting polymers can offer a high degreee of transparency. Polyaniline, polypyrrole, and polythiophene have been the predominant conducting polymers of interest for ESD protection. Doped polyaniline in the form of a dispersible powder (Versicon, a trademark of Allied-Signal Corp.) has been blended with a number of thermoplastic and thermoset resins to achieve excellent ESD properties [56].

Of particular interest for ESD protection of electronic component packages are coating formulations. The coating can be applied directly onto already fabricated packages by spray-coating, or they can be applied onto plastic sheets that are subsequently thermoformed into a package. One type of coating is based on dispersions of doped polyanilines. Versicon coatings based on soluble polyanilines have also been produced. One such system is a curable water-based coating reported by IBM [24]. An aqueous coating based on a poly(3,4-ethylene-dioxythiophene)/polystyrenesulfonic acid blend is reported to exhibit antistatic properties [57].

2.4.2.3.6 *Other Applications*

Development of electrically conducting polymers that can be processed as thermoplastics and possess the advantageous mechanical behavior and corrosion stability of plastics would open up entirely new fields of applications. Some of the many applications in which electrically conducting polymers would find a large market are cables and conducting electricity, wall papers that allow electrical heating, materials for antistatic equipment, and conducting fabrics.

The polymerization and doping of conducting polymers on such fabrics as nylon, polyester, and cotton to produce fabrics with a range of conductivities under the trade name Contex has been achieved by Milliaken and Co., Spartanburg, S.C. These materials are under intensive study for use in aircraft and tanks of the future. They could also be used for heated clothing, protecting computers against EMI, static dissipation in high speed missiles, and conveyor belts to handle static-sensitive electronic and flammable goods.

Electrically conducting polymers are also regarded as promising candidates for use as passivating layers with respect to photocorrosion of photoelectrodes. For example, films of polypyrrole have been tested as coating for certain semiconductor electrodes.

Structures similar to muscle fibers have been proposed by researchers in Japan and Italy. They are based on the property that a conducting polymer in a fiber from undergoes dimensional changes due to expansion and contraction along the fiber's length as a result of electrochemical doping and undoping.

Other possible applications include conductive paints, toners for reprographics and printing, and as components for aircraft where the combination of light weight, mechanical strength, and moderately high conductivity are required.

Recently, electrically conductive fiber composites have been prepared from polypyrrole-engineered pulp fibers [58]. To prepare such fibers, FeCl$_3$ solution is first dispersed into pulp, which is disintegrated by kneading and stirring. The dopant (anthraquinone-2-sulfonic acid, sodium salt) slurry and pyrrole solution are then added to start the polymerization, the molar ratio of FeCl$_3$ to pyrrole and that of dopant to pyrrole being 3:1 and 1:3, respectively, for the optimum effect. Paper composites can be prepared both directly from the modified pulp fibers and by adding the modified fibers as conductive-fiber fillers into the paper making stock. For the latter method, less monomer (i.e., conductive polymer) is needed to achieve the same level of conductivity while a higher tensile strength in the paper is attained, as compared to paper obtained exclusively from treated fibers.

2.4.3 Photoconductive Polymers

The enhanced flow of current under the influence of an applied electric field that occurs when a semiconductor is exposed to visible light or other electromagnetic radiation is known as photoconduction. Poly (*N*-vinyl carbazole) (VII) and various other vinyl derivatives of polynuclear aromatic compounds such as poly(2-vinyl carbazole) (VIII) and poly(vinyl pyrene) (IX) have high photoconductive efficiencies. The excellent photoconductivities of these polymers are believed to be due to their helical condormation with successive aromatic side chains lying parallel to each other in a stack along which electron transfer takes place relatively easily.

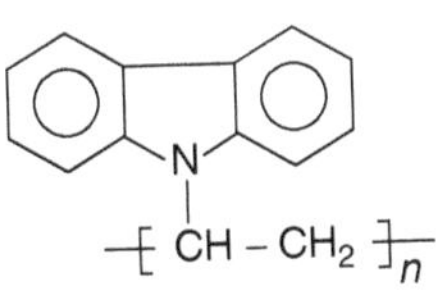

(VII)

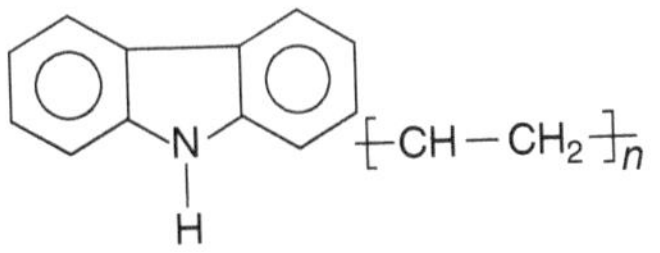

(VIII)

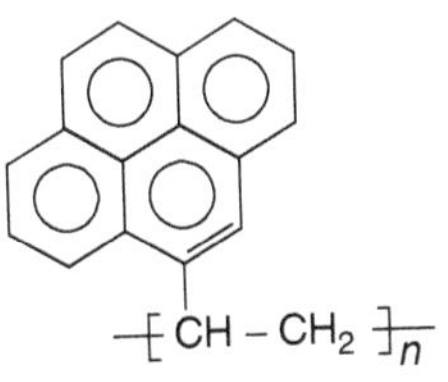

(IX)

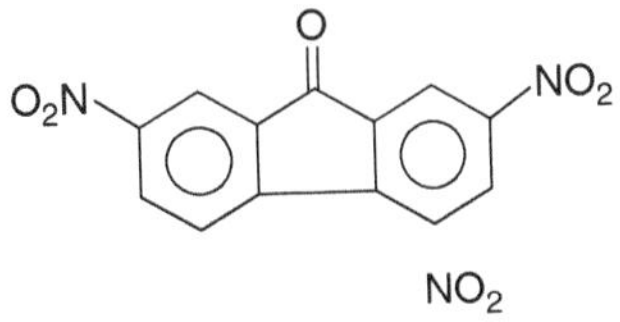

(X)

Poly(*N*-vinyl carbazole) absorbs light in the 360 nm region to undergo electronic excitation and ionization in the electric field. The photogeneration efficiency of the polymer can be greatly enhanced by the addition of an equimolar amount of 2,4,7-trinitrofluorenone (TNF) (X), which shifts the absorption of poly(*N*-vinyl carbazole) into the visible range by the formation of a charge transfer state, rendering it photoconductive at 550 nm. While the polymer alone is a hole conductor, the addition TNF creates electron carriers and the conduction mechanism actually becomes electron dominated.

Photoconduction forms the basis of electroreprography. In this photocopying process, or xerography as it is sometimes known, a photoconductive material is coated onto a metal drum and uniformly charged (sensitized) in darkness by a corona discharge. The drum is then exposed to the bright image of the object to be copied whereupon the illuminated areas of the photoconductive material become conductive and dissipate their charge to the metal drum (earthed). The photoconductive material in the dark areas is still charged and so is able to attract a positively charged black resin-coated toner powder forming a latent image of the object. The latent image is then transferred to a piece of negatively charged paper which is heated to fuse and fix the resin, thereby making the image permanent.

Early photoconductive materials were based on selenium compounds such as As_2Se_3. Such a material is difficult to handle, because it needs to be applied by vacuum sublimation and it is rather brittle. Poly (N-vinyl carbazole)-based materials are now replacing the selenium-based compounds.

Much work has been reported on the development of carbazole derivatives and also noncarbazole photoconductive polymers. An example of the latter group is poly(bis(2-naphthoxy)phosphazene), which is intrinsically an insulator with a very low photosensitivity toward both UV and visible radiation, but when doped with trinitrofluorenone in a 1:1 molar ratio, it is a strong photoconductor.

2.4.4   Polymers in Fiber Optics

Polymeric optical fibers (POFs) combine the benefits of all optical fibers with additional amazing simplicity in handling. This is mainly due to their relatively large diameter and acceptance angle (or numerical aperture). In spite of the outer diameter being typically in the range of 1 mm, the fiber remains flexible because of the polymer material used, mainly acrylics PMMA. These benefits make POF attractive for a wide variety of under-water, indoor and outdoor lighting, data transmission, and sensor application. The fiber optic cable or tube employs the phenomenon of total internal reflection (see Piped Lighting Effect, Chapter 2 of *Plastics Fundamentals, Properties, and Testing*) to carry the light throughout the length of the cable.

In construction, POF cable is made up of a light carrying solid polymer core with a thin protective coating, or covering, called the cladding. The light entering the fiber optic tube is trapped within the core and is continually reflected as it moves down the path, the interface between the core and the cladding of the tube wall acting like a mirror. The light comes out of the other end or is diverted. A less common type of fiber optic cable, called a side-emitting fiber, distributes the light along the entire length of the tubing like neon lighting.

In its simplest form, a fiber optic lighting system consists of an illuminator or light source, and a number of fiber optic cables or light guides that carry the concentrated beam of light produced by the light source (Figure 2.28) [59]. The illuminator can use a filter to remove most of the lamp's infrared and ultraviolet energy that may damage and fade the colors in textiles, paintings and graphic art pieces.

Originally conceived as a means of lighting pools and fountains, the fiber optic technology has since been embraced by lighting architects for several indoor uses due to its low heat production. As the fiber optic lighting does not create electromagnetic fields, this lighting technology can be used in areas with EMF-sensitive electronic equipment, such as magnetic resonance imaging (MRI) room.

As every other optical fiber, POF has a core-cladding structure as shown in Figure 2.29. In most cases, the core material (diameter 980 μm) is highly purified PMMA with a typical refractive index, $n_1 = 1.49$, coated with a cladding (thickness 10 μm) of fluorinated polymer with typical refractive index, $n_2 = 1.42$, while the polymers used for sheath/jacket are PE, HDPE, PVC, and nylon. The numerical aperture (NA) thus has a value of 0.50. With this high NA the bandwidth is limited to theoretical values of some 10 Mbit/s. Since the bandwidth is related to the fiber NA, the transmission capacity can be increased by lowering the NA. However, lowering NA causes increased bending sensitivity. A trade-off has therefore to be found for the optimum POF for a specific application.

As large core POFs suffer from sensitivity to bending, these losses can be reduced by using multiple small cores [Figure 2.30(a)] instead of a single large one. The problem can also be overcome by using the so-called double-step-index fibers [Figure 2.30(b)] as an intermediate step to multi-step-index fibers.

The best solution achieved so far for obtaining high bandwidth with large core diameter is using a graded index profile [Figure 2.30(c)]. Since more than 15 years a lot of progress has been made in shaping the profile by different methods. Until 1996, mainly PMMA has been used as the core material which had

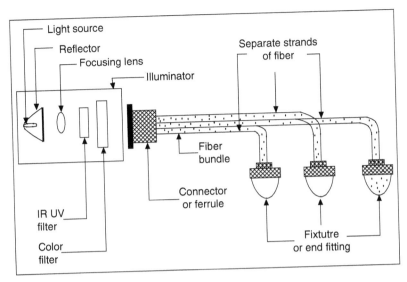

FIGURE 2.28   Schematic representation of a fiber optic lighting set-up. (After Knisley, J. 2002. Fiber optic lighting dries out, ecmweb.com/mag/electric_fiber_optic_lighting.)

to be doped or polymerized differently. However, a new product has now reached the market. It is Lucina (120 μm core diameter) fiber made out of CYTOP, offering low attenuation combined with high bandwidth, but giving up the specific benefits of large core diameter [60].

The application of POF for data communication over short distances is quite old. Thus, cables connecting HiFi set-ups or in industrial production lines are well known where the main reason for using POF was its complete immunity to EMIs. Much progress was made in 1980s when low loss polymers were developed and drawn into fibers starting with Step-index (SI) profile, followed by first attempts to

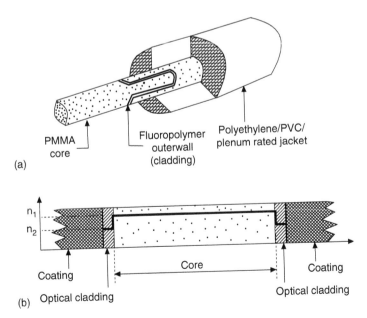

FIGURE 2.29   Standard PMMA step-index polymer optical fiber. (a) Structure of cable. (b) Refractive index (n) profile. (After Knisley, J. 2002. Fiber optic lighting dries out, ecmweb.com/mag/electric_fiber_optic_lighting.)

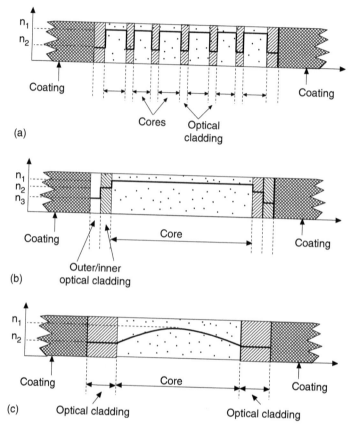

FIGURE 2.30 Refractive index (n) profiles of optical fiber cables: (a) multi-core step index fiber; (b) multi-step index fiber; (c) graded index fiber. (After Knisley, J. 2002. Fiber optic lighting dries out, ecmweb.com/mag/ electric_fiber_optic_lighting.)

produce graded index (GI) profiles. From then on more rapid progress was made in the application of POF, as bandwidth and range increased every year and with the development of CYTOP fibers, attenuation values in the range of high quality silica fibers could be reached.

The attenuation limit of PMMA, generally used as the core material of commercially available step-index polymer optical fiber (SI-POF), is approximately 100 dB/km in the visible region. This high attenuation of POF compared to the silica-based fiber (Figure 2.31) has limited the POF data link length, even when the bandwidth characteristics are improved by the graded index-POF (GI-POF) [61]. However, the perfluorinated (PF) amorphous polymer base GI-POF has a low loss wavelength region (Figure 2.31) from 500 to 1300 nm [62]. The experimentally observed total attenuation of PF polymer-based GI-POF decreases to 40 dB/km even in the near infrared region, while the theoretical limit is still lower (see Figure 2.31).

While silica based single mode optical fiber, because of its high bandwidth, is widely used in the long distance trunk area for giga bit per second transmission and beyond, the use of silica-based multi-mode fiber is a recent trend in the area of local area networks (LANs) and interconnection, because the large core diameter of the multi-mode fiber of 50 and 62.5 μm relaxes the tolerance required for connection compared with the single mode fiber whose core diameter is only 5–10 μm. However, even with the multi-mode silica fiber, an accurate alignment in the connection is still required. Compared to this, the large core diameter (100–1000 μm) of POF enables the use of inexpensive polymer connectors, prepared by injection molding, because even a displacement of ±30 μm in the connection does not seriously

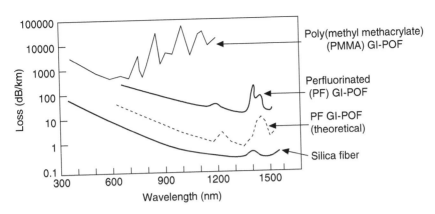

FIGURE 2.31 Attenuation of graded index polymer optical fiber in comparison with that of silica fiber.

influence the coupling loss [63]. Moreover, a large core of more than 100 μm could reduce the modal noise, which disturbs systems with multi-mode silica fibers [64].

There are many uses for POF, ranging from telecommunications to consumer electronics, in addition to conventional illumination systems based on fiber optic lighting [65]. POF is an ideal medium for short range networking, as it is easier to connect and has lower installation costs. With its ability to sustain smaller bend radius, POF is more user friendly and is able to withstand rigorous installation tasks such as pulling the fiber through walls and plenums. POF has potential applications in the aerospace industry since with its lighter weight it can replace the heavy copper wire or silica fiber now being used. Efforts are also being made to use fiber optics to channel light to instrument panels for increased brightness and reduced weight. Light weight, ease of use, and data transmission capabilities also make POF a perfect choice for consumer electronics and automobile applications where internet capability and other high data transfer features have started to make their entry. POF's tight bend radius, complete EMI immunity and low cost make it uniquely suitable for several image-transfer applications in medical industry. The primary advantages of POF over copper are increased bandwidth and durability. These properties make POF a cost effective replacement. Additionally, copper is prone to tapping and it is not a secure medium, whereas optical fiber is extremely difficult to tap.

2.4.5 Polymers in Nonlinear Optics

Materials that have nonlinear optical properties are of considerable interest because of their potential for optical switching and waveguiding applications in telecommunication [66–68]. Until recently, nearly all of the nonlinear optical materials commonly in use were inorganic solids. However, there is currently a growing interest in organic and polymeric solids because of their exceptionally large nonlinear, second-order optical properties and the larger variety of asymmetrical crystal structures available. Possible applications of these materials include amplifiers, frequency doublers, waveguides, Q-switches (used for building up power output in lasers) and fitters.

Nonlinear second-order optical properties include second harmonic generation and the linear electro-optic effect. Second harmonic generation is the ability of a material to double the frequency of light passing through it, and the linear electro-optic effect is a change in the refractive index of a material under the application of a low-frequency electric field. For these effects to occur, the material must not have a center of symmetry, while for maximum second harmonic generation, a crystal should posses propagation directions where the crystal birefringence cancels the natural dispersion, leading to the condition of equal refractive indices at the fundamental and second harmonic frequencies (phase matching).

Phase-matched second harmonic generation in single crystal polymers was first observed in 2-methyl-4-nitroaniline substituted diacetylene polymers. Subsequently, a number of other diacetylene structures have been synthesized that exhibit orders of magnitude greater phase-matched second harmonic generation than lithium iodate.

Besides single crystals, disubstituted diacetylene polymer thin films can be obtained for waveguide applications by evaporation, solidification, or deposition of the diacetylene monomers by the Langmuir–Blodgett approach, which allows film thicknesses to be controlled at the molecular level. These diacetylene films can be patterned using selective polymerization techniques such as developed for UV, x-ray, and electron beam lithography.

2.4.6 Langmuir–Blodgett Films

The preparation of ordered thin films is of considerable interest in the construction of electronic devices and model membrane systems. A method that has gained much popularity in this area is the Langmuir–Blodgett (LB) technique [67]. The technique is based on the premise that molecules with a hydrophilic head and a hydrophobic tail can form a monolayer at an airwater interface and can be transferred onto a solid surface.

The process of film formation can be carried out either by dipping a glass slide (or some other substrate) vertically into a trough containing a monolayer on the surface, as shown schematically in Figure 2.32, or alternatively by using a rotating substrate to effect a horizontal transfer. Either way can be used to build up monolayers by one passage or multilayers by repeated passage of the substrate.

To make polymeric films one selects a molecule with a polymerizable unit (double or triple bonds) so that the mono- or multilayer films can be subsequently converted into a polymer structure by thermal treatment or exposure to UV or γ radiation. Several possible monomer structures are shown in Figure 2.33.

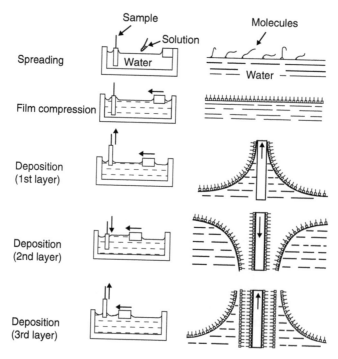

FIGURE 2.32 Schematic representation of the formation of mono-, bi-, and tri-layers of molecules from a Langmuir–Blodgett trough. (After Chemla, D. S. and Zyss, J. eds. 1987. *Nonlinear Optical Properties of Organic Molecules and Crystals*, Vol. 1, no. 4, Academic Press, San Dieo, CA.)

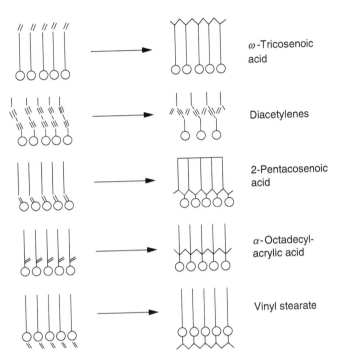

ω-Tricosenoic acid

Diacetylenes

2-Pentacosenoic acid

α-Octadecyl-acrylic acid

Vinyl stearate

FIGURE 2.33 Schematic of possible monomer structures that can be polymerized in thin films formed using the Langmuir–Blodgett method.

Polyacetylenes containing suitable amphiphilic groups can be used to produce thin films in this way, e.g., heptadeca-4,6-diyne- 1 -ol and the corresponding acid. It has been found advantageous to use the neutralized form of acid derivatives. For example, diacetylene monocarboxylic acids generally form much more stable films if the cadmium salt is used initially; in this case, Cd^{2+} ion remains in the aqueous phase as a counterion.

Ordered thin films prepared by the Langmuir–Blodgett method have found application in nonlinear optics. The method has also been applied in nanolithography. The miniaturization of integrated circuits requires high resolution, and for this purpose electron beams are used. To improve pattern definition on the resist (discussed later) used in this application, it is necessary to employ much thinner resist films and shorter exposure times. In this respect, thin LB films are superior to conventional spin coating since the latter technique does not always guarantee that the resist film will be free from defects such as pin holes that can spoil the subsequent pattern. Improved resolution has been obtained, for example, from resists prepared using polymerized ultrathin (45 nm) LB films of ω-tricosenoic acid, $CH_2=CH(CH_2)_{20}COOH$, and α-octadecyl acrylic acid $CH_2=C(C_{18}H_{37})COOH$. The LB technique holds great promise in the area of molecular electronics, where precise control of the molecular structure is of paramount importance.

2.4.7 Piezo- and Pyroelectric Polymers

Piezoelectricity is defined as an electric polarization that occurs in certain crystalline materials at mechanical deformation [69]. The polarization is proportionate to the deformation and the polarity changes with change in deformation. In reverse, electric polarization produces mechanical deformation in piezoelectric crystals. The piezoelectric materials also possess pyroelectric properties, i.e., electric polarization is generated at temperature change.

If a polymer has a larger dipole moment in its molecule than can be aligned to form a polar crystal, there is every likelihood it will exhibit strong piezo- and pyroelectricity. However, though many polar polymers such as poly(vinyl chloride), polyacrylonitrile, and nylon-11 have relatively large dipole moments, they exhibit only weak piezoelectric activities because the dipoles cannot be aligned very well by an electric field.

A major advance was made in 1969 when a strong piezoelectric effect was discovered in poly (vinylidene fluoride) (PVDF). The effect is much greater than for other polymers. In 1971, the pyroelectric properties of PVDF were also first reported, and as a consequence, considerable research and development has continued during the last two decades.

Poly(vinylidene fluoride) is a semicrystalline thermoplastic polymer consisting of $+CH_2CF_2+$ units in long chains. The crystallinity is normally about 55%, and the polymer forms lamellar crystals typically 10 mm thick by 100 nm long that are dispersed in an amorphous phase that has a glass transition temperature of about 40°C. The molecule, which may extend through several crystalline and amorphous regions, can have three morphological forms known as α, β, and γ-PVDF.

The preferred from for piezoelectric activity is the β-PVDF, which has all-trans conformation (Figure 2.34) with dipoles essentially normal to the molecular axis. The α-PVDF, on the other hand, has trans-gauche-*trans*-gauche ($tg^+ tg^-$) conformation (Figure 2.34) having components of the dipole moment both parallel and perpendicular to the chain axis. When packed in the respective unit cells, the dipoles of the chain interact in form β giving a large dipole moment, but are antiparallel in form α. In order to obtain strongly piezoelectric (pyroelectric) PVDF, the nonactive α-form must therefore be converted to the active β-form. This may be done by special treatment during the extrusion process.

In order to impart stronger piezoelectric properties to PVDF, a number of processing steps are undertaken. The material is uniaxially stretched at 150°C and annealed and then subjected to *poling* a

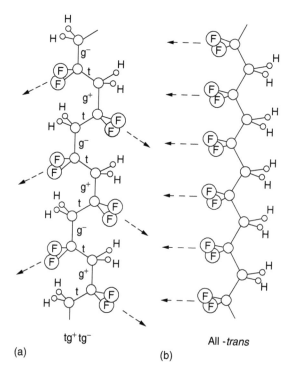

tg⁺tg⁻   All -trans
(a)   (b)

FIGURE 2.34 Schematic presentation of the two most common crystalline chain conformations in PVF₂: (a) $tg^+ tg^-$ (α form) and (b) all-trans (β form). The arrows indicate projections of the -CF₂ dipole directions on planes defined by the carbon backbone. The $tg^+ tg^-$ conformation, where t=trans, and g=gauche, and g⁻=gauche minus, has components of the dipole moment both parallel and perpendicular to the chain axis. The all-trans conformation has all dipoles essentially normal to the molecular axis.

term used to describe methods of inducing polarization), either thermally or by a corona discharge procedure.

In the case of thermal poling, electrodes are evaporated on to the PVDF and a field of 0.4–1.4 MV/cm is applied at temperatures of 90–100°C for 1 hr. Cooling with the field maintained stabilizes the polar orientation of the crystallites and produces permanent polarization (poling).

For corona poling, which is a faster method used industrially, the nonmetallized or half-metallized PVDF film is subjected to a corona discharge from a needle electrode a few centimeters away. The resulting charge buildup on the film generates a field in the sample that is sufficient to align the polar chains, even at room temperature. A very high piezoelectric activity can be obtained by a combination of stretching and corona poling.

The parallel dipolar alignment in PVDF film treated as above leads to a residual polarization, which is the basis for several proposed mechanisms that explain the piezoelectric and pyroelectric behavior of the material. Although various observations support the suggestion that PVDF is also ferroelectric, its Curie point has not been detected. (Most ferroelectrics show a phase transition at a temperature known as the *Curie point* above which they become paraelectric.)

PVDF film, as produced from the melt, is largely in the nonpolar α-form, the β phase only being obtained after subsequent processing operations, as described above. If however, vinylidene fluoride is copolymerized with as little as 7% by weight of trifluoroethylene, a copolymer is formed with crystallites completely in the β-form. This obviates the need for stretching after synthesis and the copolymer can be processed by way conventional routes, such as injection molding. Moreover, unlike PVDF, copolymers of vinylidene fluoride and trifluoroethylene have been shown to demonstrate the ferroelectric to paraelectric transition. For a copolymer with a composition of 55% vinylidene fluoride and 45% trifluoroethylene, a phase transition is observed near 70°C, and with 90% vinylidene difluoride, a phase transition at 130°C.

2.4.7.1 Applications

Technological applications exist for both the piezo- and pyroelectric phenomena, but piezoelectric applications are the most common. Piezoelectric materials are of practical importance because they permit conversion of mechanical energy into electrical energy and vice-versa. Compared to inorganic piezoelectric material, the great advantages of piezoelectric polymers lie in their workability, mechanical flexibility, high shock resistance, low cost, low density, chemical stability, and availability in large sizes.

A large number of applications have been proposed for piezoelectric polymers. The types of applications can be grouped into five major categories: sonar hydrophones, ultrasonic transducers, audio-frequency transducers, pyroelectric sensors, and electromechanical devices. The principal polymers of interest in these applications are PVDF and copolymers of vinylidene fluoride and trifluoroethylene.

The greatest use of PVDF is as a source of pressure. The fact that PVDF has an acoustic impedance that is close to that of water or the human body makes it especially suitable as a source of pressure in underwater technology and medicine. PVDF also has an extreme band width, i.e., a large frequency range.

With its low acoustic impedance, extreme bad width, high piezoelectric coefficient, and low density (only one-quarter the density of ceramic materials), PVDF is ideally suited as a transducer for broad band underwater receivers in lightweight hydrophones. The softness and flexibility of PVDF give it a compliance 30 times greater than ceramic. PVDF can thus be utilized in a hydrophone structure using various device configurations, such as compliant tubes, rolled cylinders, discs, and planar stacks of laminated material.

A 100 element 360° scanning sonar transducer utilizing PVDF has been developed by Marconi to provide a 360° view of the acoustic scene on a radar type display. Hydrophones for submarine reconnaisance are of great interest to the military. A comparison between PVDF and a ceramic material PZT-4 commonly used in sources of pressure is shown in Table 2.12. The so-called *g-d* product, which is considered to be the standard of suitability for hydrophone applications, is 2.5 times greater for PVDF as

TABLE 2.12 Comparison of the Properties of PVDF-Film (Kureha 19) with Typical Values of PZT-4 Ceramic for Hydrophone Applications

Property	Units	PVDF Polymer	PZT-4 Ceramic
Piezoelectric constant—field/stress	10^{-3} Vm/N	174	11.1
Piezoelectric constant—charge density/stress	10^{-12} m/V	20	123
g-d product	10^{-12} m^2/N	3.48	1.37
Piezoelectric coupling factor	—	0.102	0.334
Relative dielectric constant	—	13	1300
Elastic compliance	10^{-12} m^2/N	330	10.9
Density	10^3 kg/m^3	1.78	7.5

Source: Wirsen, A. 1987. *Electroactive Polymer Materials.* Technomic, Lancaster, PA.

compared to the corresponding value for PZT-4. PVDF is thus superior for acoustical recording in water. Basic constructions of hydrophones with various sensitivity and pressure resistance are shown in Figure 2.35.

There is much interest in the application of PVDF in medical imaging because of its close acoustic impedance match with both tissues. Monolithic silicon-PVDF devices have been produced in which a sheet of PVDF is bonded to a silicon wafer containing an array of metal oxide semiconductor field effect transistor (MOSFET) amplifiers arranged in such a way that when an acoustic wave is detected, the electrical signal resulting from the piezoelectric action in the PVDF appears directly on the gate of an MOS transistor. The device is therefore known as a piezoelectric oxide semiconductor field effect transistor (POSFET).

PVDF is generally an ideal material for transducers operating at frequencies above 0.5 MHz in hydrophones and pulse echo probes for medical and nonmedical testing. A 64-element linear array transducer has been produced that, operating at 5 MHz, offers a wide-bandwidth pulse response, sharp ultrasonic field distribution, and a high energy conversion efficiency.

While PVDF has piezoelectric properties similar to those of piezoelectric ceramic materials, the pyroelectric coefficient is too low to be useful except for specialized applications where the main

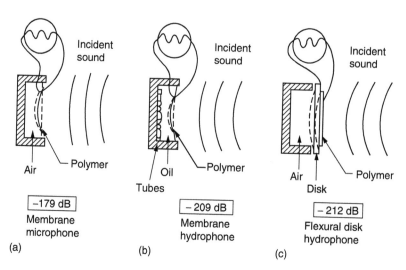

FIGURE 2.35 Cross-sections of cylindrical piezoelectric polymer receivers. (a) Membrane microphone formed by fixing a taut film over an air-filled cylindrical chamber. (b) Membrane hydrophone with an oil-filled chamber containing plastic compliant tubes to provide compliance while withstanding hydrostatic pressure. (c) Flexural disk hydrophone backed by air. The disk both excites the polymer film and provides strength against hydrostatic pressure.

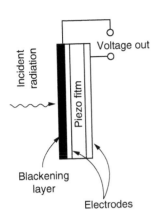

FIGURE 2.36 A pyroelectric detector made with piezo film.

requirements may be speed and resolution, and sensitivity is a minor consideration. PVDF, for example, has been used to form an array of 50 detectors for the energy profile determination of various wavelength laser beams. PVDF has also been employed in a pyroelectric vidicon to obtain a high resolution picture, because its low thermal conductivity reduces thermal spreading.

Pyroelectric detectors made with piezo films can be used in a number of specialized applications. A pyroelectric detector shown in Figure 2.36, consists of a thin semitransparent electrode placed on the front of the piezo film and a highly reflective rear electrode. A blacking layer added to the front electrode results in far more efficient broadband layer added to the front electrode results in far more efficient broadband absorption of incident radiation. The incident radiation passes into the piezo film, which absorbs strongly in the wavelength range of 8–11 μm. Pyroelectric detectors made with piezo film are used in intrusion detection systems and energy management systems. Pyroelectric detectors can also be used for humidity monitoring and certain gas detection, a good example being carbon monoxide.

Much interest has been shown in the use of PVDF for elements in telephone handsets. Plessey Telecommunications have produced bimorph microphone units using PVDF. It has also been employed in high fidelity tweeters and headphone transducers for audio applications.

2.4.8 Polymeric Electrolytes

The organic solvents currently used the most, tetrahydrofuran and propylene carbonate, have clear limitations in regards to stability [70–73]. The potential at which p-doping (indiffusion of negative ions) of polyacetylene electrode occurs is close to the anodic limit for propylene carbonate and higher than the limit for tetrahydrofuran. Thus one of the major problems with polymeric batteries is the problem of slow electrochemical breakdown of organic solvents at actual doping potentials.

An alternative to employing liquid electrolytes is to use a solid electrolyte to produce an all solid battery. This may have several advantages—an absence of electrolyte leakage or gassing, the likelihood of an extremely long shelf-life, and the capability of operating over a wide temperature range.

Polymeric materials, because of their low weight and ease of fabrication, have been investigated as electrolytes for lithium batteries. Early investigations included ion-exchange membranes, such as polystyrene sulfonic acid. However, it was found that in the dry state, the conductivity of these materials was extremely low (10^{-12}–10^{-15} ohm^{-1}-cm^{-1}), and the addition of aprotic solvents, such as propylene carbonate, did not appreciably loosen the ionic clusters formed within these membranes.

Polymer gels based on polymers such as poly(vinylidene fluoride), polyacrylonitrile, and aprotic solvents containing added alkali metal salts, gave appreciable room-temperature conductivity. However, solvent volatility and voltage stability of the electrolyte were serious problems.

A new perspective in polymer electrolytes was obtained in 1978 when Armand [70] suggested the use of PEO-alkali metal salt complexes for alkali metal rechargeable batteries. Poly(ethylene oxide) (PEO) and poly(propylene oxide) (PPO) form ion complexes with, for instance, NaI, NaBF$_4$, LiClO$_4$, LiCF$_3$SO$_3$, and others. Perhaps the most important advantage of such polymer electrolytes is the ability of the complex to form a good interface with solid electrodes, thereby permitting faster kinetics at the ion transfer between electrode and electrolyte.

Figure 2.37 shows the structure of the PEO-salt complexes proposed by Armand. The regular PEO helix is filled by metal ions (M$^+$) solvated by ether oxygens, and the counterions (X$^-$) are expelled from

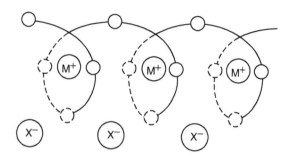

Low - temperature form

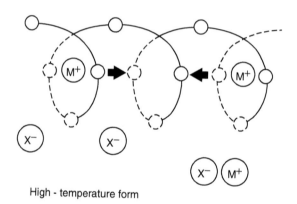

High - temperature form

FIGURE 2.37   Proposed helical structure of PEO-salt complexes.

the strands of the parallel helices. (Pure PEO has a helical structure in the crystalline state with seven CH_2CH_2O units in two turns of the helix.)

Initial measurements carried out on PEO-alkali metal salt complexes indicated that the observed conductivities were mostly ionic with little contribution from electrons. It should be noted that the ideal electrolyte for lithium rechargeable batteries is a purely ionic conductor and, furthermore, should only conduct lithium ions. Contributions to the conductivity from electrons reduces the battery performance and causes self-discharge on storage. Salts with large bulky anions are used in order to reduce ion mobility, since contributions to the conductivity from anions produces a concentration gradient that adds an additional component to the resistance of the electrolyte.

Though the conductivity of the PEO-alkali metal complexes (10^{-3} ohm^{-1}-cm^{-1} at 140°C) is fairly low in comparison with inorganic solid electrolytes such as β-alumina(Na), $RbCu_{16}I_7Cl_{13}$ and $RbAg_4I_5$ at the same temperature, this can be compensated for by the facile production of thin films typically 25–500 µm thick. A cell may thus consist of a lithium or lithium-based foil as anode, an alkali metal salt-PEO complex, such as $(PEO)_9$ $LiCF_3SO_3$ (25–50 µm), as the electrolyte, and a composite cathode (50–75 µm) containing a vanadium oxide (V_6O_{13}) as the active ingredient. The vanadium oxide is one of a number of "insertion" compounds that permits the physical insertion lithium ions reversibly into their structure and thus allows recharging of the cell.

To function as an effective polyelectrolyte, the polymer should have a low glass transition temperature to allow the freedom of molecular movement necessary for ion transport. While the advantages of PEO-alkali metal complexes are their good electrochemical stability as well as faster kinetics at the ion transfer between electrode and electrolyte, the disadvantage is the low conductivity of the polymeric electrolyte at normal operating temperatures. Thus, in order to obtain high-power densities, it is necessary to work at relatively high temperatures ($\sim 120°C$). Although this temperature of operation may be suitable for certain application, for example, vehicle traction, quite clearly it would be unsuitable for other applications, such as consumer products.

Recently, a new polymeric electrolyte consisting of polyester-substituted polyphosphazene [71] has been developed. This polymer, which is designated MEEP, forms complexes with a large number of metallic salts, the complexes having a higher conductivity at room temperature than earlier polymer electrolytes. For example, MEEP-LiF$_3$SO$_3$ has a conductivity of 10^{-4} ohm^{-1}-cm^{-1}, which is sufficient for battery use, since the polymeric electrolyte can be shaped as films between the electrodes. This holds promise of very light batteries with potentially very great energy density.

While it is possible that future polymeric batteries will be all-polymeric solid-state batteries, it is predicted, however, that the most promising solid state batteries will combine polymeric electrolytes with nonpolymeric electrode materials such as TiS$_2$, V$_6$O$_{13}$, Li or LiAl, the specific capacity of which surpasses that of polyacetylene electrodes.

As a great number of combination possibilities exist for electrode as well as electrolyte materials, the probability of developing rechargeable batteries with considerably increased performance may be considered to be high. For potential use in electric cars and other electrically operated vehicles, which may become an environmental requirement in densely populated areas, such developments in polymeric batteries are eagerly awaited.

Drastic improvement in rechargeable batteries is also of great interest militarily. One example is submarine batteries. Compared to lead-acid batteries, the polymeric batteries enjoy potential advantages in this area: faster charge, smaller volume, and great freedom in shaping the batteries to available space. Though the energy density (KWh/kg) of polymeric batteries are, in current estimation, comparable to that of the lead-acid battery, charging of the batteries, which is the most critical routine operation of a diesel-powered submarine and requires about two hours to complete, can be reduced drastically with the use of polymeric batteries.

Polymer electrolyte membrane fuel cell (PEMFC) technology has been receiving increased attention due to its high energy efficiency and environmentally friendly nature [74,75]. Among the different technologies developed, PEMFCs which operate at temperatures above 150°C have certain advantages that result in better overall performance and simplification of the system. The polymer electrolyte membranes (PEMs) in high-temperature PEMFCs must enable proton conduction and, at the same time, exhibit mechanical, thermal, and oxidative stability under operating conditions. The state-of-art material used in high-temperature PEMFCs is polybenzimidazole (PBI) which can be doped with various strong acids. In the case of phosphoric acid, PBI exhibits high acid uptake, resulting in highly conductive materials [76]. However, PBI has several drawbacks, such as moderate mechanical properties, reduced oxidative stability, limited availability, and high cost.

Among the alternative polymeric structures developed recently for PEMFC application, mention may be made of aromatic polyethers containing polar pyridine units in the main chain [77,78]. Novel poly (aryl ether sulfone) copolymers containing 2,5-biphenylpyridine and tetramethyl biphenyl moieties have been synthesized [78] by polycondensation of 4-fluorophenyl sulfone with 2,5-(4′, 4″-dihydroxy biphenyl)-pyridine and tetramethyl biphenyl diol (see Figure 2.38). These polymers exhibit excellent film-forming properties, mechanical integrity, high modulus up to 250°C, high glass transition temperatures ($> 280°C$) as well as high thermal stability up to 400°C. In addition to the above properties required for PEMFC application, this polymer shows high oxidative stability and acid doping ability, enabling proton conductivity in the range of 10^{-2} S.cm^{-1} above 130°C [78].

2.5 Polymers in Photoresist Applications

Polymers and polymeric systems that can undergo imagewise light-induced reactions are of great technological importance [79–84]. Photoresists are polymers or polymeric systems (polymer binders containing dispersed or dissolved photoactive compounds) which, applied as a surface coating to an

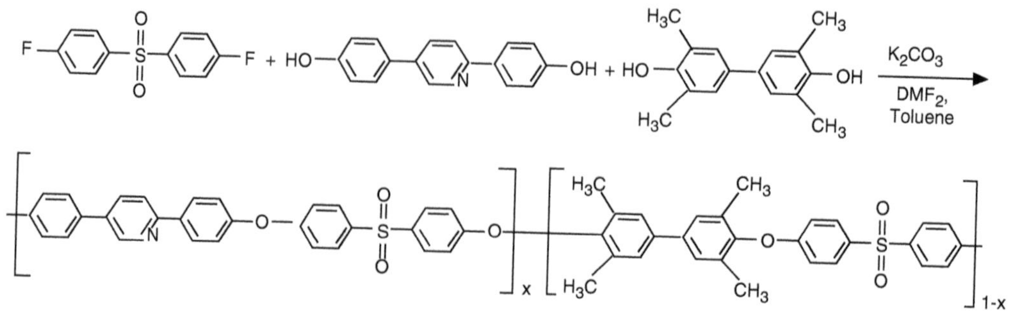

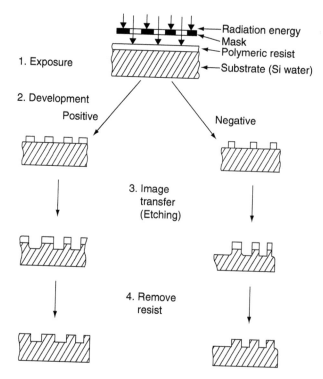

FIGURE 2.38 Polycondensation of 4-fluorophenyl sulfone with 2,5-(4′,4″-dihydroxy biphenyl)-pyridine and tetramethyl biphenyl diol for the preparation of poly(aryl ether sulfone) copolymers containing 2,5-biphenylpyridine and tetramethyl biphenyl moieties. (After Pefkianakis, E. K., Deimede, V., Daletou, M. K., Gourtoupi, N., and Kallitsis, J. K. 2005. *Macromol. Rapid Commun.*, 26, 1724. With permission.)

underlying substrate, undergo some type of reaction upon exposure to visible or ultraviolet radiation such that the differences in solubility of the exposed and unexposed areas can be exploited to obtain selective removal of either the exposed or unexposed area of the film (a process known as development). Thus, by providing suitable radiation exposure of the polymer-coated surface through a pattern or mask, such as a negative, an image can be generated on the substrate. The image can then be transferred permanently to the substrate by etching, deposition, etc. Since such images are raised above the substrate on which the polymer has been coated (see Figure 2.39), they are called relief images. In order for the image to be transferred to the substrate, the polymer must "resist" the chemical activity of the particular

FIGURE 2.39 Positive and negative modes for a photoresist film.

etchant, which can be either a wet (solvent) or dry (plasma) process. Thus, they polymers are called photoresists.

There are a number of types of photoresist that function by a variety of mechanisms, such as:

1. Cross-linking of a linear polymer backbone by the light-induced decomposition of a photo-sensitizer to generate active species (e.g., cyclized rubber with azide type photosensitizer).
2. Cross-linking of a polymer containing photosensitive groups within its own structure [e.g., poly (vinyl cinnamate)].
3. Polymerization of a monomeric material to yield a reduced solubility polymer (e.g., methylene bis-acrylamide monomer and benzoin photoinitiator in polyamide as the carrier polymer).
4. Enhancement of solubility by a photoinduced molecular rearrangement (e.g., naphthoquinone diazide in a novolac binder resin).
2. Enhancement of solubility by reduction in molecular weight by bond scission (e.g., poly(methyl methacrylate) deep-UV resist).
6. Depolymerization by high energy irradiation to allow a kind of ablative imaging (e.g., by irradiation with laser or an electron beam).

Figure 2.39 shows schematically the positive and negative modes of lithographic processes with a photoresist film. When exposure of the photoresist film to radiation produces an image that is less soluble in the developer, a negative image results. Negative images are generally produced by photochemical reactions that lead to crosslinking and network formation in the polymer film in the exposed area. Such network polymers are insoluble in the developer solvent. A second means of generating a negative image is to produce a photoinduced change which makes the exposed areas less soluble in a selective developer.

A positive image results when the areas of the film exposed to radiation are more soluble than the unexposed areas in the developer (Figure 2.39). There are two classes of chemical reactions that can lead to enhanced solubility of the exposed areas. The first is a photochemical change that converts the exposed area to a different polarity as compared to the unexposed areas. A selective dissolution of the exposed areas can then be obtained with the proper choice of a developer. The second chemical change that can occur is a backbone cleavage of the resist polymer such the exposed area of the polymer film is degraded to a low molecular weight (or all the way to monomer) that has a dissolution rate significantly higher than the rate for the unexposed higher-molecular-weight area in the developer.

There are two important aspects of the performance of a photoresist: imaging (lithography) and image transfer (resistance). An image must be formed in the resist polymer and then the image must be transferred permanently into the substrate or to other materials by etching, deposition, etc. The term photolithography is also used to refer to both the processes. Traditionally, photolithography has had great commercial impact on the printing industry, photomachining of fine parts, and so on.

More recently, the revolutions in the field of electronics have been made possible by the photopolymers (microresists) used to delineate the tiny features that make up modern integrated circuits. The microlithographic applications of photopolymers have been the driving force behind most of the technological and scientific activity in the area over the last three decades. Most of the devices that have been manufactured using microlithography have employed one of two resist systems. Negative working resists based on cyclized rubbers and bisazides were the original mainstays of this industry. Although these resists are still in use today, novolac/naphthaquinone-diazide-based positive photoresists are used for most features less than 3 μm in size. The smallest details defined by photolithography in factories are about 2 μm wide.

The drive to smaller features sizes in microlithography has brought in its wake a trend toward shorter wavelengths for the exposure radiation. To a very rough approximation, the resolution limit of an optical exposure source is give by 2λ, where λ is the wavelength. As the wavelength decreases, the potential resolution thus becomes smaller, i.e., better. This, however, requires that are sensitive at these shorter wavelengths and also exposure sources that are capable of sufficient output at these wavelengths.

Almost all commercial exposure sources use super-high-intensity Hg or Hg/Xe lamps with high-intensity outputs at 436, 405, 365, 330, and 313 nm. The first three wavelengths are normally grouped and called the near-UV. The next two wavelengths are called the mid-UV. Radiation from these lamps and other sources with wavelengths below 280 nm comprises the deep-UV. With the development of excimer lasers high-output deep-UV sources have become readily available.

Though the majority of current microlithography for semiconductor processing is carried out using radiation in the near-UV, it is resolution-limited by its relatively long wavelength. The submicron resolution that is required for increasing sophistication and degree of integration is more readily achievable by going to shorter wavelengths, and there is currently much interest in using not only deep-UV but also x-rays, ion beams, and electron beams for exposing resist. Each of these technologies has specific resist requirements, and various polymeric resist systems have been developed and are under development.

The other change in resist processing involves the use of multilayer schemes in place of a simple photoresist coating. Microimaging using multilayer materials and technology has revolutionized the electronics industry. Many silicon chips used in calculators and computers are produced in some variation of the following sequence of operations:

A silicon wafer that has one surface oxidized to a controlled depth is coated (on the oxide surface) with a photoresist, such as poly(vinyl cinnamate), to produce a thin and uniform coating several micrometers thick when dry. Exposure to UV light through a mask insolubilizes part of the polymer. The uncross-linked polymer is washed off solvents. The bare substrate parts that thus reappear are etched through the oxide layer down to the silicon layer by a fluoride solution in water or by a plasma that contains reactive ions.

Operations are then performed to alter the chemical composition of the etched regions. The operations may include ion implantation to introduce dopants that make semiconductors of the diffused-base transistor type and depositing a layer of aluminum to act as a conductor or a layer of other materials to act as insulators. After removing all of the remaining polymer by solvents, plasmas, or baking, the wafer is recoated and a new pattern is imposed and processed. The sequence may be repeated as needed to produce integrated circuits with many such layers and with amazing complexity. The minimum line widths on chips produced by photolithography are usually in the 5-μm range. However, to produce large-scale integrated circuit (LSI) devices, minimum line widths must be decreased to pack more circuits on a chip.

2.5.1   Negative Photoresists

Most negative-working resist systems are based on the fact that the dissolution rate of a polymer decreases as the molecular weight increases, and the polymer ultimately becomes insoluble in any solvent as a crosslinked network is formed.

A system widely used in silkscreen printing takes advantage of the photosensitization reaction of the dichromate/poly(vinyl alcohol) or dichromate/gelatin system. The mechanism of these dichromate resists is obscure, but it is thought that the formation of the chromate ion from the dichromate ion is an important initial step:

$$Cr_2O_7^{2-} \xrightarrow{h\nu} CrO_3 + CrO_4^{2-} \tag{2.10}$$

This is followed by the transition of chromium from Cr(VI) to a low state of oxidation, and an interaction process with the hydrophilic polymer to form a chromium/polymer complex in which binding between the polymer and photoreleased chromium compound involves either primary forces or physical forces of absorption. The resulting complex causes a solubility decrease in the aqueous system used for development.

The dichromate resists as also the diazo resist systems developed later suffer from limited stability and limited exposure wavelength sensitivity and are very slow photographically. These have therefore been displaced for resist applications by better-characterized negative photoresists using photodimerization (cinnamate type resist), photocrosslinking (bisazide resists), and photoinitiated polymerization (free-radical or cationic) to yield insoluble polymer networks.

The simplest photodimerizable resist is poly(vinyl cinnamate), made by esterification of poly(vinyl alcohol) with cinnmoyl chloride. Upon exposure, the cinnamate groups can dimerize to yield truxillate or truxinate. If the two cinnamate groups involved in this reaction are from two chains, the cyclic product represents a cross-link (Figure 2.40).

There are many cinnamate resins, including derivatives of poly(vinyl alcohol), cellulose, starch, and epoxy resins. It is the epoxy resins that have found the most applications in lithographic materials.

As cinnamate resins are water-insoluble, a suitable solvent, such as trichloroethylene, is used to achieve the solubility differential for image development. An emulsion of the solvent dispersed in an aqueous phase of gum arabic and phosphoric acid is generally used. The light-exposed regions of the coating are rendered insoluble due to cross-linking and form the printing image.

Poly(vinyl cinnamate) itself is only weakly absorbent above 320 nm. Its photoresponse is generally of the order of a tenth of that of dichromated colloids, but the rate can be accelerated by the use of photosensitizers, such as nitroamines (increase of the order of 100 times), quinones (increase of the order of 200 times for specific ones), and aromatic amino ketones (increase of the order of 300 times for specific ones). A commonly used aromatic ketone is 4,4'-bis(dimethylamino)-benzophenone, also known as Michler's ketone.

The Kodak photoresist based upon the poly(vinyl cinnamate) system is particularly suitable for printed circuit manufacture. It also finds application in some invert halftone photogravure processes and photolithographic plates. The property of superior adhesion to metal of cinnamic esters of particular epoxy resins has resulted in the preferred use of these resins for platemaking. Very few of the cinnamate-type resists have, however, been used in micro applications.

A negative-working resist system that has seen extensive use in microlithography and continues to be used where high resolution is not required, is that based on the photochemistry of bisazide as the photoactive compound in a cyclized or partially cyclized polyisoprene rubber as the binder resin. Since poly(*cis*-isoprene) has a low glass-transition temperature and is too soft for use in lithography, partial cyclization is performed to yield a polymer with good adhesion and film forming properties. The cyclization of polyisoprene catalyzed by acid produces a complex mixture of structures that increase the T_g of the rubber while reducing the solution viscosity so that high solids coatings are possible:

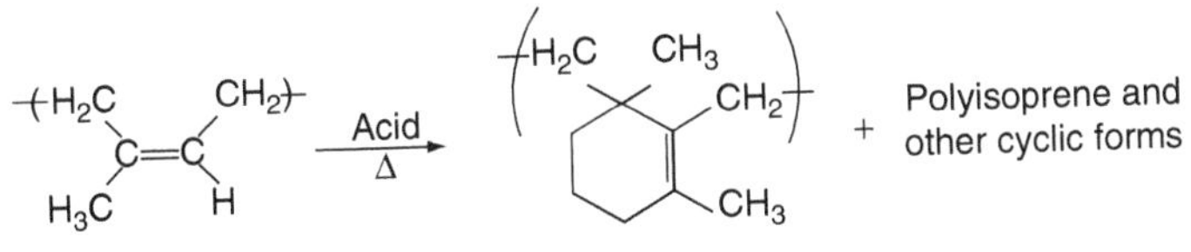

The partially cyclized rubber also has improved film-forming properties.

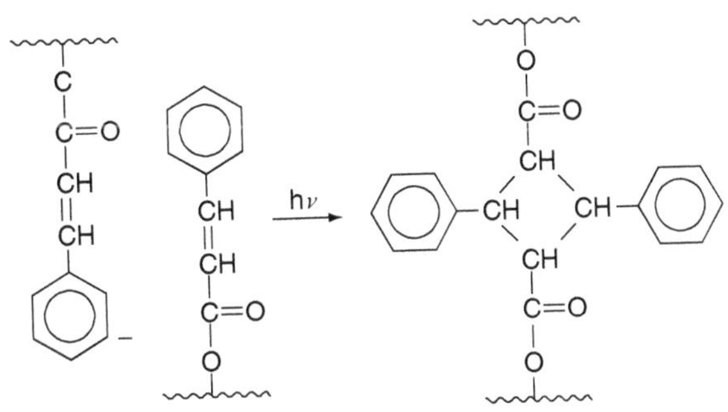

FIGURE 2.40 Photodimerization of poly(vinyl cinnamate).

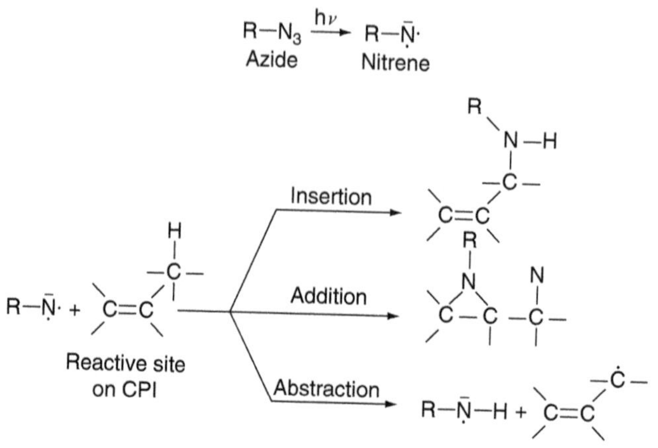

FIGURE 2.41 Reaction of nitrene formed by photodecomposition of azide, with cyclized polyisoprene (CPI) polymer.

The azide-type photoactive compounds used with the above type of resist have good thermal stability and decompose efficiently upon UV irradiation from mercury lamps to give intermediates known as nitrenes. The nitrenes being highly reactive undergo several reactions, e.g., (1) insertion reactions with carbon-hydrogen double bonds to yield secondary amines, (2) addition reactions with carbon-carbon double bonds to produce aziridines, and (3) abstraction reactions that generate radicals. These chemistries are shown in Figure 2.41. When a bisazide is used, a bisnitrene is formed on irradiation which then cross-links the polyisoprene by reacting with the double bonds or the allylic hydrogens in the exposed regions to produce an insoluble matrix (see Figure 2.42). Resists that use bisazides are mostly near UV (436 nm) resists.

The drawback of the bisazide-based cyclized rubber resist becomes manifest during development. A good solvent must be used to remove the unexposed high polymer. However, this solvent also penetrates and swells the images that are to be left behind. Though with large (>10 μm) spaces between images, the image returns to its original position, with small (<3 μm) spaces, two neighboring images may swell,

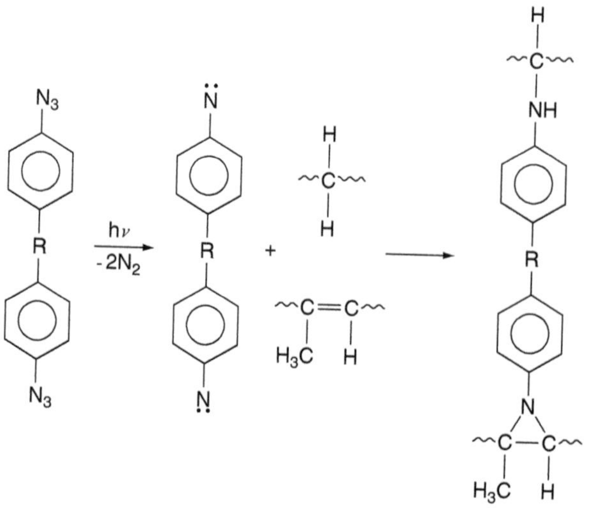

FIGURE 2.42 Cross-linking of polyisoprene by irradiation of a mixture of polyisoprene and a bisazide.

touch, and coalesce to leave a polymeric bridge between the two images. Such a bridge will finally translate into a short in the electronic circuit.

A novel microlithographic application for photoinitiated polymerization involves the polymerization of a monomer and the locking-in of a plasma-sensitive host polymer so that plasma techniques can be used to carry out all-dry development, thus avoiding the problems of swelling and resolution limitation associated with standard resists. Some plasma-developable resists are described later in this section.

2.5.2 Positive Photoresists

Positive photoresists have been increasingly important because of their higher resolution capability and better thermal stability. They become more soluble in exposed areas by increasing their acidity or by undergoing bond scission and degradation of polymer chains.

2.5.2.1 Near-UV Application

At near-UV wavelengths it is not possible to enhance solubility by photoinduced bond scission because the light is not energetic enough, and consequently an alternative mechanism is used. Positive resists with

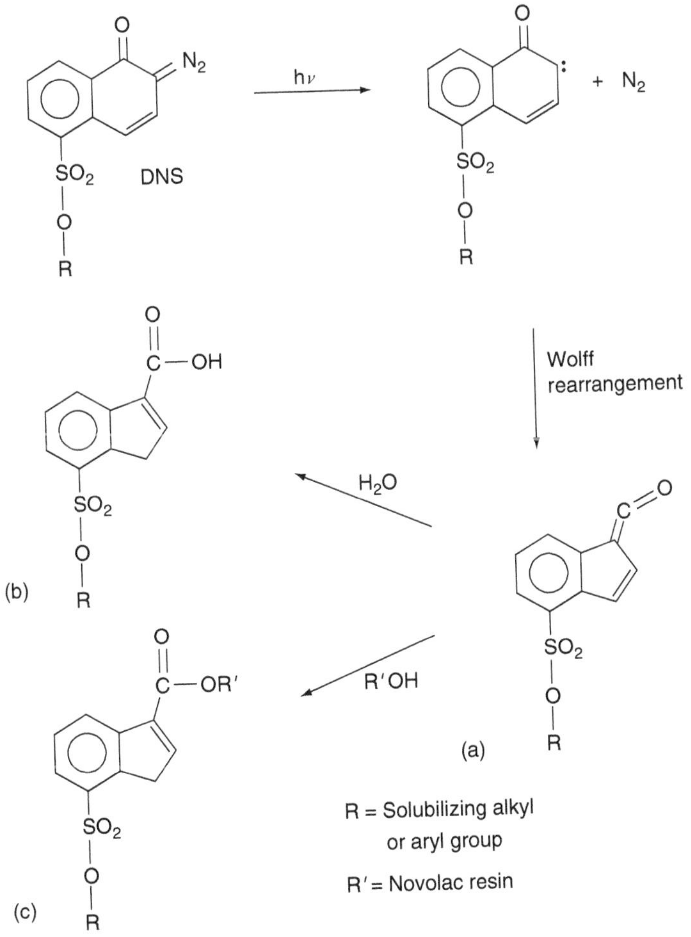

FIGURE 2.43 Photolysis of diazonaphthoquinone sulfonate esters (DNS) followed by (a) Wolff rearrangement to a ketene intermediate, which (b) reacts with water to form an indene carboxylic acid or (c) reacts with phenolic hydroxyl groups (in absence of water) to form ester linkages.

sensitivity in the near-UV usually comprise a large photoactive molecule that is insoluble in basic solvents and also sufficiently bulky to inhibit dissolution of a base soluble polymer. On exposure, the photoactive compound breaks down to form a base-soluble product and the exposed region can then be washed out using a basic solvent. The base soluble polymeric binder is typically a novolac resin of relatively low molecular weight.

The most commonly used photoresists for resolving features of 5 μm and below are based on the photolysis of diazonaphthoquinone sulfonate esters (DNS), which are used as the photo-sensitive compound in a novolac binder resin. Upon exposure at near-UV wavelengths, the photosensitive compound undergoes a Wolf rearrangement. Nitrogen is given off in this reaction producing a ketene intermediate (Figure 2.43a), which reacts with water in the film to form an indene carboxylic acid (Figure 2.43b). Under normal ambient working conditions, enough water is present in the novolac binder to ensure the formation of the carboxylic acid. However, in the absence of water, the ketene intermediate has been shown to react with the phenolic hydroxyl groups on the novolac to form pendant ester linkages (Figure 2.43c).

The unexposed DNS in the masked areas acts as a dissolution inhibitor for the novolac binder, presumably due to its hydrophobic nature, while the indene carboxylic acid that is formed from the photolysis reaction in the exposed areas is a dissolution accelerator for the novolac. The rate of development in an aqueous base is much higher for the exposed areas, which promotes good developer discrimination between the exposed and unexposed areas leading to high-contrast, high-resolution images. It has been proposed, however, that this simplistic picture of hydrophobic-to-hydrophilic switch may not represent all of the factors that enable the high discrimination and high contrast obtainable in this type of resists. The enhancement in dissolution rate may also be a result of the gaseous nitrogen that is formed during the DNS photolysis (see Figure 2.43). It is hypothesized that the nitrogen evolution causes microvoids and stresses in the novolac film and that this promotes the diffusion of the developer into the polymer film leading to an increase in the dissolution rate.

Novolacs produces by cresol-formaldehyde condensation have been the polymers of choice for the DNS photochemistry. These novolac resins are usually synthesized from commercial cresol mixtures, which contain about 60% *m*-cresol, 30% *p*-cresol and 10% of various other aromatic phenols.

These polymers are soluble in spin-coating solvents, show good film-forming and coating properties, provide excellent adhesion to most substrates, and exhibit enough balance in the needed binder properties to be used almost exclusively in all commercial DNS containing positive photoresist formulations.

With the increase in demands on photoresist formulations due to smaller imagery and new image transfer steps, such as ion implantation and plasma etching, novolacs are, however, found to fall short in two areas. First, a specific chemical composition and molecular size is difficult to reproduce because the exact composition of the cresol starting materials can vary from lot to lot and because the condensation polymerization of the cresol with formaldehyde is difficult to control, which results in a relatively low-molecular-weight polymer with M_n around 1000 and with a broad dispersity (about 20–40). These variations can lead to irreproducible lithographic performance of the high-resolution resists. Second, novolacs have low glass transition temperatures (T_g), generally in the range of 20–120°C depending on molecular weight, which often lead to unacceptable image distortions when the image is subjected to plasma etching or ion implantation process steps.

For several years, many of the commercial positive photoresists have contained multifunctional, principally trifunctional, DNS derivatives, some examples of which are shown in Figure 2.44. They contribute to enhanced image stability during the image transfer step in addition to higher contrast and faster development in the microlithographic process.

As stated earlier, the developer used to dissolve or develop out the exposed regions of the DNS-based positive photoresist coatings is an aqueous base. There are two general types of these aqueous-base developers. The first type, the metal-ion containing developers, is based on metallic hydroxides. The need to eliminate metal ion contaminants in semiconductor processing has led to the development of a second category of metal-ion free developers. The most important members in this category are quaternary

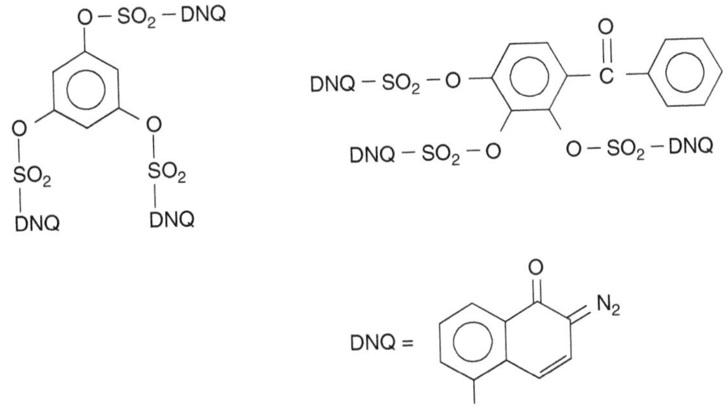

FIGURE 2.44 Typical examples of multifunctional diazonaphthoquinone sulfonate esters used in photoresists.

alkylammonium hydroxides, such as tetramethylammonium hydroxide, tetraethylammonium hydroxide, etc., used in aqueous solutions.

The DNS/novolac-based positive photoresist can also be used to obtain negative images by an image reversal process involving post-exposure treatment of the photoresist film with a reactive amine. This amine-treated film is heated and then flood-exposed. Development with a conventional aqueous base at this stage produces a high resolution negative image of the original positive image. This process is shown in Figure 2.45. An amine (e.g., monazoline, imidazole, triethanolamine) reacts with the indene carboxylic acid in the originally exposed areas to form an ammonium salt that decomposes on heat treatment to indene. Subsequent flood exposure gives indene carboxylic acid in the previously unexposed areas that are then washed out in the aqueous base developer. The hydrocarbon indene from the original exposure and subsequent process protects the novolac just as the original DNS would do.

2.5.2.2 Mid- and Deep-UV Photoresists

Though conventional positive resists have performed well at 436, 405, and 365 nm wavelength (common outputs of commercial exposure sources), the potential resolution gain has been small. This has given a push toward shorter wavelengths for achieving higher resolution.

A DNS-based positive photoresist can be used for deep-UV exposures if the binder resin in the photoresist itself does not absorb. Since novolacs made from the condensation of formaldehyde with pure *p*-cresol (instead of commercial cresol mixtures) are found to have very transparent windows at about 250 nm, a DNS/*p*-cresol novolac-based photoresist gives a positive image after deep-UV exposure and development with aqueous base. A copolymer of styrene and maleimide is also used in place of novolac as a binder for DNS-based deep-UV positive resists.

The photochemistry of *o*-nitrobenzyl esters (Figure 2.46) has been used as the basis for another dissolution inhibitor approach to making deep-UV resists. In this case, the polymer binders are base-soluble copolymers containing methacrylic acid and the dissolution inhibitor is an *o*-nitrobenzyl ester of a large organic molecule such as cholic acid. The ester cleaves, upon exposure, to form products (Figure 2.46) that no longer interfere with the resin dissolution, and thus a positive relief image is obtained after development.

Deep-UV resists are used in two principal modes: as surface resist in single layer mode, or as the thick planarizing layer in portable conformable mask (PCM) bilayer mode. In the single layer mode, the use of deep-UV resists provides improved resolution, while in the PCM mode, a deep-UV resist is used in combination with a conventional resist to make use of the wavelength selectivity of the two types of resists. The PCM process (Figure 2.47) first exposes and develops the conventional novolac-based thin top resist using a 436 nm masked exposure. A second full exposure is then made in the deep-UV, which is

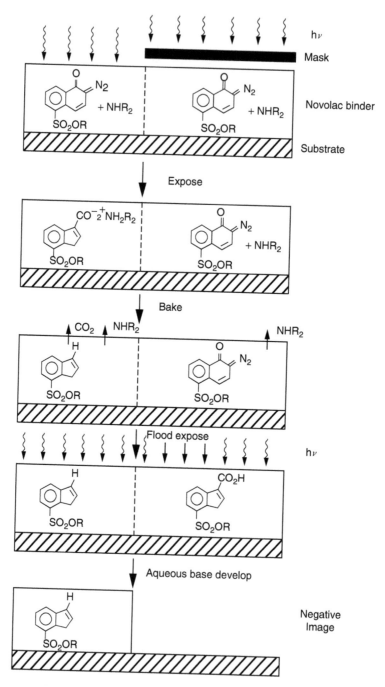

FIGURE 2.45 Schematic of image reversal process of DNS-novolac positive photoresist.

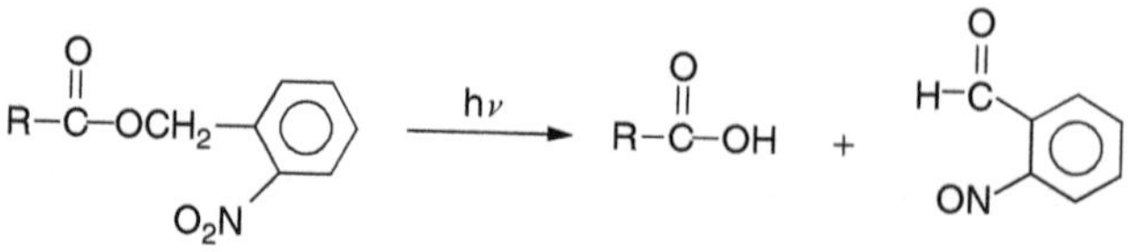

FIGURE 2.46 Photochemistry of *o*-nitrobenzyl ester.

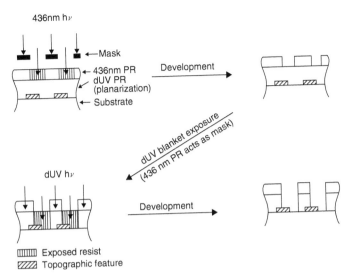

436nm hν

←Mask
←436nm PR
←dUV PR
   (planarization)
←Substrate

Development

dUV blanket exposure
(436 nm PR acts as mask)

dUV hν

Development

▥ Exposed resist
▨ Topographic feature

FIGURE 2.47 Schematic of portable conformable mask (PCM) bilayer process. (PR, photoresist; dUV, deep-UV).

masked by the novolac-based resist image of the 436 nm exposure. The unmasked (exposed) positive-working deep-UV resist is developed to give a thick, high-resolution image for subsequent processing.

The use of a photolabile acid generator to deprotect imagewise a functional group or depolymerize a polymer is the best-documented example of photochemical change systems that are known to proceed by a chain reaction mechanism. A new term, chemical amplification, has been coined to describe any of several such systems. Many photolabile acid generators are possible, but the most widely used have been iodonium and sulfonium salts, which have their natural sensitivity between 200 and 300 nm but can be substituted and sensitized out into the visible.

Polymers with acid-labile protecting group such as *t*-butyl carbonate and *t*-butyl esters have been doped with iodonium and sulfonium salts, which produce acid on radiation exposure in resist matrix. Imagewise exposure of these resist formulations to generate acid and then heating produces a relief image (Figure 2.48a). Either positive or negative images may be developed depending on the solvent chosen.

The design and synthesis of polymers with carbonate units in the backbone that allow a facile degradation of the polymer by the chemical amplification scheme with photoacid have been reported. The exposed areas of such resists can be developed by a solvent giving high-resolution positive images. On the other hand, polymers containing precursors to more volatile fragments such as copolycarbonate (Figure 2.48b) may be dry developed by thermal treatment alone. Thus, after exposure of the copolycarbonate in the presence of a photoacid and thermal treatment of 60°C for 2 min, complete image cleanout was observed.

Polymers capable of facile depolymerization upon irradiation may be suitable as "self-developing" positive resists. One polymer where depolymerization has been successful is polyphthalaldehyde (Figure 2.48c). End-capped polyphthalaldehyde is stable to a temperature of 150°C, although its ceiling temperature is −40°C. When polyphthalaldehyde is irradiated with a high-powered excimer laser, it is blown away as monomer during the exposure. This type of ablative imaging is possible because of the high-energy output of the laser. Moreover, in the presence of an onium salt, the depolymerization can be catalyzed imagewise.

The excimer laser mentioned above filled a long-standing need for a workable exposure source for deep-UV lithography. The excimer laser can deliver large amounts of light at several deep-UV wavelengths, depending on the mixture in the laser gas chamber. Several of the possible outputs of this tunable tool according to the excited complex that emits light are listed in Table 2.13.

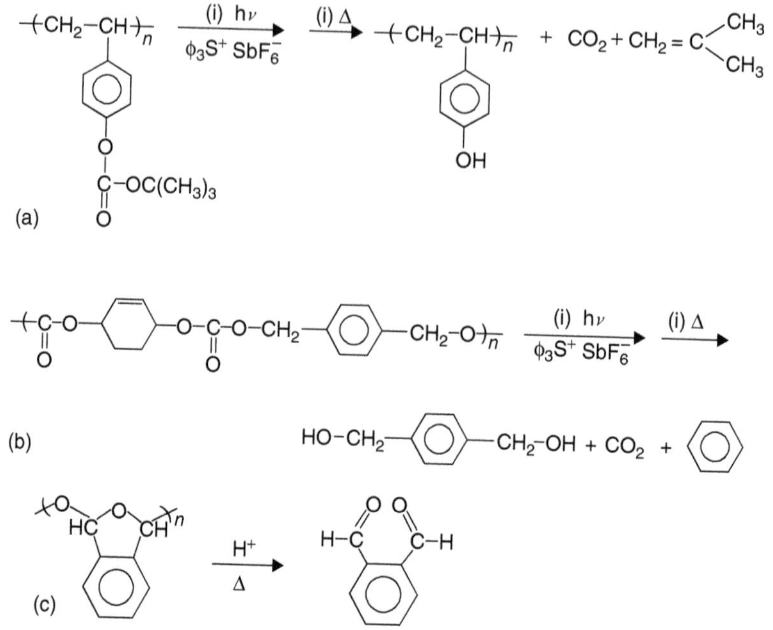

FIGURE 2.48 Photochemistry of poly(4-*t*-butoxycarbonyloxy styrene) doped with (a) triphenylsulfonium hexafluoroantimonate, (b) a dry developable polycarbonate, and (c) self-developing polyphthalaldehyde.

The extremely high-energy output of the laser also allows the potential of the above kind of ablative imaging with other polymers. For example, when PMMA is used with an ArF (193 nm) laser, the exposure produces a large amount of scissioning in the top surface. The fragments are blown out of the exposed hole by the kinetic energy of the photopulse, while continued pulses drill through the resist. Generally, the energy output is so great that almost any polymer can be ablated.

2.5.3   Electron Beam Resists

With the demand for greater and greater device integration, as in the technology of random access memories where the number of devices to be accommodated on a piece of silicon has increased dramatically with no concomitant increase in the device area, the geometry of the features on a chip has continued to fall and routine production is now in the submicron range for some advanced devices. Imaging such small geometries call for new lithographic processes. One of the most important of these new processes uses an electron beam to irradiate the resist and is known as electron beam lithography. Focused beams of electrons or focused ion beams are the fundamental methods of generating patterns with dimensions far into the submicron range.

TABLE 2.13   Excimer Laser Outputs

Complex	Wavelength (nm)
Ar–F	193
KrCl	222
KrF	248
XeBr	282
XeCl	308
XeF	351

An electron beam is of much shorter wavelength than the radiation used in standard microlithography and, therefore, provides greater resolution possibilities. For resist exposure a beam of electrons can be used whose position is controlled by a computer-driven beam deflector, thus obviating the need for a mask. Both positive- and negative-working resists are used for electron beam lithography.

Since an electron beam having shorter wavelength and higher energy can cause covalent bond scission relatively easily, polymer degradation by exposure to the beam forms the basis for a positive photoresist system. Poly(methyl methacrylate) and its analogues, such as poly(hexafluorobutyl methacrylate), have been widely studied as positive-working electron resists. Exposure to an electron beam causes extensive main chain scission in these polymers, but depolymerization back to monomer does not occur. Moreover, PMMA is not very sensitive and so other materials have been employed. Polyolefin sulfones, for example, are extremely sensitive positive-working electron resists, undergoing chain scission back to the olefin and SO_2.

Polymers containing reactive cross-linking groups such as epoxy or allyl are commonly used for negative-working electron beam resists. A typical example is a copolymer of glycidyl methacrylate and ethyl acrylate. Co-polymers of glycidyl methacrylate and 3-chlorostyrene have been shown to have high sensitivity, good adhesion, and plasma resistance. These materials are commercially available.

One drawback in electron beam lithography arises from the fact that much of the interaction between the polymer and the electron beam occurs as a result of low-energy secondary electrons being produced in the film. These are scattered beyond the definition of the exposing beam and produce unwanted reactions in areas adjacent to the primary exposure. This effect, known as the proximity effect, can cause undercutting and merging of closely spaced features. However, in x-ray and ion-beam lithography, the secondary electrons produced have lower energies and shorter path lengths, and consequently the proximity effect is less pronounced. X-ray and ion-sensitive resists may thus find greater use in the future.

2.5.4 Plasma-Developable Photoresists

One important area of resist research in recent years is the development of plasma-developable resist systems. The aim of plasma developable resists is to use nonsolvent, all dry development methods to avoid the problems of swelling and consequent resolution limitation associated with conventional resists. Much of the semiconductor fabrication process now utilizes plasma techniques as they are capable of providing high resolution images. An important consideration in this is that the plasma-developable resist images should stand up well to the plasma etching treatments.

Plasma resistance of organic polymers can increase significantly due to the presence of aromatic and heteroatomic groups. So if a resist could be made such that after exposure, the exposed area contained more of these groups than the unexposed area a differential plasma etch rate would be achieved leading to selective development (Figure 2.49).

A process known as *photolocking* based on the above method was developed at Bell Laboratories. The process used a resist consisting of a plasma degradable acrylic polymer such as poly(dichloropropyla-crylates) intimately mixed with a readily polymerizable volatile monomer such as *N*-vinyl carbazole or acenaphthylene. Upon exposure to UV radiation through a mask, the monomer only in the exposed areas polymerizes and becomes locked in the plasma sensitive host polymer. After exposure, the monomer in the unexposed regions is removed by application of vacuum leaving the plasma-sensitive poly (dichloropropylacrylate), while in the exposed regions the photolocked polymers of *N*-vinyl carbazole or acenaphthylene provides plasma resistance, thus facilitating selective development. This type of photolocking system is particularly suitable for x-ray lithography.

In photolocking systems, *N*-vinyl carbazole, which is rather toxic, was subsequently replaced by a number of silicon containing monomers. A novel feature of such silicon-based monomers (XI) is that, on oxygen plasma treatment for final resist removal, they deposit a thin layer of silicon dioxide, which is itself a useful protective dielectric layer.

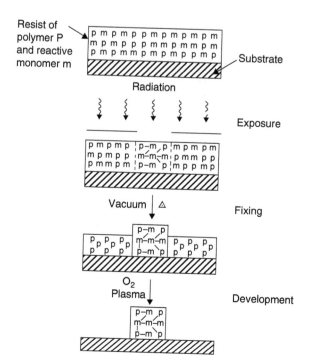

FIGURE 2.49 Schematic of photolocking process used to bind etch-resistant molecules imagewise in a polymer film.

$$CH_2=CHCOO\,(CH_2)_n - \underset{\underset{CH_3}{|}}{\overset{\overset{CH_3}{|}}{Si}} - O - \underset{\underset{CH_3}{|}}{\overset{\overset{CH_3}{|}}{Si}} - (CH_2)_n - OOCCH=CH_2$$

(XI)

2.6   Photoresist Applications for Printing

Photoresists find applications in several of the printing processes. The major areas of interest include printing plates, photoengraving, silkscreen printing, printed circuits, collotype, and proofing systems.

2.6.1   Printing Plates

Photopolymerization by actinic radiation is used extensively in the preparation of printing plates. The kind of plates may be categorized into three groups: (1) relief or raised image; (2) planographic, photolithography; and (3) gravure or intaglio-photogravure. Photopolymerization, however, finds use primarily in (1) and (2).

2.6.1.1   Relief or Raised-Image Plates

The traditional copper, zinc, and magnesium plates used by the bulk of letterpress printers and platemakers are now being replaced by photopolymer platemaking, because the latter is a more rapid and simpler process, and permits faster processing. Photopolymer plates can be made accurate to the negative. The principle of relief plate making is depicted in Figure 2.50. Press life for photopolymer plate systems is satisfactory, generally up to about half a million impressions, though current systems claim longer.

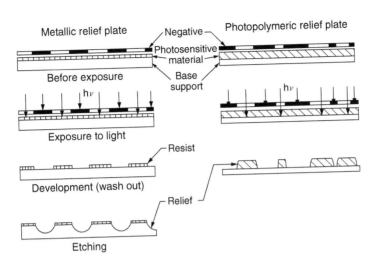

FIGURE 2.50 Principle of relief plate making.

Photosetting in conjunction with photopolymer plates has a vast potential in three chief areas of printing, namely, newspaper production, flatbed cylinder letterpress, and rotary letterpress commercial printing.

Photopolymer relief plates fall into two principal categories: liquid types and solid types. Presensitized liquid and dry raised-image or relief plates are used for letterpress, dry offset, and flexography.

The first liquid photopolymer system was the letterflex plate. The process relies on the cross-linking reaction of a polyurethane possessing allylic unsaturation and a polythiol, i.e.,

$$\sim\!\!\sim R\!-\!CH_2\!-\!CH\!=\!CH_2 \;+\; HS\!-\!R'\!\sim\!\!\sim$$

Actinic
radiation
+ Benzophenone

$$\sim\!\!\sim R\!-\!CH_2\!-\!CH_2\!-\!CH_2\!-\!S\!-\!R'\!\sim\!\!\sim$$

The liquid photosensitive resin is employed as a top coating on an aluminum support. A negative is then used and exposed to actinic radiation from a Xenon lamp. The unexposed areas are removed with a solution of surface-active agents using an ultrasonic washing technique.

Solid photopolymer plates for printing were pioneered by Dupont with the introduction of Dycril in 1957. The photosensitive layer in Dycril is a 0.3–1.0 mm thick binder composed of a water-soluble cellulose derivative such as cellulose acetate succinate, a monomer of the divinyl type such as triethylene glycol diacrylate, and a photo-initiator such as 2-ethylanthraquinone and a thermal stabilizer (inhibitor) of the *p*-methoxyphenol type. The mixture is applied on an aluminum support or steel with an adhesive sheet and exposure to actinic radiation from a carbon or mercury arc lamp is performed through a negative placed over the photosensitive layer. Postexposure treatment involves washing with a 0.2–0.5% caustic soda solution that removes the unexposed area and leaves a relief on the support. Dycril is fairly cheap, has a fairly high resolving power, and is durable.

Nyloprint, of BASF is a solid type of photosensitizer compound coated plate, around 1967. The solid photopolymerizing material is a combination of an alcohol-soluble polyamide (such as a polyamide copolymer of hexamethylene diammonium adipate and ε-caprolactam), a vinyl monomer (such as bis-acrylamide, hexamethylene bis-acrylamide, triethylene glycol diacrylate, etc.), a photo-initiator such as benzoin methyl ether, and an inhibitor (such as hydroquinone) and maleic anhydride to function as a compatibilizer for polyamide with vinyl monomers. Post-exposure treatment of Nyloprint, however, involves washing with alcohol.

Dyna-flex, a solid photosensitive plate marketed by Dynaflex, has a photosensitive emulsion on an aluminum backing or carrier plate and is thought to be based on poly(vinyl alcohol) and dichromate chemistry.

2.6.1.2 Photolithography/Planographic Plates

These are lithographic plates and find application on offset presses. The photosensitive materials used include diazo monomers, diazo polymers, and diazo compounds, in conjunction with additive film formers such as epoxy resins, phenolics, poly(vinyl acetals), polyamides, azide compounds, unsaturated chalcone, cinnamate, stilbene, and vinyl polymers and dichromated colloids. Three kinds of metallic photolithographic plates are used. These are surface plates, deep-etch plates, and bimetallic or trimetallic plates.

Surface plates are either negative- or positive-working and direct use is made of the coating material or the resultant compounds of photochemical change as the image bearer. The other two types of plates may be regarded as "reversal" positive working processes in which the coating acts as a hydrophilic stencil, and the image is in essence composed of the plate surface backed up by the lacquer application.

Deep-etch plates and trimetallic plates require a positive transparency for their preparation, whereas bimetallic plates are generally processed from photographic negatives. A bimetallic plate has an oleophilic metal, such as copper or zinc, in the image area and another one, such as chromium or iron, in the non-image areas. For trimetallic plates, backing occurs from a third metal for support. Etching is done with ferric nitrate solution, which dissolves the copper but does not attack the metal underneath. The photoresist is removed subsequent to etching, the plate is then inked, and desensitization of the nonimage areas is carried out.

Commercial presensitized plates are formulated on two chief groups of photoresponsive compounds, namely, (1) diazo resins and diazo oxides, and (2) cinnamate resins, some of which have been previously described.

2.6.1.3 Photogravure

In photogravure, ink is transferred from recesses in a metal plate or cylinder to the surface to be printed on. In contrast to the halftone processes (see below), depth of color variations are produced by varying the depth of the recesses and, consequently, the ink quantity transferred to the printed surface.

Gelatin or poly(vinyl alcohol)/dichromate photoresist may be used to sensitize the paper on which a mesh-like pattern or hardened gelatin enclosing small squares of unhardened gelatin (corresponding to the cells that will finally constitute the printing areas) is created by exposure underneath a screen made of two sets of transparent parallel lines arranged orthogonally. A positive transparency is then used to expose the paper so that every square is hardened to a depth dependent on the incident actinic radiation intensity. The paper is then transferred to a copper plate or cylinder with the metal against the gelatin. The backing paper is then removed and the image is developed by washing with water to remove the unhardened gelatin. A photoresist image remains, consisting of squares of varying depths partitioned by walls of completely hardened gelatin.

Ferric chloride is used as the etching agent to prevent gas production, as this might dislodge the resist. The greatest depth of etch occurs in the least hardened areas and, conversely, the least depth in the most hardened areas, thus producing shadow and highlight regions. The resist remaining is removed prior to printing.

2.6.2 Photoengraving

Photoengraving is the process by which blocks for illustrations are made. The illustrative material uses either line drawings with lines of differing thickness representing various tones or photographs reproduced as halftone blocks. For halftone work, the continuous tones of the original photographs are transformed to a uniform pattern of varying sized dots by either a crossline screen placed a short distance in front of the negative or a contact screen placed in contact with the negative prior to exposure.

Clean copper, zinc, or magnesium plates are used, with the fresh photosensitive coating applied over the surface such that on drying it is about 1–4 μm thick. The plate is put under a negative in a vacuum printing frame and exposed to actinic radiation from a carbon arc lamp. After exposure, the plate is washed in running water and the unexposed film areas removed with cotton-wool. The plate is electrolytically or chemically etched after coating the reverse side of the plate with an acid-resistant lacquer such as an alcoholic shellac solution. Zinc or magnesium plates are chemically etched with dilute nitric acid, whereas copper plates are etched with ferric chloride solution. Electrolytic etching is often carried out with copper plates, where they are rendered as the anode in a cell with an iron cathode. A solution of ammonium and sodium chloride is used as an electrolyte. In both types of etching, it is necessary to prevent undercutting of the lines and dots by lateral etching.

2.6.3 Printed Circuits

The use of printed circuits reduces significantly the time necessary to wire electronic apparatus. The design of the circuit ensures that contacts between the components may be provided by a metal coating on an insulating base that is covered with a thin copper layer. Poly(vinyl alcohol)/dichromate or any usable photosensitive system is applied to the copper. The resist is hardened, and the copper regions not forming part of the circuit are removed by etching with aqueous ferric chloride or peroxodisulfate solution.

2.6.4 Collotype and Proofing Systems

Collotype is similar to photolithography as the image areas are made by photocuring dichromated gelatin, with the exception that it has a nonplanar printing surface that is covered with microscopic reticulations. The use of halftones is dispensed with in collotype printing, and the process approaches continuous tone production more closely than any of the other printing methods.

With the advent of proofing systems, photoresist technology allows the graphic arts industry to approximate final full-color print. Photoresist coatings on polyester clear film in the three primary colors, yellow, magenta, and cyan and also in black, can be contacted with color-separation negatives. After exposure, these plastic-supported resists are developed in a manner similar to offset plates and, on superimposition on each other in close registration, afford a good approximation to the final full-color work.

2.7 Optical Information Storage

Optical memory may be defined as any information storage device in which writing and reading of the elements of stored information (pixels) are performed by use of electromagnetic radiation in the visible, infrared, or ultraviolet parts of the spectrum. The case for the use of optical memories in electronic systems rests on general advantages of optical storage systems which are (1) large information capacity (an optical disc, for example, might store 25,000 pages of compressed binary black and white images); (2) high resolution with multivalued pixels; (3) high degree of parallelism and megapixel arrays; (4) high speed resulting from large bandwidth×channels product and fast optical processes; and (5) no interference of incoherent beams and no EMI.

Polymeric materials can be used for optical information storage and some are ideally suited for the manufacture of optical video and digital audio discs. The information is normally transferred to the polymer using a monochromatic laser by one of four possible methods: (1) pitting of thin films (ablation, hole burning); (2) bubbling of thin films; (3) texture change in liquid crystal film; and (4) bilayer alloying. Of these, technique (1) is of greatest commercial interest. The information is stored by creating, on the polymer surface, a series of small pits that have different lengths and frequency of spacing.

The information can be retrieved by measuring the intensity and modulation of laser reflected from the pattern of pits on the disc surface.

The discs themselves must be fabricated from materials that have the following characteristics: (1) dimensional stability; (2) optical clarity; (3) isotropic expansion; and (4) low birefringence (see Chapter 2 of *Plastics Fundamentals, Properties, and Testing*). In addition, the surface should be free of contaminating particles and occlusions that would interfere with the information retrieval process.

The structures that have evolved for ablative-mode optical discs make use of interference effects to minimize the reflectance (R) of the disc in the absence of a hole. A typical ablative-mode optical disc has the structure shown in Figure 2.51. The substrate is an optically transparent material such as polycarbonate, poly(methyl methacrylate), poly(ethylene terephthalate), or poly(vinyl chloride), topped by a subbing layer to provide an optically smooth (to within a fraction of a nanometer) surface for the recording layer. A metal reflector (typically aluminum) is then incorporated next to a transparent dielectric medium such as spin-coated poly(α-methyl styrene) or plasma-polymerized fluoropolymers. This dielectric spacing layer serves both to satisfy the quarter-wave (λ/4) antireflection conditions and to insulate thermally the Al reflector from the top absorbing layer where the information pits are created.

A verity of materials have been considered for the absorbing layer in ablative-mode optical discs. Metal/polymer composites offer greater flexibility in optimizing the desired combination of parameters since the functions of energy absorption and ablation can be separated, with the metal absorbing and the polymer ablating. For example, in the proprietary material Drexon, silver particles (formed from silver halide) are dispersed in gelatin. The gelatin matrix ablates at 200°C. It is claimed that the dehydration accompanying ablation means that there is no buildup of material at the edges of the pit (Figure 2.51), and consequently a higher signal-to-noise ratio, since the buildup of material on the rim of the pit is a factor generating noise.

Dye molecules are an obvious alternative energy absorber, and, like metals, may be used either as thin films or in a polymer matrix. For example, squaryllium dyes (e.g., R=CH$_3$, R'=OH; XII) may be used as IR absorbers with GaAs or AlGaAs lasers.

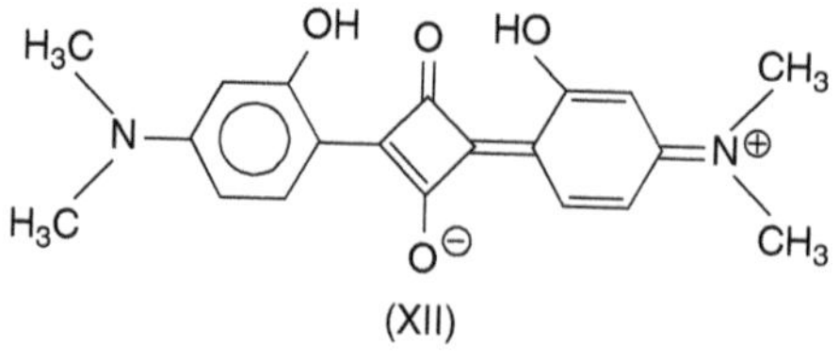

(XII)

The absorbing layer is then protected by a transparent overcoating of crosslinked poly(dimethyl siloxane) elastomer. This will produce a direct read after write, or DRAW, disc that is non-erasable.

2.8 Polymers in Adhesives

An adhesive is a substance capable of holding materials together by surface attachment or adhesive bonding which has many advantages to offer as compared to any other method of attachment [85–88]. When all applications of adhesives are taken into account, adhesive bonding must be considered as the most widely used method of holding various materials together.

The importance of the surface polarity and the surface characteristics for polymer adhesion has been considerably discussed in scientific literature [87]. A useful generalized theory of adhesion, however, can be built upon the basis of electrical attractions. The electrical attractions, resulting from uneven surfaces, which are not normally considered to be electrical, participate easily in attractive interactions if adhesives can be found that will wet them. Thus the reason that polyethylene and poly(tetrafluoroethylene) are difficult to bond is simply that available adhesives are thermodynamically more stable if their molecules attract one another than if they interact with low energy surfaces. The solution to this problem would

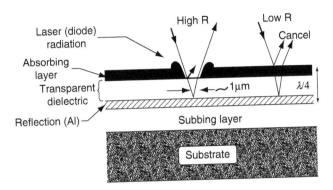

FIGURE 2.51   Trilayer structure of a typical ablative-mode optical disc.

therefore be to modify the surfaces of difficult-to-bond materials so that they become capable of strong electrical attractions. A number of processes are routinely used industrially for this purpose.

In general, for an adhesive to work it must start out in a fluid state to enable adequate wetting of the substrates to take place, and then the fluid state must be replaced by a solid state to enable the adhesive bond to gain strength. There are many types of adhesives varying in the way the liquid state is provided and/or the method of attaining the final solid state. In practice, there are, however, two ways of applying macromolecular adhesives. The adhesives may be applied as a nonpolymerized liquid (monomer, oligomer) that reacts after wetting the surface to form a polymer. This approach is typical of epoxies, phenolics, polyesters, and cyanoacrylates (see Chapter 1). The other technique is to apply the fully polymerized product, and this requires that the macromolecular compound be applied above the melting range, above its glass transition temperature, or in solution. Bond strength develops as the adhesive solidifies.

Adhesives currently on the market and used in significant quantities can be classified into four categories: solvent-based, water-based, 100% solids (hot melt), and radiation cured. Not all categories offer the same performance or processing benefits.

2.8.1   Solvent-Based Adhesives

Solvent-based adhesives are made by dissolving suitable solid materials in appropriate organic liquids. The solution is the liquid phase, which carries out the wetting of the substrates; then the solvent is removed by evaporation, leaving a solid adhesive film.

The progress of technology over the last 100 years has made available a number of organic liquids in large quantities, as well as a range of synthetic solids that can be dissolved in them. Various combinations of these materials form a substantial part of the adhesives industry, the choice of the solvent depending on what is to be dissolved, availability, cost, and health and environmental conditions.

In the U.S. converting industry, one- and two-component solvent-based adhesives are still the most widely used, accounting for nearly 80% of adhesive formulations. Many of these adhesives are typically polyester-or polyether-based urethanes (see Chapter 1) with isocyanate functionality that cure by reacting either with atmospheric moisture or a cross-linker. Polyether-based urethanes, while less expensive than polyester-based, do not provide the same strength and do not perform well in some processing environments or end uses.

The choice exists between using a one-part or a two-part formulation. One-part formulations are based on polyurethanes of very high molecular weight that are still soluble in petroleum-ester mixtures. Two-part adhesives are normally used where high cohesive strength and improved heat resistance are required and these comprise with the prepolymer terminated in some active group (such as polyether or polyester polyols) and as the second part either a multifunctional isocyanate (e.g., MDI) or an isocyanate terminated polymer (e.g., PMDI) of rather lower molecular weight. These materials can be formulated to allow a variety of potlife times, with a typical maximum of two hours.

Polyurethane adhesives are most widely used for bonding elastomers, fabrics, and thermoplastics, and have found major uses in boot and shoe industry. Polyurethane adhesives continue to represent an excellent choice for bonding metal to plastics or rubbery adherents and are superior to many other adhesives at cryogenic temperatures, maintaining their shear strength and toughness at temperatures far below those that cause serious embrittlement of other adhesives. Polyurethane two-part adhesives also find high-performance engineering applications. High-performance uses, in fact, are a fast-growing segment because of the industry's desire to eliminate standard mechanical bonding methods such as nuts and bolts, screws, and welding.

The adhesive properties of polyurethanes may be attributed to the polar nature of the macromolecular compounds used in their formulation. Furthermore, the isocyanates present in the polyurethane compositions may react with any active hydrogen present in the adherent or with the films of water often present on the surfaces of materials such as ceramics, glass, and metals. These types of reactions result in enhanced anchoring of the adhesive and attainment of high bond strength.

There exists a large number of other solvent-based adhesives derived from synthetic polymers that are soluble in organic solvents. For example, substituted nylons (aliphatic polyamides)—typically, *N*-methoxymethyl nylons—are soluble in some alcohols and alcoholic mixtures and have been used to provide solution adhesives with good rust resistance. Such soluble nylons may be compounded with many thermosetting polymers to improve their properties to yield such outstanding characteristics as their resilience and peel strength.

Polyamides of enhanced solubility also result from the use of acids or amines containing large bulky side groups that prevent close packing of molecules (e.g., branched diamines and cyclohexane dicarboxylic acid) and copolymerization to give irregular structures (e.g., copolycondensation in acetic acid medium of caprolactam with ω-amino-undecanoic acid). Strong and versatile protective adhesives, and coatings are produced by mixing solutions of some polyamides with epoxy resins. The coatings are tough, hard, flexible and strongly adhering to a wide variety of surfaces such as wood, concrete, steel, glass, and many different kinds of plastics.

Many of the organic solvent-based adhesives are based on rubbery polymers, the main ones being natural rubber, polychloroprene, butadiene-acrylonitrile, styrene-butadiene, and polyisobutylene. In a typical method of manufacture, the rubber is placed in a heavy-duty mixer and solvent is added slowly till a smooth solution is formed. In some cases, the rubber is milled beforehand to reduce viscosity and produce smoother solutions. Depending on the rubber, the solvent can either be petroleum hydrocarbon of differing boiling ranges, aromatic hydrocarbon (usually toluene) or oxygenated solvents such as ketones or esters. Chlorinated solvents are used when nonflammability is required. Various resins are added to improve adhesion to different substrates. Antioxidants are added to improve ageing characteristics which fillers are often added to improve performance.

These organic solvent-based adhesives are very easy to apply and find wide application in general industry. Whenever a variety of different materials has to be stuck together quickly and easily, organic solvent-based adhesives will often be the answer. The footwear industry uses them for sticking soles to uppers and many other operations. The automotive industry uses them for stick-on trim and a variety of other jobs. Foam-seat manufacturers use these adhesives in brushable and sprayable form. Paper converters employ solvent-based permanently tacky adhesives in the manufacture of labels and sticky tapes.

A special type of organic solvent-based adhesive is the so-called contact adhesive. This makes use of the fact that certain elastomeric or rubbery solids (e.g., polychloroprene) have the property of autohesion, i.e., they can stick readily to themselves, especially if compounded with resins and containing small amounts of solvents. The bonding takes place by a diffusion process, the adhesive being applied to both surfaces to be bonded. Thus substrates may be coated with a contact adhesive, the adhesives can be allowed to dry till most of the solvent has evaporated (the dry adhesive film at this point will contain residual solvent, often up to 7–10%, but the film will not adhere to another substrate) and the two adhesive surfaces can then be brought together, the bonding taking place immediately and without much pressure. Because of this property, contact adhesives find wide use in industry for easily sticking many different materials together without the need for clamping.

Although natural rubber was the first polymer used in contact adhesives, it has been almost entirely displaced in favor of synthetic polymers. The first to achieve prominence was polychloroprene, but this has later been joined and, for some purposes, displaced by polyurethane and acrylic copolymers.

It may be noted that contact adhesives are different from pressure-sensitive adhesives. The latter generally combine a high degree of inherent tackiness, for instantaneous bonding, with a high cohesive strength to facilitate removal of tape from a surface without leaving any residue. Typically, the adhesive is based on a film-forming elastomeric material such as SBR, polychloroprene, polybutadiene, and other normally tacky polymers, such as polyvinyl ethers and polyacrylates. Commonly, blends and combinations of various materials are needed to optimize initial tack, cohesive strength, and, of course, the actual adhesive properties of the substrate. Polyterpene resins and rosin esters are often used as tackifiers in many formulations. Pressure-sensitive adhesive tapes, for example, are composed of (1) the adhesive, (2) a primer coating that is applied to the plastic or metal foil surface, (3) a release coating on the backing to allow unrolling of the tape, and (4) a suitable backing material such as plastic, metal foil, paper, and nonwoven textiles.

2.8.2 Water-Based Adhesives

This class of adhesives contains some of the most traditional materials, for example animal glues and gums derived from natural sources, which are long-chain molecules, are soluble in water, and are either proteins or polysaccharides. Modern technology has brought other water-soluble materials into use; for example, poly(vinyl alcohol) adhesives, and a host of resins based on phenol, urea, and formaldehyde (see Formaldehyde Resins in Chapter 1).

Water-based adhesives also encompass emulsion and dispersion adhesives in which polymer chains are dispersed and stabilized into tiny particles or spheres in aqueous medium with the aid of surfactants. Within this category there are low-cost poly(vinyl acetate) and acrylics, and more expensive higher-performance urethanes, which are typically made of two components.

Because water-based adhesives contain little or no organic solvents, they offer manufacturers a strategy for compliance with volatile organic compound (VOC) regulations. The use of water-based systems is therefore growing quickly, particularly in applications that do not require high-performance properties.

Probably the most widely used industrial emulsion or dispersion adhesives are those based on poly(vinyl acetate), commonly referred to as PVA. These product are normally manufactured by emulsion polymerization whereby, basically, vinyl acetate monomer is emulsified in water with a suitable colloid-emulsifier system, such as poly(vinyl alcohol) and sodium lauryl sulfate, and, with the use of water soluble initiator such as potassium persulfate, is polymerized. The polymerization takes place over a period of four hours at 70°C. Because the reaction is exothermic, provisions must be made for cooling when the batch size exceeds a few liters. The presence of surfactants (emulsifiers) and water-soluble protective colloids facilitates the process resulting in a stable dispersion of discrete polymer particles in the aqueous phase.

Plasticizers and thickeners are two of the materials most commonly used in the formulation of PVA-based adhesives, but many other modifiers may be needed, producing a very wide range of adhesives to meet specific requirements of a wide variety of applications.

PVA-based adhesives exhibit excellent adhesion properties and adhere to a host of different substrates such as metal, porcelain, wood, paper and various textiles. The adhesives are quick-setting, a property which shortens press and assembly times in manufacturing processes.

While a substantial proportion of emulsion adhesives are based on PVA, copolymer emulsion adhesives based on vinyl acetate ethylenes (VAEs) are becoming of increasing significance and importance in the market place. These are produced by the copolymerization of vinyl acetate and 10–20% ethylene, the resulting polymer base possessing some superior properties over the PVA-based emulsions referred to above. These superior properties relate principally to the increased inherent flexibility of the dry VAE film, which enhances adhesion to many "difficult" surfaces. There also exist many alternate specialized emulsion adhesive systems for which a number of other polymers and copolymers are used as the formulating basis.

Generally, water-based adhesives are used where at least one of the substrate is porous, or permeable to moisture vapor. The two substrates to be joined are laminated while the adhesive is still wet, and the water escapes either by initial absorption into one of the substrates or by evaporation due to permeability.

Because of the need to remove water, most applications for aqueous adhesives are with materials that would allow transmission of water vapor. Hence the major uses are with paper, wood and fabric, either as binders or laminating adhesives. Typical applications of emulsion adhesives include the following: remoistenable gummed tape, tube winding, box manufacture, plywood manufacture, woodworking, bookbinding, abrasives manufacture, and as textile adhesives.

2.8.3   Hot Melt Adhesives

Hot melt adhesives are thermoplastics that when molten are applied to the substrates [89]. Setting is achieved rapidly, normally in a few seconds or less, leading to high production speeds, and for this reason melt adhesives are attractive for a number of applications. They are 100% solids, containing no solvents; this means no evaporation and no VOC emissions. Other advantages are indefinite shelf life, much reduced space requirements due to lower warehouse volume, and avoidance of drying areas.

Melt adhesives are based on thermoplastics, but usually contain a number of other components. The most commonly used melt adhesives are based on ethylene-vinyl acetate (EVA) copolymers, but polyethylene, polyesters, polyamides, and thermoplastic rubbers (e.g., styrene-butadiene block copolymers) are also used (see Adhesive Bonding of Plastics in Chapter 1 of *Plastics Fabrication and Recycling*).

Depending on the properties required, the vinyl content in the EVA copolymer may be varied and varying quantities of tackifying resins (e.g., rosin esters) and waxes are incorporated. Tackifying resins affect the adhesion at elevated temperatures, the time the adhesive takes to harden and the quality of final adhesion achieved. The main purpose of the waxes is to reduce the melt viscosity and thereby improve the wetting of the substrates.

The widespread use of hot melt adhesives includes bookbinding (soft cover binding of books and magazines), component assembly in electronics, consumer durables, automotive fields, and various applications in the packaging industry (e.g., carton sealing, bag making, labels), footwear manufacture (counters and toe puffs), woodworking, and the textile industry.

The main limitations of hot melt adheives are their modest service temperature and modest load-bearing ability. Polyester and polyamide systems are usually much superior to EVA systems in both these respects.

2.8.4   Radiation-Curable Adhesives

The term radiation curable is applied to adhesives that cure when exposed to ultraviolet light (UV-curable) or electron beams (EB-curable). These forms of radiation have the correct energy to initiate polymerization of low-molecular-weight unsaturated resin. The type of cross-linking obtained with EB and UV radiation is very similar but the way the curing is brought about differs. The polymerization is not directly initiated by UV radiation and a photoinitiator is used to interact with the UV light and produce the initiating species. Electron beams, however, have higher energy, which is sufficient to initiate polymerization.

Electron beams are more penetrating than UV radiation and can thus be used to bond some opaque substrates that cannot be bonded with UV-curable adhesives. However, this advantage is offset by the high capital costs of EB curing equipment, especially in comparison with UV lamps, which are both cheap and readily available.

Though potential UV-curable system [90] are many, only two have found significant commercial use in adhesives, namely, free-radical polymerization of acrylates and cationic polymerization of epoxies.

The essential ingredients of a free-radical adhesive formulation are an acrylate-terminated prepolymer and a photoinitiator. A wide range of prepolymers can be acrylated, including epoxies, urethanes, polyesters, polyethers, and rubbers. Those most commonly used in adhesive formulations are epoxy and urethane acrylates. Epoxy acrylates have properties similar to those of the parent epoxy resin, with excellent adhesion, chemical resistance, and toughness. Urethane acrylates, on the other hand, are noted for their high reactivity, good adhesion, flexibility, and tear resistance.

Acrylated prepolymers usually have high viscosities and require a diluent. This can be an acrylate monomer or a low-toxicity monomer that itself takes part in the curing reaction. The use of such reactive diluents results in changes to the viscosity, flexibility, and the speed of cure of the formulation.

Additives are commonly used to improve the performance of an adhesive. These might include adhesion promoters, fillers, light stabilizers, antioxidants and plasticizers. A drawback of acrylate-based formulations is the oxygen sensitivity, i.e., inhibition of curing by oxygen, which occurs due to quenching of the excited photoinitiator and scavenging of free radicals. The oxygen inhibition may be overcome by adding oxygen scavengers to the formulation. Other methods are nitrogen blanketing, the use of high intensity lamps, and varying the initiator type and concentration.

Cationic UV adhesive formulations typically contain an epoxy resin, a cure-accelerating resin, a diluent (which may or may not be reactive) and a photoinitiator. The initiation step results in the formation of a positively charged center through which an addition polymerization occurs (see Epoxy Resin in Chapter 1). There is no inherent termination and this may allow a high degree of post cure.

The advantages of a cationic adhesive formulation over that of a free radical one are the lack of oxygen sensitivity, less shrinkage on curing, and better adhesion. The disadvantages are that the photoinitiators are sensitive to moisture and basic materials and that the acidic species can promote corrosion. Consequently, the majority of UV formulations used in industry are acrylate based and cured by a free radical mechanism.

UV curing of adhesives offers several advantages over conventional methods of cure. These include significant productivity benefits arising from the low running costs of UV lamps and the very rapid cure of the adhesives. These adhesives also contain no solvents that are damaging to the environment. Requiring no heat, they are suitable for heat-sensitive substrates and are single component.

The disadvantages of UV adhesives are that one transparent substrate is normally required; they suffer from oxygen inhibition and only a limited depth of cure can be achieved. The latter problem has been tackled by the development of dual-cure adhesives. In these systems, two independent curing mechanisms are incorporated into a single system. Thus the adhesive is cured first to a chemically stable form by UV irradiation and subsequently led to full cure by a second means, for example, thermal cure.

UV curable adhesives are being used in a large number of industrial applications. Some of the most significant of them include plastics and glass bonding, as well as applications in electronics. These applications demand high-performance adhesives which bond to difficult substrates.

2.9 Degradable Polymers

Although there certainly has never been a great incentive for making unstable polymers, the idea of making degradable polymers has long existed due to its environmental significance, and quite a bit of effort has gone into research along these lines. There are two important ways of making a degradable polymeric material. One is to make the polymer sensitive to sunlight, which fractures its chemical bonds and breaks it down by photodegradation; the other is to make a polymer out of a material that is biodegradable.

To make plastics photodegradable, a material can be implanted that will absorb sunlight and becomes sufficiently reactive to attack the polymer molecules from which the material is composed. One other means of photodegradation is to incorporate in the polymer backbone suitable groups, e.g., carbonyl groups that absorb the ultraviolet (UV) component of the sunlight to form excited states energetic enough to undergo bond cleavage. Such processes (referred to as Norrish type II reactions) occur as follows:

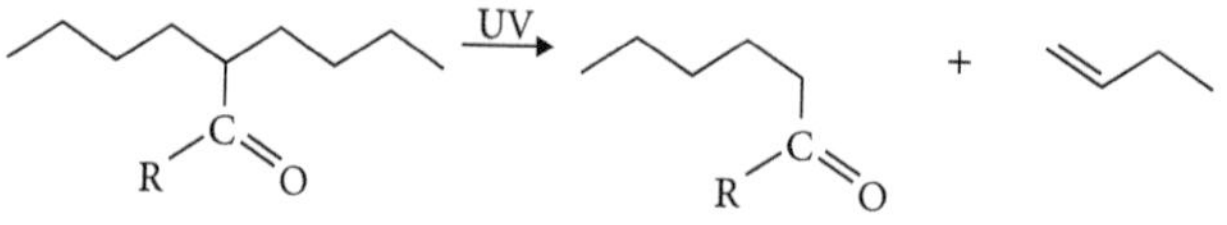

$$(2.11)$$

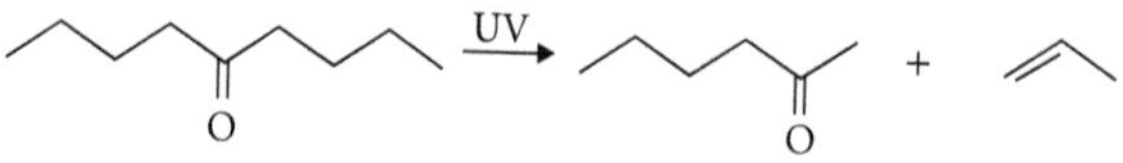

$$(2.12)$$

The main problem in making a photodegradable polymer is that it is hardly possible to combine rapid degradation upon exposure to light in a landfill after use with a good light-stability of the polymer during service. This contradiction is probably the reason why this method never really caught on.

The difference between photodegradation and biodegradation lies in the possibility to create an environment (as in a landfill) completely different from that encountered under normal storage conditions; e.g., microorganisms that can destroy organic polymers may be added to a landfill. In spite of the fact that substantial research time was spent on studies in this field, it is claimed [91] that surprisingly little is understood about the molecular-level interaction between polymers and microorganisms. For polyesters, however, a number of interesting data are available. Esterases (ester-hydrolyzing enzymes) and some microorganisms are known to biodegrade polyester at a reaction rate depending upon the polyester structure. While many aliphatic polyesters, specifically poly(hydroxy fatty acids), are suited for biodegradation, the aromatic polyester (e.g., PET) do not possess this property.

Poly(vinyl alcohol) is the most readily biodegraded, compared to other vinyl polymers, e.g., polystyrene, polyethylene, and polypropylene. Respirometric assays with mixed culture activated sludge have shown that poly(vinyl alcohol) is mineralized [92]. Extensive studies have been reported on the purification, characterization, and mechanism on degradation of poly(vinyl alcohol) by enzymes isolated from pseudomonas. The polymer is degraded as the sole carbon source by thee organisms; the mechanism of action is summarized in Figure 2.52.

Poly(ethylene-*co*-vinyl alcohol) is a thermoplastic used extensively in laminates for food containers due to its excellent film forming and oxygen barrier properties. Limited evidence for the disappearance of the polymer was provided [93]. Whether or not the copolymer is biodegradable is apparently related to the size and distribution of the ethylene blocks.

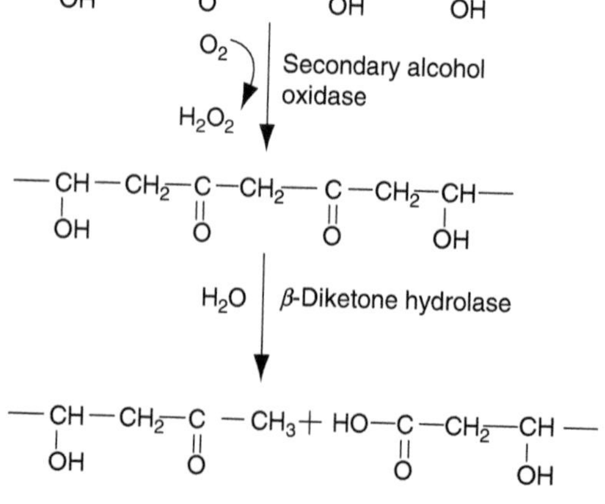

FIGURE 2.52 Mechanism of enzymatic depolymerization of poly(vinyl alcohol).

$$\cdots \overset{\displaystyle OH \quad\quad\quad OH}{\underset{\displaystyle \mid \quad\quad\quad\quad\; \mid}{\left[CH_2-CH-CH_2-CH-(CH_2)_n\right]_n}}$$

Poly(vinyl acetate) and poly(ethylene-*co*-vinyl acetate) (EVA) are slowly biodegraded [94], particularly where there is a high percentage of vinyl acetate in the copolymer. Growth of fungi on EVA copolymers correlate directly with vinyl acetate content [95], which seems to suggest that a pathway wherein enzymatic hydrolysis of side-chain ester groups occurs instead of the oxidative chain cleavage required for polyolefins may play a role.

2.9.1 Packaging Applications

The polymer poly(hydroxybutyrate) (PHB) is a natural polymer made by a wide range of microbes, such as the bacterium Alicaligenes eutrophus, as a convenient way of storing energy (in the same way that human beings store energy as fat). PHB is a crystalline, thermoplastic polyester of 3-hydroxybutyric acid [$CH_3CH(OH)CH_2COOH$]. It has been commercialized by ICI under the trade name Biopol. Often it is copolymerized with hydroxyvaleric acid $CH_3CH_2CH(OH)CH_2COOH$, to produce poly (3-hydroxybutyrate-*co*−3-hydroxyvalerate). The copolymerization increases the percentage of amorphous regions that are readily attacked by hydrolytic degradation, thereby increasing degradation rates.

The properties of the copolymer can be tailored to make it suitable either for molded articles such as shampoo bottles, or thin films for plastic envelopes or carrier bags. However, the polymer is costly, a container made of Biopol being about seven times more expensive than polyethylene. This polymer is now in production and used for packaging, agricultural products, and disposable items of personal hygiene.

Another approach to making biodegradable plastics for packaging consists of mixing small amounts of biodegradable polymers (e.g., starch) with a regular polymer (e.g., polyolefin), in order to make the end product destroyable as well. Starch is of interest as a biodegradable material because of its low cost, its availability as agricultural surplus raw material, and the thermoprocessability of the blend using conventional plastics-processing equipment.

In the early 1970s, work was in progress at two noncommercial centers to create a biodegradable plastic material that would be acceptable to the packaging industry. One such center was the USDA laboratories in Peoria, Illinois, as part of a program of work aimed at finding nonfood markets for farm products. The Peoria project approached its target by adopting starch as the matrix natural polymer, gelatinizing the starch by the action of heat and water, and then seeking synthetic polymeric additives that would make the mix processable on familiar plastics machinery. The products thus had starch gel as the continuous phase.

The other center, at Brunel University in the United Kingdom, developed the complimentary technique of using polyolefin polymers as the continuous phase with particulate starch additives as a filler. The starch component was usually below 10%. Microbes digest the starch in such product, leaving a flimsy plastic lace that disintegrates mechanically. The starch being in small amounts hardly alters the properties of the original polymer and the product can be processed on existing machinery without costly alterations.

One blend of polyethylene and starch incorporates an auto-oxidant that reacts with metal salts in soils or other environments to form peroxide radicals. The radicals degrade the polyethylene polymer chains into smaller oligomers susceptible to mineralization [96]. The starch is treated with a silane coupling agent for compatibility with polyethylene, and an unsaturated ester such as soya or corn oil is used as the auto-oxidant. Later technology has considered temperature triggers based on a composting regime [96].

The USDA technology mentioned above produces blends of gelatinized starch with nonbiodegradable components, such as ethylene and acrylic acid [97]. Blown films containing up to 60% starch are made and ammonia is added in the process. Many companies are commercializing thermally processed starch blends containing high percentages of starch for a wide variety of applications (e.g., Novon, Novamont).

Mixtures of starch with other polymers, including PET, have been studied [98], but no commercialization of the latter mixture is known.

Polyethylene can be made photodegradable by copolymerizing ethylene with a small percentage of carbon monoxide so that carbonyl groups are built into the long chain $-\!(CH_2CH_2COCH_2CH_2\,)\!-$. Polymers containing 1% of carbonyl groups lose their strength after two days in the sun. The polymer is thus decomposed into short chains [Equation 2.12] that microbial enzymes can digest.

Another approach to making photodegradable plastics is to use photosensitizers such as iron and nickel dithiocarbamates. These compounds can absorb photons to produce free radicals that trigger chain reactions leading to polymer degradation. The stability of a polymer film can be regulated by the amount of photosensitizer used. Polyethylene film used for mulching the soil in vegetable growing can thus be timed to decompose and disappear at the end of the time of harvesting.

2.9.2 Medical and Related Applications

Biodegradability is often an important consideration in the development of biomedical, pharmaceutical, and agricultural products for a number of applications. Biodegradable polymers have been formulated for uses such as controlled release and drug-delivery devices, surgical sutures, scaffolds for tissue regeneration, vascular grafts and stents, artificial skin, and orthopedic implants.

2.9.2.1 Synthetic Polymers

Degradable, synthetic polymers that are commonly known in the medical field include poly(α-hydroxy esters), poly(ε-caprolactone), poly(ortho esters), polyanhydrides, poly(3-hydroxy butyrate), polyphosphazenes, polydioxanones, polyoxalates, and poly(amino acids).

Poly(glycolic acid) (PGA) and poly(lactic acid) (PLA) are examples of poly(α-hydroxy esters.) PGA is a highly crystalline, hydrophilic, linear aliphatic polyester (XIII). As such, it has a high melting point and a relatively low solubility in most common organic solvents. PGA degrades primarily by bulk erosion. This occurs through random hydrolysis of its ester bonds.

The degradation of PGA is bimodal with the first phase of degradation occurring by diffusion of water to the amorphous regions followed by hydrolysis. The second phase begins as water penetrates and hydrolyzes the more crystalline regions. For PGA surgical sutures, mass loss occurs primarily during the second phase, completing the entire process between 4 and 12 weeks. The rate of hydrolysis in vitro can controlled by varying the pH. A more basic or acidic environment drives hydrolytic cleavage.

(XIII)   (XIV)   (XV)

The linear aliphatic polyester, PLA (XIV) is chemically synthesized by condensation polymerization of the free acid or catalytic ring-opening polymerization of the lactide (dilactone of lactic acid). The ester linkages in the polymer are sensitive to both enzymatic and chemical hydrolysis. PLA is hydrolyzed by many enzymes including pronase, proteinase K, bromalin, ficin, esterase, and trypsin. The degradation rate of PLA also varies with varying pH. The amount of lactic acid released during the course of PLA degradation is very small but increases rapidly as PLA is broken down to low molecular weight oligomers.

PLA is primarily considered for medical implants and drug delivery, but broader applications in packaging and consumer goods are also targeted. An attractive feature of this material is the relatively low cost of the monomer, lactic acid, which can be derived from biomass (fermentation), coal, petroleum, or natural gas.

Copolymerization of poly(α-hydroxyesters) provides a means to control physical and mechanical properties of the product. The degradation rates are highly dependent on the relative amount of each monomer. Copolymers with high or low ratios of comonomers are much more stable to hydrolytic attack than copolymers with a more equimolar ratio, due to their greater crystallinity.

Multiple uses of poly(lactic acid), poly(glycolic acid) homopolymers, and poly(lactic-*co*-glycolic acid) copolymers have been described including sutures, vascular grafts, drug carriers, and scaffolds for tissue engineering (discussed below). This is due in part to the FDA approval of these polymers for use as implantable materials.

Poly(ε-caprolactone) (PCL) is a semicrystalline aliphatic polyester (XV), synthesized by ring-opening polymerization of ε-caprolactone. PCL has been shown to degrade by random hydrolytic scission of its ester groups, and under certain circumstances by enzymatic degradation. Chemical hydrolytic rates of PCL homopolymers are very slow, particularly when compared with polyglycolic acid and poly(glycolic-*co*-lactic acid). PCL has a low glass transition temperature of $-62°C$, existing always in a rubbery state at room temperature, and a melting temperature of $57°C$. It has been postulated that these properties lead to a high permeability of PCL for controlled release agents.

A number of amino acid derivatives have been synthesized primarily for biomaterial applications as implants and for controlled release. These polymers include amino acid-based polyanydrides and polyesters. Poly(amide-anhydride) of structure (XVI), for example, is formed from β-alanine and sebacoyl chloride (that is, amino acid and diacid chloride) under anhydrous conditions in the presence of an acid acceptor. The premise upon which this class of polymers is designed is that the anhydride linkage will hydrolyze rapidly at the surface of a material and more slowly toward the center, providing for controlled release. The primary mechanism for depolymerization is surface erosion via chemical hydrolysis rather than enzymatic depolymerization. The polyanhydrides of aromatic diacids have advantages when compared with aliphatic polymers, including longer release and degradation times and higher mechanical strength and stability.

(XVI) (XVII)

Poly(ortho esters) (XVII) contain acid-labile linkages in the polymer backbone. As with the polyanhydrides discussed above, poly(ortho esters) are a class of polymers that can degrade heterogeneously by surface erosion. These polymers lose material from the surface only, while retaining their original geometry. As such, their primary use is in drug delivery.

2.9.2.2 Controlled Release Agents

Controlled release refers to the use of polymers containing agents of agricultural, medicinal, or pharmaceutical activity, which are released into the environment of interest at relatively constant rates over prolonged periods [99,100]. In the agricultural field, degradable mulches to promote crop growth are composed of combinations of natural polymers (which degrade readily in the presence of soil microorganisms) and synthetic polymers. Examples are starch-graft-poly(methyl acrylate) and block copolymers of amylose or cellulose with polyesters. At the end of the growing season, such mulches may be plowed directly into the soil along with crop residues.

Another application involves binding of agricultural chemicals in polymer formulations for slow release at a rate effective for their intended purpose avoiding the risk of the reagents being washed away by rain or irrigation. For example, the herbicide 2,4-dichlorophenoxyacetic acid (2,4-D) has been incorporated into polymers either as a chelate with iron (XVIII) or as a hydrolysable pendant ester group on a vinyl polymer (XIX).

(XVIII)

(XIX)

A different strategy for controlled release is based on polymer permeability other than degradation. The active reagent may be encapsulated within a polymeric membrane or in a strip, as shown in Figure 2.53. Ideally the reagent is contained in the reservoir as a saturated solution with excess in suspension. This allows diffusion through the membrane at constant rate without loss of activity. Alternatively, the reagent may be dispersed in a polymer matrix and released to the environment by diffusion or extraction. A variety of membrane and matrix devices are commercially available [99]. Pheromone release strips for insect control and household fly and cockroach strips for release of insecticide are also in commercial use.

Encapsulated pharmaceuticals have been available for many years. Reservoir strips called transdermal patches are used to release drugs through the skin; for example, nitroglycerin to treat angina or scopolamine to combat motion sickness. Implanted polymeric matrix devices, including some made with degradable polymers to release the drug through surface erosion, are available. Degradable polyesters, described above, are used as disappearing surgical sutures. The inorganic polyphosphazene, $-[-N=PCl_2-]_n-$, with amino acids, esters, and steroids in place of the chlorine atoms, are of interest for slow release of the steroid, the other degradation products (amino acid, ester, phosphate, and ammonia) being nontoxic.

2.9.2.3 Tissue Engineering

An interesting application of biocompatible and biodegradable polymers is in the field of tissue engineering, which involves the creation of natural tissue with the ability to restore missing organ function.

The problem of donor scarcity precludes the widespread utilization of whole organ transplantation as a therapy to treat many diseases. Not only is finding a suitable donor difficult, but once found it is a costly coordinated effort to harvest and transport the required organ. Scientists and physicians have therefore sought a feasible alternative to organ transplantation. Materials such as metals, ceramics, and polymers have been used extensively to replace the mechanical function of hard tissues such as cartilage and bone. Devices constructed primarily of polymers, such as artificial hearts and membrane oxygenators, have also been used to partially replace organ function, at least on a temporary basis.

Tissue engineering is a potential alternative to existing therapies, as it would allow the restoration of organ function by creating complete natural tissue, with the required mechanical or metabolic features in vivo [59]. Though still in its experimental stages, the use of tissue engineering techniques to regenerate organs has the potential to affect the quality and length of many lives. The technique can be applied for

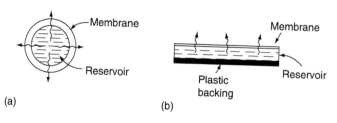

(a)   (b)

FIGURE 2.53   Membrane-controlled release devices: (a) microencapsulation and (b) strip.

virtually any tissue and organ in the human body, e.g., in bone repair, cartilege regeneration, wound healing, ocular diseases, mandible reconstruction, and andiogenesis.

The creation of natural tissue may be a achieved either by transplanting cells seeded into a porous material (Figure 2.54a and b) or, in some cases, by relying on in-growth of tissue and cells into such a material (Figure 2.54c). The process by which regeneration is effected by in-growth from surrounding tissue is known as tissue induction. The process can be facilitated or enhanced by the release of chemotactic agents from the porous material which attract cells to the area of regeneration (Figure 2.54d). Transplanted cells may be autogeneic (from the same genotype), allogeneic (from the same species but with a different genotype), or xenogeneic (from a different species). Autogeneic cells transplantation is the most preferable choice since it avoids many of the problems associated with immune rejection of foreign tissue.

There has been a great deal of research involving cell culture on biodegradable materials and development of techniques to manufacture degradable polymer scaffolds. Tissue engineering using transplantation of autologous (i.e., from the same patient) cells has been studied to restore the function of tissues such as cartilage, bone, skin, nerve, kidney, and liver, as this method of restoring organ function offers a number of advantages over whole organ trnsplantation, the two most important of these being overcoming the problem of donor scarcity and the reduced risk of rejection. Extensive research has also been performed with encapsulated xenogeneic pancreatic islets to treat insulin deficiencies in diabetics. In this case, the polymer acts both as a cell substrate and as a permanent protective barrier to the immune system.

Stem cell therapy is one very interesting strategy for tissue regeneration because it combines the potential of stem cells to develop into a particular functional tissue with the possibility of donor cell coming from the same patient, thus eliminating adverse immune response. Adult human stem cells are of particular interest, e.g., mesenchymal stem cells, which can develop into bone cartilege tissue and others [102]. Autologous stem cell treatment is carried out by the extraction of a small number of stem cells followed by their in vitro cultivation to obtain a sufficient large number, which can then be re-implanted with a scaffold and stimuli. The actual development is strongly dependent on a variety of biological regulation systems, e.g., the presence of growth factors, nutrients, cell-cell interaction, cell-matrix interaction, pH, and electrolytes [103].

The scaffold material should possess a number of properties like biocompatibility, biodegradability, mechanical strength, porosity, potential of entrapment and release of pharmaceutically active agents, and an easy processability for the clinician. Since safety is a critical issue and similar to new pharmaceuticals, regulatory agencies (e.g., FDA) need to approve the use of the scaffold. Materials already approved for scaffolds include poly(DL-lactide) (PLA), poly)(DL-lactic acid-*co*-glycolic acid) (PLGA) and its block copolymers with poly(ethylene glycol) (PEG) [104]. However, the in vitro systems do not require biodegradability, but well defined material properties are desired during the whole cell cultivation process. Hydrogels can provide all these properties when choosing the right chemistry. They can provide softness, permeability for water, nutrients, metabolites or pharmaceutical active agents, and they show certain mechanical strength, which enables them to simulate living tissue. Therefore, they are excellent scaffolds for tissue engineering as well as drug delivery devices [105]

Recent research in polymeric biomaterials for engineering skeletal tissues and/or soft tissues such as muscles or lungs is focused on development of three-dimensional (3D) biodegradable polymer scaffolds [106]. A candidate for construction of such 3D polymer scaffolds is a fiber mesh consisting of individual polymer fibers woven or knitted into 3D patterns of variable pore size [107].

In order to enhance the bioactivity and potential osteoconductivity of bioresorbable polymer scaffolds, bioactive glass becomes a candidate material to be used as a coating, effectively forming a polymer/glass composite structure. Another reason to combine a bioactive glass with biodegradable polyesters is to control undesirable hydrolytic degradation characteristics of the polyesters such as rapid internal degradation and degradation-induced morphological and compositional changes. These considerations have led to the development of a new hybrid composite material concept based on using biodegradable polymer scaffolds coated with tailored bioactive glass layers [108]. A commercial bioactive glass is named

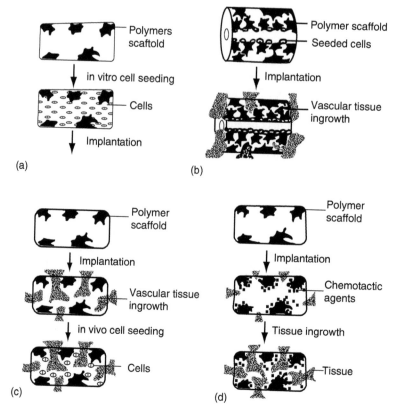

FIGURE 2.54 Schematic diagram of various regeneration techniques: (a) cell transplantation into porous scaffolds to regenerate tissues such as cartilage; (b) cell seeding into scaffolds with an annular space and subsequent tissue in-growth to regenerate tubular tissues; (c) prevascularization of scaffolds and subsequent cell seeding to regenerate metabolic organs such as liver; (d) tissue induction via release of chemotactic agents that attract the desired cells into a porous scaffold. (After Thomson, A. C., Wake, M. C., Yaszemski, M. J., and Mikos, A. G. 1995. *Adv. Polym. Sci.,* 122, 265. With permission.)

Bioglass®, which has the following composition by weight: 45% SiO_2, 24.5% Na_2O, 24.5% CaO, and 6% P_2O_5. When in contact with body fluids, this bioactive glass rapidly (2 h) forms a thin surface layer of calcium phosphate that stimulates cellular infiltration and osteointegration of the bioactive glass to the surrounding tissue by processes of osteoconduction, thus promoting bone regeneration [109].

2.10 Ionic Polymers

Ionic polymers are polymers containing chemically bound ions within their structure [110–116]. These are specialist materials, some of which have limited, though important, commercial applications. Those containing few ions are melt-processable thermoplastics called ionomers; those containing many ions are either water-soluble polymers called polyelectrolytes or crosslinked polymers containing ionic groups called ion-exchange resins. Polyelectrolytes and ion-exchange resins are, in general, intractable materials and not processable on conventional plastics machinery.

Ionic polymers may be classified according to the type of the bound ion, its position within the structure, and the amount of bound ions present along a given length of polymer chain. They may be further classified according to the nature of the counterion or the nature of the supporting polymeric backbone.

The bound ion is usually either an anion such as sulfonate $(-SO_3^-)$, phosphonate $(-PO_3^{2-})$ or carboxylate $(-CO_2^-)$ or a cation such as quaternary $(-NR_3^+)$ or amine $(-NH_3^+)$. But both types can occur together and the polymer is then said to be ampholytic. In each case, the polyion may be strong or weak according to the degree to which it ionizes. For example, polysulfonates and quaternary type polyions are strong, polycarboxylates and amine type polyions are weak, while polyphosphonates are intermediate in nature.

The bound ion can be pendant to the polymer's covalent backbone, as in a polysulfonate (XX) or it can be integral or enchained, as in an ionene (XXI)

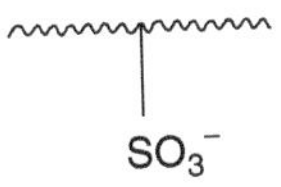

(XX) 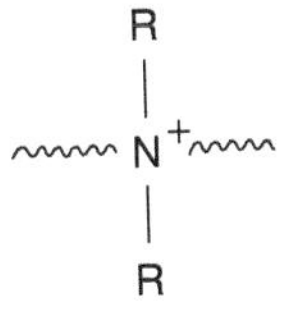   (XXI)

Almost any conventional polymer can be modified to form an ionic polymer of the pendant type. However, ionic polymers of the integral type, like the ionenes, are specialized structures whose backbones can exist only in the ionic form.

Ionic polymers contain counterions that neutralize the charges on the bound ions. The counterions may be grouped into three types: (1) univalent, (2) di- or trivalent and (c) polymeric. Polymers with polymeric counterions are often called polysalts, polyelectrolyte complexes, polyion complexes, simplexes, or coacervates.

The methods used for synthesis of ionic polymers can be divided mainly into three types: (1) direct synthesis, (2) post-functionalization of a standard preformed polymer, and (3) post-functionalization of a special preformed polymer.

Direct synthesis is used for polyelectrolytes such as carboxy polymers, e.g., poly(acrylic acid) and poly(methacrylic acid), which are probably the most common ionic polymers of all. Post-functionalization of pre-formed polymers is a frequently used synthetic method, provided that the polymer is sufficiently reactive. Common examples are the sulfonation of polystyrene and the grafting of thioglycollic acid on to polybutadiene using free radicals.

Special post-functionalizable copolymers have also been used to derive acid ionomers by hydrolysis, thus avoiding the difficulties of copolymerizing ionic and nonionic monomers. To this end there are many examples where carboxylic acid polymers are formed by hydrolyzing copolymers containing acrylate esters, acrylonitrile, or maleic anhydride. As described later, a sulfonic acid ionomer, Nafion, is formed by hydrolysis of tetrafluoroethylene copolymerized with a sulfonyl fluoride.

All of the aforementioned ionomers are anionic polymers. Cationic polymers are less common, though equally important. Pendant cations are usually of the quaternary ammonium type and made by a multi-step post-functionalization in which a precursor containing labile chloride is reacted with a tertiary amine:

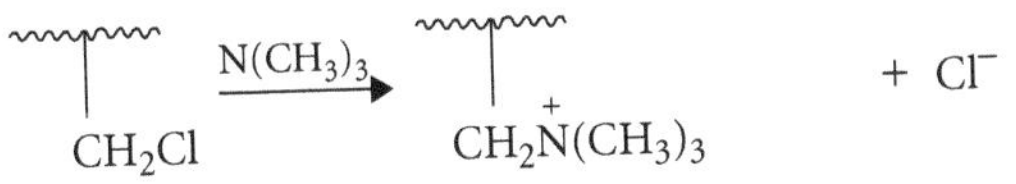

$$+ Cl^- \qquad (2.13)$$

Sometimes, the inverse of this is also done when a pre-formed polyamine is quaternized by reacting with an alkyl halide:

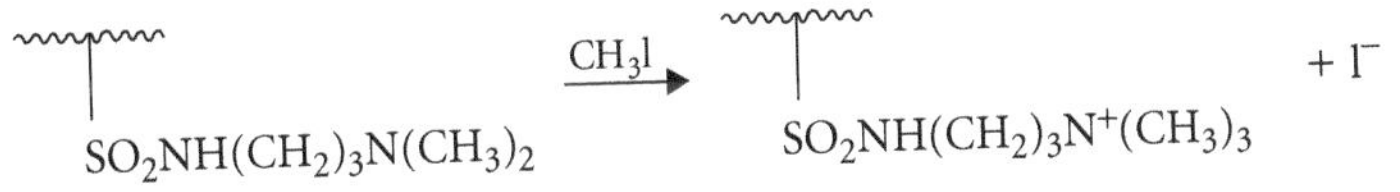

$$+ I^- \qquad (2.14)$$

2.10.1 Physical Properties and Applications

Synthetic ionic polymers have three distinctive properties that dictate their usage: (a) ionic cross-linking, (b) ion-exchange capability, and (c) hydrophilicity. Major applications for organic ionic polymers in relation to their ion content and dominant property are summarized in Table 2.14.

2.10.1.1 Ionic Cross-Linking

Since the ions in ionic polymers are held by chemical bonds within a low dielectric medium consisting of a covalent polymer backbone material with which they are incompatible, the polymer backbone is forced into conformations that allow the ions to associate with each other. Because these ionic associations involve ions from different chains they behave as crosslinks, but because they are thermally labile they reversibly break down on heating. Ionomers therefore behave as cross-linked, yet melt-processable, thermoplastic materials, or if the backbone is elastomeric, as thermoplastic rubbers. It should be noted that it is with the slightly ionic polymers, the ionomers, where the effect of ion aggregation is exploited to produce meltprocessable, specialist thermoplastic materials. With highly ionic polymers, the polyelectrolytes, the ionic cross-linking is so extreme that the polymers decompose on melting or are too viscous for use as thermoplastics.

Two types of ionic aggregates are found in ionomers, called multiplets and clusters, which coexist in equilibrium within the matrix of covalent backbone material. The multiplets, very small and very numerous, are associations of ion-pairs up to eight in number that are purely ionic and are devoid of trapped segments of the covalent backbone, while the clusters are large associations of these multiplets between which are trapped segments of the backbone. Evidence for this theory of microphase separation into multiplets and clusters is provided by many types of physical measurement, the important techniques being small-angle x-ray scattering, and infrared and Raman spectroscopy.

2.10.1.2 Ion-Exchange

Ionic polymers contain two types of ions, namely bound ions, which are part of the structure, and the counterions, which are free. In a medium in which the ionic polymer is insoluble, the counterions are exchanged for similar ions from the surrounding medium and an equilibrium is established, the kinetics of the process being dependent on factors such as the physical form of the insoluble polyion, its porosity, and surface area. The fundamental fact in this exchange is that the counterions have free movement into

TABLE 2.14 Applications of Organic Ionic Polymers in Relation to the Amount of Ion Present

Ion Content (equiv/kg)	Designation	Dominant Property	Application	Example
<0.5	Ionomer	Ionic cross-linking	Thermoplastic	Ethylene-acrylic acid copolymer
1–1.5	Ionomer	Ion-exchange	Membranes (electrodialysis, etc.)	PTFE copolymer (Nafion, Flemion)
1–1.5	Ionomer	Hydrophilicity	Membranes (reverse osmosis	Sulfonated polyether-sulfones
>4	Polyelectrolyte	Water solubility	Thickeners, dispersants, flocculants and sizes	Polyacrylic acid or copolymer
>4	Polyelectrolyte (cross-linked)	Ion-exchange	Ion-exchange resins	Sulfonated polystyrene, aminated polystyrene
14	Polyelectrolyte	Ionic cross-linking	Dental cements	Polyacrylic acid

and out of the polymer, while ions of the same type of charge as the bound ion do not. This barrier to ion movement is known as Donnan exclusion.

The property of ion-exchange has important consequences and three different types of application depend on it. They are (1) ion-exchange resins, (2) ion-exchange membranes, and (3) heterogeneous catalysis.

Ion-exchange resins are insoluble beads of an ionic polymer. Their bestknown use is in deionization or demineralization of tap water to distilled water. In this process, the water is contacted with two types of resin, one a polyacid and the other a polybase, so that the net effect is to exchange M^+ and X^- from the tap water for H^+ and OH^-, i.e., H_2O, from the polyion beads:

$$\text{(P)}SO_3^-H^+ + M^+X^- \rightleftharpoons \text{(P)}SO_3^-M^+ + H^+X^- \tag{2.15}$$

$$\text{(P)}NR_3^+OH^- + H^+X^- \rightleftharpoons \text{(P)}NR_3^-X^+ + H_2O \tag{2.16}$$

To be effective, ion-exchange resins should have high ionic content. They are therefore polyelectrolytes and, since polyelectrolytes are inherently water-soluble, they are chemically cross-linked during manufacture to make them insoluble. When immersed in water they undergo swelling, which facilitates rapid ion-exchange. The degree of swelling is, however, inversely related to the density of the cross-links.

A polyion in the form of a thin membrane is used as ion-exchange membrane in another application of the ion-exchange phenomenon. When exposed to an electrolyte, an ion-exchange membrane will allow counterions to pass through it, but will act as a barrier to the complementary ion, and is therefore said to be permselective. Thus a polyanionic membrane will allow passage of cations and a polycationic membrane that of anions, so that under the influence of an electric current, continuous fluxes of cations and anions, respectively, can be set up across these membranes. This principle is exploited in electrodialysis and in chlor-alkali cells as described later.

The dialyzer used for desalination of water by electrodialysis (Figure 2.55) is an electric cell divided into a series of sub-cells that are separated by alternate polyanion and polycation membranes. While the ions move under the influence of the applied electric potential, the complimentary permselectivities of the ion-exchange membranes insure that salt is removed from alternate sub-cells and concentrated in the others. Desalination dialyzers used in practice have up to 100 such sub-cells.

Other applications of dialyzers include concentration of brine and the desalting of cheese whey. (It is important not to confuse electrodialysis with ordinary dialysis, which is not an ion-exchange process but

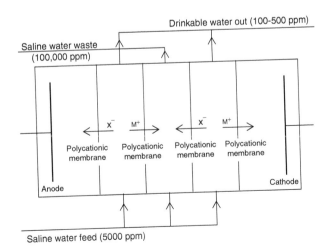

FIGURE 2.55 Desalination of water by electrodialysis.

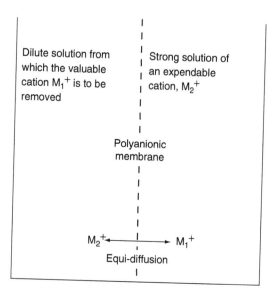

Dilute solution from which the valuable cation M_1^+ is to be removed

Strong solution of an expendable cation, M_2^+

Polyanionic membrane

$M_2^+ \longleftarrow\longrightarrow M_1^+$

Equi-diffusion

FIGURE 2.56　Ion-exchange dialysis.

is a separation based on differences in size and hence diffusivity between large and small molecules through a membrane, which, generally speaking, is not ionic.)

A related application of ion-exchange membrane is in ion-selective electrode. Thus when an electrode with an ion-exchange membrane encasing is immersed in an ionic medium, it will develop a potential proportional to the activity of the ion to which the membrane is permselective.

A high ion content is not necessary for an ion-exchange membrane, since its function is not governed by its exchange capacity. Though a high ion content can be advantageous as it decreases the electrical resistance, it can in fact be counterproductive if the membrane is so highly swollen by water that its permselectivity is reduced. The ion content of an ionexchange membrane is thus often intermediate between that of an ionomer and an ion-exchange resin.

While chemical cross-linking is used for ion-exchange resins to limit this swelling or to prevent the resin from dissolving, this method is generally not desirable for membranes, because cross-linking interferes with the process of membrane fabrication.

Ion-exchange dialysis (or Donnan dialysis) like ordinary dialysis is a diffusion-controlled separation process, but unlike the latter it involves ionexchange and so needs an ionic membrane (though without an applied current). If a polyanionic membrane is used, as shown in Figure 2.56, the cations diffuse each way across the membrane but the anions cannot so that, in effect, cations are separated as they swap over, but the amount of electrolyte on each side of the membrane remains constant. This type of dialysis, using continuous cells, is being developed as a means of stripping and concentrating radioactive ions from dilute solutions of radioactive wastes.

2.10.1.3　Hydrophilicity

Ionic polymers are hydrophilic. Those of moderate ion content swell on contact with water and those of high ion content dissolve, unless they are cross-linked. With moderately ionic polymers, which swell without dissolving, the hydrophilicity has the advantage of making the structure more permeable to ions for ion-exchange. The hydrophilicity of moderately ionic polymers leads to another type of membrane application, that of reverse osmosis.

The method of reverse osmosis involves application of pressure to the surface of a saline solution, thus forcing pure water to pass from the solution through a semipermeable membrane that does not permit passage of ions. Since the natural osmotic pressure tends to force the water form the region of low ionic concentration to that of the high, to achieve a flow in the opposite direction, as in reverse osmosis, a pressure has to be applied exceeding the osmotic pressure.

In the process used for desalination, a saline, brackish water or sea water is forced at very high pressure through a hydrophilic membrane resulting in water of drinkable quality (Figure 2.57). Reverse osmosis is also utilized for recovering waste water for paper mill operations, pollution control, industrial water treatment, chemical separations, and food processing.

The compartments indicated in Figure 2.57a are a schematic representation of reverse osmosis process. In practice, the reverse osmosis process is conducted in a tubular configuration system (Figure 2.57b). Raw waste water flows under high pressure (greater than osmotic pressure) through an inner tube made

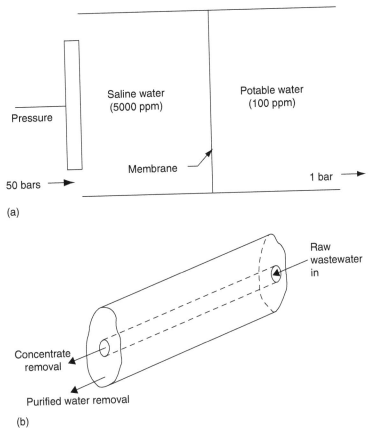

FIGURE 2.57 (a) Schematic representation of reverse osmosis process. (b) Tubular configuration system for wastewater treatment by reverse osmosis.

of a semipermeable membrane material and designed for high pressure operation. Purified water is removed from the outer tube that is at atmospheric pressure and is made of ordinary material.

Although the membrane used for reverse osmosis may be ionic and ion-exchange can occur, there is, however, no overall transmission of salt because of Donnan exclusion (and unlike electrodialysis no electric current is applied). Thus, while an essential requirement of a membrane for reverse osmosis is hydrophilicity, it need not be ionic. In fact, the most successful reverse osmosis membranes developed are made of nonionic cellulose acetate. Certain ionic polymers, such as the sufonated polyaromatics, has been used because of their greater chemical stability and resistance to biological degradation.

Several mechanisms have been proposed to explain reverse osmosis. According to the preferential sorption-capillary flow mechanism of Sourirajan [114], reverse osmosis separation is the combined result of an interfacial phenomenon and fluid transport under pressure through capillary pores. Figure 2.58a is a conceptual model of this mechanism for recovery of fresh water from aqueous salt solutions. The surface of the membrane in contact with the solution has a preferential sorption for water and/or preferential repulsion for the solute, while a continuous removal of the preferentially sorbed interfacial water, which is of a monomolecular nature, is effected by flow under pressure through the membrane capillaries. According to this model, the critical pore diameter for a maximum separation and permeability is equal to twice the thickness of the preferentially sorbed interfacial layer (Figure 2.58b).

From an industrial standpoint, the two basic parameters characterizing reverse osmosis systems for a given separation are: (1) rejection factor, which involves the choice of the appropriate chemical nature of the film surface, and (2) water flux, which depends on methods for preparing films containing the largest

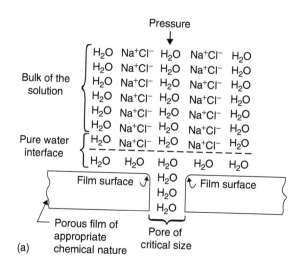

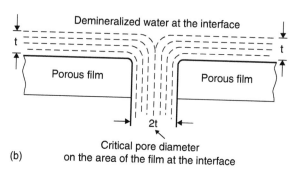

FIGURE 2.58 (a) Schematic representation of preferential sorption capillary flow mechanism. (b) Critical pore diameter for maximum separation and permeability. (After Agarwal, J. P., and Sourirajan, S. 1969. *Ind. Eng. Chem.*, 61, 62. With permission.)

numbers of pores of the required size. This approach is the basis of the successful development of the Sourirajan-Loeb type of porous cellulose acetate membranes for desalination and other applications.

To achieve high flux in reverse osmosis, the membrane should have a high surface area and be very thin (0.02–1.0 µm), but to withstand the high applied pressure it also needs to be very strong. To meet these requirements, which are in conflict, the semipermeable membrane is used as a skin mounted on a support, which is another membrane that is very porous, like a filter paper, and that is nonselective, and very thick (>100 µm). In some such bilayers each layer is of the same polymer and the membrane is then called integral or asymmetric, but in others the layers are of different polymers and the membrane is called a composite. Composite membranes have the advantage of each layer being individually optimized—the skin for its permeability and the support for its strength—but they are harder to make them integral membranes.

Polyelectrolytes, except those which are covalently cross-linked, are soluble in water, and their applications are based on this property. Basically, these applications depend on the polyelectrolyte altering the fluid properties of an aqueous medium, or modifying the behavior of particles in aqueous slurries or colloidal suspensions.

Polyelectrolytes raise the viscosity of aqueous solutions thus acting as thickeners, and the magnitude of the effect increases with the polymer's molecular weight. Naturally occurring gums and acidic polysaccharides have been traditionally used as thickening agents in food stuffs and pharmaceutical products, but, more recently, synthetic polyelectrolytes have been used in these roles. Much use is also made of polyelectrolytes to modify the characteristics of latex paints and similar proprietary fluids.

Polyelectrolytes can also stabilize particles in aqueous suspension and so act as dispersants. In this reaction, the hydrophobic backbones of polyelectrolytes are absorbed onto the surface of the particles by van der Waals attraction, while their ions form a hydrophilic surface and interacts with water. In suspension or bead polymerization, for example, a hydrophobic vinyl monomer is dispersed in water by agitation to form droplets that are stabilized by a polyelectrolyte. Each stabilized droplet of the monomer acts as a bulk polymerization system and on polymerization gets the shape of a bead.

Polyelectrolytes, depending on their ionic charge, can interact with colloidal particles and neutralize the stabilizing hydrophilic charges, thus acting as flocculating agents. They have been used in this way to coagulate slurries and industrial wastes.

There are other applications of polyelectrolytes that depend on their behavior in water in various ways. They are thus used as sizes in the textile industry and in paper manufacturing, and as additives to drilling muds and to soil for conditioning purposes.

A very different kind of application of polyelectrolytes is their use as dental cements. Here a divalent cation is added to an aqueous solution of a polyanion to form a highly cross-linked precipitate of great strength.

2.10.2 Ionomers

2.10.2.1 Polyethylene Ionomers

Ionomers based on polyethylene were described earlier (Chapter 1). They are mainly copolymers with pendant carboxylate groups in which the polyethylene backbone is their major component ($>90\%$). Ethylene is directly copolymerized with methacrylic acid in a continuous process developed from the high-pressure method used to make low-density polyethylene by free-radical initiation. The comonomers are mixed in appropriate proportions, allowing for the greater reactivity of the methacrylic acid, and introduced with a peroxide initiator to a reactor at a pressure of about 30,000 psi (207 MPa) and a temperature of 250–280°C. Because methacrylic acid reacts disproportionately rapidly and has to be replenished frequently, the conversion to polymer is kept low per pass, and at 15–20% conversion the residual monomer is removed, then recycled with fresh monomer. This ensures that the methacrylic acid units, and hence the ionic cross-links formed later, are randomly distributed and uniform in composition.

The distinctive properties of the ionomers become manifest only when the carboxylic acid groups are neutralized. The neutralization is a kind of post-treatment that is performed by melting the polyacid in a mill at 150°C and adding a base or basic salt in powder form or in solution. In these neutralizations, the melt starts as a soft, fluid, opaque mass of polyacid and then becomes stiff, rubbery, and transparent as the ionomer is formed.

The loss in fluidity, or gain in melt strength, is a significant factor determining the usefulness of ionomers. That the effect is due to the formation of strong ionic cross-links is clearly shown by a comparison of the melt viscosity of an ionomer with its acid precursor (Figure 2.59). The melt viscosity of a copolymer of ethylene and methacrylic acid containing, for example, 2 mole per cent of acid is increased by only about 50% compared to an otherwise equivalent polyethylene homopolymer, but when neutralized, and therefore ionized, the melt viscosity increases twenty-fold. (The slight increase with the acid is attributed to the weak hydrogen-bonded cross-links.) The melt-strength of an ionomer is such that its molten film can be drawn over the sharp edges of a nail without it being punctured or torn.

Though, as just explained, the viscosity of an ionomer at low shear is very much greater than its acid precursor, at high shear, however, the two are nearly the same (Figure 2.60), which is attributed to breaking down of the physical bonding of the cross-links. The high shear sensitivity of melt viscosity is an important characteristic of ionomers and is a useful feature in a number of melt-fabrication processes.

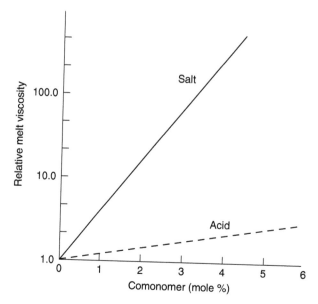

FIGURE 2.59 Melt viscosity of poly(ethylene-co-methacrylic acid) and its sodium salt relative to polyethylene.

Thus the high melt-strength at low shear makes ionomers useful for extrusion or blow-molding, or for any process where the melt is partially supported.

A striking property of the ethylene ionomers is that they are transparent, unlike their acid precursors or polyethylene itself, which are not. The haze in polyethylene is due to the fact that the polymer is partially crystalline, and minute crystallites within it agglomerate into spherulites, which are of a size to scatter light. Ionomers also contain crystallites, and the crystallite formation is, in fact, helped by the ions, the domains of which serve to nucleate them; but the crystallites are unable to agglomerate because of the high viscosity of their surroundings. Thus, microcrystallinity is enhanced by the ions, but macro-crystallinity, which causes haze, is inhibited. The transparency of ionomers is useful in packaging applications.

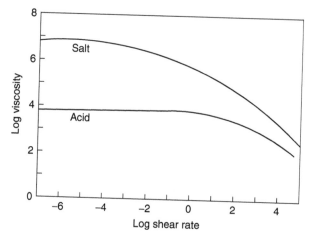

FIGURE 2.60 Melt viscosities (160°C) at various shear rates of poly(ethylene-co-methacrylic acid) with 3.5 mole% comonomer and its sodium salt.

Some physical properties of polyethylene ionomers are compared with those of polyethylene and the acid copolymer, poly(ethylene-co-methacrylic acid) in Table 2.15. Ionomer is generally tougher and, as shown in Table 2.15, relative to the acid copolymer, its tensile strength is increased by 27–53% and its stiffness is nearly tripled.

Polyethylene ionomers are described as flexible and tough with good impact toughness at low temperature, and with good resistance to grease and solvents, to stress-cracking, and to abrasion. They are blow-molded into films, sheets, bottles, and blister packs, and injection-molded into various objects. Their resistance to grease and solvents has made them useful in meat packaging. They are very effective as external coatings on glass bottles to contain breakages.

2.10.2.2 Elastomeric Ionomers

A number of ionic polymers exist that have a recognized elastomer as the covalent backbone and have a small ionic content, so they may be called elastomeric ionomers. The ions provide at least a part of the cross-links in these polymers. Those elastic ionomers that are cross-linked exclusively by their ions have, however, the useful feature of being thermoplastic.

Carboxylated polybutadiene ionomers, which are close relatives of the polyethylene ionomers described above, have an essentially polybutadiene backbone that contains some acrylonitrile and styrene to adjust its flexibility and toughness, and, in addition, up to 6% by weight of acrylic or methacrylic acid. Like the polyethylene ionomers, they are usually made by direct copolymerization with the carboxylic acid monomer using, however, emulsion methods. Typically the monomers are slurried in water with sodium dodecylbenzene sulfonate as the emulsifier and potassium persulfate as the free-radical initiator. The tendency of the carboxylic acid monomer to dissolve in the aqueous phase instead of remaining in the butadiene-rich phase is suppressed by making the aqueous phase acidic so that the monomer remains in the nonionized form.

The carboxylated polybutadienes, when neutralized, undergo ionic cross-linking, producing the effect of vulcanization. The neutralization can be done by treating with aqueous sodium hydroxide then heating, or by heating directly with zinc oxide. The ionic cross-link formed with the sodium ion of moderate strength at room temperature and dissociates at 100°C. With zinc, the ionic cross-link formed is much stronger, although the polymer is capable of substantial flow at higher temperatures.

Some properties of sodium and zinc vulcanizates are compared with those of the acid precursor in Table 2.16. High tensile strength is a characteristic of ionic vulcanizates. As a comparison, a standard polybutadiene elastomer vulcanized with sulfur gives a strength of 1.9–5.8 MPa (275–841 lbf/in.^{2}), whereas an equivalent copolymer containing 1.5 mole% methacrylic acid, and vulcanized with

TABLE 2.15 Comparative Physical Properties of Polyethylene Ionomers and Their Acid Precursor

			Ionomer	
Property	PE	Copolymer[a] ($-CO_2 H$)	Na$^+$	Zn^{2+}
Appearance	Hazy	Hazy	Transparent	Transparent
Melt index (g/10 min)[b]		5.8	0.03	0.09
Yield srength				
(10^3 lbf/in^2)	1.20	0.88	1.91	1.93
(MPa)	8.3	6.1	13.2	13.3
Elongation (%)	600	553	330	313
Tensile strength				
(10^6 lbf/in^2)	1.8	3.4	5.2	4.3
(MPa)	12.4	23.4	35.9	29.6
Stiffness (relative)		1.0	2.8	3.0

[a] Poly(ethylene-co-methacrylic acid) with 3.5 mole% methacrylic acid comonomer.
[b] ASTM-D-1238-57T.

magnesium oxide, gives 29.0 MPa (4.2×10^3 lbf/in.^{2}). They also respond differently to fillers such as carbon black—sulfur vulcanizates are reinforced, while ionic vulcanizates are weakened.

Carboxylated polybutadienes have not been used much as thermoplastic elastomers, principally because they have poor compression set, high stress relaxation, and poor performance at higher temperatures. Carboxylated polybutadienes, supplied as lattices, are used mainly in dipping and coating processes, applications, which often do not involve ionized carboxylate, at least directly. The applications include adhesives, paper coating, glove-dipping, carpet-backing, binding nonwoven fabrics, and also shrink-proofing woolen garments where the carboxyl groups are believed to react with pendant amino groups in the proteins of the wool. A wide range of carboxylated lattices are available as well-developed items of commerce; some of these are styrene-butadiene rubbers (SBRs) and others are acrylonitrile-butadiene rubbers (NBRs).

Mixed vulcanizations, such as with zinc oxide and sulfur or zinc oxide and peroxide, are used for carboxylated NBRs in order to combine the advantages of the ionic method with the conventional methods. The mixed vulcanizates have high tensile strength and notable resistance to abrasion, oil, and fuel. They are used in applications such as industrial rollers and wheels and shoe soles. Dry NBRs are available in fewer grades than lattices. Examples are Krynac by Doverstrand and Hycar by Goodrich.

Elastomeric ionomers have also been developed from ethylene-propylene-diene ternary copolymers known as EPDM rubbers. The diene is commonly ethylidene norbornene. Du Pont made carboxylated ionomers by free-radical grafting of maleic anhydride (0.5–5%) onto the diene moiety of the polymer and neutralized the product with rosin salt. The ethylidene norbornene can also be sulfonated, thus:

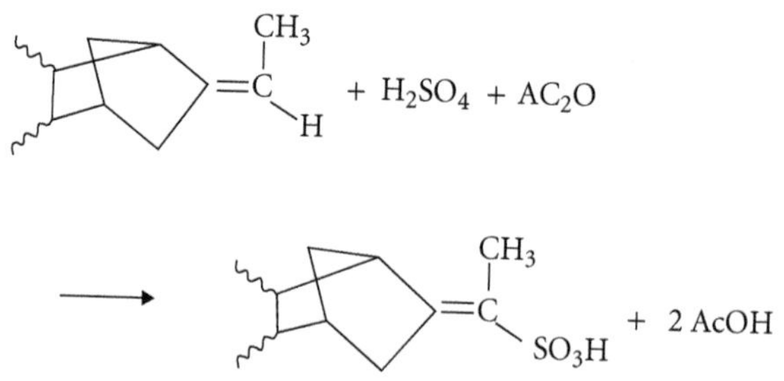

$$(2.17)$$

Ionomer properties of sulfonated EPDM are said to develop only when the acids are neutralized. The best properties are given by zinc salts, particularly when they are plasticized by zinc stearate. (See the section on thermoplastic elastomers in Chapter 1 for properties of ionic elastomers.)

2.10.2.3   Ionomers Based on Polytetrafluoroethylene

Polymers with a polytetrafluoroethylene (PTFE) backbone and pendant perfluorosulfonate or perfluorocarboxylate groups have become commercially important materials, although they are expensive. The sulfonates were introduced as Nafion by Du Pont in the early 1970s and the carboxylates as Flemion by Asahi Glass in 1978. They are made by free-radical copolymerization of tetrafluoroethylene and perfluorovinyl monomers giving precursor copolymers, (XXII) and (XXIII), which can be post-functionalized by hydrolysis to generate sulfonic and carboxylic acid groups. The perfluorovinyl

TABLE 2.16   Vulcanization of Carboxylated Polybutadienes

Form	Tensile Strength		Elongation (%)
	10^3 lbf/in.^{2}	MPa	
Acid copolymer	0.1	0.7	1600
Sodium vulcanizate	1.7	11.7	900
Zinc vulcanizate	6.0	41.4	400

Poly(butadiene-co-methacrylic acid) with 6.7 mole% methacrylic acid.

monomers themselves are made from tetrafluoroethylene by multi-step synthesis using hexafluoropropylene oxide.

$$PTFE - CF_2 - CF \sim\sim\sim$$
$$|$$
$$O$$
$$|$$
$$CF_2$$
$$|$$
$$CF \cdot CF_3$$
$$|$$
$$O(CF_2)_2\, SO_2\, F$$

Nafion precursor

(XXII)

$$PTFE - CF_2 - CF \sim\sim\sim$$
$$|$$
$$O$$
$$|$$
$$CF_2$$
$$|$$
$$CF \cdot CF_3$$
$$|$$
$$O(CF_2)_3\, COOCH_3$$

Flemion precursor

(XXIII)

Although originally developed for use as a membrane in fuel cells, Nafion has found more useful applications in various other electrolytic separation processes. In these applications, use is made of its cation-exchange properties and its ability to survive in extremely aggressive chemical environments. An outstanding example is the use of Nafion as a membrane in the chlor-alkali cell where it is gradually replacing the traditional diaphragm and mercury cells. In the chlor-alkali process a cell is partitioned by the polyanionic membrane into an anode compartment, to which brine is added, and a cathode compartment to which water is added (Figure 2.61). On application of electric potential, Na^+ passes from the brine around the anode and through the membrane to the water around the cathode where it forms sodium hydroxide. The membrane keeping the brine and water separate allows transfer of Na^+ ions by ion-exchange and acts as a barrier to Cl^- and OH^-.

The Nafion membranes have ionic contents of 0.66–0.91 Eq/kg. A bilayer of these two extremes of composition is made and also a bilayer with a perfluorocarboxylate (Flemion) polymer. The bilayered membranes are made by melt processing the precursor copolymer, which then, in membrane form, is hydrolyzed.

Perfluorocarboxylates, the Flemions, were introduced with the idea that a carboxylate at a given ion content would be less hydrophilic than a sulfonate and so would be of help in balancing the ionic content of an ion-exchange membrane with its permselectivity and hydrophilicity as required. The bilayered Nafions also resulted from similar thinking and at tempt to combine certain advantages of the two types.

The efficiency with which power is consumed in the electrolysis and the highest concentration of uncontaminated sodium hydroxide that can be produced determine the membrane performance in the

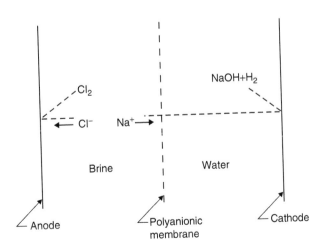

FIGURE 2.61   Chlor-alkali cell.

chlor-alkali process. Rapid progress in membrane performance has been made since the introduction of carboxylate membranes, leading to an efficiency level of 95% and a concentration of 33%.

Bilayer carboxylate membranes can be produced by surface modification of Nafion-type membranes. In a process used by Asahi Chemical the sulfonate on a Nafion-type surface is reduced to sulfinic and sulfenic acids, then oxidized to a carboxylate layer of 2–10 μm thickness:

$$\text{~~~OCF}_2\text{CF}_2\text{SO}_2\text{Cl} \xrightarrow{\text{Reduction}} \text{~~~OCF}_2\text{CF}_2\text{SOOH}$$

$$\downarrow \begin{array}{l} -\text{SO}_2 \\ \text{and} \\ \text{oxidation} \end{array}$$

$$\text{~~~OCF}_2\text{COOH}$$

(2.18)

Tokoyama Soda makes Neosepta F membranes by treating a Nafion-type precursor with an alcohol and the oxidizing the surface in air.

Nafions in their acid forms are super-acidic, i.e., stronger than sulfuric acid and so have high catalytic power. Nafion powders in acid form have therefore been used as catalysts in many types of reactions, such as esterification and Friedel-Crafts reactions at low temperatures.

2.10.2.4   Ionomers Based on Polysulfones

Ionomers of commercial polysulfones, principally sulfonates, are being developed as membranes, particularly for purifying water by reverse osmosis. In this application, they are superior to conventional membrane materials because they are resistant to oxidation by the chlorine used in water treatment, to harsh chemical cleaning operations, to biological fouling, and to compaction under high operating pressures. The traditional polymer, cellulose acetate, is less strong, while Nafion, though very inert chemically, is too costly and difficult to fabricate in a form suitable for reverse osmosis. The sulfones, having completely amorphous backbones and solubility, can be easily cast into membranes from solvents. The optimum ion content is about one eq/kg, which provides a balance of properties between a high flux of the permeant (water) and a low leakage of the rejected species (the dissolved salts).

Commercial polysulfones (see Table 1.30) such as Udel (Union Carbide) and Victrex (ICI) are sulfonated in a post-functionalization step to make ionomers. When dissolved in dichloroethane and treated with a complex of sulfur trioxide and triethyl phosphate, Udel (XXIV) becomes monosulfonated on the rings marked **R** (as the rings connected to the sulfone group are deactivated by the sulfone group).

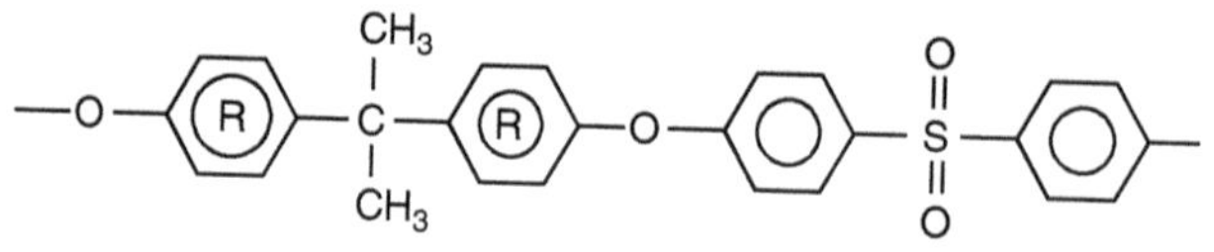

(XXIV)

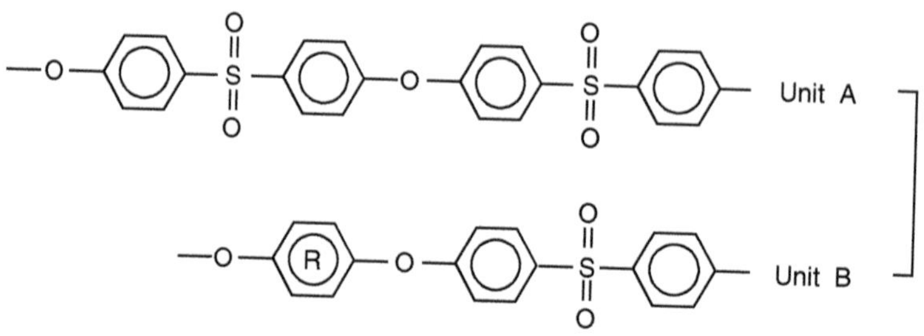

(XXV)

Victrex is a random copolymer of two units (XXV) in which the ring marked **R** of unit **B** becomes monosulfonated by the simple process of dissolving it in sulfuric acid. The other aromatic rings and the whole of unit A are inert. The extent of the reaction is predetermined by the proportion of unit B in the copolymer. The strength of the ionic associations in sulfonated Victrex is indicated by the increase in T_g, which occurs linearly with the content of unit B despite the already high value of 230°C for the parent polymer.

2.10.3   Polyelectrolytes

2.10.3.1   Ion Exchangers

The most important class of ion-exchangers are the organic ion-exchange resins. Their framework, the so-called matrix, consists of an irregular, macromolecular, three-dimensional network of hydrocarbon chains. The matrix carries ionic groups such as $-SO_3^-$, $-COO^-$, $-PO_3^{2-}$, and AsO_3^{2-} in cation exchangers and $-NH_3^+$, $\rangle NH_2^+$, $\rangle N^+ \langle$, and $-\rangle S^+$ in anion exchangers. Ion-exchange resins thus are cross-linked polyelectrolytes. The first completely synthetic ion-exchange resins were prepared by B. A. Adams and E. C. Holmes in England in 1935. Today more than 100 synthetic ion-exchange resins are marketed throughout the world by various companies. Some of the largest manufacturing companies and the trade names of the resins are listed in Table 2.17.

Ion-exchange resins are supplied as insoluble, water-swellable beads that have either a dense internal structure (gel-type) or a porous, multichannelled one (macroporous-type or macroreticular). The gel-type PS resin was the first to be introduced (1947) and the macroporous-type came later (1959).

Ion-exchangers are broadly classified as cation exchangers and anion-exchangers. Carriers of exchangeable cations are called cation exchangers; they have acidic functional groups bound to the resin matrix. Carriers of exchangeable anions are called anion exchangers; they have basic functional groups bound to the resin matrix (Table 2.18). Chelating ionexchange resins contain chemically bound chelating functional groups, which sorb metal ions by chelation.

Polystyrene (PS) cross-linked with divinyl benzene (DVB) is the matrix on which most of the commercial ion-exchange resins are based; the ionic groups are introduced by post-functionalization of the cross-linked polymer matrix. By varying the DVB content, the degree of cross-linking can be adjusted in a simple and reproducible manner. The nominal DVB content, which refers to mole% of DVB in the polymerization mixture, is used to indicate the degree of cross-linking. General-purpose ion-exchangers contain between 8 and 12 mole% DVB, the most common being 8 mole%. For special purposes, resins with as little as 0.25% DVB and as much as 25% DVB have been prepared. Resins with low DVB content swell strongly and are soft and gelatinous, while those with very high DVB content swell very little and are tough and mechanically more stable.

TABLE 2.17   Trade Names and Manufacturers of Ion-Exchange Resins

Trade Name	Manufacturer
Amberlite	Rohm and Haas Co., Philadelphia, PA
De-Acidite	The Permutit Co. Ltd., London, England
Dowex	Dow Chemical Co., Midland, MI
Duolite	Chemical Process Co., Redwood City, CA
Imac	Industrieele Mij. Activit N. V., Amsterdam, Netherlands
Ionac[a]	Ionac Co. Ltd., New York, NY
Lewatit	Farbenfabriken Bayer, Leverkusen, Germany
Nalcite[b]	National Aluminate Corp., Chicago, IL
Permutit	Permutit Co., New York, NY
Wofatit	VEB Farbenfabrik Wolfen, Wolfen, Kr. Bitterfeld, Germany
Zeo-Karb	The Permutit Co. Ltd., London, England

[a] Product of Permutit Co., marketed by Ionac Co.
[b] Product of Dow chemical Co., marketed by National Aluminate Corp.

TABLE 2.18   Types of Ion-Exchange Resins

Active Group	Structure
Cation-exchange resins	
Sulfonic acid	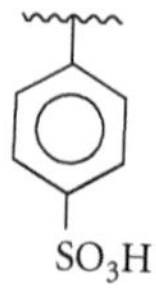 (benzene ring) SO$_3$H
Carboxylic acid	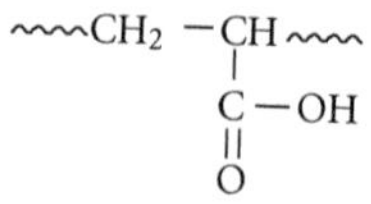 $\sim\sim\sim CH_2 - CH \sim\sim\sim$ with $C - OH$, $\parallel$ O
Phosphonic acid	 (benzene ring) OP(OH)$_2$
Anion-exchange resins	
Quaternary ammonium salt	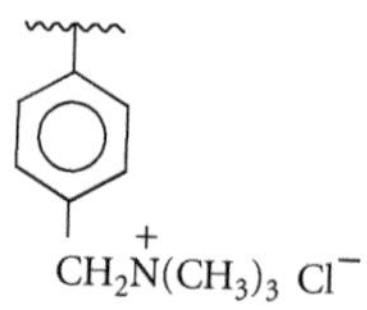 (benzene ring) CH$_2\overset{+}{N}(CH_3)_3$ Cl$^-$
Secondary amine	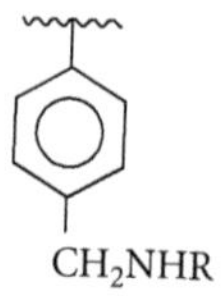 (benzene ring) CH$_2$NHR
Tertiary amine	 (benzene ring) CH$_2$NR$_2$

The styrene-DVB copolymer beads are prepared by suspension (pearl) polymerization technique. The monomers are mixed and a polymerization catalyst such as benzoyl peroxide is added. The mixture is then dispersed into small droplets in a thoroughly agitated aqueous solution that is kept at a temperature required for polymerization (usually 85°C–100°C). A suspension stabilizer (gelatin, polyvinyl alcohol, sodium oleate, magnesium silicate, etc.) in the aqueous phase prevents agglomeration of the droplets. The size of the droplets depends chiefly on the stabilizer, the viscosity of the solution, and the agitation, and it can be varied within wide limits. As polymerization takes place, the droplets are transformed into polymer beads. For most purposes, a bead size of 0.1–0.5 mm is preferred, but beads from 1 μm to 2 mm in diameter can be prepared without much difficulty.

The above method gives beads of gel-type polymer matrix. Porous beads can be made by incorporating a component (for example, styrene homopolymer) that is soluble in the monomer mixture. After polymerization this component is removed from the matrix with, for example, toluene, thus leaving pores within the structure in the final product.

Highly porous, so-called macroreticular beads can be prepared by a variation of the conventional pearl polymerization technique. An organic solvent is added in which the mixture of monomers is soluble, but the polymer, when it is formed, is insoluble. Thus, as polymerization progresses, the solvent is squeezed out by the growing polymer regions. In this way, one can obtain spherical beads with large pores (several hundred angstrom units), which guarantee access to the interior of the beads even when nonpolar solvents are used.

Styrene-DVB copolymer beads are sulfonated to produce the most widely used strong-acid type cation-exchange resins. To make them, the copolymer precursor beads are dispersed in about 10 times their weight of concentrated sulfuric acid and heated slowly to 150°C. The sulfonic acid group is normally introduced into the para position. Though the reaction is very simple in principle, it involves delicate operations and close control of parameters in order to achieve beads of suitable structure and durability. A fully mono-sulfonated, polystyrene has a theoretical ion content of 5.1 equivalents/kg (dry) but many commercial resins have about 4.4–5.2 eq/kg. Amberlite IR-120, Dowex-50, Nalcite HCR, Permutit Q, Duolite C-20 and C-25, and Lewatit S-100 are resins of this type.

Weak-acid, cation exchange resins are prepared by copolymerization of an organic acid or acid anhydride and a cross-linking agent. As a rule, acrylic or methacrylic acid is used in combination with divinyl benzene, ethylene dimethacrylate, or similar compounds with at least two vinyl groups. The pearl polymerization technique described above can be used if esters instead of the water-soluble acids are polymerized. The esters are hydrolyzed after polymerization. The final products have ionic contents of 9–10 eq/kg (dry). Resins of this type are Amberlite IRC-50, Duolite CS-101, Permutit H-70, and Wofatit CP-300.

Strong-based anion-exchange resins, also very common, are made (Figure 2.62) by chloromethylating the styrene-DVB copolymer, then aminating the product with a tertiary alkyl amine. The quaternization of chloromethylated resins with tertiary alkyl amines takes place smoothly and quantitatively. The two most common strong-base resins are made with trimethylamine and contain the quaternary amine groups—$N(CH_3)_3$ and—$N(CH_3)_2CH_2OH$. The first type of resin is classed as a Type I strongbased anion exchanger and the second type as a Type II exchanger. Dowex-1, Amberlite IRA-400, Permutit S-1, Nalcite SBR, Duolite A-42, and De Acidite FF are Type 1 resins. Dowex-2, Amberlite IRA-410, Permutit S-2, Nalcite SA-R, and Duolite A-40 are Type II resins. The resins Amberlite IRA-401 and 411 differ from

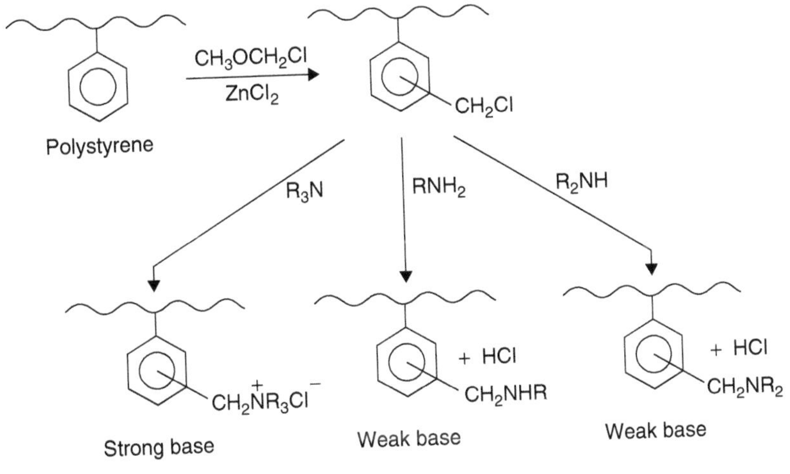

FIGURE 2.62 Reaction scheme for the preparation of weak-base and strong-base resins starting with polystyrene.

the standard types 400 and 410 only by having a lower DVB content. The ion contents of strong-based resins are 3–4 eq/kg (dry).

Type I resins have better thermal and oxidative stability and maintain the integrity of the quaternary groups over a long period of time. Type II resins, when used in the hydroxide form, are limited to a maximum temperature of approximately 40°C and should not be used under oxidizing conditions. Type II resins are often used because of lower operations costs. They regenerate somewhat more easily and have higher operating capacities than the Type I products. The resins have a useful industrial life of 3–5 years.

Weak-base resins are made by treating a chloromethylated intermediate with primary or secondary alkyl amines (Figure 2.62). The treatment with secondary amines leads to monofunctional weak-base resins having tertiary amino groups. By treatment with primary alkyl amines, however, polyfunctional weak-base resins having secondary and tertiary amino groups are obtained, the latter being formed by the reaction of the primary amine with two chloromethyl groups:

$$\begin{array}{c} \cdots-CH_2Cl \\ \\ \cdots-CH_2Cl \end{array} + H_2NR \longrightarrow \begin{array}{c} \cdots-CH_2 \\ \diagdown \\ \diagup \\ \cdots-CH_2 \end{array} NRH^+Cl^- + HCl \qquad (2.19)$$

Additional cross-linking results if the chloromethyl groups belong to different chains (Occasionally, polyamines such as tetramethylenepentamine are used; they can react in a similar way with two or more chloromethyl groups.) The resins Amberlite IR-45, Dowex-3, Nalcite WBR, and Duolite A-14 are polyfunctional weak-base anion exchangers. De-Acidite G is a monofunctional resin with tertiary amino groups.

Deionization by ion-exchange is usually confined to relatively dilute solutions, i.e., concentrations less than 0.03N (1500 ppm $CaCO_3$). Since ion-exchange resins have finite capacities, economics are most favorable for dilute solutions, although concentrations up to 0.1N (5000 ppm $CaCO_3$) can be treated quite satisfactorily.

Cation exchange resins having strongly acidic sulfonic acid groups are the major cation exchangers used for deionization purposes. When dilute solutions are passed through beds of such resins, either in acid form or sodium salt form, cation exchange takes place according to the following reactions (shown for $MgCl_2$):

$$(1/2)MgCl_2 + \textcircled{P}SO_3^-H^+ \rightleftharpoons \textcircled{P}SO_3^-(Mg^{2+})/2 + HCl \qquad (2.20)$$

or

$$(1/2)MgCl_2 + \textcircled{P}SO_3^-Na^+ \rightleftharpoons \textcircled{P}SO_3^-(Mg^{2+})/2 + NaCl \qquad (2.21)$$

The above reactions are reversible, and the exhausted resin can be regenerated with moderate concentrations of strong acids or NaCl. The relative affinities for the resin of various cations have an important influence on the efficiency of exchange. In general, the order of ease of replacement of common cations is $Li^+ > Na^+ > K^+ > Mg^{2+} > Ca^{2+} > Al^{3+} > Fe^{3+}$.

Strongly basic anion-exchange resins can act as acid neutralizers or can split salts in the same manner as the sulfonic acid cation exchangers. The following are typical reactions:

$$\textcircled{P}NR_4^+OH^- + H^+X^- \longrightarrow \textcircled{P}NR_4^+X^- + H_2O \qquad (2.22)$$

or

$$\textcircled{P}NR_4^+OH^- + M^+X^- \longrightarrow \textcircled{P}NR_4^+X^- + M^+OH^- \qquad (2.23)$$

The resins show marked affinity relationships depending on ion size and valene. The order of affinity for common anions is: $SO_4 > Cl^- > HCO_3^- > F^- >> HSiO_3^-$. Reaction (23) can be reversed by regenerating with moderate concentrations of sodium hydroxide solutions. However, since the reactions are not as easily reversed, regeneration levels of 150%–200% of the stoichiometric requirements are frequently employed.

In addition to the commonly employed sulfonic acid cation exchanger, resins based on the carboxylic acid groups are sometimes employed under special conditions. These resins are effective for exchange reactions in neutral and alkaline solution. For example, the effluent from a saltsplitting reaction, Equation 2.23, can be treated with a carboxylic exchanger according to the reaction:

$$\textcircled{P}COOH + MOH \longrightarrow \textcircled{P}COOM + H_2O \qquad (2.24)$$

The reaction is reversed very effectively by the hydrogen ion, and regeneration can be accomplished readily at efficiencies approaching 100%.

The role of weak-base anion-exchange resins in deionization is often confined to acid neutralization, as shown by the following equation:

$$RNH_3OH + HX \rightarrow RNH_3X + H_2O \qquad (2.25)$$

The reaction can be reversed by addition of some alkaline reagent such as $NaOH$, Na_2CO_3, NH_3, and $NaHCO_3$. The regeneration is accomplished readily at efficiencies approaching 100%. The ease with which the hydroxyl ion can be replaced by other anions is $SO_4 > Br^- > F^- > CH_3COO^- >> HCO_3^-$.

The weak-acid and weak-base resins do not totally dimineralize the water like their strong counterparts. However, the water they produce, containing 100–200 ppm of dissolved solids, is adequate for many purposes. The regeneration of the weak resins is also accomplished more readily than the strong resins.

A relatively recent development is a thermally regenerable ampholytic resin containing both weak acid and a weak-base functionality within the one bead (Sirotherm by ICI), which absorbs significant quantities of salt at ambient temperatures and releases salt on heating to 70–90°C. Thus, when the resin is fully exchanged or exhausted, it can be rinsed in hot water and reused. The adsorption step involves the transfer of protons form carboxylic acid groups to amino groups to form the cation and anion exchange sites:

$$\textcircled{P}\begin{matrix} \diagup COOH \\ \diagdown NR_2 \end{matrix} + Na^+ + Cl^- \rightleftharpoons \textcircled{P}\begin{matrix} \diagup COO^- Na^+ \\ + \\ \diagdown NR_2H\ Cl^- \end{matrix} \qquad (2.26)$$

The equilibrium is temperature sensitive, with both types of groups showing weaker electrolyte behavior on heating. The large increase in the ionization of water that occurs on heating, about 30-fold from 25°C to 85°C, releases additional protons and hydroxyl ions, which suppress the ionization of the weak electrolyte resins. The hot water can thus be looked on as providing the acidic and basic regenerants that usually have to be added separately.

Operating data indicate that Sirotherm TR-20 reins can produce waters of salinities as low as 50–100 ppm dissolved dalts, the economic upper range of salinities in feed water being restricted to 2000–3000 ppm. The resins are expected to find application in the demineralization of mildly brackish surface and underground waters for industrial and municipal use and as a roughing stage in the production of high-quality boiler feed water.

Some polystyrene resins (cross-linked with DVB) are specially modified to have chelating functional groups bound to the matrix so as to make them selective towards certain ions. Such resins with iminodiacetic acid groups are marketed under the trade names Dowex A-1 (Dow Chemical) and Chelex 100 (Bio-Rad Laboratories). The complex (XXVI) formation constants with metal ions of the chelating resin are so large that the resin absorbs metal ions equivalent to the iminodiacetic acid groups (used in sodium salt form), i.e., the efficiency of metal ion adsorption is near 100%. A particular metal ion can be removed by controlling the pH of aqueous solution. For example, at pH 2, mercury and copper ions are preferentially adsorbed, while zinc, cobalt, and cadmium ions are little adsorbed.

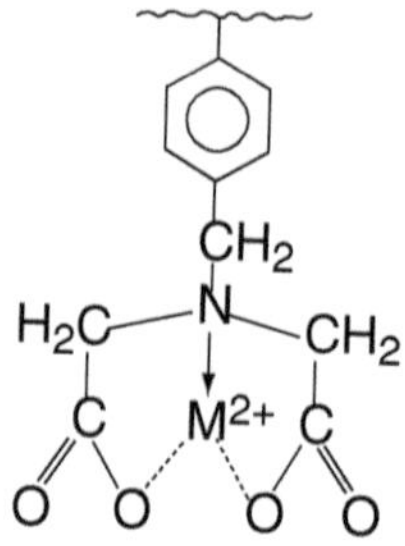

(XXVI) 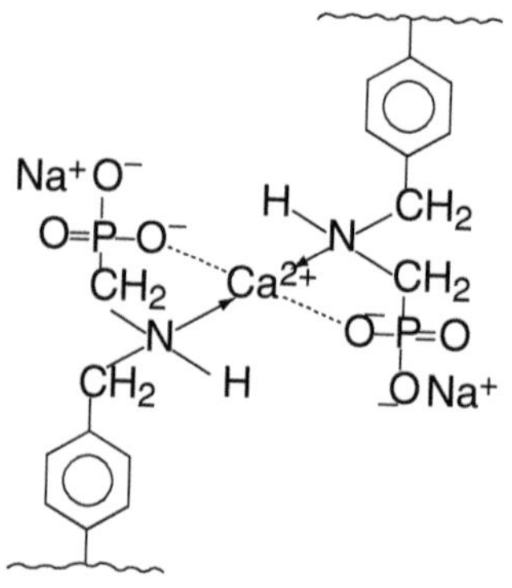   (XXVII)

Polystyrene resins that have aminophosphonate chelating groups are highly selective towards calcium ion. Such a resin, e.g., Duolite ES467 (Rohm and Haas), when added to a strong brine solution (25%) contaminated with 10 mg/liter of calcium ion will selectively remove the calcium ion, forming calcium amino-phosphonate complex (XXVII), until only 0.02 mg/liter remains. This process is particularly useful for purifying the brine used in the chlor-alkali cell described earlier.

Many other specific resins have been prepared. To give only a few examples, resins with hydroxamic acid groups (XXVIII) are specific for Fe^{3+} ions and those with mercapto groups (XXIX) prefer Hg^{2+} ions. Resins containing chlorophyll and haemin derivatives or similar compounds form extremely strong chelates with ions such as Fe^{3+}. In fact, the chelated counterions are held so strongly that they can hardly be displaced. Another undesired consequence of such strong association is that the mobility of the counterion in the resin is greatly reduced. Hence for any application one should choose the resin carefully, seeking a reasonable compromise between selectivity, ease of regeneration, and rate of ion exchange.

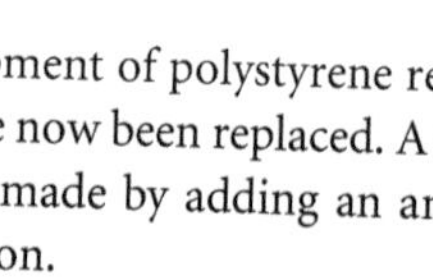

(XXVIII) 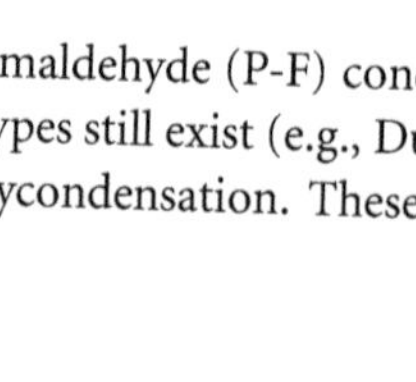   (XXIX)

Prior to the development of polystyrene resins, phenol-formaldehyde (P-F) condensates were used as matrices, but they have now been replaced. A few weak-base types still exist (e.g., Duolite ES562 of Rohm and Haas), which are made by adding an amine during polycondensation. These P-F condensates are used for enzyme fixation.

2.10.3.2 Applications

Applications of ion-exchange resins are extremely varied, ranging from water-softening to purification of chemicals and therapeutic applications. An extremely useful industrial development of the ion-exchange technique is the production of demineralized water rivaling that of distilled water in purity.

Most ion-exchange reactions for industrial applications are done with columns of resin in which the ions in solution (say B) are depleted, proceeding from top to the bottom of the column, by exchange with ions (say A) in the resin. With the progress of ion exchange the resin bed shows an exhausted portion at the top, an ion-exchange zone in the middle and a regenerated portion at the bottom (Figure 2.63).

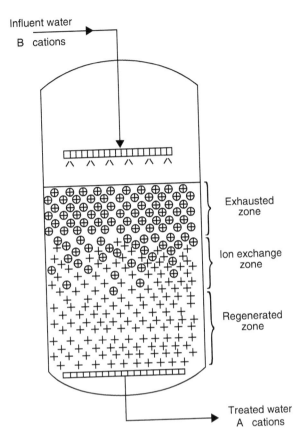

Influent water
B cations

Exhausted zone

Ion exchange zone

Regenerated zone

Treated water
A cations

FIGURE 2.63 Ion-exchange column in service: ($+$, resin containing A cations; $\oplus$, resin containing B cations).

As the ion-exchange zone moves down through the resin column, the **B** ions eventually reach the outlet, at which time breakthrough occurs.

In most ion-exchange installations, the vertical column is the most commonly employed unit. The system may consist of a single column with one type of resin (single column system), two or more columns containing a variety of cation and/or anion exchange resins (multiple-bed system), or a single column containing a mixed bed of two (or more) resins (monobed or mixed-bed system).

Much of the appeal of the ion-exchange process stems from the simplicity of the single-column system. Water softening by ion-exchange is the most widely used example of this system (Figure 2.64). The hardness ions, calcium and magnesium, are exchanged for sodium as the hard water flows down through a column of cation exchange resin used in the sodium form [see Equation 2.21]. Conversion of the exhausted resin back to the sodium form [i.e., reverse of Equation 2.21] is accomplished in a regeneration step by contacting the column with an excess of sodium chloride solution. Water softeners range in size from small household units (e.g., 20 cm in diameter and 60 cm deep) to large industrial units (e.g., 360 cm in diameter and 150 cm deep).

It will be noted that in reactions (20) and (22) the overall result of ion-exchange is the complete elimination of dissolved salts with the formation of an amount of water equivalent to the amount of electrolyte removed. Deionization requires contacting the water with both cation and anion exchange resins, which can be done in a multiple-bed system.

Two-bed deionization of water is widely practiced. Such systems generally use a strong-acid cation exchange followed by either a weak-base or a strong-base anion exchanger. Reactions similar to equations

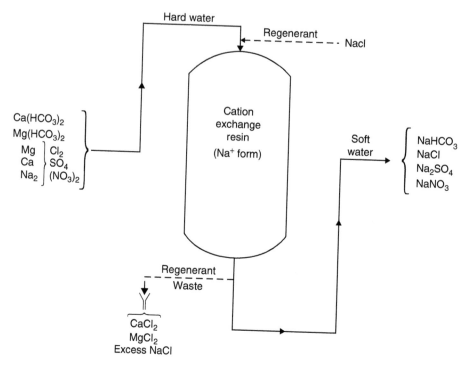

FIGURE 2.64 Water softening by ion-exchange in single-column system.

(20), (22), or (25) occur. Three beds or more are used either to achieve operating economy or a greater degree of deionization. For example, a three-bed system of strong acid cation exchanger/weak-base anion exchanger/strong-base anion exchanger offers economy in regeneration. In service, the weak-base resin removes the mineral acids coming from the cation exchanger and the strong-base resin removes principally carbonic and silicic acids.

For installations requiring very high effluent water quality, monobed or mixed-bed deionization has been used widely, yielding a demineralized water in one operation. The standard mixed-bed is a special case of a single column. It usually consists of an intimate mixture of a strong-acid cation exchange resin and a strong-base anion exchange resin in the hydrogen and hydroxide forms, respectively. When water is passed through a fixed bed of such a mixture, the ions in solution are alternately exposed to numerous contacts with cation and anion exchange sites, the effect being similar to that obtained with a very large number of multiple beds, with the result that almost complete deionization occurs. The process follows the reaction:

$$MX + \left\{\begin{array}{l} \textcircled{P}SO_3^-H^+ \\ \textcircled{P}NR_3^+OH^- \end{array}\right\} \longrightarrow H_2O + \left\{\begin{array}{l} \textcircled{P}SO_3^-M^+ \\ \textcircled{P}NR_3^+X^- \end{array}\right\} \qquad (2.27)$$

Following exhaustion of a mixed bed and prior to regeneration, the resins in the bed are separated by applying backwash at a flow rate sufficiently high to fluidize the bed. By virtue of that fact that anion exchange resins have a lower density than cation exchange resins, hydraulic separation into two layers occurs. The regenerant caustic soda then contacts the upper layer of anion resin and the acid regenerant flows through the lower layer of cation resin. Then following a rinse with relatively pure water, the resins are air-mixed prior to the next ion-exchange run.

Ion-exchange methods are established for treating various effluents arising from the metal finishing processes such as plating and anodizing. The use of strong-base resins for decolorizing sugar liquors is widely practiced. The coloring bodies are organic anions that are sorbed by weakly cross-linked strong-base gel resins.

Even wines are sometimes treated by column cation exchange. Potassium hydrogen tartrate, which causes an unpleasant precipitate in wines, is converted to the more soluble sodium salt by treatment with polystyrene sulfonic acid resin in the sodium (Na^+) form.

Ion-exchange resins are used for metal recovery from low-grade ores and dilute leach liquors. One of the best examples is the recovery of uranium.

Ion-exchange chromatography is well known for separating mixtures of ions in solution. Possibly the best-known organic analytical ion-exchange application is the chromatographic separation and isolation of amino acids. Commercially, the most significant application is the recovery of antibiotics such as streptomycin and neomycin. The fermentation broth containing the impure antibiotic is treated with a polyacrylic weak-acid resin on which the antibiotic is sorbed to the exclusion of other organic impurities. The product is recovered by elution with dilute mineral acid.

2.10.3.3   Polycarboxylates

Polyacrylate-type homopolymers are polyelectrolytes and are the most ionic of the organic polymers. They dissolve in water giving aqueous solutions with unusual and useful physical properties. They are generally made by free-radical polymerization in aqueous solution. Very-high-molecular weight (e.g., 4×10^6) polymers can be obtained that give very viscous solutions. Polyacrylic (XXX), poly-methacrylic (XXXI), and polyitaconic (XXXII) acids are the three main types having theoretical capacities of 13.9, 11.6, and 15.4 eq/kg, respectively. Aqueous solutions or dry powders of these materials are commercially available.

Versicols (Allied Colloids) and Texigels (Scott Bader) are homopolymers of acrylic or methacrylic acids or their copolymers with acrylamide. They are used as stabilizers, and protective colloids and thickeners for aqueous dispersions, binders, and flocculants. Carbopols (B. F. Goodrich) are different grades of polyacrylic acid of varied molecular weight having excellent suspending, thickening, and gel-forming properties. Carbosets (B. F. Goodrich) are acrylic copolymers and have a similar carboxylic content but generally can be dissolved in alkaline solutions. They are used mainly in coating applications. In some applications, they are covalently cross-linked with epoxides and so on, but some applications use ionic cross-links made with zinc ions.

A notable application of polyacrylic acid is for cements in dentistry. These are made by mixing an aqueous solution of the polymer with zinc oxide when the zinc salt precipitates as a highly cross-linked gel that rapidly sets to a hard mass under oral conditions. In a variation of this reaction, the zinc oxide is replaced with a tooth-colored glass powder that releases Al^{3+} and Ca^{2+} ions. These cements, called ASPA (aluminosilicate polyacrylic acid) or glass ionomer, set very rapidly, bond well to tooth enamel, and are compatible with living tissue.

2.10.3.4   Integral Polyelectrolytes

Polyelectrolytes having bound ions integrated in the polymer backbone are called *ionenes*. Some ionenes have been studied for their bacteriostatic and bactericidal activity. Ionenes with segments of polypropylene oxide in the backbone have been evaluated as thermoplastic elastomers.

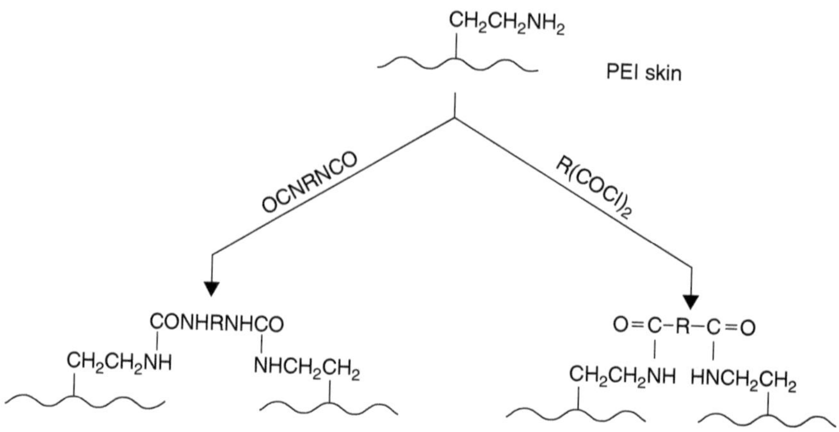

FIGURE 2.65   Insolubilization of polyethylenimine by treatment with toluene diisocyanate or phthaloyl dichloride.

    Polyethylenimine (PEI) is an integral polyelectrolyte that is available commercially, e.g., Polymin (BASF). It is formed by the ring-opening polymerization [reaction (28)] of ethyleneimine (aziridine). The resulting polyamine has about 50% of the expected secondary-amine functionality and about 50% primary and tertiary due to branching:

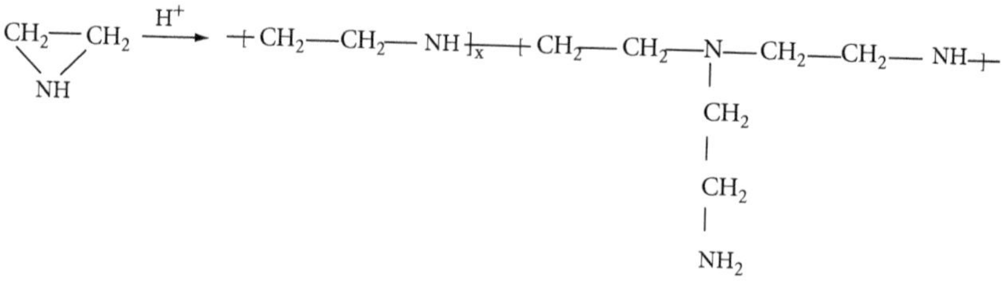

$$(2.28)$$

    PEI has typical polyelectrolyte properties; it is a highly viscous hygroscopic liquid, completely miscible with water and lower alcohols, insoluble in benzene, and reactive toward cellulose. PEI is mainly used as a size, flocculating agent, or protective colloid, notably in the paper and textile industries, because of its ability to bind to cellulosic fibers.

    Membranes based on PEI were introduced for use in reverse osmosis to desalinate water. These membranes, known as NS100 and NS101, are made by forming a PEI skin on polysulfone support and insolubilizing it by treatment with toluene di-isocyanate or phthaloyl dichloride to produce a polyurea or polyamide (Figure 2.65).

2.11   Scavenger Resins

The field of organic chemistry has seen the most extensive use of polymeric materials as aids in effecting chemical transformation and product isolation. Insoluble polymer supports have been used as handles to facilitate these functions. As chemical reagents can be bound to an insoluble polymer carrier and used in organic synthesis [117,118], polymer-bound reagents can also be used to assist in the purification step of solution-phase reactions [119,120]. The latter are known as scavenger resins. These are added to the reaction mixture upon completion of the reaction in order to quench and selectively bind to the unreacted reagents or by-products. The polymer-bound impurities are then removed from the product by simple filtration to obtain pure compounds. For example, aminomethylated poly(styrene-*co*-divinyl benzene) can be used to remove acid chlorides, sulfonyl chlorides, isocyanates, thiocyanates, and proton. Similarly, 2-Chlorotrityl resins have been developed for the attachment of carboxylic acids, alcohols, phenols, and amines under mild conditions. Several such commercially available examples of scavenger resins are listed in Table 2.19.

2.12 Synthetic Polymer Membranes

A membrane can be described as a thin barrier that permits selective mass transport. This property is described as permselectivity in order to distinguish the membrane from thin nonpermeable film or layer. The mode of permeation and separation is dictated by its morphology. The basic morphologies are isotropic (dense or porous) and anisotropic with a tight surface extending from a highly porous wall structure [121]. While the transport rate through a dense membrane is inversely proportional to the membrane thickness, membrane permselectivity is independent of thickness. Thus anisotropic membranes permit both high transport rates and excellent separation, eliminating the mechanical integrity problems associated with the handling of ultrathin membranes. This concept has led further to the development of thin-film composite membranes which consist of ultrathin semipermeable layers on highly porous substrates as support offering minimum resistance to the permeates. In this way, materials that exhibit semipermeable properties but cannot form self-supporting membranes can be deposited on other porous substrates. For example, a polysulfone porous-support matrix is coated with cross-linked polyethyleneimine (PEI)-toluene diisocyanate (TDI) or furan resin as dense layers of 0.1–1 μm thickness to produce composite membranes. The permeability of such membranes depends on the porosity of the substrates and the complex composition of the deposited permselective layer [121].

As permselective barriers, synthetic membranes have been employed in a variety of applications, which include dialysis, mirofiltration, ultrafiltration, reverse osmosis, pervaporation, electrodialysis, and gas separation. Synthetic membranes also find special applications as permselective barriers for ion-specific electrodes, biosensors, controlled release, and tissue-culture growth. Some commercial polymer membranes are listed in Table 2.20.

2.12.1 Membrane Preparation

Most of the membranes listed in Table 2.20 are formed through phase separation processes, i.e., melt extrusion or coagulation of a polymer solution by a nonsolvent. In melt extrusion, a polymer melt is extruded into a cooler atmosphere which induces phase transition. The melt extrusion of a single polymer usually gives a dense, isotropic membrane. However, the presence of a compound (latent solvent) that is miscible with the polymer at the extrusion temperature but not at the ambient temperature, may lead to a secondary phase separation upon cooling. Removal of the solvent then yields a porous isotropic membrane. Anisotropic membranes may result from melt extrusion of a dope mixture of polymers containing plasticizers.

Membranes are prepared from polymer(s) dissolved in a solvent using either a dry process or a wet process. In the dry process, a volatile solvent is used for dissolving the polymer(s) and the extruded polymer solution is transferred into an evaporation chamber to yield a porous, isotropic or anisotropic membrane. In the wet process, on the other hand, the extruded mixture is coagulated by exposing the mixture to a nonsolvent in the form of vapor or liquid. The latter process is often referred to as the phase inversion process.

2.12.1.1 Wet-Extrusion Process

Membranes used in microfiltration, reverse osmosis, dialysis, and gas separation are usually prepared by the wet-extrusion process, since it can be used to produce almost every membrane morphology. In the process, homogeneous solutions of the polymers are made in solvent and nonsolvent mixtures, while phase inversion is achieved by any of the several processes, such as solvent evaporation, exposure to excess nonsolvent, and thermal gelation. In most formulations, polymer solutions of 15–40 wt% concentration are cast or spun and subsequently coagulated in a bath containing a nonsolvent (usually water).

TABLE 2.19 Some Commercially Available Scavenger Resins

Resin Name	Structure	Reacts With
Poly(Styrene-*co*-divinyl benzene), aminomethylated	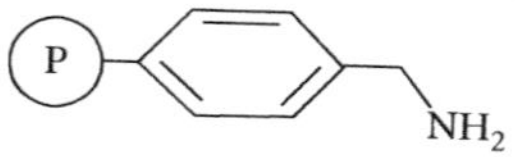	$RCOCl$, RSO_2Cl, $RNCO$, $RNCS$, H^+
Ethylenediamine polymer-bound	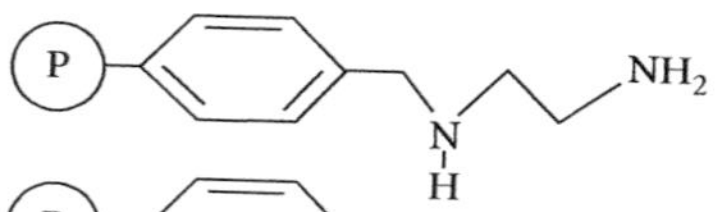	$RCOCl$, RSO_2Cl, $RNCO$, $RNCS$, H^+
Morpholine, polymer-bound	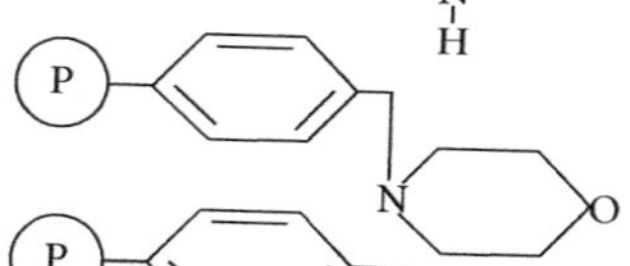	H^+
Piperidine, polymer-bound	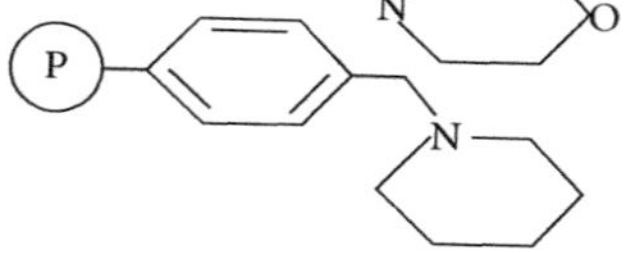	H^+
Sulfonly chloride polymer-found	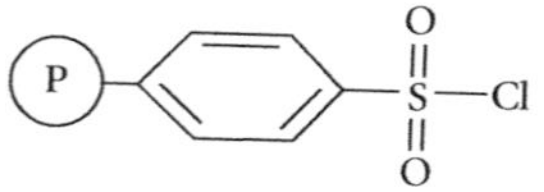	ROH, RNH_2
p-Toluenesulfonic acid, polymer-bound	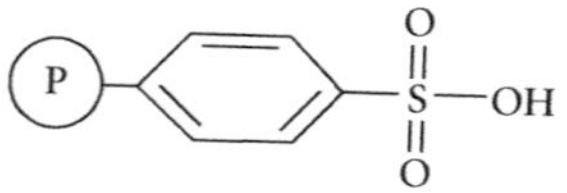	ROH, RNH_2
Isocyanate, polymer-bound	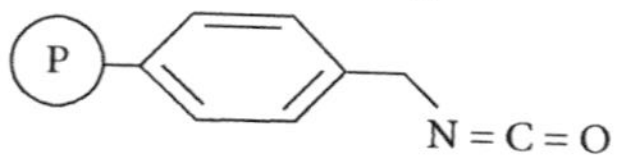	RNH_2, $RNHNH_2$, RO^-
2-Chlorotrityl chloride, polymer-bound	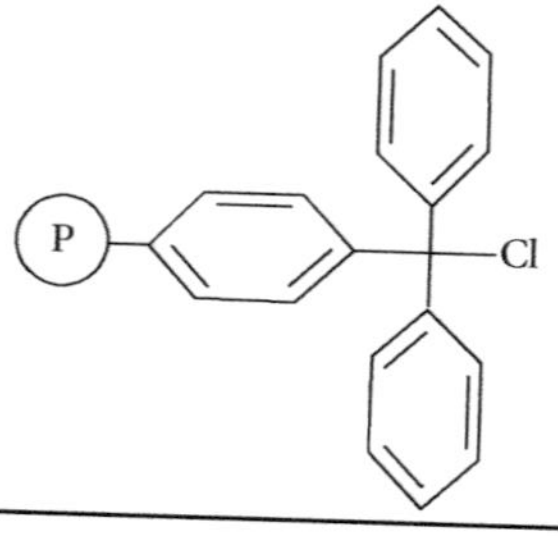	RCO_2H, ROH, $PhOH$, RNH_2

Source: Catalog of Sigma-Aldrich Corporation, Milwaukee, WI, U.S.A.

TABLE 2.20 Some Commercial Polymer Membranes and Their Applications

Material	Applications[a]
Cellulose acetate (CA)	MF, UF, RO, D, GS
Cellulose triacetate (CTA)	MF, UF, RO, GS
CA-CTA blend	RO, D, GS
Cellulose esters, mixed	MF, D
Cellulose, regenerated	MF, UF, D
Polyamide, aromatic	MF, UF, RO, D
Polyimide	UF, RO
Polyacrylonitrile	UF, D
Polysulfone	MF, UF, D, GS
Polytetrafluoroethylene	MF
Poly(vinylidene fluoride)	MF, UF
Polypropylene	MF
Polydimethylsiloxane	GS

Source: Cabasso, I., 1987. *Encyclopedia of Polymer Science and Engineering, Vol. 9,* J. I. Kroschwitz, ed., Wiley-Interscience, New York.

[a] MF, microfiltration; UF, ultrafiltration; RO, reverse osmosis; D, dialysis; GS, gas separation.

The polymer concentration in the solution is adjusted depending on the viscosity. Higher solution viscosities are required for the production of hollow-fiber membranes, as compared to flat sheet production, because the fiber fabrication is performed without a casting surface.

The morphology of the membrane is strongly influenced by the concentration gradient of the permeating coagulant (nonsolvent) within the cast layer. If the concentration profile is flat, coagulation occurs virtually the same time over the entire layer, yielding an isotropic porous membrane. This happens, for example, when the membrane is precipitated by exposing the cast layer to the coagulant vapor phase or if solvents with low vapor pressure are used. On the contrary, if the concentration gradient of the permeating coagulant is steep, as when membranes are coagulated in a nonsolvent bath, an anisotropic porous membrane forms, which can be used for ultrafiltration, reverse osmosis, and gas separation.

Thermally induced phase separation (TIPS) in polymer solutions is one of the most versatile and widely used methods for the production of microporous membranes [122]. In the TIPS process, a homogeneous solution is formed by the dissolution of a polymer in a diluent at a high temperature and phase separation is then induced by cooling the polymer solution. The compatibility between the polymer and the diluent is one of the key factors affecting the morphology of the membrane. In many cases, polyolefin has been used as the polymer material to prepare microporous membranes [123,124]. A high-density polyethylene hollow-fiber membrane has been prepared by polymer crystallization via the TIPS process [125]. Poly(ethylene-*co*-vinyl alcohol) hollow fiber membranes with 44 mol% ethylene content, showing better pore connectivity and rejection of $\sim$20 nm diameter solute, has been prepared by TIPS, using a mixture of 1,3-propanediol and glycerol (50:50) as diluent [126].

Polymers containing ethylene oxide units are of considerable interest since ether oxygen linkages lead to flexible polymer chains and specific interactions with metal ions, polar molecules such as H_2O and H_2S, and quadrupolar molecules such as CO_2. Thus rubbery membrane materials have been made for the removal of acidic gases such as CO_2 and H_2S from natural gas (mainly CH_4) using a highly branched, cross-linked PEO hydrogel (see below). Unlike conventional size-sieving membrane materials, which achieve high permeability selectivity mainly via high diffusivity selectivity, these polar rubbery membrane materials exhibit high CO_2 permeability and high CO_2/CH_4 mixed-gas selectivity due to high gas diffusivity and high CO_2/CH_4 solubility selectivity [127].

In a typical method of preparation of the aforesaid rubbery membrane material [127], a prepolymer solution is prepared by adding 0.1 wt% initiator (e.g., 1-hydroxy-cyclohexyl phenyl ketone) to poly (ethylene glycol) diacrylate (PEGDA, 743 g/mol) or mixtures of PEGDA and poly(ethylene glycol) methyl ether acrylate (460g/mol). After mixing and sonicating to eliminate bubbles, the solution is sandwiched between two quartz plates separated by spacers to control thickness and polymerized by exposure to 312 nm UV light for 90 s at 3 mW/cm².

Compared to flat membranes, hollow-fiber membranes have much wider applications at the commercial scale because they provide a higher membrane area per unit membrane module volume.

2.12.1.2 Hollow-Fiber Membranes

Integrally skinned asymmetric asymmetric membranes used for gas and liquid separations consist of a thin skin layer supported by a porous substructure. The skin layer determines the permeability and selectivity of the membrane, whereas the porous substructure functions primarily as a physical support for the skin. Both layers are composed of the same material and are integrally bonded. The skin layer usually has a thickness on the order of several hundred to several thousand angstroms.

At present, hollow-fiber membranes used for gas and liquid separations are mostly prepared from amorphous polymers by means of a phase inversion technique. In this technique, a polymer is dissolved in a suitable solvent or solvent mixtures and spun into a coagulation bath, where solvent exchange occurs between the extruded fiber and coagulant, yielding the asymmetric membrane structure.

Technology development on the fabrication of asymmetric membranes with an ultrathin dense layer has received much attention due to the fact that the thinner the dense layer is, the higher is the productivity. The fabrication of a hollow fiber with a desirable pore-size distribution and performance is not a trivial process as many factors influence fiber morphology during the phase inversion.

The controlling factors for hollow fiber spinning are not only complicated, but also quite different from those for flat membranes. For example, two coagulations take place in hollow fiber spinning (at internal and external surfaces), while there is only one major coagulation surface for an asymmetric flat-sheet membrane. Moreover, whereas there is usually a waiting period for an asymmetric flat membrane before immersing it into a coagulant, the internal coagulation process for a hollow fiber, if liquids are used as *bore* fluids, starts immediately after extrusion from a spinneret and the fiber then goes through the external coagulation. In addition, the spinning dope suitable for fabricating hollow fibers generally has a much greater viscosity and elasticity than that for flat membranes. Furthermore, the phase

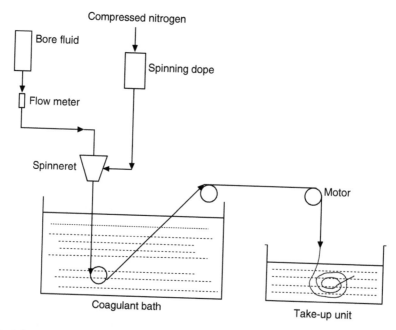

FIGURE 2.66 Schematic diagram of a hollow-fiber spinning process. (After Chung, T. S. and Hu, X. 1997. *J. Appl. Polym. Sci.*, 66, 1067. With permission.)

inversion process for hollow-fiber formation usually takes place under tension or elongational stress, while flat membranes do not experience this type of stress [128].

A typical hollow-fiber spinning set-up is shown in Figure 2.66. For poly(ether sulfone) (PES) hollow fiber [128] a spinning solution (dope) of PES (30%) in N-methyl-2-pyrrolidone (NMP) is extruded under a nitrogen pressure of 10 psi through a spinneret having dimensions (typically) of 800 and 550 μm for outer and inner diameters, respectively. The bore fluid of composition NMP/water (40/60) is conveyed by the gravity force through a flow meter. The spinning dope and the bore fluid, flowing at the rate of 0.317 and 0.083 cm³/min, respectively, meet at the tip of the spinneret, then pass through an air gap ranging from 0 to 14.4 cm before entering the coagulation (water) bath. Once a hollow fiber is formed, it is stored in water bath for several days and then transformed to a tank containing methanol to remove the residual NMP completely.

The air gap distance between the spinneret and coagulation bath (see Figure 2.66) plays a very important role on fiber performance. An increase in air-gap distance may result in a significant decrease in permeance. This may arise from the fact that different precipitation paths arise during the wet-spinning and dry-jet wet-spinning processes. As illustrated in Figure 2.67, the nascent hollow fiber experiences vigorous and rapid coagulation at its internal and external surfaces simultaneously in a wet-spinning process. However, in a dry-jet wet-spinning process the nascent as-spun fiber experiences two different coagulation paths (see Figure 2.67) before entering the coagulation bath, viz., a vigorous coagulation at its internal surface and a nonvigorous coagulation at its external surface. A big air-gap distance may thus produce a greater orientation and tighter molecular packing at the outer surface, causing a decrease in permeance. The difference between the coagulations of the two spinning processes is further explained below.

Since in a wet-spinning process (Figure 2.68), the as-spun fiber is immersed in the non-solvent coagulation bath immediately after exiting from the spinneret, the coagulations at both inner and external surfaces are vigorous and rapid with the result that the extended and randomly oriented polymer chains contract suddenly and almost instantaneously, thereby entrapping a significant amount of non-solvent and solvent in the contracted chains [128]. Both the outer and inner skin layers formed may, therefore, have a long-range random and entangled structure with some macro/microporosity or free volume.

In the case of dry-jet wet-spinning process (Figure 2.68) with a certain air-gap distance, the moisture-induced precipitation process slows the speed of chain contraction and provides contracting chains with time needed for conformation rearrangement [128]. The external surface layer of a hollow fiber made by this process may thus have compact short-range random arrangement of polymer chains in circumferential and lateral directions, and a slightly oriented and stretched structure in the axial direction, producing a skin morphology that has less micro- and macroporosity or free volume. The dry-jet wet-spun fibers thus have more compact structure than wet-spun fibers.

2.12.2 Membrane Modules

Mainly four types of membrane modules are used: plate-and-frame, spiral-wound, tube-in-shell, and hollow fiber. The plate-and-frame module consists of a series of membranes (10–500 μm thick) sandwiched between spacers that act as flow channels (Figure 2.69). (The membranes are often laminated on a porous support that offers no flow resistance.) The feed flows in one set of channels and the permeate, with or without carrier fluid, flows in alternate channels. Plate-and-frame modules find use in ultrafiltration and dialysis applications which include hemodialysis and electrodialysis.

In the spiral-wound system, membranes glued together at the ends and separated by spacers (that provide flow channels to the feed and permeate) are layered and wound around a central porous tube several times, thus forming a multilayered and cylindrical module. The feed mixture flows axially into the channels and the permeate flows spirally into a central porous tube and out of the system. A spiral-wound module has at least twice the packing density of a plate-and-frame module. Several spiral-wound modules can be lined up in series forming a cartridge system.

The tube-in-shell modules usually consist of porous tubes with the membrane attached either inside or outside depending on the application and are especially suited to handle liquids that contain suspended

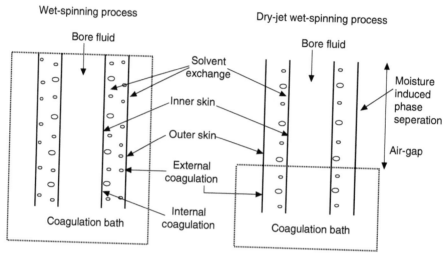

FIGURE 2.67 Comparison of precipitation in the wet-spinning and dry-jet wet-spinning processes. (After Chung, T. S. and Hu, X. 1997. *J. Appl. Polym. Sci.*, 66, 1067. With permission.)

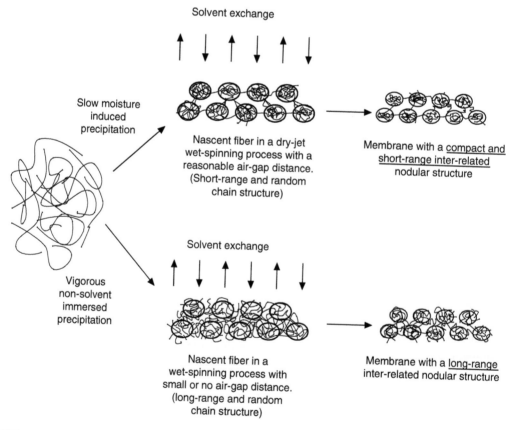

FIGURE 2.68 Schematic of skin morphologies in the wet-spinning and dry-jet wet-spinning processes. (After Chung, T. S. and Hu, X. 1997. *J. Appl. Polym. Sci.*, 66, 1067. With permission.)

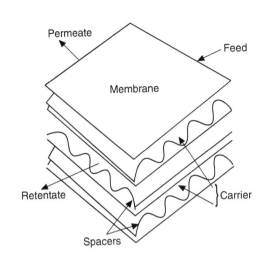

FIGURE 2.69 Schematic of a plate-and-frame type multilayer membrane module.

solids and colloids. When the feed liquid contains a high concentration of suspended solids, membranes with higher tube diameters are often used sacrificing packing density. Tube-in-shell modules are suited for high flux operations in ultrafiltration and reverse osmosis where deposits resulting from concentration polarization tend to clog other configurations rapidly. In these modules, the deposits on single tubes can be easily removed and tube replacement is inexpensive.

Hollow fibers have two advantages over flat sheet or tubular membranes, viz., they provide higher productivity per unit volume and they are self-supporting. The higher productivity results from the fact that surface-to-volume ratio varies inversely with fiber diameter. Thus with hollow fibers of 100 μm diameter, a 0.3 m^3 device can offer 500 m^2 of effective membrane area, as compared to about 20 m^2 for flat-sheet membrane and about 5 m^2 for membrane in tubular form. As hollow fibers are self-supporting, the hardware requirement for the fabrication of a hollow-fiber unit is simple. Unlike flat-sheet membranes which must be assembled (see Figure 2.69) with spacers, porous supports, etc., a hollow fiber module can be made by simply potting a bundle of hollow fibers into a plastic or metal tube (Figure 2.70). The hollow-fiber unit is, however, vulnerable to fouling and clogging by suspended particulates in the feed. Thorough pretreatment of the feed is therefore necessary.

The most important applications of hollow-fiber modules are in hemodialysis (Figure 2.70), reverse osmosis, and gas separation units. Modules up to 50 cm diameter containing hundreds of thousands of fibers are used in gas separation.

2.12.3 Applications

In hemodialysis, low molecular metabolic waste such as urea, creatinine, and other toxic substances (solutes up to 6000 mol wt) are removed from the blood of uremia patients by diffusive transport, which is driven by a concentration gradient of blood solutes being dialyzed against a physiological solution. A complimentary process is hemofiltration, in which solutes up to 20,000 mol wt are removed via an ultrafiltration membrane, the transport being caused by a convective transmembrane flux generated by mild hydraulic pressure differences across the membrane.

Several classes of polymeric materials are found to perform adequately for blood processing, including cellulose and cellulose esters, polyamides, polysulfone, and some acrylic and polycarbonate copolymers. However, commercial cellulose, used for the first membranes in the late 1940s, remains the principal material in which hemodialysis membranes are made. Membranes are obtained by casting or spinning a dope mixture of cellulose dissolved in cuprammonium solution or by deacetylating cellulose acetate hollow fibers [121]. However, polycarbonate-polyether (PC-PE) block copolymers, in which the ratio between hydrophobic PC and hydrophilic PE blocks can be varied to modulate the mechanical properties as well as the diffusivity and permeability of the membrane, compete with cellulose in the hemodialysis market.

Low-density polyethylene and polypropylene in the form of flat-sheet and hollow-fiber membranes are used in plasmapheresis and as oxygenators in the heart-lung machine. Other materials commonly used in plasmapheresis are cellulose acetate, polycarbonate, and polysulfone [129].

Membrane technology is employed in the separation of gases, e.g., H_2 from N_2, CO, and CH_4; CO_2 and water vapor from natural gas. It finds use in H_2 recovery from ammonia production plants. The

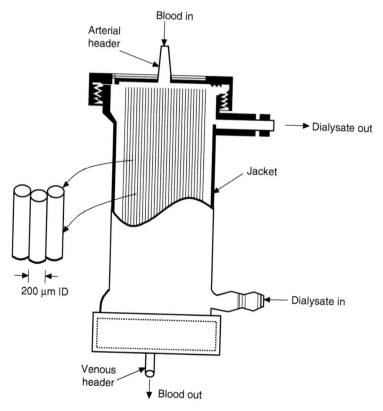

FIGURE 2.70   Schematic of a hollow-fiber membrane cartridge used for blood dialysis.

separation of gas components by polymer membranes is based on chemical affinities and kinetic (sieving) diameters of gases, e.g., He 0.26, H_2 0.289, CO_2 0.33, O_2 0.346, N_2 0.364, CH_4 0.38 nm [130]. At low pressures, the membrane selectivity is closely related to the separation factor, $\alpha_{AB} = (S_A/S_B)(D_A/D_B)$, where S_A/S_B represents the preferential sorption or solubility and D_A/D_B is the ratio of diffusion coefficients (mobilities) of the components A and B. In general, high diffusions and solubilities are associated with the rubbery polymer as it has high mobility of chain segments and higher rate of increase of free volume with temperature, as compared to the glassy polymer.

Oily water wastes constitute a major environmental problem in many industries. Stable oil/water emulsions, which cannot be broken by mechanical or chemical means, require more sophisticated treatment to meet the effluent standards. Various physical methods including microfiltration, ultra-filtration, nanofiltration, centrifugation, air flotation, and fiber or packed bed coalescence have been applied in oil-surfactant-water separation [131]. Among these physical methods, membrane technology is by far the most widely used.

In an effort to improve resistance to fouling, hydrophilic and low surface charge membranes have been developed because hydrophilic materials are less sensitive to adsorption than hydrophobic ones. Using polyetherimide as the membrane material and polybenzimidazole and poly(ethylene glycol) as the additives, hydrophilic hollow-fiber membranes have been prepared for oil-surfactant-water separation [132], showing rejection rates of 51%–79%, 83%–93%, and more than 99% for surfactant, total organic carbon, and oil, respectively. Polyetherimide hollow-fiber membranes have also been prepared using polyvinylpyrrolidones as additives for separation of oil-surfactant-water emulsion systems [133], achieving the corresponding rejection rates of 76%–80%, 91%–93%, and more than 99%. Figure 2.71 shows the flow diagram of a typical oil-surfactant-water membrane separation system.

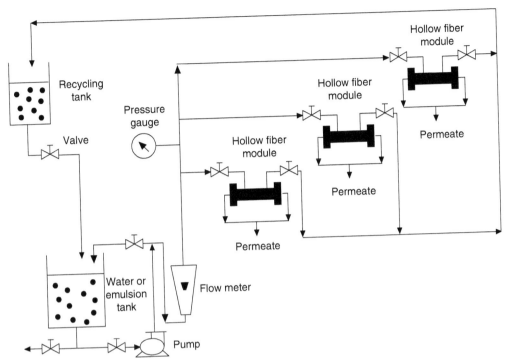

FIGURE 2.71 Flow chart of a separation unit using hollow fiber membranes for oil-surfactant-water emulsion systems. (After Xu, Z. L., Chung, T. S., Loh, K. C., and Lim, B. C. 1999. *J. Appl. Polym. Sci.*, 158, 41. With permission.)

2.13 Hydrogels and Smart Polymers

Hydrogels are three-dimensional hydrophilic polymer networks that absorb water but do not dissolve in water. The extent of volume change due to water absorption varies with the degree of ionization of the gel and for a superabsorbent it may be as large as 500 fold. The volume change is understood as a phase transition which is a manifestation of competition among three forces on the gel—the positive osmotic pressure of counterions, the negative pressure due to polymer-polymer affinity, and the rubber elasticity of the polymer network [134]. The balance of these forces varies with changes in temperature or solvent properties.

Partially hydrolyzed and lightly cross-linked polyacrylamide is a typical superabsorbent hydrogel. The polyacrylamide gels are prepared by free-radical polymerization. In a typical procedure [134], acrylamide (linear constituent), N,N'-methylene-bisacrylamide (tertrafunctional cross-linking constituent), and N,N,N',N'-tetramethyl ethylene diamine (TEMED) (accelerator) are dissolved in (degassed) water at 0°C to which ammonium persulfate (initiator) is added to initiate polymerization. The polymerization is typically done in micropipettes, although gels have also been alternatively synthesized between glass plates with spacers. (In many other systems, other comonomers, preferably with ionic groups, are added to achieve desired properties.) After gelation, the cylindrical samples from capillaries or sliced pieces from a macroscopic product are dialyzed with water to remove residual monomers and then partially hydrolyzed to convert a part of the amide groups to carboxylic acid groups:

$$-CONH_2 \rightarrow -COOH \rightarrow -COO^- + H^+$$

The possibility of using superabsorbent hydrogels as water managing materials for the renewal of arid and desert environments has attracted great attention [135]. These materials can reduce irrigation water consumption, improve fertilizer retention in soil, lower the plant death rate, and increase the plant growth rate [136]. However, the application of superabsorbents in this field has been limited because most polymeric superabsorbents are based on pure poly(sodium acrylate) and so they are too expensive and not suitable for saline water and soils [137]. In order to reduce production costs and improve salt resistance, superabsorbent composites have been made by incorporating mineral powders into hydrogels [138,139].

Fertilizers are as important as water in agriculture and horticulture. However, about 40%–70% fertilizer is lost to the environment and cannot be absorbed by corps and trees when mixed with soil directly, resulting in large resource losses and serious environmental pollution [140]. Incorporating fertilizers into a superabsorbent polymeric network may thus be an effective way of increasing the utilization efficiency of both water and fertilizer [141]. Thus a multifunctional superabsorbent composite based on poly(acrylic acid-*co*-acrylamide) (PAA-co-AAm) has been obtained by incorporating sodium humate (SH), which can perform a number of functions such as regulate plant growth, accelerate root development, improve soil cluster structures, and enhance the absorption of nutrient elements. In a typical procedure, 3.6 g of AAm, 4.3 g of AA (50% neutralized with 2M sodium hydroxide solution) and 8.8 mg N,N'-methylenebisacrylamide (cross-linking agent) are polymerized with ammonium persulfate as initiator for 3 h at 50°C, in the presence of an appropriate amount of SH dispersed in the mixed solution. The product is washed with water and ethanol and then dried in an oven at 70°C.

The superabsorbent composites containing SH show release of the fertilizer over 10–40 days, depending on the SH content (5 wt% to 30 wt%). The release rates into water in the initial period are higher since the SH existing on the surface or freely incorporated in the composite network are dissolved more readily in water. The SH bonded with the polymeric network needs more time to diffuse from the hydrogel granule and dissolve in water. Figure 2.72 shows schematic structures of a PAA-*co*-AAm/SH

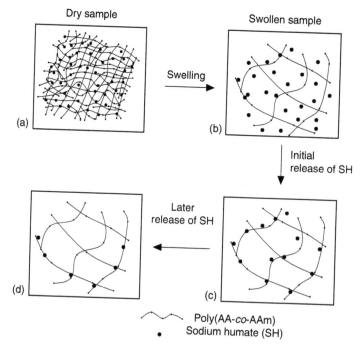

FIGURE 2.72 Schematic structures of a PAA-AM/SH superabsorbent composite (see text): (a) in dry state; (b) in swollen state; (c) after release of SH which is on the surface or freely incorporated in the polymer network; and (d) after release of SH that is bonded with the polymer network. (After Zhang, J., Liu, R., Li, A., and Wang, A. 2006. *Ind. Eng. Chem. Res.*, 45, 48. With permission.)

superabsorbent composite in the dry state, in the swollen state, and after the release of SH. The superabsorbent composite also improves the water-retention capacity of the soil. Thus, compared to sand soil without the superabsorbent composite, 33.45 wt% water was still retained on the 20th day when the sand soil was mixed with 1.0 wt% composite containing 30 wt% SH [141].

2.13.1   Smart Polymers

Stimuli-responsive hydrogels which exhibit volume changes or phase transitions in response to differences and variations in the surrounding environment, such as temperature [142], pH [143], pressure [144], electricity [145], or light [146] are given the term smart or intelligent because their strange properties can be put to use in a wide range of applications. Two of the most studied polymers in the intelligent materials stage are poly(N-isopropylacrylamide) or PNIPAAm and poly(vinyl methylether) or PVME.

$$-(CH_2-CH)_n- \qquad -(CH_2-CH)_n-$$

C=O	O
NH	CH_3
CH(CH_3)_2	

(PNIPAAm)   (PVME)

Both PNIPAAm and PVME exhibit unique thermo-shrinking properties. Thus, as an aqueous solution is heated beyond a certain point, the polymer shrinks and a phase separation occurs. This temperature is commonly referred to as the lower critical solution temperature (LCST). For PNIPAAm, it lies between ca. 30 and 35°C, the exact temperature being a function of the detailed microstructure of the macromolecule. Below LCST, the polymer is soluble in the aqueous phase, as the chains are extended and surrounded by water molecules. Above the LCST, the polymer becomes insoluble and phase separation occurs. Because of the abrupt nature of these transitions and their reversibility (which allows repeated thermal switching) these polymers have stirred up particular interest in the field of science and engineering since their first appearance in the open literature in 1956.

PNIPAAm has been synthesized from N-isopropylacrylamide (NIPAAm) by a variety of techniques, the most widely used being free-radical initiation of organic solutions [147] and redox initiation in aqueous media [148]. Redox polymerization of NIPAAm in aqueous media typically uses ammonium persulfate or potassium persulfate as the initiator and either sodium metabisulfite or N,N,N'N'-tetramethylethylenediamine (TEMED) as the accelerator. In addition, the solutions are usually buffered to constant pH since in the absence of buffer much greater polydispersity is obtained. Whether one polymerizes NIPAAm in organic or aqueous solution also affects polymer properties [149].

One of the very useful applications involving thermo-shrinking polymers is a polymeric tool (a gel hand) made of three layers of PNIPAAm, PAAm, and an inert spacer (Figure 2.73). This gel hand can be used like a tweezer to pick up a target compound in aqueous solution by simply raising the temperature above LCST and to release the compound below LCST.

Thermoresponsive polymers can also be used in many biotechnological applications. The use of such polymers for in vitro studies allows the control of surface properties and thus the stimulation of cell adhesion and detachment through temperature changes. For example, normally an enzyme would be used to detach cultured tissue cells from culture dishes for further culturing, but the process is very inefficient because many of the detached cells are damaged by the enzyme. This problem has been overcome by using PNIPAAm which aids cell adherence by exhibiting hydrophobicity above its LCST, but below this temperature the polymer exhibits hydrophilicity and binds to water thus pushing the cells away. Therefore, by lowering the temperature just below the LCST of the polymer, almost all of the cells can be detached for culturing. This process is simple, efficient, inexpensive and polymer-grafted dishes can be re-used many times.

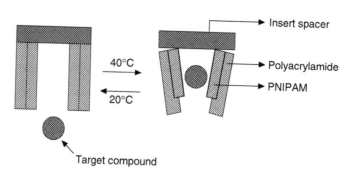

FIGURE 2.73　A schematic illustration showing how thermo-shrinking polymer PNIPAAm can be used as a gel hand to isolate a target compound.

In an interesting biochemical application, PNIPAAm has been attached near the receptor site of the protein Streptavadin, which binds to the ligand biotin with a very large association constant. Below LCST ($\sim 32°C$) of PNIPAAm, the polymer is a soluble random coil and allows the binding of the ligand to the protein. However, above the LCST, the polymer molecule collapses onto the receptor site, preventing the binding of the ligand to the protein. This reaction can thus be controlled by small changes in the temperature. Since protein-ligand interactions are used in many assay procedures, such as immunoassays, the above ability to control ligand binding can find many applications in biotechnology and medicine.

In recent years, many kinds of temperature-responsive PNIPAAm and its copolymer hydrogels with other acrylic monomers have been synthesized [142]. Besides being used for hydrogels, NIPAAm monomer can be grafted on to polymer substrates by electron beam, irradiation or UV-initiated graft polymerization to achieve special modification of polymer surfaces. Thus NIPAAm has been grafted on porous polymer films such as LDPE, PP, or polyamide films in order to prepare novel films for pervaporation of liquid mixtures or separation membranes [150,151].

LCSTs have been reported [152] for a series of poly(N-alkyl acrylamides) which proceed from complete solubility to insolubility as the size of the alkyl side group increases. NIPAAm has also been copolymerized with other N-alkyl acrylamides [152,153] over the entire composition range. Typically, continuous changes in the LCST either up or down are observed when the comonomer (with the exception of N-n-butyl acrylamide) has respectively a smaller or larger N-alkyl group than N-isopropyl. Thus the LCST of PNIPAAm-like polymers can be shifted from <0 to >100°C by varying the copolymer composition. This provides excellent flexibility in tailoring transitions for specific uses. One can switch off solubility at biologically relevant temperatures as well as room temperature. Moreover, the same LCST can be obtained with a small amount of a very hydrophobic comonomer (such as N-decyl) or a high fraction of a less hydrophobic comonomer (such as N-t-butyl). This permits further modulation of interactions with cosolutes in the system.

PNIPAAm gels have been applied to a myriad applications which exploit the change in gel dimensions to modulate the differential diffusion of species in a medium. It has thus been possible to selectively remove [154,155] and deliver [156–158] consolutes with thermal-switching control. A few of the processes are described below.

A research group of Cussler at the University of Minnesota did considerable applied research on separations with PNIPAAm gels. Figure 2.74 shows a flow chart of their separation process. It is based on the premise that if a gel is placed into an aqueous solution of a mixture of large cosolute A, such as macromolecules, and small cosolute B (or clean water) (Steps 1 and 2), when the gel swells, A will be excluded from entering the gel pores by steric hindrance and B will freely enter the gel pores. On removing the swollen gel (Step 3), the solution left behind (raffinate) is now more concentrated in A. The removed gel, on the other hand, can be placed into a warm solution to obtain B free of A and the gel

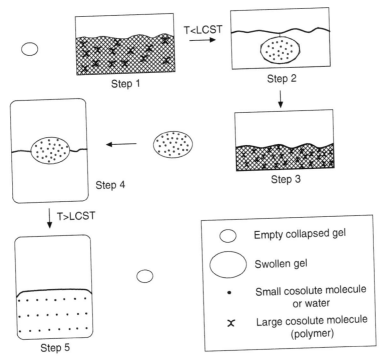

FIGURE 2.74 Steps in separation process using PNIPAAm gel. (After Schild, H. G. 1992. *Prog. Polym. Sci.*, 17, 163. With permission.)

for recycling. The efficiency of the separation process is defined as

$$\text{efficiency} = \frac{\Delta C_R V_R}{V_S^o}$$

where ΔC_R is the measured increase in concentration in raffinate relative to feed, V_R is the volume of raffinate and V_S^o is the initial solution volume. Thus, if the concentration doubles when the volume is halved, then the efficiency is 100%. On the other hand, if the original and final concentrations are the same, then the efficiency is 0%. This separation system has been termed [159] as size selective extraction solvent. It serves as an alternative to ultrafiltration techniques [142].

As would be expected, PNIPAAm gels with higher cross-linking densities yield increased efficiency of separation. However, as cross-links are randomly distributed, there is no abrupt cut-off at an exact molecular weight. Nonetheless, PNIPAAm gel particles have commercial potential as they are able to absorb up to 30 times their dry weight in water [154] and such systems have been patented [155] as an inexpensive alternative to ultrafiltration. The gel separation process has been used for removing low molecular weight contaminants from soy protein [160] and water from gasoline or fuel oils [161]. Polymers in the PNIPAAm class have also been cross-linked into fibrous materials for use as a urine absorbent [162].

Prior to PNIPAAm, the Cussler group [159] used pH-sensitive hydrolyzed polyacrylamide (PAAm) anionic gel, which swells at high pH (and thus picks up small cosolute B); to recover the gel for re-use, the swollen gel is removed and placed in water at low pH when it collapses and squeezes out B. High separation efficiencies are found for species greater than 3 nm in diameter such as proteins and synthetic polymers. However, only negatively charged or neutral species can be separated. Cationic species such as lysozyme cannot be separated as they precipitate on the gel. This is a limitation of the PAAm anionic gel.

Japanese workers [163] have made PVME gels typically through cross-linking with gamma ray irradiation. These exhibit LCST behavior and have been applied in wastewater sludge dewatering, much

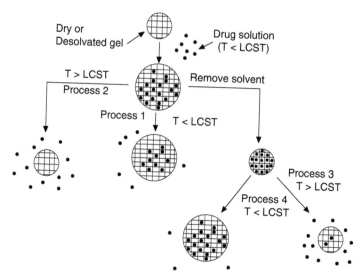

FIGURE 2.75   Various processes of drug delivery using PNIPAAm gel. (After Schild, H. G. 1992. *Prog. Polym. Sci.*, 17, 163. With permission.)

in the same way as PNIPAAm discussed above. The water absorbing rate of the gel varied with the properties and solid concentration of the sludge, as would be expected. Sludges were dewatered by repeated use of the gel.

The application of hydrogels for biomedical applications including drug delivery devices has been of increasing interest. Changes in swelling states of PNIPAAm gels can influence the diffusion of solutes from within the gels to the outside aqueous media [142,156–158,164]. Several different schemes can thus be envisioned for drug delivery application, as shown schematically in Figure 2.75. The LCST of PNIPAAm-type gels can be adjusted to near body temperature (37°C) by copolymerization, by varying the length of the alkyl chain, and by use of additives [142], thus tailoring the drug delivery as desired and making them viable for in vitro applications. Some delivery systems, such as processes 1 and 3 (Figure 2.75), depend on Fickian-type diffusion, while others, such as process 2 (Figure 2.75), may rely on the pressure generated during gel collapse to squeeze out the drug. In process 2, however, after the initial release of drug, there may be increasing retardation in the system due to the formation of a surface skin layer and closure of surface pores as the surface region is the first to contact the solution and undergo the phase transition.

PNIPAM-based interpenetrating networks (IPNs) have been developed [158,165]. As the presence of the second polymer (essentially inert) network has little effect on the LCST, it provides the ability to control only the degree of swelling, as was demonstrated [158] by a poly(ethylene oxide–dimethylsi-loxane—ethylene oxide)/PNIPAAm system. In contrast, the copolymerization method influences both the degree of swelling and the LCST.

For numerous applications, it is desirable to maximize the volume change with a minute pH change near the phase transition point. A promising candidate [143] for such a system is segmented poly(amine urea) with alternating polar (N,N'- diethylenediamine) and apolar (diethylenephenylene) units in the main chain. In a typical method of preparation, an end-isocyanated polyamine was prepared by the 2:1 addition reaction of 4,4$'$-methylenediphenyldiisocyanate (MDI) to telechelic polyamine having secondary amino groups at both ends and the isocyanated polyamine was then extended by the reaction with equimolar ethylenediamine to form a segmented poly(amine urea) (SPAU). The SPAU exhibited reproducible swelling/deswelling in response to pH changes, with a discontinuous 270-fold change in the swelling degree at a critical pH indicating a phase transition. The regulation of solute transport across

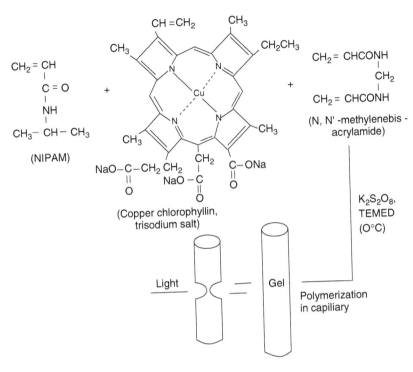

FIGURE 2.76 Preparation of PNIPAAm gel containing light-sensitive chromophore, trisodium salt of copper chlorophyllin. Chlorophyllin has a double bond through which it enters the polymer network forming covalent bonds. The formed gel collapses under illumination. (After Suzuki, A. and Tanaka, T. 1990. *Nature*, 346, 345. With permission.)

the SPAU matrix responding to a pH jump in the external milieu was demonstrated using vitamin B_{12} as a model solute.

Hydrogels sensitive to ultraviolet light [166] as also those sensitive to visible light [146] have been reported. The ultraviolet light initiates an ionization reaction in the gel, creating internal osmotic pressure which induces swelling, while in the absence of this light, the equilibrium tends towards the neutral polymer system and the gel collapses. This transition process is slow, as it depends on the photochemical ionization and subsequent recombination of ions. It is, however, technologically desirable that the transition be induced by faster mechanisms.

In the visible light induced polymeric gel systems [146], the transition mechanism is due only to the direct heating of the network polymers by light which is an extremely fast process. Such systems might be used as photoresponsive artificial muscles, switches, and memory devices. In a typical method of preparation by free-radical copolymerization (Figure 2.76), 7.8 g N-isopropylacrylamide (NIPAAm), 0.72 g trisodium salt of copper chlorophyllin (light-sensitive chromophore), 0.67 g N,N'-methylene bisacrylamide (cross-linker), and 240 μl of tetramethylethylenediamine (TEMED) (accelerator) were dissolved in 100 ml water at 0°C, to which 0.2 g of ammonium persulfate was added to initiate the polymerization. Capillaries (~100 μm diameter) were immersed into the solution. After gelation, the cylindrical samples were removed from the capillaries and dialyzed with water to remove unreacted monomers. Light shrank the gel (Figure 2.76) over the entire temperature range, but the largest effect was seen in the transition region at 31.5°C. For 1 μm diameter gels the response time is expected to be about 5 ms.

In partially hydrolyzed (ionic) acrylamide gels, phase transition can also be induced by the application of an electric field [167]. The electric forces on the charged sites of the network produce a stress gradient along the electric field lines in the gel. There exists a critical stress below which the gel is swollen and above which the gel collapses. The volume change at the transition is either discrete or continuous, depending on the degree of ionization of the gel and on the solvent composition. The discrete volume transition of the gel induced by an electric field can be used to make switches, memories, and mechanochemical trasducers. For example, ionic gels controlled by coordinated signals from a microcomputer may be used as an artificial muscle. Furthermore, two- or three-dimensional images may be stored by using the local collapse and swelling of the gel [167].

One of the problems faced in applying PNIPAAm hydrogels is that the response rate to temperature changes is very slow, which restricts their wider applications, such as on-off valves and artificial muscles. According to the Tanaka-Fillmore theory [168], $\tau \approx R^2/D$, where τ, R, and D are the characteristic time for gel shrinkage, the size of the gel, and the cooperative diffusion coefficient, respectively. It can be observed from this equation that the response time of the gel is proportional, and hence the response rate is inversely proportional, to the square of the size of the gel. Consequently, micrometer-sized gel particles exhibit a more rapid response rate. However, in many applications, bulk hydrogels are needed. Many approaches have therefore been proposed to prepare fast-responsive macroscopic PNIPAAm hydrogels, such as producing phase-separated microstructures in gels [169], preparing comb-type PNIPAAm hydrogels [170], incorporation of silica microparticles, followed by treatment to remove silica [171], polymerization/cross-linking of NIPAAm in solid state [172], polymerizing NIPAAm in salt [173] or saccharide [174], and simple freeze-drying tretament of PNIPAAm hydrogels [175].

Some of the above methods have the limitation that they require the use of strong acids or organic solvents in the gel. A novel, simple, and effective strategy to achieve fast-responsive bulky PNIPAAm hydrogels involves incorporation of PNIPAAm nanoparticles into PNIPAAm networks to form composite hydrogels [176]. PNIPAAm nanoparticles (~ 500 nm) synthesized by emulsion polymerization [176] are dispersed in an aqueous solution of NIPAAm and N,N'-methylenebisacrylamide which are polymerized at room temperature using ammonium persulfate and N,N,N', N''-tetramethylenediamine (TEMED) as initiator and accelerator, respectively. When the polymerization is complete, the generated PNIPAAm chains may pass through the PNIPAAm particles, which would be immobilized in the PNIPAAm networks. In comparison with conventional PNIPAAm hydrogels, the PNIPAAm nanoparticle-incorporated PNIPAAm hydrogels thus prepared exhibit much faster response rates as the temperature is raised above the LCST. These improved response properties are a result of the superfast shrinkage of PNIPAAm nanoparticles which generate pores for water molecules to be quickly squeezed out of the bulky PNIPAAm gels (see Figure 2.77) [176].

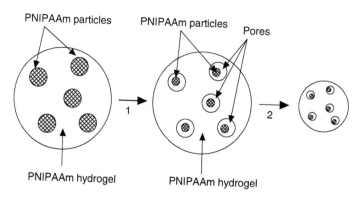

FIGURE 2.77 Schematic illustration of the shrinking process: (1) PNIPAAm particles shrink and generate pores; (2) bulky gels then shrink further at a rapid rate. (After Zhang, J.-T., Huang, S.-W., Xue, Y.-N., and Zhuo, R. X. 2005. *Makromol. Rapid Commun.*, 26, 1346. With permission.)

2.14 Dendritic Polymers

A little more than 30 years after Flory [177] theorized about synthesizing condensation polymers from multifunctional monomers, the first papers on synthesis of dendritic polymers (dendron, Greek for tree) appeared [178,179], revealing a number of very unique and different properties of these polymers, compared to their linear analogs. For instance, at high molecular weights the dendritic polymers were found to be globular and, in contrast to linear polymers, they behaved more like molecular micelles [180]. The descriptors starburst, dendrimers, arborols, cauliflower, cascade, and hyperbranched used for such polymers all describe specific geometric forms of structure.

Following the first papers of Tomalia [178] and Newkome et al. [179] in 1985 dealing with dendrimers, a large number of dendrimers have been presented in the literature ranging from polyamidoamine, aromatic polyethers and polyesters, aliphatic polyethers and polyesters, polypropyleneimine, polyphenylene to polysilane. Copolymers of linear blocks with dendrimer segments (dendrons) and block copolymers of different dendrons have been described.

Dendrimers, as shown in the generalized form in Figure 2.78(a), are obtained when each ray in a star molecule is terminated by an *f*-functional branching from which (*f*–1) rays of the same length again emanate. A next generation is created when these *f*–1 rays are again terminated by the branching units from which again rays originate, etc. In recent years, the chemistry of preparing dendrimers has become very successful, although the synthesis of perfect monodisperse dendrimers is time-consuming and painfully cumbersome. Another serious drawback is the space filling due to which it has not been possible to prepare more than five generations. Either the reaction to a higher generation stops completely or, as it happens in practice, the outermost shells develop imperfections.

Though perfect monodisperse dendrimers have very interesting material properties, for use as engineering materials they are far too complicated and costly to produce. This was soon realized by several researchers at DuPont Experimental Station working on dendritic polymers as rheology control agents and as spherical multifunctional initiators. The need to obtain the material rapidly and in large quantities forced them to develop a route for a one-step synthesis of dendritic polymers. These polymers

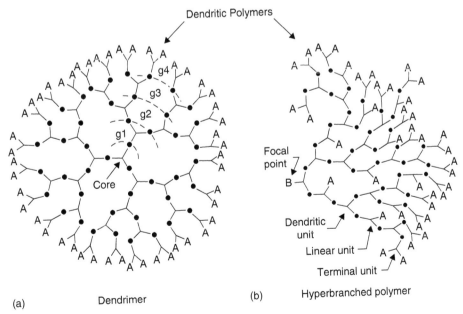

FIGURE 2.78 Schematic representation of dendritic polymers comprising dendrimers and hyperbranched polymers (g level indicates generation number). (After Hult, A., Johansson, M., and Malmström, E. 1999. *Adv. Polym. Sci.*, 143, 1. With permission.)

were, however, polydisperse and had defects in the form of linear segments between branch points but they were highly branched dendritic molecules. Kim and Webster [181] named them hyperbranched polymers. Ever since, a wide variety of hyperbranched polymers have been synthesized and reported in the literature [182].

While in a perfectly branched dendrimer derived from A_xB monomer only one type of repeat unit can be distinguished apart from the terminal units carrying the chain ends [see Figure 2.78(a)], a hyperbranched polymer of A_xB (at high conversion of B) will have three different types of repeat units as illustrated in Figure 2.78(b); these are dendritic units (fully incorporated A_xB monomers), terminal units having two A groups unreacted, and linear units having one A group unreacted. The linear segments are generally described as defects. Since the hyperbranched polymers are allowed to contain some linearly incorporated A_xB monomers, the synthesis of such polymers, unlike that of dendrimers, does not require the use of protection/deprotection steps. The most common synthesis route thus follows a one-pot procedure [183] where A_xB monomers are condensed in the presence of a catalyst. Such one-step polycondensations result in highly branched polymers even though they are not as idealized as the generation-wise constructed dendrimers.

Tedious and repetitive synthetic procedures for dendrimers result in expensive polymers and limit their availability. On the other hand, the one-step process, mentioned above, allows production of hyperbranched polymers on a large scale, giving them an advantage over dendrimers in applications requiring large amounts of material, although the properties of hyperbranched polymers are only intermediate between those of dendrimers and linear polymers. Hyperbranched polymers are also unique in that their properties are easily tailored by changing the nature of the end groups. For some areas, such as coating resins and tougheners in epoxy resins, hyperbranched polymers are envisaged to play an important role. Various other applications of these novel materials have been suggested. One reason for the emerging interest in hyperbranched polymers is their different properties compared to conventional linear polymers.

A step change in hyperbranched polymer synthesis occurred when self-condensing vinyl polymerization was introduced by Fréchet et al. [184], extending hyperbranched methodologies to vinyl monomers with end group control and living polymerization features. Hyperbranched polymer synthesis based on vinyl monomers was developed further by several research groups [185]. An inherent limitation with many of these methodologies, however, is the need for designed functional monomers, which are costly to manufacture and would, therefore, render many cases economically nonviable for large-scale production. In a significant development in this context, Sherrington and co-workers [186,187] have recently demonstrated a new facile route for the synthesis of branched polymers based on commercially available vinyl monomers and the addition of chain transfer agents (CTAs). The stoichiometric balance between the cross-linker [multifunctional monomer (MFM)] and the CTA is the most important parameter that determines whether the polymer remains soluble throughout the reaction or gels and forms a cross-linked network. Thus, hyperbranched poly(methyl methacrylate) has been synthesized using a facile one-step batch solution polymerization reaction, which is essentially a linear polymerization doped with appropriate amounts of MFM and CTA [188].

2.14.1 Applications

Among the various applications suggested for hyperbranched polymers are surface modification, additives, tougheners for epoxy-based composites, coatings, and medicines.

It has been demonstrated that hydrophobic, fluorinated, hyperbranched poly(acrylic acid) films can passivate and block electrochemical reactions on metal surfaces thus preventing surface corrosion [189]. Hyperbranched films can be synthesized on self-assembling monolayers on the metal surface via sequential grafting reactions to obtain thick and homogeneous films.

The lack of mechanical strength makes hyperbranched polymers more suitable as additives in thermoplast applications. Hyperbranched polyphenylenes, for example, have been shown to act successfully as rheology modifiers when processing linear thermoplastics. A small amount added to polystyrene thus results in reduced melt viscosity [190].

The use of epoxidized hyperbranched polyesters as toughening additives in carbon-fiber reinforced epoxy composites has been demonstrated [191]. The use of hyperbranched polymers as the base for various coating resins has been described in the literature. For example, a comparative study [192] between an alkyd resin based on a hyperbranched aliphatic polyester and a conventional (less branched) high-solid alkyd showed that the former had a substantially lower viscosity and much shorter drying time than the conventional resin of comparable molecular weight.

An important application of dendritic polymers being explored in medicine is in advanced drug delivery systems. However, most applications within this field, described in the literature, deal with dendrimers and not with hyperbranched polymers. In a study on the effect of dendrimer size when used inside the human body, it was found [193] that large dendrimers (M_w ca. 87,000) were excreted into the urine within two days, whereas smaller dendrimers (M_w ca. 5,000) accumulated mostly in the liver, kidney and spleen with no urine excretion. Hyperbranched polymers, being mostly polydisperse, are thus unsuitable in vivo applications.

A special feature of dendritic polymers is the possibility to combine an interior structure having one polarity with a shell (end groups) having another polarity, e.g., a hydrophobic inner structure and hydrophilic end groups. Thus hyperbranched polyphenylenes with (anionic) carboxylate end groups have been described [194], where the carboxylate end groups make the polymer water soluble while the hydrophobic interior hosts non-polar guest molecules. In another example [195], hyperbranched aromatic poly(ether ketone)s having acid end groups have been used to solubilize hydrophobic molecules in water. In such a case, a critical micellar concentration (CMC) is not observed and instead there occurs a steady increase in solubility of the hydrophobic compound with polymer concentration. Such dendritic polymers have been described as unimolecular micelles. In a recent review [196], the guest-host possibility is described for various dendritic polymers considered suitable for medical applications such as drug delivery.

Hyperbranched polymers allow a wide range of variation in properties, which depend on several parameters, the most important however being the backbone and end-group structure in combination. The glass transition temperature, for example, can be shifted 100°C simply by changing the polarity of the end groups, while keeping the backbone structure unchanged.

Though most hyperbranched polymers are considered to be amorphous, some examples of crystalline and liquid crystalline hyperbranched polymers have been described in the literature. The possibility of crystallinity has further expanded the application potential of these polymers. In some cases, in application as toughening additives, the polarity of the hyperbranched polymer relative to the thermosetting matrix resin can be adjusted to give a reaction-induced phase separation in composites, resulting in a dramatic increase in toughness but still retaining the overall good mechanical properties (such as high modulus) of the system. This has been demonstrated for hyperbranched aliphatic polyesters added to epoxy/amine thermoset system. Use as toughening additives for composites represents a successful application of hyperbranched polymers.

Self-assembly, which is a supramolecular approach relying on complementary non-covalent interactions, such as electrostatic interactions and hydrogen bonds, is an incredibly powerful concept in modern molecular science. It offers an attractive option by which small, synthetically accessible, relatively inexpensive dendritic systems (dendrons) can be simply assembled into highly branched complex nanoscale assemblies having wide range of novel properties, that depend on the reversibility and specificity of the assembly process as well as on the branching inherent within the dendritic building blocks. It has been shown [197] that such self-assemblies of dendritic building blocks in solution have potential applications in diverse areas, including controlled release, nanoscale electronics, gel-phase and liquid crystalline materials, and biotechnology.

Dendronized polymers, i.e., polymers with dendritic side chains, are currently under intense investigation with respect to various applications, including the synthesis of hierarchically structured materials, catalysis, applications in the biosciences, such as ion channel mimics and DNA compactization, as well as photoelectronics applications [198].

Though numerous applications have been suggested for hyperbranched polymers, few have reached the stage of full commercial exploitation. Hyperbranched polymers is, however, a young and growing area in the domain of macromolecules. The special properties of these polymers are now receiving greater attention and a number of interesting applications of the hyperbranched polymers are expected to bring them to the market place. The future thus looks bright for these novel materials.

2.15 Shape Memory Polymers

When a polymer is deformed at a temperature above its glass transition temperature (T_g) and quickly cooled to a temperature below T_g, the deformed shape becomes frozen. An ordinary polymer does not restore its original shape when the temperature is again raised above T_g. But a shape memory polymer (SMP) is known to recover the original shape when it is warmed above T_g. Because of this property, SMPs have drawn wide attention from various fields during the past decade. Figure 2.79 illustrates typical shape memory behavior of a commercially available SMP product. As functional polymers they find applications in a broad range of temperature sensing elements.

SMPs basically consist of two phases [199], viz., frozen phase or fixed points and reversible phase. Thus, shape memory effects have been observed with such polymers as *trans*-polyisoprene (TPI), styrene-butadiene polymer (SB) and segmented polyurethanes (PUs). In these materials, the crystalline soft domains (e.g., crystalline phase of TPI, crystalline polybutadiene segments of SB, and crystalline soft segments of PU) form the reversible phase, with their crystalline melting temperature being the shape recovery temperature (T_s), and hard domains (e.g., cross-links in TPI, styrene blocks in SB, and hard segments of PU) become the fixed points or frozen phases. During the second shaping process, which normally is done at a temperature higher than T_s, frozen phases remain intact. The reversible phases soften above T_s and harden when cooled below T_s.

Basic principles of shape memory behavior of polymers can be best described in terms of their modulus (E)-temperature (T) behavior, as shown in Figure 2.80, where T_s is the crystalline melting temperature of soft segment and T_h is the softening-hardening transition temperature of fixed phase, while T_l and T_u are the typical loading and unloading temperatures, respectively. During the primary processing, such as injection molding, the materials are heated above T_h, where the previous memories are completely erased. In the mold, as the temperature decreases below T_h, fixed phases emerge and the formation is completed at T_s. As the specimen is cooled further below T_s, soft segments crystallize and the material is frozen to its glassy state. The shape of this molded specimen is referred to as the "original shape" for the shape memory experiment.

In the second shaping process (e.g., extension, compression, and transfer molding), if the shape of the specimen is changed by deforming at a temperature T such that $T_s < T < T_h$, this new shape becomes fixed on subsequent cooling below T_s under constant strain. Thereafter on warming to a temperature above T_s, the specimen returns to the original shape giving an exhibition of shape memory of the polymer. The driving force of the shape recovery is the elastic strain generated during the deformation.

A high glassy state modulus (E_g) provides the material with high shape fixity during simultaneous cooling and unloading, while a sharp transition from glassy state to rubbery state makes the material sensitive to temperature variation. Thus, a high elasticity ratio (E_g/E_r)—preferably, a difference of two orders of magnitude—allows easy shaping at $T > T_s$ and great resistance to deformation at $T < T_s$. To satisfy all these properties, systematic designs should be carried out to relate the structural varieties with the properties. While the stress-strain-temperature behavior of the material, as explained above, is of basic importance with regard to shape fixing and shape memory effect evaluations, the shape memory element is often subjected to cyclic deformation, such as for an actuator. Hence the cyclic characteristics are also of practical importance in evaluating the durability of the shape memory element.

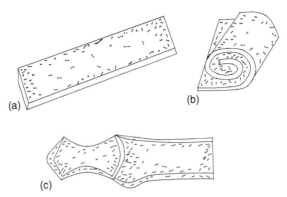

FIGURE 2.79 Illustration of shape memory behavior of a commercial shape memory polymer Veriflex™ (CRG Industries). (a) Coupon of Veriflex having a rectangular memory shape. When heated above its transition temperature, it becomes elastic and can be manipulated into different shapes such as (b) and (c) and then cooled to maintain the new shape in a rigid state. When reheated above its transition temperature, it returns to its memorized shape.

As compared to metallic compounds used as shape memory materials, shape memory polymers have low density, high shape recoverability, easy processability, and low cost. Since the discovery by Mitsubishi in 1988, polyurethane SMPs have attracted a great deal of attention due to their unique properties, such as a wide range of shape recovery temperatures ($-30°C$ to $70°C$) and excellent biocompatibility, besides the usual advantages of plastics. A series of shape memory polyurethanes (SPMUs), prepared from polycaprolactone diols (PCL), 1,4-butanediol (BDO) (chain extender), and 4,4'-diphenylmethane diisocyanate (MDI) or toluene diisocyanate (TDI) have recently been introduced [200–202].

SMPUs are basically block copolymers of soft segments, which are polyols, and hard segments built from diisocyantes and chain extenders (see Figure 1.31). Depending on the types and compositions of soft and hard segments, and preparation procedures, the structure-property relationships of SMPUs are extremely diverse and easily controlled, and hence shape recovery temperature can be set at any temperature between $-30°C$ and $70°C$, allowing a broad range of applications. They can be molded using conventional processing techniques, including extrusion, injection and blow molding, which allow versatility of shaping. In the future, the shape memory effect can be applied to areas like smart fabrics that can control moisture permeability or smart materials with damping capability.

Until the present time, most SMPUs have been prepared from linear PUs, which have physically cross-linked segments. However, these linear SMPUs cannot endure repeated changes in shape memory. In fact, some studies have found that the shape retention and shape recovery of SMPUs decrease dramatically after the first cycle [200]. To improve cyclic shape memory retention and shape memory recovery, SMPUs have therefore been chemically cross-linked by excess MDI or by glycerin [203].

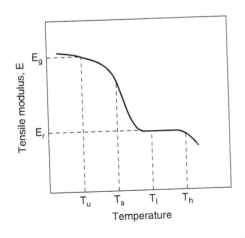

FIGURE 2.80 Typical modulus-temperature curve of a shape memory polymer. (After Kim, B. K., Lee, S. Y., and Xu, M. 1996. *Polymer*, 37, 5781. With permission.)

2.16 Microencapsulation

Microencapsulation [204–207] is a process by which small solid particles, liquid droplets, or gas bubbles are enveloped within a second (coating) material for the purpose of shielding them from the surrounding environment and to release at need. Microcapsules range in size between 1 and 1000 µm. (Capsules greater than 1000 µm can be called macrocapsules, and those smaller than 1 µm are termed nanocapsules.) The coating material used to form the capsule may be an organic polymer, hydrocolloid, sugar, wax, fat, or inorganic oxide. Capsules release their contents at a later time by a mechanism appropriate to the application. The four typical mechanisms by which the encapsulated material is released are mechanical rupture of the capsule wall, dissolution of the wall, melting of the wall, and diffusion through the wall [208]. Less common release mechanisms include ablation (slow erosion of the shell) and biodegradation.

Two well-known examples of microencapsulated products that depend on shell rupture to release the core contents are the scratch-and-sniff perfume advertisements and the carbonless copy paper. The former have tiny perfume-filled microcapsules coated onto the magazine page so that, when scratched, the shell wall ruptures, releasing the perfume. In the case of carbonless copy paper (see Figure 2.81), the underside of the top sheet is coated with small capsules (1–20 µm diameter) containing a dye precursor (2%–6% solution of a leuco dye in a high boiling organic solvent) which is colorless, but darkens when in contact with an acidic component (such as attapulgite clay or phenolic resin). The top of the lower sheet is coated with this acidic component so that when the capsules break due to high local pressure beneath a pen point in writing on the top sheet, the leuco dye react with the acid sites to give an image on the lower sheet. Initially, a mixture of crystal violet lactone and *N*-benzoyl leucomethylene blue was used as the leuco dye. Subsequently, fluoran and phthalide leuco dyes, as also metal chelate color systems were developed. Solvents used as dye carriers are high boiling (>200°C) organic liquids which include benzylated ethylbenzene, benzyl butyl phthalate, isopropylbisphenyl, and diisopropylnaphthalene.

Microencapsulation has been applied widely in the detergent industry. Some powder detergents contain protein reactive enzymes such as protease, encapsulated in a water soluble polymer such as

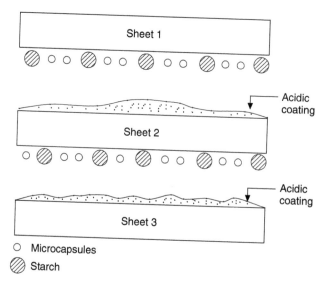

FIGURE 2.81 Cross-section of a three-part business form made with carbonless paper. (After Thies, C. 1989. *Biomaterials and Medical Applications*, Encyclopedia Reprint Series, J. I. Kroschwitz, ed., pp. 346–367. John Wiley, New York.)

polyethylene glycol. Released upon shell dissolution in the washing machine, the enzymes help remove blood stains by attacking the blood protein.

Encapsulated ingredients are used in many packaged baking mixes to delay chemical reactions until proper temperatures are reached. Sodium carbonate is a baking ingredient as it reacts with food acids to produce leavening agents. It is encapsulated in a fat that is solid at room temperature but melts at a temperature of about 125°F. This delays the release of the core material until the proper temperature is reached.

Microencapsulation is a common means of providing sustained release of a medication and is frequently used in the pharmaceutical industry for this purpose. For example, aspirin which provides effective relief for fever, inflammation, and arthritis, can cause peptic ulcers and bleeding in direct doses. Therefore, it is sometimes encapsulated in ethyl cellulose or hydroxypropyl methylcellulose and starch; tablets are produced by pressing together these microcapsules. Instead of being released all at once, the encapsulated aspirin diffuses through the shell producing a slow and sustained release.

A wide range of materials in addition to those mentioned above have been encapsulated. These include pharmaceuticals, vitamins, living cells, catalysts, flavor oils, agrochemicals, adhesives, and water. The advantages of encapsulation are many. Thus unpleasant odor or taste can be effectively masked in a food product, toxic products can be safely handled, sensitive materials can be protected, liquids can be handled as solids, and drug delivery can be controlled and targeted. Microencapsulation has thus found increasing applications in many areas. Usually organic polymers are employed as shell materials. Waxes and fats also find use, especially in food and drug applications for which the U.S. Food and Drug Administration specifications are to be met.

2.16.1 Processes for Microencapsulation

Many processes for making microencapsulated product have been reported in the literature. They encompass a broad range of scientific and engineering disciplines. Some are used in high volume applications as in carbonless copy paper, others are in low volume specialty applications; many are used in pilot plant scale.

2.16.1.1 Complex Coacervation

Conceived in the 1930s by Barrett Green at the National Cash Register Corporation, complex coacervation was the first process used to make microcapsules for carbonless copy paper. The process occurs in aqueous media and is used to encapsulate water-immiscible liquids or water-insoluble solid. The substance (liquid or solid) to be encapsulated is first dispersed as tiny droplets or particles in an aqueous solution of a polymer such as gelatin by mechanical agitation. A second water soluble polymer, such as gum arabic, is then added to this emulsion and mixed. As the pH is adjusted to 4.0–4.4 by the addition of acetic acid, there occurs spontaneous formation of two incompatible liquid phases—one phase, called the coacervate, having relatively high concentrations and the other phase, called the supernatant, having low concentrations of the two polymers. If the materials are properly chosen, a complex coacervate preferentially adsorbs on the dispersed droplets (or particles of water-insoluble substance) to form shell cover of microcapsules. The shells are usually hardened, first by cooling to 5°C, and then by cross-linking with formaldehyde or glutaraldehyde. The capsules are isolated and dried, or used directly as an aqueous dispersion. Since with aldehyde cross-linking the shells remain hydrophilic and hence prone to swelling in water, drying may lead to aggregation. Therefore, to reduce hydrophilicity, such capsules are treated with urea-formaldehyde at a low pH (~ 2).

Any pair of oppositely charged polyelectrolytes capable of forming complex coacervate can be used in the above process of encapsulation by coacervation. Gelatin, a positively charged polyion, forms complex coacervates with a number of polyanions, including alginate, polyphosphate, carrageenan and ethylene-maleic acid copolymers. A wide variety of microcapsules containing water-insoluble or water-immiscible materials can be made by complex coacervation. The capsule size ranges from 5 to 1000 μm with typical loading of 80%–95%.

2.16.1.2   Polymer–Polymer Incompatibility

Microencapsulation based on polymer-polymer incompatibility utilizes the phenomenon that when two chemically different polymers are dissolved in a common solvent (usually organic) they separate spontaneously into two liquid phases with each phase containing one of the polymers predominantly. When a substance (active agent to be encapsulated) which is insoluble in the solvent is dispersed in such a system it is engulfed by the polymer in one of the two phases due to preferential adsorption, resulting in the formation of embryo microcapsules. The coating polymer is then insolubilized by chemical cross-linking or addition of a nonsolvent and microcapsules are separated. The active agents encapsulated are typically polar solids possessing some degree of water solubility and ethyl cellulose is the most commonly used coating polymer. Aggregation of capsules during isolation is often encountered in this type of process.

2.16.1.3   Interfacial and In Situ Polymerization

The first step in all interfacial polymerization processes for encapsulation is to form an emulsion. This is followed by initiation of a polymerization process to form the capsule wall. Most commercial products based on interfacial or in situ polymerization employ water-immiscible liquids. For encapsulation of a water-immiscible oil, an oil-in-water emulsion is first formed. Four processes are schematically illustrated in Figure 2.82. In Figure 2.82(a), reactants in two immiscible phases react at the interface forming the polymer capsule wall. For example, to encapsulate a water-immiscible solvent, multi-functional acid chlorides or isocyanates are dissolved in the solvent and the solution is dispersed in water with the aid of a polymeric emulsifier, e.g., poly(vinyl alcohol). When a polyfunctional water-soluble amine is then added with stirring to the aqueous phase, it diffuses to the solvent-water interface where it reacts with acid chlorides or isocyanates forming the insoluble polymer capsule wall. Normally some reactants with more than two functional groups are used to minimize aggregation due to the formation of sticky walls.

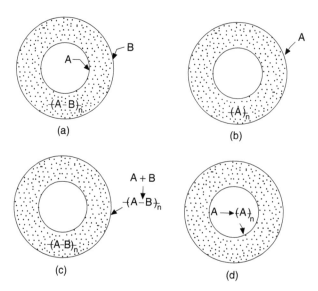

FIGURE 2.82   Microcapsule formation by interfacial and in situ polymerization. A and B are reactants, while—$(A–B)_n$—and—$(A)_n$—are polymeric products. See text for explanation. (After Thies, C. 1989. *Biomaterials and Medical Applications*, Encyclopedia Reprint Series, J. I. Kroschwitz, ed., pp. 346–367. John Wiley, New York.)

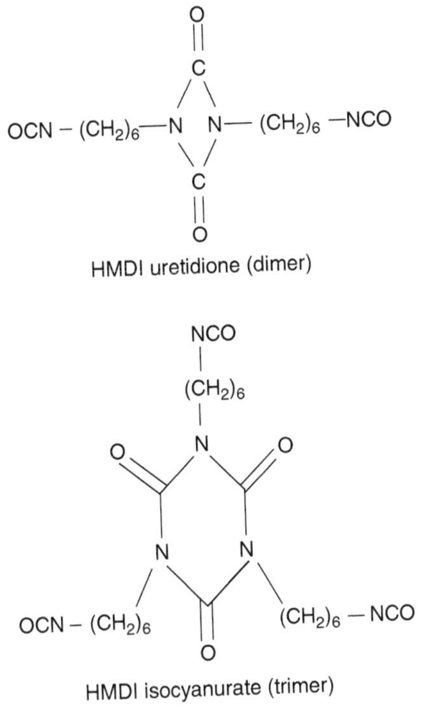

FIGURE 2.83 Chemical structures of HMDI ureti-dione and HMDI isocyanurate. (After Takahashi, T., Taguchi, Y., and Tanaka, M. 2005. *J. Chem. Eng. Japan*, 38(11), 929. With permission.)

Acid chlorides commonly used include sebacoyl chloride, terephthaloyl chloride, and trimesoyl chloride. The last named acid chloride has three functional groups and hence serves as a cross-linking agent. Among the various isocyanates used, mention may be made of polymeric isocyanates based on p,p'-diphenylmethane diisocyanate and other isocyanate adducts based on toluene diisocyanate and hexamethylene diisocyanate. Amines, such as ethylenediamine, hexamethylene-diamine, and triethylenetetramine are added to the aqueous phase, the last named amine being a cross-linking agent. A base is added to serve as an acid scavenger if acid chlorides are used.

Microencapsulation by interfacial polyconden-sation is a useful method to microencapsulate a liquid core material. Especially, polyurea and polyurethane microcapsules have been extensively investigated in various industries [209]. For example, aliphatic hexamethylene diisocyanate (HMDI) and aliphatic ethylene diamine (EDA) have been used to prepare polyurea microcapsules containing insecticide called diazinon [210]. A urea linkage is formed immediately by the reaction between an amine and an isocyanate group (see Figure 1.31), and a polyurea is synthesized by the reaction between an amine with two or more amine groups and an isocyanate with two or more isocyanate groups.

In the presence of a basic catalyst, HMDI can dimerize to HMDI uretidione and trimerize to HMDI isocyanurate (Figure 2.83). Polyurea microcapsules prepared from HMDI uretidione or isocyanurate (by reacting with EDA) are stronger than those obtained from HMDI. Moreover, HMDI uretidione and isocyanurate do not have bad smell like the HMDI monomer. So these may be preferred for the industrial application. The preparation of polyurea microcapsules containing a pyrethroid insecticide by the reaction between HMDI uretidione/isocyanurate and EDA has been reported [211]. In a typical method of preparation, as shown in Figure 2.84, given amounts of insecticide and the mixture of HMDI uretidione/isocyanurate (Desmodur N3400) are mixed and dispersed in water with poly(vinyl alcohol) as a stabilizer (4 wt%) to form oil-in-water (O/W) emulsion to which a given amount of EDA aqueous solution is then added with agitation to facilitate interfacial polymerization. Single cored microcapsules, 0.2–6.0 μm in size, have been obtained by this method in about 98% yield [211].

Figure 2.82(b) represents capsule wall formation by direct polymerization of a monomer (A) such as *n*-alkyl cyanoacrylate at the water-solvent interface. In this case, water is dispersed in a water-immiscible solvent with the aid of an emulsifier and *n*-alkyl cyanoacrylate is added to the solvent phase from where it diffuses to the solvent-water interface and polymerizes to poly(*n*-alkyl cyanoacrylate), forming capsule wall membrane.

In a variation of the above process, a water soluble monomer A is adsorbed on water-immiscible solvent droplets dispersed in the aqueous solution of A and then polymerized to form closed shell structure. For example, cross-linked polysaccharide capsules with diameters ranging from 200 nm to several microns and wall thicknesses of several tens of nanometers have been fabricated by interfacial polymerization of methacrylated *N,N*-diethylaminoethyl dextran (DdexMA) [212]. In a typical method of preparation of DdexMA (see Figure 2.85), a mixture of *N,N*-diethylaminoethyl dextran (Ddex),

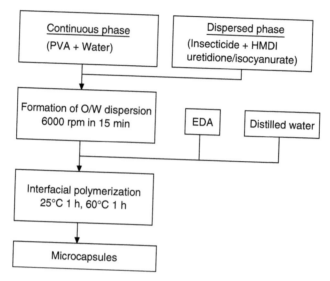

FIGURE 2.84   Flow chart of the process of preparation of polyurea microcapsules. (After Takahashi, T., Taguchi, Y., and Tanaka, M. 2005. *J. Chem. Eng. Japan*, 38(11), 929. With permission.)

methacrylic acid (MAA), and water soluble carbodiimide (1-ethyl-3-dimethylaminopropyl carbodiimide, EDAC) is incubated at 25°C in phosphate buffered saline (pH 7.4) for 24 h. After filtration the synthesized macromonomers are dialyzed in distilled water to remove the excess MAA and EDAC.

In a typical method of encapsulation (see Figure 2.86), droplets of chloroform or chloroform solution of the drug to be encapsulated are dispersed in water containing DdexMA with the help of an emulsifier (e.g., Triton X-100). The polymerization of DdexMA is initiated by the addition of redox initiator $K_2S_2O_8/NaHSO_3$ under nitrogen atmosphere at 40°C. After polymerizing for 8 h, chloroform is

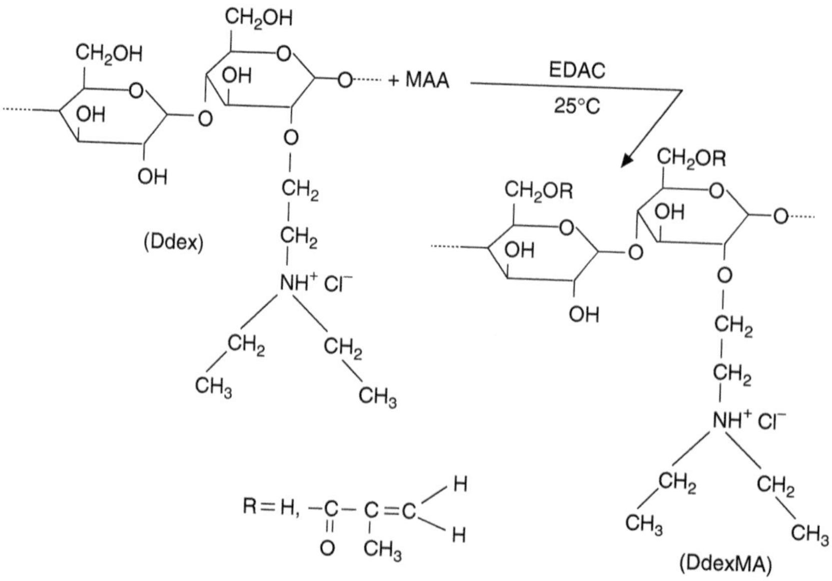

FIGURE 2.85   Synthesis of methacrylated *N,N*-diethylaminoethyl dextran (DdexMA) (see text for description). (After Jiang, B., Hu, L., Gao, C., and Shen, J. 2006. *Acta Biomaterialia*, 2, 9. With permission.)

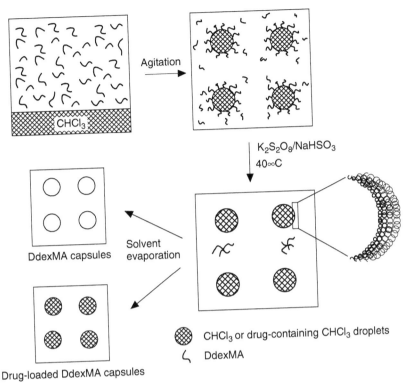

FIGURE 2.86 Schematic description of the process of capsule formation by interfacial polymerization of DdexMA (see text). (After Jiang, B., Hu, L., Gao, C., and Shen, J. 2006. *Acta Biomaterialia*, 2, 9. With permission.)

evaporated to obtain empty DdexMA capsules or drug-loaded DdexMA capsules. The loaded drug can be released from the capsules in a sustained manner.

In the encapsulation process depicted in Figure 2.82(c), polymerization is initiated in the water phase of oil-in-water emulsion. As the molecular size of the polymer increases, it deposits at the water-oil interface where it continues to grow forming a cross-linked polymer capsule wall. In a typical example, methylol urea or methylolmelamine is added to the aqueous phase along with an ionic polymer. The pH is adjusted to 3.5–4.5 and the mixture allowed to react for 1–3 h at 50–60°C. The ionic polymer in the aqueous phase assists deposition of the aminoplastic at the water-oil interface.

The process represented in Figure 2.82(d) is similar to that in Figure 2.82(c), except that polymerization is initiated in the water-immiscible solvent phase. A vinyl monomer, e.g., styrene, methyl methacrylate, or vinyl acetate, is dissolved in a water-immiscible solvent together with an initiator. The solution is emulsified in water using an emulsifier and heated to initiate free-radical polymerization. The resulting polymer deposits at the solvent-water interface forming the capsule wall.

2.16.1.4 Spray Drying

Spray-dry encapsulation is a low-cost process capable of producing a range of microcapsules in good yield. It is the earliest commercial encapsulation process and has been improved considerably over the years as a means of forming soluble microcapsules, with active agent payloads up to 60 wt%, though 20–25 wt% is more common. Gum arabic, starch derivatives, maltodextrins, hydrolyzed gelatin, and mixtures of these materials with sucrose or sorbitol are commonly used as the water soluble carrier (wall material) for microcapsules, especially those containing food derivatives. While gum arabic produces spray-dried capsules that are dispersible in cold water, hydrolyzed gelatin yields capsules that are soluble in hot as well as cold water.

In the process, the active agent (usually water-immiscible liquid or oil) is dispersed in aqueous solution of the carrier material with the aid of a suitable emulsifier and the resulting oil-in-water emulsion is sprayed into a heated drying chamber where they remain for 30 s or less. The product of this operation is a dry microcapsule powder consisting of small, spherical particles with a typical diameter of 10–40 μm in which the oil phase is dispersed throughout as 1–3 μm size droplets. A small amount of the active agent in the powder, however, remains free or unencapsulated; it can only be minimized but not totally eliminated.

Various waxes, fatty alcohols, fats and fatty acids are also used as coating (carrier) material. In these cases, the active agent is dispersed in the molten carrier (using an emulsifier, if necessary) and the dispersion atomized through heated nozzles into a cooling chamber where the carrier solidifies to form the coating.

2.16.1.5 Fluidized-Bed Coating

Fluidized-bed coating is more suitable for encapsulation of solid particles. It can also be used for liquids, if they are first absorbed on a porous solid. The solid particles to be coated are placed in a coating chamber and suspended by an air stream in a cyclic flow (Figure 2.87) past a nozzle (at the bottom of the chamber) which sprays the liquid coating phase onto the particles. Carried up into the chamber by the air stream, the coated particles dry by evaporation of the solvent, fall back to the chamber bottom to be carried again by the air stream past the spray nozzle and up into the coating chamber. The cycle is repeated until the coating attains the desired thickness.

Though submicron solid particles can be coated by this process, the particles agglomerate to form capsules of 50 μm or more. For example, in the Wurster process of fluid-bed coating [213], the production is limited to finished particle sizes of at least 150 μm diameter. A large variety of coating materials can be used including hydrocolloids, solvent or water-soluble polymers, sugars, waxes, and fats, applied as solutions, dispersions, or hot melts. Increasing emphasis on reduction of volatile emissions has, however, led to the preference for aqueous dispersions of polymers such as ethyl cellulose and acrylic resins.

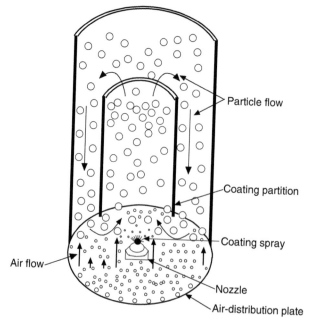

FIGURE 2.87 Schematic of fluidized bed coating chamber (Wurster process). (After Thies, C. 1989. *Biomaterials and Medical Applications*, Encyclopedia Reprint Series, J. I. Kroschwitz, ed., pp. 346–367. John Wiley, New York.)

Coated fertilizers are physically prepared from granules of conventional fertilizers coated with materials that reduce their dissolution rate. Various materials are found to be suitable for coating purposes. The most important of these include wax and sulfur and organic polymers such as polyethylene [214] and polyacrylamide [215]. In a typical process described for coating urea [214], urea prills (size range 0.5–2 mm) are fluidized and sprayed with LDPE solution (10 wt%) in toluene using a dosing pump and atomizing air, followed by drying. The solvent vapors are collected by a tubular condenser. The dissolution of urea in water decreases with increasing coating percentage. This is due to the reduction in the number of pinholes which are the primary path for urea release from LDPE-coated urea.

2.16.1.6 Co-Extrusion Capsule Formation

In this novel encapsulation process, liquid core and shell materials are pumped through concentric orifices in the form of a fluid jet, with the core material flowing through the central orifice and the shell material through the outer annulus. The fluid jet breaks up forming compound drops because of disturbances on the jet surface, the size distribution of the compound drops being thus related to the frequencies of these disturbances. As the compound drop is composed of core fluid encased by a layer of shell fluid, capsules are formed by hardening the shell by appropriate means, such as chemical cross-linking in the case of polymers, solvent evaporation in the case of solution, and cooling in the case of fats or waxes. The size of the capsules formed and the quantity of the core material contained within each capsule depend, in a complex way, on a host of property and process variables, such as density, viscosity, interfacial tension, flow rate and temperature of the fluids, diameters of the inner and outer orifices of the nozzle, and the amplitude and frequency of the vibrational disturbances imparted to the fluid jet. A novel machine vision-based control system (Figure 2.88) under development at the Southwest Research

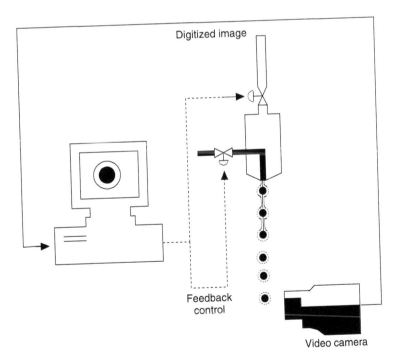

FIGURE 2.88 Schematic of a co-extrusion process of microencapsulation being developed at Southwest Research Institute (SWRI) using machine vision-based control system that allows inflight capsule inspection. The desired capsule size and sphericity are obtained by manipulation of process input variables, such as flow rate, temperature, vibrational frequency, and amplitude via a feedback control loop. (After Franjione, J. and Vasishtha, N. 1995. *Technology Today*, Southwest Research Institute Publications.)

Institute (San Antonio, Texas) allows inflight capsule inspection, while the process input variables including flow rate, temperature, vibrational frequency and amplitude are manipulated via a feedback control loop to obtain desired capsule size and sphericity [207]. Due to the involvement of so many factors and their complex interrelationships, co-extrusion processes are designed and operating conditions determined usually on a case-by-case basis.

2.16.1.7 Other Processes

In addition to the processes described above, there are many other encapsulation processes [Thies, 204] which are in use or are under various stages of development. These include phase inversion in polymer solution, pan coating, solvent evaporation, gelation, electrostatic encapsulation, and vapor deposition.

The phase inversion technique (wet method), in which a polymer is precipitated from its solution with a nonsolvent, can be used to form coatings on solid particles. This technique has been used to encapsulate fertilizers to reduce fertilizer losses and to minimize environmental pollution. For example, polysulfone has been used as a coating for soluble NPK granular fertilizer in controlled-release formulations. In a typical procedure [216], the granular (2–5 mm) fertilizer is added to a 10%–20% solution of polysulfone in *N*,*N*-dimethylformamide (DMF) and the granules covered with a polymer solution layer are subsequently dropped into a water bath, where the gelation process takes place (the time of the gelation being about 5 s). The coated granules are separated from the water after a maximum of 1 min and then dried in a drier (104°C) to a constant mass. A double coating can be achieved by immersion of the single-coated fertilizer into an adequate polymer solution followed by precipitation in water and drying.

The polymer concentration in the film-forming solution has influence on the physical properties (porosity) of the coatings and the release rate of nutrients from coated granules. Thus for polysulfone-coated NPK fertilizer with coating having 38.5% porosity (prepared from 13.5% polymer solution) 100% of NH_4 was released after 5 h test, whereas only 19.0% of NH_4 was released after 5 h for the coating with 11% porosity [216].

In the pan coating process, a coating material is sprayed onto a particulate ($\geqq 800$ μm) mass tumbled in a mixer, forming coated particles. The process is used by the pharmaceutical industry to produce controlled-release products.

In solvent evaporation, the active agent is dispersed or dissolved in a water-immiscible and volatile solvent (such as methylene chloride), and the mixture is dispersed in water allowing evaporation of the solvent to produce microspheres that can be isolated. The technique has been used in the pharmaceutical industry to produce injectable microspheres.

The gelation technology employs chemical interactions to cause liquid droplets to gel, forming microcapsules or microspheres. This technique is used by the pharmaceutical industry to encapsulate active agents and also to immobilize live cells and organisms. In one process, live cells are first entrapped in gel matrix beads produced by the reaction of sodium alginate with calcium ions. The outer layer of the beads is then hardened by treatment with a polycation to form a polyelectrolyte complex, while the interior of the beads is solubilized by treating with sodium nitrate to form a soluble complex.

In the electrostatic encapsulation process, a core material and an immiscible liquid coating phase are converted into oppositely charged aerosols so that the core phase (having higher surface tension) is surrounded by a shell of the coating phase, which is then hardened by a suitable means, forming small microcapsules.

Coating by vapor deposition has been used to form microcapsules, the coating being done on solid particles. The latter may be droplets of frozen liquid or liquid encapsulated by some other process and the vapor deposition may be caused by pyrolysis of di-*p*-xylylene in vacuuo [see Poly(*p*-xylylene) in this Chapter]; the reactive *p*-xylylene radical formed by the pyrolysis polymerizes on the solid particles to be encapsulated, forming a thin coating of poly(*p*-xylylene).

2.16.2 Applications

By far the single largest application of microcapsules is in carbonless copy paper used to make multipart business forms. The principle on which the carbonless paper is based is illustrated by the three-part business form depicted in Figure 2.81, which shows the bottom and/or top faces of the sheets coated with a layer of microcapsules and the acidic component, respectively. The capsules and the acidic material may also be coated on the same side and the product is then called self-contained carbonless copy paper.

Small size microcapsules on the paper give sharper image, but are less easily broken by a normal writing instrument. In practice, the upper size limit is ca 20 μm and the lower ca 1–2 μm. Multipart business forms contain up to seven sheets, i.e., one original and up to six copies. For optimum color formation in such a form, sizes of 3–6 μm are recommended.

Another application of microencapsulation in graphic arts is the Sanders light-sensitive imaging system, illustrated in Figure 2.89. Here the top of a paper sheet is coated with microcapsules containing a light-sensitive monomer, a photoinitiator, and a leuco dye. As a pattern or object is imaged on the sheet by a light source, the monomer in those capsules that are exposed to light polymerizes, causing the capsules to solidify (harden), while the liquid in the unexposed capsules remain unchanged. The exposed sheet next passes through a set of rollers whereupon the unexposed capsules collapse and release the dye, thereby forming an image. Color images result from the use of three different dye precursors and three different sensitivities in three different capsules. This technology can be used to reproduce color pictures and documents.

Microencapsulation is common in pharmaceutical industry, particularly when sustained release of a medication is required. Ethyl cellulose is a common coating material. Most capsules are formed by solvent evaporation, polymer-polymer phase separation, or fluidized-bed coating process. Common examples of encapsulated drugs include aspirin, acetaminophen, ampicillin, and potassium chloride. Orally administered capsules serve to conceal an unpleasant taste and reduce gastrointestinal irritation that can be caused by oral unencapsulated drug.

Prolonged release formulations injected intramuscularly, subcutaneously or intravenously and targeted to specific body sites are based on small microcapsules with polylactide or lactide-glycolide copolymers used as wall materials. Some examples of formulations, including those at various stages of clinical trials, are microencapsulated fertility-control drug, luteinizing hormone-releasing analogue, and chemotherapeutic agents.

A number of food additives and ingredients are available in encapsulated form, which include active agents such as citrous flavors, citric acid, ascorbic acid, spice extracts, and vegetable extracts. Encapsulation serves to enhance stability, reduce loss by volatilization and oxidative degradation, increase shelf-life, and impart better handling properties. Water-soluble carriers in common use for food additives are gum arabic, modified starches, maltodextrins and hydrolyzed gelatins. Encapsulation

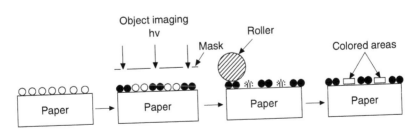

○ Unexposed microcapsule (soft)

● Exposed microcapsule (solidified)

FIGURE 2.89 Mechanism of Sanders imaging system. (After Thies, C. 1989. *Biomaterials and Medical Applications*, Encyclopedia Reprint Series, J. I. Kroschwitz, ed., pp. 346–367. John Wiley, New York.)

is done by extrusion, for which coating formulations are prepared as concentrated solutions. Capsules obtained by extrusion are larger than those obtained by spray drying. Treatment of the capsules with cold and hot water is required for the release of the active ingredients.

Common water-insoluble carriers for food ingredients are low-melting, hydrogenated vegetable oils, e.g., cottonseed or soybean oil. Solid ingredients are coated by the Wurster air suspension process, while liquid ingredients are encapsulated by spray chilling or spray cooling. The capsules are heated above the melting point of the carrier to release the active agents.

Encapsulation of pesticides and herbicides serves to prolong effectiveness and reduce oral toxicity, dermal toxicity, and phytotoxicity, besides concealing any objectionable odor. There are many commercial examples of encapsulated pesticides and herbicides, some of which are methyl parathion, diazinon (highly effective against cockroaches), thiocarbamate herbicides, fonofos (soil insecticide), methoprene (mosquito-growth regulator), pyrethins (effective against crawling insects, e.g., cockroaches), and warfarin (rodenticide). All these products are encapsulated by interfacial or in situ polymerization.

2.17   Polymer Nanocomposites

Nanotechnology is one of the evolving and promising technologies for the new millennium. It is believed to harbor the potential to become a key technology, especially since two different promising major approaches can be observed in the field—first, miniaturization beyond the micrometer (10^{-6} m) size to nano (10^{-9} m) size (which is about 10,000 times finer than a human hair) and second, the exploitation of new effects that arise from nanostructured materials.

A nanocomposite material can be defined as one that consists of two or more different material components, at least one of which has a domension (i.e., length, width, or thickness) below 100 nm. There are many types of nanocomposites presently under research and development including polymer/inorganic particle, polymer/polymer, metal/ceramic, and inorganic-based nanocomposites. However, the first named one, commonly called polymer nanocomposite (PNC) and defined as the combination of a polymer matrix resin (continuous phase) and inclusions having at least one dimension less than 100 nm, is the only type of nanocomposite to date that has seen any significant commercial activity.

PNCs have emerged as a very efficient strategy to upgrade properties of synthetic polymers, exceeding largely the performance of the conventional fiber- or mineral-reinforced polymer composites. One additional advantage is that relatively small amounts of filler, typically 2–10 wt%, are required as a result of the nanometric scale dispersion of the filler in the matrix. They therefore avoid many of the costly and cumbersome fabrication techniques common to conventional composites. Instead they can be processed by techniques like extrusion, injection molding, and casting normally reserved for unfilled polymers. Furthermore, they are adaptable to films, fibers, as well as monoliths. Because of the small amount of filler used in PNCs, nanocomposite plastic parts offer a 25% weight savings on average over highly filled plastics.

One of the most promising PNC systems is the one based on organic polymers and inorganic clay minerals with layered structure, which belong to the general family of 2:1 layered silicates. Compared to their micro and macro counterparts and the pristine polymer matrix, polymer-clay nanocomposites (PCNs) exhibit higher tensile strength and moduli, lower thermal expansion coefficients, lower gas permeability, greater swelling resistance, enhanced ion conductivity, and greater thermal stability and flame retardance. Moreover, since the length scale involved minimizes scattering, all these favorable properties can be generated without losing transparency. For nanocomposites, therefore, one could also use the term: composites with transparent fillers.

The nanoscopic fillers, as mentioned above, have at least one characteristic length that is of the order of nanometers. Uniform dispersion of these nanoscopically sized particles or nanoelements can lead to ultra-large interfacial area between the constituents (approaching 700 m^2/cm^3 in dispersions of layered silicates in polymers) and also to ultrasmall distance between the nanoelements (approaching molecular dimensions at extremely low loadings of the nanoparticles).

Considering, for instance, a system containing 1 nm thick plates, 1μm in diameter, the distance between plates would approach 10 nm at only 7 vol% of plates [217]. The behavior of PNCs can be rationalized as follows. The proliferation of internal inorganic-polymer interfaces means the majority of polymer chains reside near an inorganic surface. Since an interface restricts the conformations that polymer molecules can adopt, and since in PNCs with only a few volume percent of dispersed nanoparticles the entire matrix polymer may be considered as nanoscopically confined interfacial polymer, the restrictions in chain conformations will alter molecular mobility, relaxation behavior, and the consequent thermal transitions such as glass transition temperature of the composites [217].

The dimensions of the added nanoelements also contribute to the characteristic properties of PNCs. Thus, when the dimensions of the particles approach the fundamental length scale of a physical property, they exhibit unique mechanical, optical and electrical properties, not observed for the macroscopic counterpart. Bulk materials comprising dispersions of these nanoelements thus display properties related to solid-state physics of the nanoscale. A list of potential nanoparticulate components includes metal, layered graphite, layered chalcogenides, metal oxide, nitride, carbide, carbon nanotubes and nanofibers. The performance of PNCs thus depends on three major attributes: nanoscopically confined matrix polymer, nanosize inorganic constituents, and nanoscale arrangement of these constituents. The current research is focused on developing tools that would enable optimum combination of these unique characteristics for best performance of PNCs.

2.17.1 Preparation of Polymer Nanocomposites

Clay minerals have a layered structure with unit layers about 1 nm thick and sizes ranging from several nanometers to several micrometers. Montmorillonite (MMT) is a clay most commonly used as nanoparticle reinforcement in PNC preparation. It is a crystalline 2:1 layered silicate clay mineral consisting of several hundred individual nanoscale layers held together by electrostatic forces. Its platelets consist of octahedral aluminum sheets sandwiched between two tetrahedral silica sheets. As there are also some magnesium cations (Mg^{2+}) replacing some of the aluminum cations (Al^{3+}), the layers have a net negative charge which is balanced by hydrated inorganic cations positioned in the spacing (galleries) between the aluminosilicate layers [217].

The two main morphologies that are found in PNCs are intercalated morphology and exfoliated morphology (see Figure 2.90). The best performances are commonly observed for the exfoliated nanocomposites. However, the two extreme situations can coexist in the same material. In the intercalated morphology, the polymer chains are located between the clay nanolayers and though the layer spacing may be increased (the amount of separation being determined by the thermodynamic interactions of the clay layer, the cations residing between these layers, and the matrix polymer), the attractive forces between the nanolayers cause them to be in regularly spaced stacks or tactoids.

On the other hand, in exfoliated morphology, individual clay layers are delaminated and dispersed, regularly or disorderly, in the polymer matrix. The objective of exfoliation in PNC fabrication, however, is to uniformly disperse and distribute the layered inorganic (initially comprising aggregates of the nanolayers) within the polymer. In a fully exfoliated morphology (true exfoliates), the individual nanolayers are uniformly (randomly) dispersed throughout the composite. The achievement of full exfoliation in PNC fabrication, i.e., transformation of an initially microscopically heterogeneous system to a nanoscopically homogeneous system, is the goal of many research activities because of the good effect it has in exhibiting new and improved performance properties of the nanocomposites.

The main techniques that can be used to prepare polymer/clay nanocomposites are: (a) melt mixing the layered clay with polymer, (b) mixing the layered clay with solution of polymer followed by solvent removal, and (c) in situ intercalative polymerization, where the monomer is first intercalated in the clay and subsequently polymerized in situ.

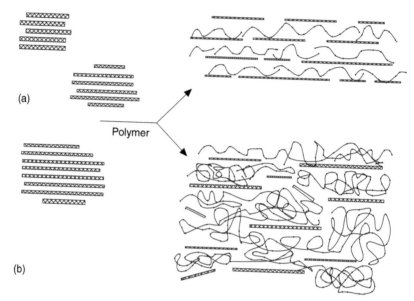

FIGURE 2.90 Schematic presentation of two extremes of composite structures that can be obtained by polymer melt intercalation of layered silicates. The rectangular bars represent individual silicate layers (1 nm thick). (a) Single polymer layers intercalated in the silicate galleries. (b) Delamination (exfoliation) of layered silicates and dispersion in a continuous polymer matrix. (After Giannelis, E. P. 1996. *Adv. Mater.*, 8(1), 29. With permission.)

Natural, unmodified montmorillonite-Na (MMT-Na) has cation exchange capacity, typically 80–90 mequiv/100 g. Although some polymers, such as polyethylene oxide or polyvinylpyrrolidone, are of sufficient polarity to be able to directly exfoliate unmodified MMT-Na, organic modification of the layered clay is usually required to render the hydrophilic surface of the clay more hydrophobic and thus more compatible with most polymers, thereby improving the wettability and dispersibility of the clay in the polymer matrix.

The polymer-clay compatibility is usually promoted by an ion-exchange reaction of the interlayer sodium cations with some cationic surfactants, most of which are alkyl amine salts or quaternary ammonium salts, such as dodecylbenzyl dimethyl alkyl ammonium (I) chloride and dimethyl 2-ethylhexyl alkyl ammonium (II) chloride, in which the alkyl group is C_nH_{2n+1} (n=14–18). Hydroxylated quaternary ammonium ions, such as methyl bis(2-hydroxyethyl) alkyl ammonium (III) have also been observed [218] to exfoliate the layered clay as easily as quaternary ammonium ions with no polar functional groups. Thus, poly(ε-caprolactam) (PCL)-based nancomposites have been prepared [219] by the method of melt intercalation by mechanically kneading PCL on a two-roll mill at 130°C with MMT organomodified by (II) or (III). The resulting PNC has interlayer spacing (called the *d*-spacing) 27–28 Å and 30–31 Å, respectively and contains small stacks of intercalated MMT as also exfoliated sheets.

The organic modifier chain length within the silicate galleries of MMT plays a crucial role in determining the dispersion behavior in nancomposites. Three types of conformation of aliphatic chains between silicate galleries are possible depending on the number of carbon atoms n [220,221], namely, a lateral monolayer (n up to 6), pseudo bilayer (n=9–12) and pseudo trilayer (n=13–18). No intercalation occurs for pseudo bilayer, while for n>12 intercalated hybrids are formed with the gallery height increasing with the increase in chain length. The increasing n value reflects in the enhanced ability to draw polymer chains between the silicate layers and the decrease in van der Waals attractive forces, leading to intercalated or exfoliated dispersions.

Polymers in Special Uses

$$H_3C-\overset{\overset{\displaystyle CH_3}{|}}{\underset{\underset{\displaystyle C_nH_{2n+1}\ (n=14\text{-}18)}{|}}{N^+}}-CH_2-\langle\!\bigcirc\!\rangle-C_{12}H_{25}$$

(I)

$$H_3C-\overset{\overset{\displaystyle CH_3}{|}}{\underset{\underset{\displaystyle C_nH_{2n+1}\ (n=14\text{-}18)}{|}}{N^+}}-CH_2-CH_2-\overset{\overset{\displaystyle C_2H_5}{|}}{\underset{}{CH}}-CH_3$$

(II)

$$H_3C-\overset{\overset{\displaystyle CH_2CH_2OH}{|}}{\underset{\underset{\displaystyle C_nH_{2n+1}\ (n=14\text{-}18)}{|}}{N^+}}-CH_2CH_2OH$$

(III)

$$H_2N-\overset{\overset{\displaystyle CH_3}{|}}{CH}-CH_2-O\left[CH_2-\overset{\overset{\displaystyle CH_3}{|}}{CH}-O\right]CH_2-\overset{\overset{\displaystyle CH_3}{|}}{CH}-NH_2$$

(IV)

$$\langle\!\bigcirc\!\rangle\!\begin{array}{l}-CH_2NH_2\\[4pt]\ \ \ CH_2NH_2\end{array}$$

(V)

Layered clay nanocomposites have been prepared by melt intercalation for a variety of polymers, including polystyrene [221], nylon-6 [222], ethylene-vinyl acetate copolymers [223], polypropylene [224], polyimide [225], poly(styrene-*b*-butadiene) [226], and PEO [227].

Clay-matrix interaction can be improved further by organic cations containing reactive functional groups which react with either the matrix group or the curing agent [228]. For instance, polyoxypropylene diamines (IV) with a long molecular chain have been used as a clay surface modifier, while it also acts as intragallery polymerization catalyst and curing agent, producing a large increase in interlayer spacing (46 Å) and a high degree of exfoliation [229]. In another instance, to induce MMT to be highly exfoliated and homogeneously dispersed in epoxy matrix [230], two different organic reagents, dodecylbenzyl dimethyl alkyl ammonium chloride (I) and *m*-xylylenediamine (V), have been used to modify MMT, since (V) has excellent miscibility with (I) and thus provides a compatible and reactive clay surface for the epoxy.

Besides melt intercalation, described above, in situ intercalative polymerization of ϵ-caprolactone (ϵ-CL) has also been used [231] to prepare polycaprolactone (PCL)-based nanocomposites. The in situ intercalative polymerization, or monomer exfoliation, method was pioneered by Toyota Motor Company to create nylon-6/clay nanocomposites. The method involves in-reactor processing of ϵ-CL and MMT, which has been ion-exchanged with the hydrochloride salt of aminolauric acid (12-aminodecanoic acid). Nanocomposite materials from polymers such as polystyrene, polyacrylates or methacrylates, styrene-butadiene rubber, polyester, polyurethane, and epoxy are amenable to the monomer approach.

An emulsion route can also be used to make nanocomposites with some polymers. Thus well-dispersed (exfoliated) PMMA nanocomposites have been synthesized via emulsion polymerization of MMA in a suspension of MMT (unmodified) or MMT modified with 2,2′-azobis(2-methyl propionamide), using sodium lauryl sulfate as the emulsifier and potassium persulfate as the initiator [232].

It should be noted that among the three different methods of PNC preparation mentioned above, the solution method and the in situ polymerization method have only limited applications because neither a compatible polymer-silicate solvent system nor a suitable monomer is always available. Moreover, they are not always compatible with current polymer processing techniques. Among all the methods to prepare PNCs, the approach based on direct melt mixing/intercalation is perhaps the most versatile and environmentally benign.

While polyolefin nanocomposites are commonly prepared by melt intercalation of MMT organo-clays with polyolefins, unmodified or modified with suitable functional groups (e.g., polypropylene with maleic anhydride or hydroxyl groups [233,234], a new promising technique, known as polymerization filling technique has been reported [235]. The process described for making polyethylene nanocomposites consists in anchoring the coordination catalytic complex (MAO/metallocene system) at the surface in the interlayer of the given non-modified layered clay (hectorite, MMT) and polymerizing ethylene from this surface (Figure 2.91). Molecular hydrogen is added as a transfer agent to control the polymer

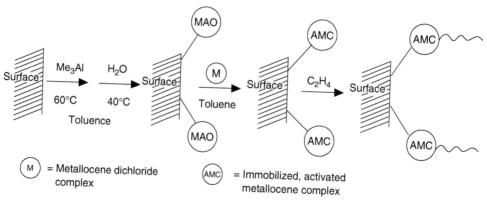

FIGURE 2.91 Schematic presentation of the process of anchoring the coordination catalytic complex, MAO/ metallocene, at the surface in the interlayer of layered clay, followed by ethylene polymerization at the immobilized catalyst site. (After Köppl, A., Alt, H. G., and Phillips, M. D. 2001. *J. Appl. Polym.*, 80 (3), 454. With permission.)

molecular weight, yielding composites with improved tensile properties, characterized by a marked increase in Young's modulus, and improved strain behavior.

Carbon nanotubes (CNTs) and carbon nanofibers (CNFs), due to their unique structure and properties, appear to offer quite promising potential for industrial application [236]. As prices decrease, they become increasingly affordable for use in polymer nanocomposites as structural materials in many large scale applications. In fact, three applications of multiwall CNT have been discussed recently: first, antistatic or conductive materials [237]; second, mechanically reinforced materials [238,239]; and third, flame retarded materials [240,241]. The success of CNTs in the field of antistatic or conductive materials is based on the extraordinary electrical properties of CNTs and their special geometry, which enables percolation at very low concentrations of nanotubes in the polymer matrix [242].

CNFs are intrinsically less conductive than either single-walled CNTs or multi-walled CNTs and CNF composites have the potential for creating inexpensive semiconducting polymers. These composites require a homogeneous dispersion within the polymer. Most well dispersed materials are made by high shear methods like twin screw extrusion [243,244]. However, the aspect ratios of the CNFs are significantly reduced, which leads to decreased mechanical properties and conductivity.

Inorganic-organic hybrid nanocomposites using sol-gel process where the inorganic phase is grown in situ is being actively pursued in many laboratories [245,246]. The main advantage of the sol-gel process, as compared to the traditional practice of mechanically blending the reinforcing fillers into a polymeric matrix, is the subtle control it provides over morphology and/or surface characteristics of the growing inorganic phase in the polymer matrix by control of various reaction parameters like pH, concentration, temperature, etc. Unlike the traditional composites, which have macroscale domain size varying from micrometer to millimeter scale, the organic-inorganic hybrids have domain size varying typically from 1 to 100 nm [247] and so are usually optically transparent, even though microphase separation may exist.

The method of sol-gel hybrid preparation is based on the growth of the inorganic phase by the hydrolysis-condensation of alkoxysilanes (Figure 2.92) like tetraethoxysilane (TEOS) in a solution containing the organic polymer. The reactions can be represented as

$$\text{Hydrolysis}: \quad \text{Si(OR)}_4 + n\text{H}_2\text{O} \rightarrow \text{Si(OH)}_n(\text{OR})_{4-n} + n\text{ROH}$$

$$\text{Condensation}: \quad \equiv \text{SiOH} + \text{RO} - \text{Si} \equiv \rightarrow \equiv \text{Si} - \text{O} - \text{Si} \equiv + \text{ROH}$$

$$\text{or} \quad \equiv \text{SiOH} + \text{HO} - \text{Si} \equiv \rightarrow \equiv \text{Si} - \text{O} - \text{Si} \equiv + \text{H}_2\text{O}$$

So far many hybrids have been prepared in this way using, e.g., poly(vinyl acetate) [248,249], poly (methyl mathacrylate) [250], PEO [251], poly(dimethylsiloxane) [252], perfluorosulfonic acid film,

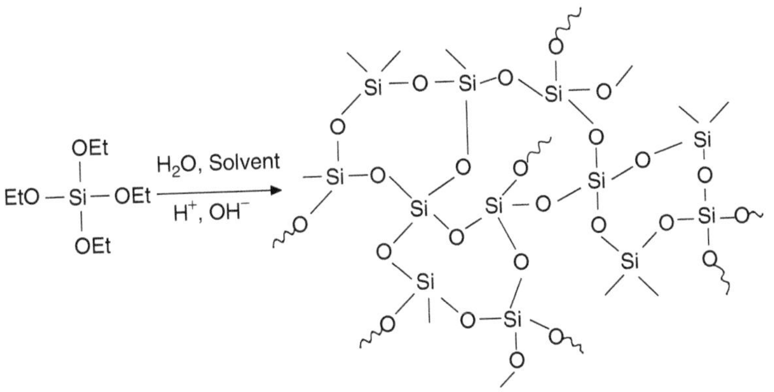

FIGURE 2.92 Overall scheme for the formation of three-dimensional silica network by hydrolysis and condensation of tetraethoxysilane (TEOS). (After Sengupta, R., Bandopadhyay, A., Sabharwal, S., Chaki, T. K., and Bhowmick, A. K. 2005. *Polymer*, 46, 3343. With permission.)

Nafion [253], poly(vinyl alcohol) [254], and polyamide-6,6 [255]. A procedure often followed for polymers soluble in tetrahydrofuran (THF) is to add TEOS to a THF solution of the polymer, followed by addition of water (4 moles based on Si) in the form of 0.15 M HCl or 0.1 M NH_4OH and allowing the reaction to take place. Films are made by casting on an inert substrate such as Teflon and drying under proper conditions. Nafion composite films are made by impregnating swollen Nafion films in alcohol solution of TEOS. The micro- or nanocomposite films made by the sol-gel process are expected to have technological opportunities in important arena of gas-liquid separations, heterogeneous catalysis, electronic materials, and ceramic precursors.

Highly electrically conductive polyaniline/graphite nanocomposites have been made via in situ polymerization of aniline in the presence of exfoliated graphite nanosheets [256]. Graphite nanosheets can be prepared via the microwave irradiation and sonication from expandable graphite powders. In a typical procedure, graphite flakes (80 mesh) are treated with a mixture of concentrated sulfuric acid and hydrogen peroxide (1: 0.08 v/v), washed and dried at 100°C. The resulting expandable graphite powders are irradiated in a microwave oven, thereby obtaining exfoliated graphite (also known as graphite worm), which is then dispersed in 50 wt% alcohol, sonicated for prolonged period, washed and dried to obtain graphite nanosheets.

In a novel procedure reported recently [257], high-performance PNCs have been prepared by growing carbon nanotubes (CNTs) on clay and dispersing the as-prepared CNT-clay hybrid in a polymeric matrix by simple melt blending. This is further elaborated below.

As has been demonstrated, iron, nickel, or cobalt nanoparticles supported on oxides are efficient catalysts for the synthesis of CNTs by chemical vapor deposition (CVD), which has proven to be a cost-efficient way of mass-producing CNTs [258]. Clay has also been used as the support for catalytic CVD growth of CNTs [258a]. While clay platelets are typical two-dimensional (2D) nanofillers for incorporation into polymeric matrices, carbon nanotubes, as one-dimensional (1D) nanomaterials, have also been considered as ideal enhancement fillers for making PNCs because of their extremely high mechanical strength and high electrical and thermal conductivity. For both 1D and 2D nanofillers, homogeneous dispersion in polymeric matrices and strong interactions with the matrices are important issues, which are addressed effectively by using CNT-clay hybrid nanofillers for making composites. This has been demonstrated by using clay-supported iron nanoparticles as a catlyst for the growth of CNTs followed by incorporation of the as-prepared CNT-clay hybrid nanofillers into a nylon-6 (PA6) matrix to make PA6/CNT-clay composites [257].

Figure 2.93 depicts the procedure for making the nanofiller and the polymer nanocomposite by the aforesaid method. Firstly, the sodium montmorillonite (Na^+MMT) is modified by impregnation with a

Fe$(NO_3)_3$ solution whereby Fe^{3+} ions are intercalated into the layers of MMT. The Fe^{3+} ions are changed to Fe$_2$O$_3$ particles by calcination. During CVD growth of CNTs, the Fe$_2$O$_3$ particles are reduced to Fe particles in situ, which serve as seeds for the growth of CNTs. The platelets of the clay are further delaminated as the CNTs grow on them, forming a 3D nanostructure consisting of a 2D nanoclay platelet and several attached nanotubes. The obtained CNT-clay hybrid has been directly used as a filler and incorporated into PA6 by melt-blending for the preparation of PA6/CNT-clay nanocomposite [257]. By transmission electron microscopy, the CNTs attached to the clay platelets have been found to be multiwalled with thicknesses ranging from several nanometers to about 20 nm.

The CNT-clay hybrid with its unique structure is expected to be an ideal filler for high-performance PNCs. This has been demonstrated [257] for the PA6/CNT-clay nanocomposite as incorporation of only 1 wt% CNT-clay hybrid is found to improve significantly the mechanical properties of PA6 (see Figure 2.94).

Organic nanofillers are relatively unknown. It is, however, reported that elastomeric nanoparticles (ENP) added to brittle plastics can significantly increase their toughness. Thus phenolic resins have been blended with nitrile butadiene and carboxylic nitrile butadiene ENPs leading to large increase in impact strength and simultaneous improvement in flexural strength and heat resistance [259]. The ENPs are special ultra-fine fully vulcanized powdered rubbers, prepared by a special irradiation technique [259a].

2.17.2 Applications of Polymer Nanocomposites

The only examples of PNCs that have achieved any commercial significance to date are the nanocomposites produced using layered clay minerals (montmorillonite, hectorite, etc.,) and carbon nanotubes (CNTs). Polyolefin nanocomposites made via melt intercalation or by the various in situ methods using olefin polymerization catalysts supported on nanofillers are a potentially promising route to new materials for automotive applications. Perhaps the most successful PNC system so far demonstrating dramatic improvement in properties is polyamide (nylon)-6/clay nanocomposite (PA6CN), which shows, as compared with nylon-6, a doubling of the tensile modulus (2.1 vs. 1.1 GPa) and strength without sacrificing impact resistance for products containing as little as 2 vol% of layered clay. In addition, the coefficient of linear thermal expansion is reduced in half (6.3×10^{-5} vs. 13×10^{-5}), a key factor in dimensional stability and an essential factor to manufacture large vehicle parts, while the heat distortion temperature increases to more than 100°C, extending the use of the PNCs to higher temperature environments, such as for under-the-hood parts in automobiles (e.g., timing chain cover for Toyota). PA6CNs are available both as films for packaging and composites for injection molding and other processes.

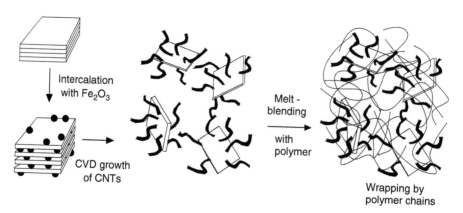

FIGURE 2.93 Schematic presentation of the procedure of growing carbon nanotubes (CNT) on clay particles and incorporation of CNT-clay hybrid filler into PA6 matrix for making PA6/CNT-clay nanocomposite. (After Zhang, W. D., Phang, I. Y., and Liu, T. 2006. *Adv. Mater.*, 18, 73. With permission.)

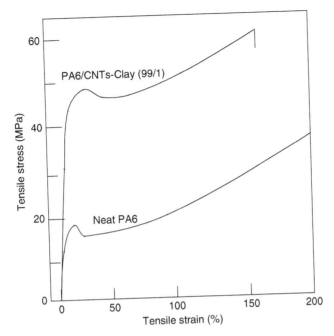

FIGURE 2.94 Stress-strain curves of neat PA6 and its nanocomposite containing 1 wt% CNT-clay nanohybrid. (After Zhang, W.D., Phang, I.Y., and Liu, T. 2006. *Adv. Mater.*, 18, 73. With permission.)

Since the large increase in stiffness and strength is realized with only a small amount of inorganic reinforcement, PA6CNs are much lighter compared to conventional polymer composites, offering a 25% weight saving on average over highly filled plastics and as much as 80% over steel. Using the PNCs only for structurally non-critical parts such as front and rear fascia, cowl vent grills, valve/timing covers, and truck beds could yield several billion kilograms of weight saved per year. This reduction in weight offers to the automotive industry and vehicle users the potential of significant energy savings, which could be further expanded by applications into structural components, interiors, and body panels [235].

The weight advantage of using PNCs in auotomotive manufacturing could have significant impact on environmental concerns besides many other potential benefits. Thus, it has been reported that wide and large scale use of PNCs by U.S. vehicle manufacturers could save 1.5 billion liters of gasoline over the life of one year's production of vehicles and the corresponding reduction in carbon dioxide emissions would be more than 10 billion pounds [217]. The use of nanocomposites is also expected to improve manufacturing speed and promote recycling.

Even minor loadings of layered clay in PNCs result in substantial reduction in flammability which has been attributed to the build-up of a protective char layer involving the clay [260]. Substantial improvement in the ablative properties of materials due to a similar char formation has also been reported [261]. These properties could lead to the development of novel fire resistant composites without the use of halogenated compounds that have negative environmental impact and hydrated inorganic oxides that require substantial loading.

It has been found that the platelet nature of exfoliated clays in polymers dramatically reduce the permeability of liquids and gases through the composite. The outstanding combination of this barrier behavior and improved mechanical properties of nanocomposites may eliminate the need for a multipolymer layer design for food and beverage packaging [217] enabling simpler process for polymer recycling. Commercial grades of PA6CN for packaging film are available from Ube and Bayer.

There are a variety of potential applications of carbon nanotube (CNT) reinforced thermoplastic composites in automotive industry and systems involving electrostatic dissipation (ESD) and EMI.

Unlike conventional carbon fillers which require high loading to provide conductivity to thermoplastics, leading to considerable deterioration in physical properties, CNT is able to provide conductivity to the composite at very low loading levels. For example, the loading requirement of CNT for electrostatic painting operations is stated [262] to be as low as 2%. CNTs have thus been commercially exploited to assist in the electrostatic paintability of automotive components molded from polyphenylene oxide/polyamide (PPO/PA) alloys. Being produced commercially by Hyperion Catalysts International, the CNTs are 10 nm in diameter and have an average of eight layers of graphite in the tube wall.

Other potential applications of PNCs include layered clays in elastomers for asphalt modification [263] and for tire applications (low air permeability) [264], and layered clay nanocomposites with epoxy [265], unsaturated polyesters [266], vinyl esters [267], and polyurethanes [268] for coatings and composites.

2.18   Wood–Polymer Composites

Wood–polymer composites (WPCs), variously known as wood-plastic composites, wood fiber-plastic composites, and green composites, are a new group of materials that are generating interest in many applications. [Here WPCs are defined as composite materials that contain wood (in various forms) and thermoplastic polymer and do not include wood flour-thermoset plastics like bakelite and particleboard products such as medium density fiberboard.] WPC products use a range of polymers such as polyethylene, polypropylene or polyvinyls in various proportions along with wood or other natural fibers to produce profiles or molded objects with the structural integrity and workability of wood and the durability of polymers. While wood flour or waste wood is mainly used as a cost-cutting alternative to mineral fillers like talc and calcium carbonate, plant fibers like flax, hemp and kenaf are currently being evaluated as environmentally friendly and low-cost alternatives for glass or carbon fibers in engineering composites.

Wood makes an excellent functional filler, but within limits. While the heat used to melt and process plastics does not affect mineral-based fillers, it does affect wood. Therefore, great care must be exercised when using wood as a functional filler. Moreover, wood absorbs moisture, while thermoplastic processing equipment has a near-zero tolerance for moisture.

Plant and wood fibers have a number of advantages over glass fibers. These are: abundant, cheap, renewable, lightweight, non-abrasive, biodegradable, can be incinerated with energy recovery, good mechanical properties as well as good acoustic and thermal insulating properties. The fibers themselves are cellulose fiber-reinforced materials as they consist of microfibrils in an amorphous matrix of hemicellulose and lignin. The hydrogen bonds and other linkages in cellulose provide the necessary strength and stiffness to the fiber, while polymers lignin and hemicellulose are responsible for most of the physical and chemical properties, such as biodegradability, flammability, moisture sensitivity, degradability by UV light etc.

Natural fibers can be classified as: seed fibers (such as cotton), bast fibers (like flax, hemp, jute, kenaf, ramie), hard fibers (like sisal), fruit fibers (like coir), and wood fibers. The chemical composition and dimensions of some common agro-fibers are presented in Table 2.21. The origin of wood fibers can be: sawmill chips, sawdust, wood flour or powder, cutter shavings, pulp or wood residues. As binders for these fibers, both thermosetting (like phenolic, epoxy, polyester) resins and thermosetting matrices [such as polyethylene (LDPE, HDPE), polypropylene (PP), poly(vinyl chloride) (PVC), polystyrene (PS)] can be used. Thermoplastic composites are, however, less expensive to process than thermosetting composites, in addition to their ability to be manufactured into complex shapes.

WPC is claimed to be superior to natural wood in several ways. Its main features can be summarized as follows: (1) the thermoplastic polymer component resists rot and insects, thus providing an inherently longer product life and less maintenance requirement than wood alone; (2) the cellulosic fiber content provides reinforcement, increases the rigidity of the product compared to the polymer component; (3) WPC products can be worked like wood using current tools and fastening techniques; (4) WPC products can be pigmented during processing for long-lasting color, or painted after installation; (5) WPC products can have virtually 100 per cent recycled content from post-consumer polymeric waste such as milk bottles and grocery sacks and wood fiber scrap from large processors such as furniture or window producers.

2.18.1   WPC Feedstocks

There is no such thing as a typical feedstock that should be used to manufacture a specific WPC and the materials selected depend very much on cost, availability, market value of the product (i.e., low-cost or high-value end of the market) and product performance requirements.

2.18.1.1   Wood

Ponderosa pine, a common species used extensively in window and door manufacture and routinely made into wood floor, is a good overall performer. Other species, such as maple, oak, and many other hardwoods and softwoods can be successfully used. The density of hardwoods can be almost twice that of softwoods and so will result is a heavier product. Moreover, some wood fibers are more durable than others depending on the environmental conditions (wet or dry). Also the modulus is generally higher for hardwoods resulting in a stiffer WPC product.

Wood fiber can be obtained by chemical treatment of the wood (Kraft process, which removes the lignin and low molecular weight waxes) or by thermo-mechanical wood treatment processes, which conserve the lignin and wax content. Wood fiber has a length to diameter ratio between 10:1 and 20:1. Wood flour is available in many sizes, from 20 mesh (coarse) to 400 mesh (extra fine), with 40 mesh being most common. Fine mesh wood flour increases stiffness but reduces impact strength, whereas longer wood fibers contribute to strength but are more difficult to bind with the polymer. For most applications, however, 40 mesh wood flour gives satisfactory performance and ease of processing [269].

Some of the most common sources of recyclate wood feedstocks suitable for composites include [269]: (a) primary wood wastes (such as wood wastes from sawmills; (b) secondary wood wastes (generated when wood products, such as furniture, cabinets and doors are made); and (c) post-consumer wood wastes (which can include anything from construction and demolition debris to packaging, crates and pallets. Primary and secondary wood wastes are the key materials used for WPC production in the U.S.A.

TABLE 2.21   Dimensions and Chemical Composition of some Common Agro-Fibers

Type of Fiber	Cellulose (%)	Lignin (%)	Fiber Dimension (mm)	
			Mean Length	Mean Width
Cotton	85–90	0.7–1.6	25	0.02
Flax	43–47	21–23	30	0.02
Hemp	57–77	9–13	20	0.022
Sisal	47–62	7–9	3.3	0.02
Bamboo	26–43	21–31	2.7	0.014
Kenaf	44–57	15–19	2.6	0.02
Jute	45–63	21–26	2.5	0.02
Bagasse	32–37	18–26	1.7	0.02
Deciduous wood	38–49	23–30	1.2	0.03
Coir	35–62	30–45	0.7	0.02

Source: Rowell, R. M. 1998. *Proceedings of the Fourth Pacific Rim Bio-based Composites Symposium*, Indonesia, Nov. 2–5, pp. 1–18.

2.18.1.2 Plastics

As wood degrades at high temperatures, plastics which can be processed below 200°C are generally used. The most commonly used polymers for WPCs using both virgin and recycled material is polyethylene (PE). All polyethylene grades (LDPE, LLDPE, HDPE) are used for the manufacture of WPC. Polypropylene (PP) is also used but requires higher levels of additives to prevent degradation. Poly(vinyl chloride) (PVC) is one of the first plastics to find commercial use in WPC and its use is still growing. The relative quantities of these polymers used in WPC manufacture are PE 70%, PP 17%, and PVC 13%. PE-based products are cheaper and have a higher heat distortion temperature than PVC-based products. But PE-based products have low surface energy which makes painting and post-treatment difficult. This is, however, not the case with PVC.

As for virgin plastics, any recycled plastic that can melt and process below the degradation point of wood (200°C) is usually suitable for manufacturing WPC materials. The choice of plastic, however, depends on the particular application requirements.

2.18.1.3 Compounded Pellets

Compounded WPC pellets are an expensive feedstock option. However, the high initial cost is offset by the fact that these can be molded with less expensive and less complex machines without the need for any additional processing and specialized handling equipment for wood flour. A number of companies have set up plants to produce compounded WPC pellets to sell on to plastics processors. In the compounding process, the filler (wood flour) and additives are dispersed in the molten polymers to produce a homogeneous blend.

2.18.1.4 Additives

Different types of additives are used in WPCs to aid processing operation (e.g., lubricants), to provide processing stability and preservation in long-term service (e.g., heat and light stabilizers) and to improve mechanical properties (e.g., coupling agents).

Lubricants and process aids can be either (a) internal, whereby they act in the resin phase to increase melt flow and throughput, prevent shear burning, and resist melt fracture (by reducing viscosity at high shear rate), or (b) external, whereby they act at the interface between resin and other materials to improve release of the composite, promote dispersion of fillers, resist melt fracture, and/or reduce friction between resin and process equipment.

Primary phenolics (free-radical scavengers) or secondary phosphite (hydroperoxide decomposer) may be used as heat stabilizers for WPC, while light stabilizers commonly used in WPC include UV absorbers (e.g., benzotriazole or benzophenone), radical scavengers or hydroperoxide decomposers, and hindered amine light stabilizers (HALS).

Compatibilizers or coupling agents are polymers containing both polar functional groups that can react or interact with the hydroxyl groups of cellulose and non-polar chain sections that are more compatible with the hydrocarbon chains of WPC polyolefin [270]. Examples of compatibilizers used as WPC additives are maleic anhydride (MAH) grafted polyethylene (PE-*g*-MAH) and polypropylene (PP-*g*-MAH), trimethoxyvinylsilane grafted polyethylene and analogous copolymers, polymethylene polyphenylene isocyanate [269], and methylol phenolic grafted polyolefins. The best approach is to use, for example, a grafted PE for a PE base resin, a grafted PP for a PP base resin, and so on. Low molecular weight silanes can be added to the wood filler by spraying or by intensive mixing before blending with the polymer. The compatibilizers promote adhesion and dispersion of the wood component in the polymer matrix, thereby improving the mechanical properties.

Microcellular foamed WPCs are lighter and feel more like real wood [271]. Both extrusion and injection molding have been used to produce foamed WPCs [272], producing materials with a density reduction of approximately 25%.

2.18.2 Manufacture of WPC Products

The processes that can be used for the manufacture of WPC products are compounding, extrusion, injection molding and pultrusion. Drying of wood is an important prerequisite of WPC production since the moisture content of wood will have a significant effect on the processing and final product quality of WPCs. In some cases, moisture is removed as part of the processing itself (discussed later), while in other cases wood drying is performed separately using additional drying equipment such as preheaters, hot air dryers and rotary tube furnace, the last named device being the most suitable.

2.18.2.1 Compounding

Many options are available for compounding which in many cases is the first stage of the WPC manufacturing process. In batch compounding systems, internal and thermokinetic mixers are used to disperse the wood based filler and additives in the molten polymer to produce a homogeneous blend, while in continuous systems kneaders or extenders are used. The batch process has the advantage that processing parameters (e.g., residence time, shear, and temperature) are easier to control, but has the problem of batch to batch quality difference. The continuous compounding extruder does not have this problem. The raw materials can either be introduced into the extruder simultaneously or the wood fiber added to the molten plastic, a section of the extruder being placed under vacuum to remove moisture from the blend. The compounded material can either be formed directly into an end product or made into pellets for processing in future. A typical WPC pellet compounding line may have a capacity of 4500–9000 kg/h. The pellets can be made with an exact fiber content required for a specific application or, alternatively, highly wood filled WPC pellets can be combined with more resins. The pellets are a boon to manufacturers who do not typically do their own compounding or do not wish to compound in-line, as for example, most single-screw profiles molding companies [273].

2.18.2.2 Extrusion

In order to produce a WPC product by the extrusion process, it is required to first melt the polymer, then mix the molten polymer with the dried wood fiber such that a homogeneous melt is achieved and any remaining moisture can be removed by vacuum venting. The extruder then compresses the homogeneous blend to pass through the die. Various extrusion configurations have been developed for WPC product manufacture [274], some of which are described below.

Single-Screw Extruder (SSE). Equipped with 24:1 or 30:1 (*L:D*) single screw, gravity hopper for material feed, in-line dryer or hopper for moisture removal, and melting/mixing by barrel heat and screw shear, the extruder used for profile extrusion has the advantages of proven technology and lowest machine cost. The SSE system, however, has a number of disadvantages, such as poor mixing, lower output, greater difficulty to vent, and greater risk of burning the wood fiber.

Co-Rotating Twin Screw Extruder (CTSE). This system is commonly used for compounding and profile extrusion in high output applications. CTSE with its high shear is, however, harder to control than the counter rotating system (see below).

Counter Rotating Twin Screw Extruder (CRTSE). This system has parallel or conical, counter rotating twin screws and uses fiber and polymer in same particle size, usually 40 mesh. The extruder is provided with vacuum vent for moisture removal and it uses barrel heat, screw mixing and shear heating for melting and mixing. The system is capable of compounding and generating the high pressure required for profile extrusion. Having shorter residence time and a narrow residence time distribution, the system is also suited to manufacture of PVC windows.

The CRTSE system has the advantages of proven technology, low rpm and low shear. But it has the disadvantages that drying system and size reduction system are required. As the polymer is melted with the fiber, there is a greater risk of burning the fiber, besides venting being more difficult.

Conical Twin Screw Extruder. Conical twin screw extruders are gaining greater acceptance in WPC processing systems. The large screw diameters in the feed section and the increasing compression along the taper of the screws make conical machines well suited for processing materials with low apparent density such as wood fiber. The slimness of the conical screws in the metering section also minimizes the shearing stress on the material which helps the wood fiber remain intact. Moreover, under conditions of identical output, a conical extruder will have more favorable metering section residence times since the melt will pass the high temperature zone much faster and this will prevent darkening of the WPCs extruded at high temperatures necessary to plasticize the polymer.

Conical Conex Extruder. The conical extruder Conex has a unique design (Figure 2.95) in which conventional single or twin screws do not exist. The material flow is divided into multiple spiral channels with a deep conical rotor ensuring that thermal degradation of polymers and additives do not take place even at relatively high melt temperatures used. Mixing is based on forcing the materials to flow through holes existing at the rotor from spiral grooves, machined onto outer stator surface, into the spiral grooves machined on the inner stator surface, which effectively grind, mix and melt the composite materials being processed in the extruder. The mechanism suits well for composite materials with high filling level of natural fibers and prevents fiber degradation [275]. As the design is self-cleaning, it avoids any stagnation of the flow.

A special feature of Conex extruder not existing in conventional extruders is its ability to grind virtually any cellulose material (such as wood sticks and solid parts, paper, textile, nut shells, straw, etc.) while processing, with adjustable flight clearance between the conical rotor and the stators (typical value being 0.5 mm) and the extruder filling level (starve feeding or full feeding) controlled by the feeding screw rpm vs. the rotor rpm to result in fibers being made on-line of the cellulosic material.

A significant feature of the Conex design is that larger plastics particles will become ground as well as wood particles in the initial part of the extruder. This permits use of feed containing even a limited amount of recycled thermosetting plastics materials, as it will become ground into non-melting plastic dust and act as fillers in composite materials [275]. The possibility of using non-dried wood materials is a notable benefit when comparing with existing technologies. Thus the extruder may be equipped with steam ejection system so that very wet wood can also be used (with the stator temperature at rear set at a high value), the released moisture being ejected as steam from the extruder.

Woodtruder. The woodtruder is a dual extrusion system for wood composite materials made by Davis-Standard Ltd. [276]. It comprises a primary parallel, 28:1 *L/D*, counter-rotating twin-screw extruder, a

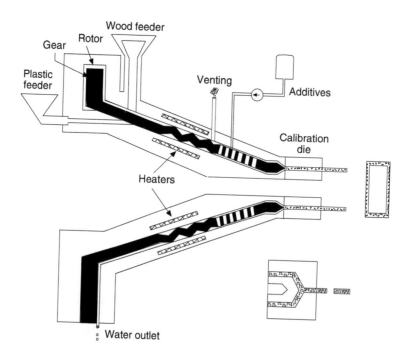

FIGURE 2.95 Schematic layout of Conex® wood extruder. (After Mali, J., Sarsama, P., Suomi-Lindberg, L., Metsä-kortelainen, S., Peltonen, J., Viekki, M., Koto, T., and Tisala, S. 2003. *Wood Fiber-Plastic Composites*, A Report dated 31.12.2003, VTT Building and Transport, Finland.)

mounted single screw, side-injection extruder, and a crammer feeder that conveys the wood fiber at a controlled rate into the primary extruder. A heating and vacuum venting system in the first section of the primary extruder serve to eliminate moisture from the wood fiber. The side injection extruder, positioned midway through the primary extruder, separately melts and mixes the polymer and then injects the molten polymer at a controlled rate into the primary extruder where the polymer encapsulates the wood fiber, resulting in a thoroughly mixed composite of up to 80% wood fiber. Typical applications include decking, fencing, and window and door profiles. HDPE, PP, PS, PVC, ABS and recycled plastics are used as the plastic component, while the wood component may come from various sources such as wood fiber, peanut shells, rice husks, wood shavings, and jute.

Putrusion. The schematic of a wood-base pultrusion system is shown in Figure 2.96. The tapered entrance section of the die design is followed by a constant geometry section (of the same dimension as the profile) where the thermoplastic polymer melts and the blend then passes through a cooling section where the polymer hardens. The main requirement of the pultrusion process is that the WPC blend is capable of flowing through the die forming a product of consistent profile [276].

Injection Molding. The WPC injection molding process [277] is similar to conventional injection molding of thermoplastics. The molding material should have good homogeneity and less than 0.5% moisture content. The processing requirements include high rate of plastification, short residence time, and small temperature range of melt.

2.18.3 Properties of WPC Products

In a properly manufactured WPC product, individual wood elements are encapsulated in a continuous plastic matrix which serves to protect the wood from the environment. Therefore care should be taken to stabilize the plastic to UV light and other environmental factors. Moreover, if individual wood elements are not adequately encapsulated, they may absorb moisture, leading to swelling, delamination, and fungal decay [278], though properly manufactured WPC products change very little with humidity variations as the plastic hinders moisture movement in the body. On the other hand, adding wood to plastics produces a beneficial effect by significantly decreasing linear thermal expansion, often by 50 per cent or more [Figure 2.97(a)]. Consequently, WPC products exhibit significantly less mold shrinkage, as compared to virgin plastics [Figure 2.97(b)].

Adding wood to plastics can increase the stiffness sufficiently for certain building applications. As shown in Figure 2.97(c), addition of wood fiber can cause 2–3 fold increase in bending stiffness as compared to unfilled plastics. However, even with these improvements, most WPCs have moduli of elasticity less than half that of solid wood [278].

Other effects of adding wood to plastics can be stated as follows. With increasing content of wood in WPC product, the tensile strength decreases [Figure 2.97(d)], flexural strength increases (up to a maximum), melt index decreases, and notched impact energy increases (while unnotched decreases).

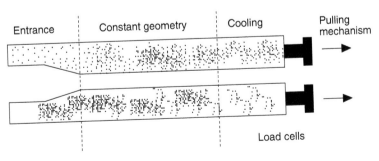

FIGURE 2.96 Pultrusion die design for wood plastic composites. (After WRAP Research Report, 2003. Wood Plastic Composites Study: Technologies and U.K. Market Opportunities. Optimat Ltd and MERL Ltd., The Waste and Resources Action Programme, The Old Academy, Banbury, U.K.)

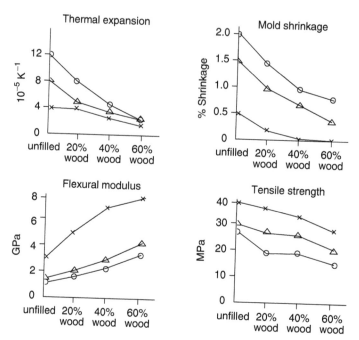

FIGURE 2.97 Effect of plastic type (polyethylene o, polypropylene Δ, polystyrene ×) and wood loading levels on properties of wood plastic composites. (After Chelsea Center for Recycling and Economic Development (CCRED), 2000. Technical Report #19, *An Investigation of the Potential to Expand the Manufacture of Recycled Wood Plastic Composite Products in Massachusetts*, Univ. of Massachusetts, Massachusetts.)

The particle size also has an effect on the property performance of WPC products. For increasing wood particle size in PP-based WPC products, both melt index and tensile elongation increase, notched impact energy increases but unnotched impact energy decreases, and flexural modulus and strength increase for particles smaller than ∼0.25 mm.

2.18.4 Applications of WPC Products

One of the first commercial use of WPCs was the use of PVC and wood flour for flooring tiles starting in the mid-50s. In 1973, the Sonnesson Plast AB Company marketed a wood-PVC composite called Sonwood. It was made by first compounding wood flour and PVC together to make a pelletized feedstock that was then extruded into thin sheets or profiles. Another profile (a shaped product) developed over thirty years ago in Italy called "woodstock" used a mix of about 50% wood flour and 50% PP. Woodstock was extruded in thin sheets, reheated and molded to produce automobile panels for Fiat automobiles. Woodstock is still widely used today.

The WPC industry is, however, only a fraction of a percent of the total wood products industry, though the use of WPCs is increasing at a substantial rate, the greatest growth potential being in building products that have limited structural requirements. Presently, the main WPC product produced in North America is decking. Polyethylene is used in 70% of all products, PVC in 18%, and PP makes bout 18%. Other areas of activity are outdoor furniture such as picnic tables, park benches, nature trails/walkways, fencing piers, boardwalks, window and door profiles, automobile components, and pallets [279]. Although WPC decking is more expensive than pressure-treated wood, manufacturers promote its lower maintenance, lack of cracking or splintering [273].

Window and door profile manufacturers form another large industrial segment that uses WPCs. Although more expensive than unfilled PVC, wood-filled PVC is gaining favor because of its balance of

Polymers in Special Uses

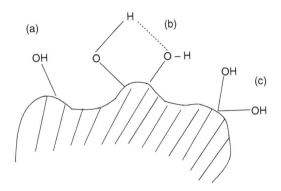

FIGURE 2.98 Surface hydroxyl groups on silica: (A) isolated groups; (B) hydrogen-bonded groups; (C) geminated hydroxyl groups (from the same Si atom). (After Hindryckx, F., Dubois, P., Jerome, R., Teyssie, P., and Marti, M. G. 1997. *J. Appl. Polym. Sci.*, 64, 423, 439. With permission.)

thermal stability, moisture resistance and stiffness [273,280]. Different approaches are being used by several industry leaders to include WPC profiles in their product lines, such as by co-extruding a wood-filled PVC with an unfilled PVC outside layer for increased durability, co-extruding a PVC core with a wood-filled PVC surface that can be painted [281] or co-extruding a wood-filled PVC and a composite with a foamed interior for easy nailing and screwing [280].

In Europe where environmental concerns are a strong driving force, there has been a high growth in the use of natural fiber-reinforced thermoplastics in automotive applications, though the growth is much slower in the U.S.A. Automotive interior applications (decorative, structural, and furniture) for WPC materials are being increasingly promoted and companies such as the Japanese firm Ein Engineering offers technology for production of profiles suitable as decorative moldings and trimmings.

Considerable growth in WPC market is expected in the near future. The phase out of chromated copper arsenate (CCA) treated wood for residential uses such as decks, playgrounds, and fencing (EPA 2002) may also help the growth of the WPC market. WPC sleepers are presently being assessed to replace wooden sleepers for railroad crossties [269]. New building products based on WPC have been developed,

FIGURE 2.99 Schematic of the method of synthesis of a bridge-anchored *ansa*-metallocene. (The indenyl rings are shown in pseudo-racemic orientation as the high isotacticity of polypropylene produced with the catalyst suggests a structural analogy to *rac*-Et(Ind)$_2$ZrCl$_2$. (After Soga, K., Kim, H. J., and Shiono, T. 1994. *Macromol. Chem. Phys.*, 195, 3347. With permission.)

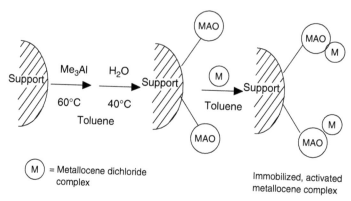

FIGURE 2.100 Synthesis of a heterogeneous metallocene catalyst with immobilized MAO. (After Köpl, A., Alt, H. G., and Phillips, M. D. 2001. *J. Appl. Polym. Sci.*, 80(3), 454.)

for example, preprimed WPC planks for front porches, siding [282] and roof shingles with A class fire rating made from natural fibers and polyethylene.

Advanced WPCs are being investigated to replace treated timber currently used to support piers and absorb the shock of docking ships. Other products include flowerpots, shims (thin washer or strip), cosmetic pencils, grading stakes, tool handles, hot tub siding, and office accessories [273].

2.19 Polymerization-Filled Composites

The industrial interest in filled polymers is steadily increasing as the addition of fillers results in an appreciable reduction of cost and provides the opportunity of producing materials with a new set of selected properties. The preparation of composites by melt blending the polymeric matrix and the filler is a straight-forward procedure but is poorly efficient when the properties of the resulting composites are concerned. In order to overcome these limitations, the polymerization-filling technique has been developed [283,284]. It consists of attaching a Ziegler-Natta-type catalyst onto the surface of an inorganic filler, so that olefin can be polymerized from the filler surface. This allows a very high filler loading (up to 95 vol.%) to be reached together with acceptable mechanical properties. Indeed, the polymer structure and molecular weight (ultrahigh molecular weight polyethylene, UHMWPE) and

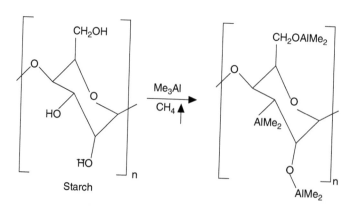

FIGURE 2.101 Reaction of trimethylaluminum with starch in toluene. (After Köpl, A., Alt, H. G., Phillips, M. D. 2001. *J. Appl. Polym. Sci.*, 80(3), 454. With permission.)

the high filling degree are the basis of completely new composite materials that cannot be produced by the standard mixing method [285].

In the early 1980s, UHMWPE was synthesized from a filler-supported catalyst and showed an unusual combination of high stiffness and high impact resistance even for a filler content as high as 60–70 wt%. In spite of these valuable mechanical properties, the interest in this UHMWPE was limited due to very poor processability, requiring either compression molding at high temperature or blending with low-molecular-weight polyethylene so as to reach an acceptable melt viscosity. This latter strategy was used in Russia for producing Norplastic composites [286].

Different methods have been considered to generate active polymerization sites on the filler surface. As a rule, the transition metal compounds have been either merely deposited on the surface and possibly within the pores of the particles, or the organometallic compounds have been reacted with some functional groups, for example, hydroxyls available on the filler surface. As far as silica is concerned, there are three kinds of OH groups on the silica surface, viz., isolated, hydrogen-bonded, and geminated hydroxyl groups (Figure 2.98) [285]. Bridged metallocenes can be covalently bound to silica through reaction of the surface silanols with reactive carbon halide or silicon halide bonds on the ligand bridge. The synthesis of such bridge-anchored *ansa*-metallocene is shown schematically in Figure 2.99. Activities of catalysts based on such surface-bound precursors are reported [287] to be greatly superior to the homogeneous analogs.

In an alternative and significantly more effective procedure [288], trimethylaluminum (TMA) is first attached to silica gel by heating the carrier material with TMA in an inert solvent such as toluene at 60°C until the gas (methane) evolution subsides. The reaction mixture is then cooled down to 40°C and a little amount of water is bubbled through it using a moist argon flow to produce a completely immobilized form of MAO:

$$Me_3Al + SiO_2/H_2O \xrightarrow{Toluene} SiO_2 - MAO + CH_4$$

Finally, a catalyst precursor (metallocene) is added to the mixture at room temperature, the amount being dependent on the desired Al:Zr ratio. The process is shown schematically in Figure 2.100.

Besides silica, numerous other materials, including alumina, zeolites, clays, and organic polymers have been used as supports for metallocene-based catalysts. In the case of nonreactive particle surfaces, a pretreatment with a reagent bearing functional groups has been used.

Cellulose, starch, and flour are excellently suited for immobilization of MAO because of the presence of OH groups as anchor points (Figure 2.101). The arrangement of the OH groups on the surface possibly serves as a template for the formation of aluminoxane structures that exhibit advantageous cocatalyst properties and the activities of the resulting supported catalysts are typically in the range of 270 kg polyethylene/g Zr-h [288].

References

1. Cassidy, P. E. 1980. *Thermally Stable Polymers*, Marcel Dekker, New York.
2. Abadie, M. J. and Silhon, B. eds. 1991. *Polyimides and Other High-Temperature Polymers*, pp. 200–79. Elsevier, Amsterdam.
3. Kalugina, E. V., Gumaragalieva, K. Z., and Zaikov, G. E. 2005. *Thermal Stability of Engineering Heterochain Thermoresistant Polymers*, VNU Science Publishers, Leiden, The Netherlands.
4. Cowie, J. M. G. 1991. *Polymers: Chemistry and Physics of Modern Materials*, Blackie, Glasgow and London.
5. Gordon, M. and Platé, N. A. 1984. *Liquid Crystal Polymers: Advances in Polymer Science, Vol. 59–6*, Springer-Verlag, New York.
6. Calundann, G. W. and Griffin, A. C. 1986. Thermoplastic polyesters, In *High Performance Polymers: Their Origins and Development*, R. B. Seymour and G. S. Kirschenbaum, eds., pp. 200–79. Elsevier Science, New York.

7. Kwolek, S. L., Morgan, P. W., and Schaefgen, J. R. 1987. In *Liquid Crystal Polymers*, H.F. Mark, N.M. Bikales, C.G. Overberger, and G. Menges, eds., In *Encyclopedia of Polymer Science and Engineering*, *Vol. 9*, pp. 200–79. Wiley-Interscience, New York.

8. Lenze, R. W. and Lin, J. I. 1986. *Polym. News*, 11, 200.

9. Morgan, P. W. 1979. *Chem. Tech.*, 316.

10. Krigbaum, W. R., Watanabe, J., and Ishikawa, T. 1983. *Macromolecules*, 16, 1271.

11. Blumstein, A. 1985. *Polymeric Liquid Crystals*, Plenum Press, New York.

12. Seymour, R. B. ed. 1981. *Conductive Polymers*, pp. 1416–79. Plenum Press, New York.

13. Wirsen, A. 1987. *Electroactive Polymer Materials*. Technomic, Lancaster, Pennsylvania.

14. Bowden, J. M. and Turner, S. R. eds. 1988. *Electronics and Photonic Applications of Polyemrs*, *Advances in Chemistry Series*, *Vol. 218*, pp. 1416–79. American Chemical Society, Washington, D.C.

15. Margolis, J. M. ed. 1989. *Conductive Polymers and Plastics*. Chapman and Hall, New York.

16. Aldissi, M. 1989. *Inherently Conducting Polymers*, Noyes Data Corp., Park Ridge, New Jersey.

17. Little, W. A. 1964. *Phys. Rev.*, 134, 1416.

18. Ito, T., Shirakawa, H., and Ikeda, S. 1974. *J. Polym. Sci.*, 12, 11.

19. Edwards, J. H., Feast, W. J., and Bott, D. C. 1984. *Polymer*, 25, 395.

20. Gill, W. D., Clarke, T. C., and Streat, G. B. 1982. *Appl. Phys. Commun.*, 2, 4, 211.

21. Chiang, C. K., Gau, S. C., Fincher, C. R., Park, Y. W., Macdiarmid, A. G., and Heege, A. J. 1978. *Appl. Phys. Lett.*, 33, 1, 18.

22. Chen, S. A. and Hwang, G. W. 1995. *J. Am. Chem. Soc.*, 117, 40, 10055.

23. Nguyen, M. T., Kasai, P., Miller, J. L., and Diaz, A. F. 1994. *Macromolecules*, 27, 3625.

24. Angelopoulos, M., Patel, N., and Shaw, J. M. 1994. *Mater. Res. Soc. Symp. Proc.*, 328, 173.

25. Ferraro, J. R. and Williams, J. M. 1987. *Introduction to Synthetic Electrical Conductors*, Academic Press, San Diego, California.

26. Scrosati, B. and Owens, B. B. 1987. *Solid State Ionic*, 23, 275.

27. Adhikari, B. and Majumdar, S. 2004. *Prog. Polym. Sci.*, 29, 699.

28. Persand, K. C. and Pelosi, P. 1985. An approach to an artificial nose, *Trans. Am. Soc. Artif. Int. Organs*, 31, 297.

29. Marsella, M. J., Carroll, P. J., and Swager, T. M. 1995. *J. Am. Chem. Soc.*, 117, 9832.

30. Torsi, L., Pezzuto, M., Siciliano, P., Rella, R., Sabbatini, L., Valli, L., and Zambonin, P. G. 1998. *Sens. Actuators B*, 48, 362.

31. Hirata, M. and Sun, L. 1994. *Sens. Actuators A*, 40, 159.

32. Unde, S., Ganu, J., and Radhakrishnan, S. 1996. *Adv. Mater. Opt. Elecr.*, 6, 151.

33. Chabukswar, V. V., Pethkar, S., and Athawale, A. A. 2001. *Sens. Actuators B*, 77, 657.

34. Li, D., Jiang, Y., Wu, Z., Chen, X., and Li, Y. 2000. *Sens. Actuators B*, 66, 125.

35. Chiang, J. C. and MacDiarmid, A. G. 1986. *Synth. Met.*, 13, 193.

36. Athawale, A. A. and Kulkarni, M. V. 2000. *Sens. Actuators B*, 67, 173.

37. Jun, H-K. , Hoh, Y-S. , Lee, B-S. , Lee, S-T. , Lim, J-O. , Lee, D-D. , and Huh, J-S. 2003. *Sens. Actuators B*, 96, 576.

38. Potje-Kamloth, K. 2002. *Crit. Rev. Anal. Chem.*, 32, 121.

39. Brahim, S. and Narinesingh, D. 2002. A.Guiseppi-Elie, *Biosens. Bioelectron.*, 17, 53.

40. Bi, J. J., Downs, J. C., and Jacob, J. T. 2004. *J. Biomater. Sci. Polym. Ed.*, 15, 905.

41. Menon, V. P., Lei, J., and Martin, C. R. 1996. *Chem. Mater.*, 8, 2382.

42. Sasha, S. K. 2002. *Appl. Phys. Lett.*, 81, 19.

43. Ikegame, M., Tajima, K., and Aida, T. 2003. *Angew. Chem. Int. Ed.*, 42, 5154.

44. Zhang, X. and Manohar, S. K. 2004. *J. Am. Chem. Soc.*, 126, 12714.

45. Li, D. and Xia, Y. N. 2004. *Adv. Mater.*, 16, 1151.

46. Fenot, A. and Chronakis, I. S. 2003. *Curr. Opin. Colloid Interface Sci.*, 8, 64.

47. Huang, Z. M., Zhang, Y. Z., Kotaki, M., and Ramakrishna, S. 2003. *Compos. Sci. Technol.*, 63, 2223.

48. Liu, H. Q., Kameoka, J., Czaplewski, D. A., and Craighead, H. G. 2004. *Nano Lett.*, 4, 671.

49. Nair, S., Natarajan, S., and Kim, S. H. 2005. *Macromol. Rapid Commun.*, 26, 1599.

50. Winther-Jensen, B., Chen, J., West, K., and Wallace, G. 2004. *Macromolecules*, 37, 5930.

51. Angelopoulos, M. 1998. Conducting polymers in microelectronics, In *Handbook of Conducting Polymers*, T.A. Skotheim, R.L. Elsenbaumer, and J.R. Reynolds, eds., pp. L14–79. Marcel Dekker, New York, Chapter 32.

52. Gottesfeld, S., Uribe, F. A., and Armes, S. P. 1992. *J. Electrochem. Soc.*, 139, 1, L14.

53. Salaneck, W. R., Lundstrom, I., and Ranby, B. eds. 1993. *Conjugated Polymers and Related Materials: The Interconnection of Chemical and Electronic Structure*, pp. 539–79. Oxford Scientific, Oxford, England.

54. Burroughes, J. H., Bradley, D. D. C., Brown, A. R., Marks, R. N., Mackay, K., Friend, R. H., Burn, P. L., and Holmes, A. B. 1990. *Nature*, 347, 539.

55. Gustavsson, G., Cao, Y., Treacy, G. M., Klavetter, F., Colaneri, N., and Heeger, A. J. 1992. *Nature*, 357, 477.

56. Kulkarni, V. G. 1993. Processing of polyanilines, In *Intrinsically Conducting Polymers; An Emerging Technology*, M. Aldissi, ed., pp. 45–79. Dordrecht, Kluwer.

57. Heywang, G. and Jonas, F. 1992. *Adv. Mater.*, 4, 2, 116.

58. Huang, B., Kang, G., and Ni, Y. 2005. *Can. J. Chem. Eng.*, 83, 896.

59. Knisley, J. Oct. 2002. *Fiber Optic Lighting Dries Out*, ecmweb.com/mag/electric_fiber_optic_lighting.

60. Naritomi, M. Dec. 1996. *CYTOP Amorphous Fluoropolymers for Low Loss POF*. Proc. POF-Asia-Pacific Forum, Tokyo.

61. Koike, Y., Ishigure, T., and Nihei, E. 1995. *IEEE J. Lightwave Technol.*, 13, 7, 1475.

62. Khoe, G. D. 14–16 July 1999. *Exploring the use of GIPOF Systems in the 640 nm to 1300 nm Wavelength Area*, Proc. 8th Intern. POF Conf., pp. 36–43, Chiba, Japan.

63. Khoe, G. D., Yoike, Y., Ishigure, T., Bennekom, P. K. v., Boom, H. P. A. v. d., Li, W., and Yabre, G. 1999. *Status of GIPOF Systems and Related Technologies*, 25th European Conf. on Opt. Comm. 99, 26–30 Sept. 1999, pp II/274-II/277.

64. Epworth, R. E. Sept. 1978. *The Phenomenon of Modal Noise in Analogue and Digital Optical Fiber Communication*. Proc. of 4th European Conf. on opt. comm., pp. 492–501. Genoa, Italy.

65. Paradigm Optics, http://www.paradigmoptics.com/pof/pofLags.html, 1989.

66. Garito, A. F. and Wong, K. Y. 1987. *Polym. J.*, 19, 51.

67. Chemla, D. S. and Zyss, J. eds. 1987. *Nonlinear Optical Properties of Organic Molecules and Crystals*, Vol. 1, pp. 6854–79. Academic Press, San Diego, California,. No. 4.

68. Prasad, P. N. and Ulrich, D. R. eds. 1988. *Nonlinear Optical and Electroactive Polymers*, pp. 6854–79. Plenum, New York.

69. Wang, T. T., Herbert, J. M., and Glass, A. eds. 1988. *The Applications of Ferroelectric Polymers*, pp. 6854–79. Blackie, Glasgow.

70. Armand, M. B., Chabagno, J. M., and Duclot, M. Sept. 1978. *Abstract 6, Second International Conference on Solid Electrolytes*, pp. 20–22, St. Andrews.

71. Blonsky, P. M., Shriver, D. F., Austin, P., and Allcock, H. R. 1984. *J. Am. Chem. Soc.*, 106, 6854.

72. Abraham, K. M., Alamgir, M., and Reynolds, R. K. 1989. *J. Electrochem. Soc.*, 136, 12, 3576.

73. Wu, H., Shy, H., and Ko, H. 1989. *J. Power Sources*, 27, 1, 59.

74. Steele, B. C. H. and Heinzel, A. 2001. *Nature*, 414, 345.

75. Yang, C., Costamagna, P., Srinivasan, S., Benziger, J., and Bocarsly, B. 2001. *J. Power Sources*, 103, 1.

76. Li, Q., He, R., Jensen, J. O., and Bjerrum, N. J. 2003. *Chem. Mater.*, 15, 4896.

77. Daletou, M. K., Gourdoupi, N., and Kallitsis, J. K. 2005. *J. Membr. Sci.*, 252, 115.

78. Pefkianakis, E. K., Deimede, V., Daletou, M. K., Gourdoupi, N., and Kallitsis, J. K. 2005. *Macomol. Rapid Commun.*, 26, 1724.

79. De Forest, W. S. 1975. *Photoresist: Materials and Processes*, McGraw-Hill, New York.

80. Roffery, C. G. 1982. *Photopolymerization of Surface Coatings*, Wiley, New York.

81. Roberts, E. D. 1985. *Chem. Ind.,* 251.
82. Allen, N. S. ed. 1989. *Photopolymerization and Photoimaging Science and Technology,* pp. 1273–79. Elsevier, Barking, Essex, England.
83. Reichmanis, E. and Thomson, L. F. 1989. *Chem. Rev.,* 89, 6, 1273.
84. Hatzakis, M., Stewart, K. J., Shaw, J. M., and Rishton, S. A. 1990. *Microelect. Eng.,* 11, 1–4, 487.
85. Patrick, R. L. 1969. *Treatise on Adhesion and Adhesives, Vol. 2,* Marcel Dekker, New York.
86. Skeist, I. 1977. *Handbook of Adhesives,* Van Nostrand, New York.
87. Lee, L. H. ed. 1984. *Adhesive Chemistry,* pp. 127–79. Plenum, New York.
88. Wake, W. C. 1976. *Adhesion and Formulation of Adhesives,* Applied Science Publishers, London, England.
89. Bateman, D. L. 1978. *Hot Melt Adhesives.* Noyes Data Corp., New Jersey.
90. Pappas, S. P. 1978. UV Curing, Science and Technology, Vol. 1–2, Technology Marketing Corp., Norwalk, Connecticut.
91. Loomis, G. 1991. *ACS polym, Preprints,* 32, 2, 127.
92. Casey, J. P. and Manly, D. G., 1976. Poly(vinyl alcohol) biodegradation by oxygen-activated sludge, *Proceedings of 3rd International Biodegradation Symposium,* Applied Science, Barkins, England.
93. Griffin, G. J. L. 1994. *Chemistry and Technology of Biodegradable Polymers,* Blackie, London, England.
94. Potts, J. E. 1984. Plastics, environmentally degradable, In *Kirk-Othmer Encyclopedia of ChemicalTechnology,* M. Grayson, ed., pp. 1022–79. Wiley-Interscience, New York.
95. Griffin, G. J. L. and Mivetchi, H. 1976. Biodegradation of ethylene/vinyl acetate copolymers, In *Proceedings of 3rd International Biodegradation Symposium,* J.M. Sharpley and A.M. Kaplan, eds., pp. 1022–367. Applied Science, Barkins, England..
96. Maddever, W. J. and Chapman, G. M. 1989. *Plastics Eng.,* 31.
97. Oley, F. H. and Westhoff, R. 1984. Starch-based films, *IEC Prod. Res. Develop.,* June.
98. Narayan, R. 1989. *Kunststoffe,* 79, 1022.
99. Paul, D. R. and Harris, F. W. eds. 1976. *Controlled Release Polymeric Formulations,* ACS Symposium Series No. 33, pp. 265–79. *American Chemical Society,* Washington, D.C.
100. Gebelein, C. G. and Carraher, C. E. eds. 1983. *Polymeric Materials in Medication,* pp. 265–79. Plenum Press, New York.
101. Thomson, R. C., Wake, M. C., Yaszemski, M. J., and Mikos, A. G. 1995. *Adv. Polym. Sci.,* 122, 265.
102. Pittenger, M. F., Mackay, A. M., Beck, S. C., Jaiswal, S. K., Douglas, R., Mosca, J. D., Moorman, M. A., Simonetti, D. W., Craig, S., and Marshak, D. R. 1999. *Science,* 284, 143.
103. Minuth, W. W., Strehl, R., and Schumacher, K. 2002. *Von der Zellkultur zum Tissue Engineering,* Pabst Science Publishers, Lengerich.
104. Han, D. and Hubbell, J. A. 1997. *Macromolecules,* 30, 6077.
105. Schmaljohann, D. 2005. Thermoresponsive polymers and hydrogels in tissue engineering. *e-Polymers* No. 021.
106. Chen, G., Ushida, T., and Tateishi, T. 2000. *Adv. Mater.,* 12, 455.
107. Chu, C.-C. 1995. In *The Biomedical Engineering Handbook,* J.D. Bronzino, ed., pp. 105–79. CRC Press, FL.
108. Stamboulis, A. G., Boccaccini, A. R., and Hench, L. L. 2002. *Adv. Eng. Mater.,* 4, 3, 105.
109. Wheeler, D. L., Montfort, M. J., and McLoughlin, S. W. 2001. *J. Biomed. Mater. Res.,* 55, 603.
110. Holliday, L. ed. 1975. Ion Polymers, pp. 62–79. Halsted Press, New York.
111. Eisenberg, A. and King, M. 1977. Ion Containing Polymers, Academic Press, New York.
112. Longworth, R. 1983. *Developments in Ionic Polymers I,* Applied Science Publishers, London, England.
113. Flett, D. S. ed. 1983. Ion Exchange Membranes, pp. 62–79. Ellis Horwood, Chichester, England.
114. Sourirajan, S. and Agarwal, J. P. 1969. *Ind. Eng. Chem.,* 61, 62.
115. Kunin, R. 1972. *Ion Exchange Resins,* 2nd ed., R.E. Krieger, Melbourne, Florida.

116. Osawa, F. 1970. *Polyelectrolytes*, Marcel Dekker, New York.
117. Hodge, P. and Sherrington, D. C. eds. 1980. *Polymer Supported Reactions in Organic Synthesis*, pp. 9326–79. John Wiley, Chichester, England.
118. Mathur, N. K., Narang, C. K., and Williams, R. E. 1980. *Polymers as Aids Inorganic Chemistry*, Academic Press, New York.
119. Zhong, H. E. et al. 1997. *J. Org. Chem.*, 62, 9326.
120. Flynn, D. L. et al. 1997. *J. Am. Chem. Soc.*, 119, 4874.
121. Cabasso, I. 1987. Membranes, In *Encyclopedia of Polymer Science and Engineering*, Vol. 9, J.I. Kroschwitz, ed., pp. 13-367, Wiley-Interscience, New York.
122. Lloyd, D. R. 1985. *Materials Science of Synthetic Membranes*, ACS Symp. Serie 269, American Chemical Society, Washington, DC.
123. Kim, S. S. and Lloyd, D. R. 1991. *J. Membr. Sci.*, 64, 13.
124. Matsuyama, H., Maki, T., Teramoto, M., and Asano, K. 2002. *J. Membr. Sci.*, 204, 323.
125. Sun, H., Rhee, K. B., Kitano, T., and Mah, S. I. 1999. *J. Appl. Polym. Sci.*, 75, 1235.
126. Shang, M., Matsuyana, H., Teramoto, M., Okuno, J., Lloyd, D. R., and Kubota, N. 2005. *J. Appl. Polym. Sci.*, 95, 219.
127. Lin, H., Van Wagner, E., Roharjo, R., Freeman, B. D., and Roman, I. 2006. *Adv. Mater.*, 18, 39.
128. Chung, T. S. and Hu, X. 1997. *J. Appl. Polym. Sci.*, 66, 1067.
129. Konstantin, P., Göehl, H., and Gullberg, C. 1981. *Artif. Organs*, 5, 691.
130. Breck, D. W. 1974. *Zeolite Molecular Sieves*, John Wiley, New York.
131. Nazzal, F. F. and Wiesner, M. R. 1996. *Water Environ. Res.*, 68, 1187.
132. Xu, Z. L., Chung, T. S., and Huang, Y. 1999. *J. Appl. Polym. Sci.*, 74, 2220.
133. Xu, Z. L., Chung, T. S., Loh, K. C., and Lim, B. C. 1999. *J. Membr. Sci.*, 158, 41.
134. Tanaka, T., Fillmore, D. J., Nishio, I., Sun, S. T., Swislow, G., and Shah, A. 1980. *Phys. Rev. Lett.*, 45, 1936.
135. Li, Y. F., Li, X. Z., Zhou, L. C., Zhu, X. X., and Li, B. N. 2004. *Polym. Adv. Technol.*, 15, 34.
136. Raju, K. M., Raju, M. P., and Mohan, Y. M. 2003. *Polym. Int.*, 52, 768.
137. Kohls, S. J., Baker, D. D., Kremer, D. A., and Dawson, J. O. 1999. *Plant Soil*, 214, 105.
138. Wu, J. H., Wei, Y. L., and Lin, S. B. 2003. *Polymer*, 44, 6513.
139. Li, A., Wang, A. Q., and Chen, J. M. 2004. *J. Appl. Polym. Sci.*, 92, 1596.
140. Jarosiewicz, A. and Tomaszewska, M. 2003. *J. Agric. Food Chem.*, 51, 413.
141. Zhang, J., Liu, R., Li, A., and Wang, A. 2006. *Ind. Eng. Chem. Res.*, 45, 48.
142. Schild, H. G. 1992. *Prog. Polym. Sci.*, 17, 163.
143. Kataoka, K., Koyo, H., and Tsuruta, T. 1995. *Macromolecules*, 28, 3336.
144. Urry, D. W., Hayes, L. C., and Parker, T. M. 1982. *Chem. Phys. Lett.*, 210, 218.
145. Tanaka, T., Nishio, I., and Sun, S. T. 1982. *Science*, 29, 218.
146. Suzuki, A. and Tanaka, T. 1990. *Nature*, 346, 345.
147. Iomata, H., Goto, S., and Saito, S. 1990. *Macromolecules*, 23, 4887.
148. Cole, C. A., Schreiner, S. M., Priest, J. H., Monji, N., and Hoffman, A. S. 1987. *ACS Symp. Ser.*, 350, 245.
149. Schild, H. G. and Tirrell, D. A. 1990. *J. Phys. Chem.*, 94, 4352.
150. Liang, L., Feng, X., and Lin, J. 1999. *J. Appl. Polym. Sci.*, 72, 1.
151. Kondo, T., Koyama, M., and Kubota, H. 1998. *J. Appl. Polym. Sci.*, 67, 2057.
152. Taylor, L. D. and Cerankowski, L. D. 1975. *J. Polym. Sci. Pt. A: Polym. Chem.*, 13, 2551.
153. Priest, J. H., Murray, S. L., Nelson, J. R., and Hoffman, A. S. 1987. *ACS Sym. Ser.*, 350, 255.
154. Freitas, R. F. S. and Cussler, E. L. 1987. *Sep. Sci. Technol.*, 22, 911.
155. Cussler, E. L. 1989. U.S. Pat. 4,828, 701; (assigned to Regents of the University of Minnesota).
156. Hoffman, A. S., Afrassiabi, A., and Dong, L. C. 1986. *J. Control. Release*, 4, 213.
157. Hoffman, A. S. 1987. *J. Control. Release*, 6, 297.
158. Mukae, K., Bae, Y. H., Okano, T., and Kim, S. W. 1990. *Polym. J.*, 22, 250.
159. Cussler, E. L., Stoker, M. R., and Varberg, J. E. 1984. *AIChE J.*, 30, 578.

160. Trank, S. J., Johnson, D. W., and Cussler, E. L. 1989. *Food Technol. (Chicago)*, 43, 6, 78.

161. Uehara, A., Kurahashi, M., and Tokunaga, Y. 1988. *Jap. Pat., Showa*, 63, 143, 907 assigned to Nippon Steel Corp.

162. Itoh, H., Nitta, A., Tanaka, T., and Kamio, H., 1986. Eur. Pat. Appl., 178, 175, (assigned to Mitsui Toatsu Chemicals, Inc.)

163. Huang, X., Akehata, T., Unno, H., and Hirasa, O. 1989. *Biotechnol. Bioeng.*, 34, 102.

164. Park, T. G. and Hoffman, A. S. 1990. *Biotech. Bioeng.*, 35, 152.

165. Okano, T., Bae, Y. H., Jacobs, H., and Kim, S. W. 1990. *J. Control. Release*, 11, 255.

166. Mamada, A., Tanaka, T., Kungwatchakun, D., and Irie, M. 1990. *Macromolecules*, 23, 1517.

167. Tanaka, T., Nishio, I., Sun, S. T., and Ueno-Nishio, S. 1982. *Science*, 218, 4571, 467.

168. Tanaka, T. and Fillmore, D. J. 1979. *J. Chem. Phys.*, 70, 1214.

169. Wu, X. S., Hoffman, A. S., and Yager, P. 1992. *J. Polym. Sci., Part A: Polym. Chem.*, 30, 2121.

170. Yoshida, R., Uchida, K., Kaneko, Y., Sakai, K., Kikuchi, A., Sakurai, Y., and Okano, T. 1995. *Nature*, 374, 240.

171. Serizawa, T., Wakita, K., Kaneko, T., and Akashi, M. 2002. *J. Polym. Sci., Part A: Polym. Chem.*, 40, 4228.

172. Xue, W., Hamley, I. W., and Huglin, M. B. 2002. *Polymer*, 43, 5181.

173. Cheng, S. X., Zhang, J. T., and Zhuo, R. X. 2003. *J. Biomed. Mater. Res., Part A*, 67, 96.

174. Zhang, J. T., Cheng, S. X., and Zhuo, R. X. 2003. *J. Polym. Sci., Part A: Polym. Chem.*, 41, 2390.

175. Kato, N., Sakai, Y., and Shibata, S. 2003. *Macromolecules*, 36, 961.

176. Zhang, J-T. , Huang, S-W. , Xue, Y-N. , and Zhuo, R. X. 2005. *Makromol. Rapid. Commun.*, 26, 1346.

177. Flory, P. J. 1953. *Principles of Polymer Chemistry*, Cornell University Press, Ithaca, N.Y.

178. Tomalia, D. A., Baker, H., Dewald, J., Hall, M., Kallos, G., Martin, J. R., Ryder, J., and Smith, P. 1985. *Polym. J.*, 17, 117.

179. Newkome, G. R., Yao, Z., Baker, G. R., and Gupta, V. K. 1985. *J. Org. Chem.*, 50, 2004.

180. Zeng, F. and Zimmerman, S. C. 1997. *Chem. Rev.*, 97, 1681.

181. Kim, Y. H. and Webster, O. W. 1990. *J. Am. Chem. Soc.*, 112, 4592.

182. Hult, A., Johansson, M., and Malmström, E. 1999. *Adv. Polym. Sci.*, 143, 1.

183. Hawker, C. J., Lee, R., and Fréchet, J. M. J. 1991. *J. Am.Chem. Soc.*, 113, 4583.

184. Fréchet, J. M. J., Henmi, M., Gitsov, I., Aoshima, S., Leduc, M. R., and Grubbs, R. B. 1995. *Science*, 269, 1080.

185. Ambade, A. V. and Kumar, A. 2000. *Prog. Polym. Sci.*, 25, 1141.

186. Costello, P. A., Martin, I. K., Stark, A. T., Sherrington, D. C., and Titterton, A. 2002. *Polymer*, 43, 245.

187. Isaure, F., Cormack, P. A. G., and Sherrington, D. C. 2004. *Macromolecules*, 37, 2096.

188. Gretton-Watson, S. P., Alpay, E., Steinke, J. H. G., and Higgins, J. S. 2005. *Ind. Eng. Chem. Res.*, 44, 8682.

189. Bruening, M. L., Zhou, Y., Aguilar, G., Agee, R., Bergbreiter, D. E., and Crooks, R. M. 1997. *Langmuir*, 13, 770.

190. Kim, Y. H. and Webster, O. W. 1992. *Macromolecules*, 25, 5561.

191. Boogh, L. Pettersen, B. Japon, S. and Manson, J.-A. 1995. *Proceedings of 28th International Conference on Composite Materials, Vol. 4*, p. 389, Whistler, Canada.

192. Pettersen, B., and Sorensen, K., 1994. *Proceedings of the 21st Waterborne, Higher Solids & Powder Coatings Symposium*, p. 753, New Orleans, Louisiana.

193. Roberts, J., Bhalgat, M., and Zera, R. 1996. *J. Biomed. Res.*, 30, 53.

194. Kim, Y. H. and Webster, O. W. 1990. *J. Am. Chem. Soc.*, 112, 4592.

195. Hawker, C. J. and Chu, F. 1996. *Macromolecules*, 29, 4370.

196. Uhric, K. 1997. *Trends Polym. Sci.*, 5, 12, 388.

197. Smith, D. K., Hirst, A. R., Love, C. S., Hardy, J. G., Brignell, S. V., and Huang, B. 2005. *Prog. Poym. Sci.*, 30, 220.

198. Frauenrath, H. 2005. *Prog. Polym. Sci.*, 30, 325.

199. Kim, B. K., Lee, S. Y., and Xu, M. 1996. *Polymer*, 37, 5781.

200. Kim, B. K., Lee, S. Y., Lee, J. S., Baek, S. H., Choi, Y. J., Lee, J. O., and Xu, M. 1998. *Polymer*, 39, 2803.

201. Jeong, H. M., Lee, J. B., Lee, S. Y., and Kim, B. K. 2000. *J. Mater. Sci.*, 35, 279.

202. Han, M. J., Boung, K. A., and Byung, K. K. 2001. *Eur. Polym. J.*, 37, 2245.

203. Hu, J., Yang, Z., Yeung, L., Ji, F., and Liu, Y. 2005. *Polym. Int.*, 54, 854.

204. Thies, C. 1989. Microencapsulation, In *Polymers: Biomaterials and Medical Applications*, J.I. Kroschwitz, ed., In *Encyclopedia Reprint Series*, pp. 346–367, Wiley, New York.

205. Benita, S. ed. 1996. *Microencapsulation: Methods and Industrial Applications*, pp. 213–79. Marcel Dekker, New York.

206. Bonnett, R. 1992. *Microencapsulation of Drugs*, CRC Press, Boca Raton, Florida.

207. Franjione, J. and Vasishtha, N. Summer 1995. *The Art and Science of Microencapsulation, Technol. Today*, Southwest Research Institute Publications.

208. Yadav, S. K., Khilar, K. C., and Suresh, A. K. 1997. *J. Membr. Sci.*, 125, 213.

209. Hong, K. and Park, S. 1999. *Matl. Res. Bull.*, 34, 963; Hong, K. and Park, S. 1999. *Matl. Chem. Phys.*, 64, 20.

210. Hirech, K., Payan, S., Karnelle, G., Brujes, L., and Legrand, J. 2003. *Powder Technol.*, 130, 324.

211. Takahashi, T., Taguchi, Y., and Tanaka, M. 2005. *J. Chem. Eng. Japan*, 38, 11, 929.

212. Jiang, B., Hu, L., Gao, C., and Shen, J. 2006. *Acta Biomaterialia*, 2, 9.

213. Hall, H.S. and Pondell, R. E. 1980. A. F. Kydonieus, ed., In *Controlled Release Technologies: Methods, Theory and Applications*, Vol. 2, pp. 630–79, CRC Press, Boca Raton, Florida.

214. Salman, O. A. 1989. *Ind. Eng. Chem. Res.*, 28, 630.

215. Rajsekharan, A. J. and Pillai, V. N. 1996. *J. Appl. Polym. Sci.*, 60, 2347.

216. Tomaszewska, M. and Jarosiewicz, A. 2002. *J. Agri. Food Chem.*, 50, 4634.

217. Krishnamoorti, R. and Vaia, R. A. eds. 2002. *Polymer Naocomposites, Synthesis, Characterization and Modeling*, ACS Symposium Series 804, pp. 2758–79, American Chemical Soc., Washington, DC.

218. Park, J. H. and Jana, S. C. 2003. *Macromolecules*, 36, 2758.

219. Lepoittevin, K., Devalckenaere, M., Pantoustier, N., Alexandre, M., Kubies, D., Calberg, C., Jerome, R., and Dubois, P. 2002. *Polymer*, 43, 4017.

220. Vaia, R. A., Jandt, K. D., Kramer, E. J., and Giannelis, E. P. 1996. *Chem. Mater.*, 8, 2628.

221. Vaia, R. A. and Giannelis, E. P. 1997. *Macromolecules*, 30, 8000.

222. Liu, L., Qi, Z., and Zhu, X. 1999. *J. Appl. Polym. Sci.*, 71, 1133.

223. Alexandre, M., Beyer, G., Henrist, C., Cloots, R., Rulmont, A., Jerome, R., and Dubois, P. 2001. *Macromol. Rapid Commun.*, 22, 643.

224. Reichert, P., Nitz, H., Klinke, S., Brandsch, R., Thomann, R., and Mülhaupt, R. 2000. *Macromol. Mater. Engng.*, 275, 8.

225. Huang, J. C., Zhu, Z., Yin, J., Qian, X., and Sun, Y. Y. 2001. *Polymer*, 42, 873.

226. Lans, M., Francesangeli, O., and Sandrolini, F. 1997. *J. Mater. Res.*, 12, 3134.

227. Vaia, R. A., Vasudevan, S., Kramiec, W., Scanlon, L. G., and Giannelis, E. P. 1995. *Adv. Mater.*, 7, 154.

228. Morgan, A. B., Gilman, J. W., and Jackson, C. L. 2001. *Macromolecules*, 34, 2735.

229. Triantafillidis, C. S., LeBaron, P. C., and Pinnavaia, T. J. 2002. *J. Solid State Chem.*, 167, 354.

230. Lu, H., Liang, G., Ma, X., Zhang, B., and Chen, X. 2004. *Polym. Intl.*, 53, 1545.

231. Messersmith, P. B. and Giannelis, E. P. 1995. *J. Polym. Sci.*, Part A, 33, 1047.

232. Badopadhyay, S., Hsieh, A. J., and Giannelis, E. P. 2002. PMMA nanocomposites synthesized by emulsion polymerization, In *Polymer Nanocomposites: Synthesis, Characterization and Modeling*, R. Krishnamoorti and R. A. Vaia, eds., In *ACS Symp. Ser.*, No. 804., pp. 137–79, American Chemical Soc., Washington, D. C.

233. Usuki, A., Kato, M., Okada, A., and Kurauchi, T. 1997. *J. Appl. Polym. Sci.*, 63, 137.

234. Kato, M., Usuki, A., and Okada, A. 1997. *J. Appl. Polym. Sci.*, 66, 1781.
235. Garcés, J. M., Moll, D. J., Bicerno, J., Fibiger, R., and McLeod, D. G. 2000. *Adv. Mater.*, 12, 23, 1835.
236. Lau, K. T. and Hui, D. 2002. *Composites B: Eng.*, 33, 263.
237. Baughman, R. H., Zakhidov, A. A., and de Heer, W. A. 2002. *Science*, 297, 787.
238. Lahr, B. and Sandler, J. 2000. *Kunststoffe*, 90, 94.
239. Sandler, J. K. W., Pregel, S., Sadek, M., Gojny, F., van Es, M., Lohmar, J., Blau, W. J., Schulte, K., Windle, A. H., and Shaffer, M. S. P. 2004. *Polymer*, 45, 2001.
240. Beyer, G. 2002. *Fire Mater.*, 26, 291.
241. Kashiwagi, T., Gulke, E., Hilding, J., Groth, K., Harris, R., Butler, K., Shields, J., Kharchenko, S., and Douglas, J. 2004. *Polymer*, 45, 4227.
242. Kilbride, B. E., Coleman, J. N., Fraysse, J., Fournet, P., Cadek, M., Drury, A., Hutzler, S., Roth, S., and Blau, W. J. 2000. *J. Appl. Phys.*, 92, 4024.
243. Sennett, M., Welsch, E., Wright, J. B., Li, W. Z., Wen, J. G., and Ren, Z. F. 2003. *Appl. Phys. A*, 76, 111.
244. Hammel, E., Tang, X., Trampert, M., Schmitt, T., Mauthner, K., eder, A., and Pötschke, P. 2004. *Carbon*, 42, 1153.
245. Armelao, L., Barreca, D., Bottaro, G., Gasparotto, A., Tondello, E., Ferroni, M., and Polizzi, S. 2004. *Chem. Mater.*, 16, 3331.
246. Kim, D. S., Park, H. B., Rhim, J. W., and Lee, Y. M. 2004. *J. Membr. Sci.*, 240, 37.
247. Wen, J. and Wilkes, G. L. 1996. *Chem. Mater.*, 8, 1667.
248. Fitzgerald, J. J., Landry, C. J. T., and Pochan, J. M. 1992. *Macromolecules*, 25, 3715.
249. Landry, C. J. T., Coltrain, B. F., Landry, M. R., Fitzgerald, J. J., and Long, V. K. 1993. *Macromolecules*, 26, 3702.
250. Landry, C. J. T., Coltrain, B. K., and Brady, B. K. 1992. *Polymer*, 33, 1486.
251. Ravaine, D., Seminel, A., Charbouillot, Y., and Vincens, M. J. 1986. *Non-Cryst. Solids*, 82, 210.
252. Sun, C. C. and Mark, J. E. 1989. *Polymer*, 30, 104.
253. Stefanithis, I. D. and Mauritz, K. A. 1990. *Macromolecules*, 23, 2397.
254. Suzuki, F., Onozato, K., and Kurokawa, Y. 1990. *J. Appl. Polym. Sci.*, 39, 371.
255. Sengupta, R., Bandopadhyay, A., Sabharwal, S., Chaki, T. K., and Bhowmick, A. K. 2005. *Polymer*, 46, 3343.
256. Du, X. S., Xia, M., and Meng, Y. Z. 2004. *Eur. Polym. J.*, 40, 1489.
257. Zhang, W. D., Phang, I. Y., and Liu, T. 2006. *Adv. Mater.*, 18, 73.
258. Andrews, R., Jacques, D., Qian, D., and Rantell, D. 2002. *T. Acc. Chem. Res.*, 35, 1008; Gournis, D., Karakassides, M. A., Bakas, T., Boukos, N., and Petridis, D. 2002. *Carbon*, 40, 2641.
259. Ma, H., Wei, G., Liu, Y., Zhang, X., Gao, J., Huang, F., Tan, B., song, Z., and Qiao, J. 2005. *Polymer*, 46, 10568; Huang, F., Liu, Y., Zhang, X., and Wei, G. 2002. *Macromol. Rapid Commun.*, 23, 2786.
260. Gilman, J. W. and Kashiwagi, T. 1997. *SAMPE J.*, 33, 4, .
261. Vaia, R. A., Price, G., Ruth, P. N., Nguyen, H. T., and Lichtenham, J. 1999. *Appl. Clay Sci.*, 15, 67.
262. Colister, J. 2002. Commercialization of polymer nanocomposites, In *Polymer Nanocomposites: Sunthesis, Characterization, and Modeling*, R. Krishnamoorti and R.A. Vaia, eds., In ACS Symposium Series No. 804, pp. 11–79, American Chemical Soc, Washington, DC.
263. Edit, Jr., C. M. July 29, 1997. U.S. Pat. 5,652,284 (Assigned to Exxon Research and Engineering).
264. Elspass, C. W. Sept 15, 1998. U.S. Pat. 5,807,629 (Assigned to Exxon Research and Engineering).
265. LeBaron, P. C., Wang, Z., and Pinnavaia, T. J. 1999. *Appl. Clay Sci.*, 15, 11.
266. Kornmann, X., Berglund, L. A., and Sterte, J. 1998. *Polym. Eng. Sci.*, 38, 8, 1351.
267. Polansky, C. A. Oct 28, 1999. WIPO Pat. Appl., WO 99/54393.
268. Wang, Z. and Pinnavaia, T. J. 1998. *Chem. Mater.*, 10, 3769.

269. WRAP Research Report, 2003. *Wood Plastic Composites Study: Technologies and U.K. Market Opportunities.* Optimat Ltd and MERL Ltd., The Waste and Resources Action Programme, The Old Academy, Banbury, U.K.
270. Hill, C. A. S. 2000. *J. Inst. Wood Sci.*, 3, 15.
271. Bledzki, A. K., Sperber, V. E., and Frank, O. 2002. *Rapra Rev. Rep.*, 8.
272. Matuana, L. and Li, Q. 2001. *Cell. Polym.*, 20, 2, 115.
273. Clemons, C. 2002. *J. Forest Products*, 52, 6, 10.
274. Gardner, D. J. 2002. *Wood Plastic Composite Extrusion-Processing Systems*, www.umaine. edu/adhesion/gardner/5502002/wpc%20ext%20pro%203-11-02.pdf.
275. Mali, J., Sarsama, P., Suomi-Lindberg, L., Metsä-Kortelainen, S., Peltonen, J., Viekki, M., Koto, T., and Tisala, S., *Wood Fiber-Plastic Composites*, A Report dated 31.12.2003, VTT Building and Transport, Finland.
276. Optimat Ltd and MERL Ltd., August 2003. *Wood Plastic Composites Study—Technologies and U.K. Market Opportunities*, WRAP, Banbury, U.K.
277. Bleier, H., Kirsch, E. and Battenfeld, May 2002. *Wood Injection Molding, Conf. Proc. Wood Plastic Composites: A Sustainable Future*, Vienna, Austria.
278. English, B. September 1996. *Proceedings No. 7286, The Use of Recycled Wood and Paper in Building Applications*, Madison, Wisconsin.
279. Chelsea Center for Recycling and Economic Development (CCRED), April 2000. Technical Report #19, *An Investigation of the Potential to Expand the Manufacture of Recycled Wood Plastic Composite Products in Massachusetts*, Univ. of Massachusetts, Massachusetts.
280. Defosse, M. 1999. *Mod. Plastics*74–79.
281. Schut, J. 1999. *Plastics Technol.*46–52.
282. DeRosa, A. Feb. 2002. Composites a hit in building industry, *Plastics News*, Crain Communications Inc., p. 1.
283. Dyachovskii, F. S. and Novokshonova, L. A. 1984. *Russian Chem. Rev.*, 53, 117.
284. Enilkolopyan, N. S. 1990. *Filled Polymers I, Science and Technology*, Springer, Berlin.
285. Hindryckx, F., Dubois, P., Jerome, R., Teyssie, P., and Marti, M. G. 1997. *J. Appl. Polym. Sci.*, 64, 423, See also 439.
286. Enilkolopyan, N. S., Fridman, A. A., Popov, W. L., Stalnova, I. O., Briekenstein, A. A., Rudakov, W. M., Gherasina, N. P., and Tchalykh, A. E. 1986. *J. Appl. Polym. Sci.*, 32, 6107.
287. Soga, K., Kim, H. J., and Shiono, T. 1994. *Macromol. Chem. Phys.*, 195, 3347.
288. Köpl, A., Alt, H. G., and Phillips, M. D. 2001. *J. Appl. Polym. Sci.*, 80, 3, 454.

3

Trends in Polymer Applications

3.1 Introduction

While the plastics industry has witnessed a spectacular growth over the last six decades, the acceleration in consumption rates of plastics has taken place in several phases since World War II. Much of the use of plastics just after the war was as a cheap substitute for traditional materials (in other cases, the material was used for its novelty value), and in many instances the result was detrimental to the industry. It required several years of painstaking work by the technical service departments of major plastics manufacturers before confidence was regained in the use of plastics. Even today the public image of plastics is not entirely positive, and the important role of plastics in contributing to raising the standard of living and quality of life is not fully appreciated.

In some areas of applications, plastics materials have been long established. A well-known example is in the electrical industries where the excellent combination of properties such as insulation characteristics, toughness, durability, and, where desired, flame-retardation capacity have led to increasing acceptance of plastics for plugs, sockets, wire and cable insulation, and for housing electrical and electronic equipment. This has lead to the increased use of the more general-purpose plastics. Plastics are also finding use in sophisticated techniques. For example, the photoconductive behavior of poly(vinyl carbazole) is made use of in photo-copying equipment and in the preparation of holographs, while the notable piezo-electric and pyro-electric properties of poly(vinylidene fluoride) are utilized in transducers, loudspeakers, and detectors.

As the plastics industry has matured, it has been realized that it is better to emphasize those applications where plastics are preferable to traditional materials. In the building industry, this approach has led to a number of uses of plastics that include piping, guttering and conduit, flooring, insulation, wall cladding, damp course layers, and window frames. In the domestic and commercial furniture industry, which forms another important outlet for plastics, uses include stacking chairs, armchair body shells, foam upholstery, cupboard drawers, and decorative laminates.

Plastics have found wide acceptance as packaging materials. Substituting plastics bottles for glass containers have been particularly appreciated in the bathroom where breakage of glass has often been the cause of many serious accidents. Small containers for medicines are also widely made from plastics. The ability of many plastics to withstand corrosive chemicals and their lighter weight compared with glass, which reduces the energy required for transportation, have been of benefit to the chemical and related industries. The wide use of plastics film for wrapping, bags, and sacks is well known. However, the quantity of plastics used in this area results in a large volume of waste plastics film, which is left around as an all-too-durable eyesore and a litter problem and has stimulated research into biodegradable plastics.

For many years the main uses of plastics in the automotive industry were associated with car electrical equipment, such as batteries, plugs, switches, and flex and distributor caps, as well as, with interior body trim including light fittings and seating upholstery. Subsequently there has been increased use in under-the-bonnet (under-the-hood) applications, such as radiator fans, drain plugs, gasoline tubing, and coolant water reservoirs. In recent years, the need for lighter cars for greater fuel economy and the emphasis on increased occupant safety have led to a substantially increased use of plastics materials for bumpers, radiator grills, and fascia assemblies. As a result, the automotive industry is now a major user of plastics, with the weight of plastics being used per car increasing every year.

Plastics are also finding increased use in the construction of vehicles for both water and air transport. There light weight, noncorrosive nature, antimagnetic characteristics, ease of maintenance, and the economy in manufacture of glass–giber reinforced plastics have led to their wide use for making boats and hulls. Plastics are particularly useful in aircraft, mainly due to their low density.

Plastics have been widely accepted in industrial equipment development as materials for many critical components and for accessories where such features as toughness, abrasion resistance, corrosion resistance, electrical insulation capability, nonstick properties, and transparency are of importance. For example, PVC, polyacetal, and PTFE are used for pipes, pumps, and valves because of their excellent corrosion resistance. Nylons are used for such diverse applications as mine conveyor belts and main drive gears for knitting machines and paper making equipment because of their excellent abrasion resistance, toughness, and low coefficient of friction.

In the medical sector, plastics uses range from medical packaging and disposables (e.g., bandages and injection syringes) to nontoxic sterilizable items (such as tubing and catheters) to spare-part surgery (such as hip joints and heart valves). A wide-range of polymeric materials are found in these applications though medical grades of PVC and polystyrene are the most commonly used. Engineering resins, however, are finding an increasing number of uses in this area.

While natural and synthetic fibers have remained the major materials for clothing, plastics have also found increasing use in this area. Footwear has provided a major outlet for plastics are they are used not only in soles and uppers, but increasingly in an all-plastics molded shoe that is in high demand in underdeveloped regions of the world. In rainwear, plastics and rubbers continue to be used for waterproof lining and in the manufacture of all plastics packable raincoat. In cold-weather apparel, polyurethane foam has found increasing use as an insulation layer and for giving "body" to clothing.

The photographic industry was one of the earliest users of plastics, which is for photographic film. There has also been widespread use of plastics in darkroom equipment. More recently, plastics have found increasing acceptance in cameras, both inexpensive and expensive, as it is recognized that well-designed camera bodies made from properly chosen plastics materials are more able to withstand rough usage and resist denting than metal camera bodies. In the audio field, while the use of plastics for types and discs is well established, plastics are now almost standard for the housings of reproduction equipment. Sports goods manufacturers make increasing use of glass- and even carbon–fiber reinforced plastics for such diverse articles as fishing rods, racquets, canoes, and so on.

While the above paragraphs indicate some of the major uses of plastics in a number of areas, the more recent trends in plastics applications in these areas form the subject of subsequent discussion in this chapter.

3.2 Polymers in Packaging

The packaging industry [1–5] is the most important user of plastics both in Europe and America, and it is expected to increase with an annual growth rate of 3%–5%. The five most common polymers in use today as packaging are polyethylene, polypropylene, polystyrene, poly(ethylene terephthalate), and poly(vinyl chloride).

LDPE, the oldest of the polyethylenes, has been used extensively in packaging, but due to its higher price and lower mechanical performance compared with LLDPE, it is being used increasingly in specialty and niche areas where its unique properties add value. In blown and slot-cast film, LDPE offers excellent clarity and easy processing. It is often used in blends with LLDPE for these reasons. The rheological characteristics of LDPE and its ease of extrusion suit it for extrusion coating of various substrates. And because of its lower melting point at a given density, it is used where ease of heat sealing is important.

LLDPE has major outlets in film and sheet markets. In film LLDPE offers greatly improved mechanical performance compared with LDPE, and it has steadily displaced the latter in many uses. However, LLDPE has significantly different processing characteristics than LDPE due to the absence of long-chain branching; this requires different extrusion equipment in order to produce film at high rates. The trend in LLDPE film resins for packaging has been consistently toward higher performance. These improved products range from butene copolymers to the higher-alpha-olefin (HAO) hexene and octene grades, and most recently the "superstrengh" HAO grades.

The penetration of LLDPE in the market of *stretch-cling wrapping* is increasing rapidly. Stretch-cling wrapping, used for consumer as well as industrial packaging, is a process in which a pre-extruded tacky film is wrapped around any product, such as bundles, rolls, and pallet loads, of any shape and size to impart a film grip to the package. The tacky characteristics of the film helps in self-adhesion of the film without any external aids.

LLDPE has all the required properties for manufacturing stretch-cling film. With excellent drawdown characteristics, the film offers high elongation in machine direction, good tear strength at cross direction, and excellent puncture resistance. A required type of tackifier is used to impart the cling effect. Depending on the end-use requirements, both HAO-grade LLDPE and butene LLDPE are commercially used, the former offering superior strength but at a higher cost. To avail cost benefit and achieve the property, processors commonly offer coextruded stretch films with one layer of HAO grade LLDPE and the other layer (or layers) of butene LLDPE. This type of films are tougher than 100% butene LLDPE films.

Very-low-density polyethylenes (VLDPEs), ranging in density from 0.880 to 0.912 g/cm^3, are ideal products for a variety of low-temperature film applications, such as ice and frozen-food packaging. The low crystallinity of the VLDPE results in energy-absorbing characteristics that yield outstanding low-temperature impact properties. The lower-crystallinity products of the VLDPE family also show increased permeability to oxygen and other gases, while their water vapor transmission rate is moderate. Thus, VLDPE offers a novel balance of barrier properties necessary for fresh produce packaging film.

Polypropylene (PP) homopolymers, in oriented and cast film, provide clarity and gloss, high tensile and tear strength, stiffness, and a wide heat-sealing range. Other features of homopolymers include good moisture-barrier properties and printability. Cast film is used in good packaging, in bags for textiles, and (in combination with other materials) for composite packaging film laminates. Oriented polypropylene film achieves its high strength, clarity, and impact resistance due to mechanical processing (orientation) that aligns the polymer molecules. It is often used in food packaging, particularly for snacks, cigarette packaging, and pressure-sensitive tapes.

Random copolymers of ethylene and propylene produced by random addition of ethylene to a propylene chain as it grows are noted for high clarity, a lower and broader melting range than homopolymer grades, and reduced flexural modulus. At refrigerator temperatures, these materials also have impact resistance greater than that of homopolymers. Orienting the material further improves its clarity and impact resistance. Applications include blow-molded bottles, cast film, and injection-molded products, such as food storage containers, which take advantage of the excellent clarity and good balance of impact and stiffness of this material. The high gloss and very broad heating-sealing range of this resin is useful in such cast film applications as trading cards and document protectors.

The melt-flow rate of ethylene–propylene random copolymers ranges from 1 g/10 min for a blow-molding grade to 30 g/10 min for an injection-molding grade. The density is about 0.90 g/cm^3 and the notched Izod impact strength of the materials ranges from under 1 to more than 5 ft.-lb/in. Special grades of copolymers with a melt-flow rate of 35 g/10 min or higher find applications in thinwall parts, usually for injection molded food packaging such as delicatessen containers or yogurt cups. Such containers have walls with a length-to-thickness ratio as high as 400:1, yet they retain the properties of top-load strength, impact resistance, and recyclability that are typical of polypropylene. Injection molders have these specialized copolymers to reduce container weights by 30% over containers produced with other materials. Meanwhile, extrusion coaters can now make use of high-melt-strength PP resins with melt-flow rates as high as 40 g/10 min. These materials resist pinholes and curling.

Polypropylene is making inroads in the packaging market at the expense of other materials, a process aided by new technologies that extend poly-propylene properties. High-crystallinity polypropylene, for example, is targeting such materials as polystyrene and PVC in packaging. High-modulus grades of PP, as produced via Himont's Catalloy polymerization technology (see Chapter 1), are expected to challenge clear plastics (such as PET) and glass in container applications. Other companies that have launched enhanced PP lines include Hoechst, Amoco, and Slovay.

Major markets of rigid PVC include bottles and packaging sheet while flexible PVC finds major packaging uses in film, sheet, blood and other bags.

New technologies are broadening the range of available thermoplastic elastomers (TPEs). These development include low-stiffness reactor grades of thermoplastic polyolefin (TPO) copolymers, which are likely to make inroads in packaging markets at the expense of PVC. Potential end products of reactor-based TPOs include packaging films and sheets, shipping sacks, medical-waste-disposal bags, and blood bags.

Major innovations in the packaging field are retortable plastics as a replacement for the can, hot-fillable rivals to glass, high-barrier-coated or laminated bottles for carbonated drinks, controlled or modified atmosphere packagings and biodegradable or ecofriendly packaging. These developments are highlighted below.

3.2.1 Retorting

Retorting is the most widely used method for processing food. In this method, the raw or partially cooked food material is placed in a container with sufficient liquid to fill it. Prior to sealing the container, it is heated under pressure to at least 121°C to kill all bacteria.

American Can's Omni retortable plastic can marks a major development in food packaging. The five layer can, used by Hormel, Del Monte, and Campbell Soup Company, consists of polyolefin structural layers, adhesive layers incorporating desiccants, and ethylene vinyl alcohol (EVOH) polymer as the barrier layer.

A newer development in this area is the production of retortable, multilayer, blow-molded containers that incorporate polycarbonate as the tough outer layer, with EVOH or poly(vinylidene chloride) (PVDC) as the barrier, and polypropylene (PP) as the food contact layer. The advantages of polycarbonate are its light weight, improved heat stability, and good optics compared to the polyolefins that are normally used.

3.2.2 Asceptic Packaging

For asceptic packaging, both the food product and the container are sterilized separately and then brought together in a sterile environment. Both low-acid foods, like yogurt and soup, and high-acid foods, such as juices and drinks, are packaged by this method.

Combibloc cartons for Bowater PK and Tetra-Brik from Tetra Pak are used for packaging high-acid pure fruit juices and give a shelf life of six months. These are multi-layer structures with a paperboard base and polyolefin and aluminum foil for barrier layers. The cartons are also used in the range of adult nutritional products for hospitals. Tetra Pak's range of asceptic packaging for pasteurized products includes Tetra-Tals, Tetra-Top, and Tetra-King, all of which feature improved opening and closing facilities.

The Connofast thermoform-label-fill-seal system of Continental Can Company can be used to produce a microwaveable asceptic container for low-acid products. The container is a coextruded tray or cup combined with a foil-free lid. LDPE provides the food contact surface while the barrier layer is either PVDC or EVOH.

3.2.3 Hot-Filling

In the hot-filling technique, which is being used as a cheap alternative to asceptic packaging, food is cooked or pasteurized by heating at a high enough temperature to kill the bacteria and then placed into the container at a temperature above 80°C to kill all the bacteria in the container. With this method flexible materials can be used to replace metal cans at a reduced cost.

In the Q-box system introduced by Mead packaging for hot-filling high-acid juices, a web made of LDPE, paperboard, ionomer, foil, polyester, and LDPE is used. Products are pasteurized at 90°C before filling. Also used for high-acid products, is an oriented polyester bottle from Monsanto Company.

Hot-fill pouches are used for both low- and high-acid applications. Heinz Company's form-fill seal Pouch Pack, which is claimed to give a shelf life of one year, has a web of 3-mil LLDPE, ethylene–acrylic acid copolymer, 35-gauge foil, ethylene–acrylic acid copolymer, and biaxially oriented nylon, from inside outward. The Pouch Pack system from Du Pont offers pouches of sizes 4 oz to 1.5 L to fill a range of high- and low-acid food applications at various temperatures, including fruit toppings at 80°C.

3.2.4 Controlled-Atmosphere Packaging

Controlled atmosphere packaging (CAP), or modified atmosphere packaging (MAP), has shown remarkable growth over the past decade. For many perishable products, exclusion of oxygen and high concentration of carbon dioxide improve the stability of the products as the growth of aerobic microorganisms is prevented. Most CAP used until recently PVDC and EVOH polymer-based oxygen barrier films to package red meat and baked goods. These films still allow the food to continue to respire after being packed, thus altering the balance of CAP gases introduced at the time of packing. However, selectively permeable films can now be used to control the amount of carbon diozide, oxygen, and moisture in contact with the produce after it has been packed. In this way, CAP can more than double the shelf life of certain fruits and vegetables.

One such film that has been successfully used to package tomatoes is DRG's Ventaflex selectively permeable film. It is a polyester and HDPE strip-laminated film that has a laminating adhesive applied in diagonal strips. The film allows the carbon dioxide produced by the fruit to pass through the HDPE layer into the atmosphere while air penetration is limited by the outer polyester film. Besides tomatoes, other products commercially packaged in Ventaflex film include cheese, coffee beans, and bean-sprouts.

3.2.5 High-Barrier Films

Resins with high-barrier properties are being developed by a number of companies in response to a big increase in the demand for high-barrier films. Selar PA (Du Pont) amorphous nylon resins offer a good oxygen and flavor/odor barrier, high clarity, good mechanical properties at elevated temperatures, and good processability. The films are used for meat wrap, snack food bags, and cereal box liners.

Selar PT amorphous poly(ethylene terephthalate) (PET) blends combined the inherent barrier properties of PET with the ability to be processed on conventional equipment without orientation or crystallization steps. Besides food packaging grades, other industrial grades of Selar PT are available that rival PP or HDPE for paint pails and other containers that require a hydrocarbon or solvent barrier.

Dow's Saran wrap, which is a copolymer of vinylidene chloride and vinyl chloride, is a useful packaging material that possesses exceptional clarity, toughness, and impermeability to water and gases. Also available from Dow is a range of Saran resin and films with improved barrier properties. For example, Saran HB monolayer lamination films are claimed to exhibit an oxygen barrier 10 times greater than the conventional Saran wrap films.

In the snack food area, metallized plastics are being increasingly used. Metallized oriented PP is particularly popular for this application.

Nylon-MXD6 is one of the crystalline polyamide resins which is produced through polycondensation of *meta*-xylylene diamine (MXDA) with adipic acid with Mitsubishi Gas Chemical Company's own technology. It is a unique aliphatic polyamide resin which contains *m*-xylylene groups in the polymer chain. It has the best gas barrier property of all nylon resins, better than EVOH in a humid atmosphere and keeps excellent gas barrier property even after retorting or boiling treatment.

Nanocomposites are new entrants to the packaging arena. Ube industries in Japan, in collaboration with Toyota, has developed nanocomposite barrier films for food packaging and other applications.

3.2.6   Oxygen Scavenger-Based Packaging

In-pack oxygen scavenging involves the use of a variety of forms of scavenger. Oxygen scavengers are capable of reducing oxygen levels to less than 0.01%, which is much lower than the typical 0.3%–3.0% residual oxygen levels achievable by modified atmosphere packaging (MAP). Oxygen scavengers can be used alone or in combination with MAP. In commercial practice, however, oxygen scavenging is combined with MAP. The presence of an oxygen scavenger reduces the need for achieving extremely low oxygen levels in MAP. The consequent reduction in gas flush time can result in accelerated line speeds and higher production rates.

In nearly all forms of packaging, oxygen is introduced to the food product from three major sources. These are head space gas, ingress of oxygen, and air entrapped in the product itself. There are primarily three types of oxygen scavengers used today. The predominant type is a metal-base scavenger, such as iron. The finely powdered iron enters the oxide state, thus binding the free oxygen, when exposed to the appropriate humidity condition. The most common form is sachet where the ground metal is packaged in a highly permeable pouch. (A disadvantage of the iron-based oxygen scavenger is that it is dependent on moisture to initiate and maintain the scavenging reaction.) The second approach uses low molecular weight organic reducing agents, such as ascorbic acid and sodium ascorbate, and enzymic oxygen scavenger systems, glucose oxidase or ethanol oxidase, which would be incorporated into sachets, adhesive labels, or immobilized onto packaging film surfaces. The third method incorporates a polymer-based oxygen scavenging resin along with a catalyst into a film structure.

The most obvious benefit of using an oxygen scavenging film versus a sachet is that it does not appear as a foreign object. Moreover, the polymer-based film, unlike the metal-based scavenger, is also completely independent of moisture and has no effect on metal detection systems.

Chevron has designed a polymer that readily scavenges oxygen without degrading into smaller undesirable compounds. Also, through the use of proprietary photoinitiator and catalyst system the polymer can remain in a non-scavenging state until triggered by exact quantities of UV light. This allows the packagers to purchase materials, store them for a reasonable period, and activate the scavenging properties during the filling process.

The scavenging system, which is extruded into a multilayer film, is a blend of two components. Comprising about 90% of the blend is a cyclohexene containing copolymer which is used to absorb oxygen. The second component (about 10%) is a masterbatch containing, in the proper ratio (dependent on application and activation needs), a photoinitiator and a cobalt salt (catalyst). For packaging applications, a multilayer structure is produced containing the oxygen scavenging polymer blend as an individual layer between a passive oxygen barrier, such as nylon, EVOH or PET, and an inside seal layer, such as LDPE, LLDPE, or ionomers. At the time of filling, the oxygen scavenging polymer layer would need to be exposed to sufficient UV light in order to trigger the scavenging mechanism.

3.2.7   Plastic Bottles

Plastic bottles for carbonated beverages are mostly made of PET as it offers high tensile strength and good barrier properties. Blow-molded PET bottles account for 33% of the total carbonated soft drink market. For packaging beer, which is oxygen sensitive, PET bottles are coated with PVDC, while the base cups that support the bottle are usually made of HDPE.

However, since PET barrier properties decrease with a decrease in size of the bottle, smaller size poses a problem. Two possible ways to solve this problem are to employ coated PET or to use coextrusion technology. As an example of the latter case, PET has been successfully injected with nylon to form a PET-nylon-PET construction, with nylon improving the barrier properties.

In addition to carbonated drinks, PET bottles are also being used in other markets, which include toiletries, cosmetic ranges, spirit miniatures, and PET aerosols for cosmetic and pharmaceutical applications.

Biaxially stretched oriented PVC (OPVC) pressure bottles have been used as an alternative to PET for packaging some carbonated drinks. In this application, the level of carbon dioxide inside the bottle and the temperature at which the bottles are kept are important. Producing OPVC by extrusion stretch blow molding results in greater clarity, increases impact resistance, and reduces cost by up to 14% compared to ordinary PVC. Fruit, squashes, and edible oils are commonly packed in PVC bottles.

American Can pioneered the concept of multilayer squeezable plastic bottles with the introduction of its Gamma bottle. Another such product is Continental Plastic Containers' Lamicon bottle. The bottles are being used for food products such as ketchup, jelly, mayonnaise, and barbecue sauce. In a new development, polycarbonate has been used by Holmicon Plast as the outer layer, rather than polyolefin, which is normally used. Polycarbonate gives the bottle a more shiny appearance.

3.2.8 Chemical Containers

Household and industrial chemical containers are increasingly being made of plastics. HDPE has almost completely replaced glass for packaging household liquids. However, for solvents and flammable liquids, a plastic is required with higher barrier properties than conventional HDPE.

Three methods can be employed to obtain higher barrier properties. These are fluorination, coextrusion, and laminar blends. Fluorination of HDPE bottles is used most commonly for packaging solvent-based products and agricultural chemicals in particular. One process consists of treating the surface of HDPE bottles with a mixture of fluorine and nitrogen gas to produce a fluorocarbon barrier layer. The process developed by Air Products Limited, creates an internal barrier layer during the blow-molding process, while Bettix Limited's fluorination process, Fluoro-seal, is post-molded and particularly suited to short runs. The fluorination process suffers from the disadvantage that it is potentially hazardous.

Coextrusion to form an effective barrier layer has been used by Plysu Containers Limited and Reed Plastic Containers, both producing tough solvent-resistant containers. The five-layer construction used by Plysu has HDPE both on the inside and outside, and a central barrier layer with adhesive layers on either side to allow lamination to HDPE. The Reedpac range of blow-molded containers developed by Reed Plastic Containers have polyethylene, polypropylene, polycarbonate, or PET combined with barrier resins such as polyamide, PVDC, and EVOH polymer.

The laminated resin blend technology using Du Pont's barrier resins, trade-named Selar, has been employed to produce barrier bottles. Blow-mocan Limited has used the Selar RB polyamide barrier resin in a one-step process that blends with polyethylene to form overlapping discontinuous plates within the wall system. The container can be used for many cleaning chemicals and agricultural products.

Metal Box has developed a copolymer polypropylene container, Poly-can, which has an anti-static additive and is suitable for water-based paints and coating products. Another development is the first bag-in-box paints from Ashwell Paints. A smooth flow is obtained in this system by the use of an open bung and screw top rather than a tap.

3.2.9 Dual Ovenables

Dual ovenable containers must be able to perform under freezable conditions as well as in both microwave and conventional ovens. Materials commonly used for dual ovenables are temperature-resistant plastics such as crystallized PET (CPET), thermoset polyesters, and polycarbonate. CPET, the dominant material, has a temperature stability up to 232°C.

Attention has been paid to the possible use of multilayer coextrusion of engineering resins for dual ovenable containers, as multilayer packaging could increase shelf stability and thus eliminate the need for freezing or refrigeration. One multilayer combination in commercial use is a sheet of Lexan polycarbonate sandwiched between layers of Ultem polyetherimide. Thermoformed trays of this material are reported to have better break resistance at lower temperatures and higher stiffness at temperatures up to 232°C, as compared to CPET. They are also insert, compatible with polyester lidding materials, and have lower oxygen permeability than PET materials.

Most plastics used in dual ovenables are two to six times more expensive than the aluminum containers they replace in frozen food applications. Less expensive plastic containers are made of molded fiber and polyester and paper laminates. However, due to the popularity of the microwave oven, in many cases microwaveable-only containers, instead of dual ovenables, are used. For example, Campbell uses a polypropylene microwaveable-only container for its Soup du Jour, thereby reducing the material costs considerably.

With the increase in the market of dual-ovenable and microwaveable trays there has been an increase in the demand for special lidding material, as the lid has to provide an oxygen barrier and a 100% hermetic steal to prevent food spoilage, besides being easy to remove. Grades of Du Pont's Mylar OL polyester lidding can be used to seal CPET, thermoset polyesters, and paper-board trays.

3.2.10   Closures

Tamper-evident closures are used to prevent customers from opening containers before they buy them. Several tamper-evident features were originally introduced including neck bands, tamper-rings, and shrink wrapping. More recently, a number of new tamper-evident and child-resistant closures (CRC) have been introduced.

Clic-loc is a CRC available from United Closures and Plastics. A specialist version is provided with a thermoset inner liner for chemical applications.

A double tamper-evident cap sealer has been introduced by Paklink. It has a ring style enclosure incorporating an inner foil seal which, once heat sealed, provides a hermetic barrier suitable for products in the food, drug, and dairy industries.

A number of CRCs, including the Childgard and Safecap, are available from Johnson and Jorgensen. The Childgard is a one-piece jigger cap style CRC used for pharmaceuticals, while Safecap is a two-piece palm and turn style CRC used for a variety of pharmaceutical and household products.

3.2.11   Biodegradable Packaging

In the 1980s and earlier, products were designed to balance cost and performance—during their useful life. Customers, end-users, legislators, and environmentalists now demand a new type of product, one that is also designed for responsible disposal. With the number of approved landfills diminishing each year, the costs of disposal at remaining sites will escalate annually. That is why recycling and composting have a useful place in an overall solid-waste strategy. This has provided impetus for the development of biodegradable, and therefore compostable, polymeric products.

Incorporation of starch, a biodegradable natural polymer, into synthetic plastic imparts biodegradability. Shopping bags containing up to 15% starch in a matrix of polyethylene are said to be biodegradable. Microbes digest the starch and leave a flimsy plastic lace that disintegrates mechanically. Ferruzzi, an Italian firm that dominates the European cornstarch industry, has launched a polymer that is more than 50% starch, and it hopes eventually to raise the content to 90%.

Novon, a biodegradable plastic made by Werner Lambert, based in New Jersey, is reported to be composed entirely from starch [3]. Through the use of different starches and nontoxic biodegradable additives, Novon specialty polymers can be formulated to provide a wide range of properties and still biodegrade completely in biologically active environments, ultimately yielding carbon dioxide, water, and minerals, and leaving no toxic, hazardous, or synthetic residues. Any consumer packaging product with a high potential of being dropped on the ground may thus be considered in Novon polymers. For example, biodegradable loosefill packaging peanuts are made of Novon 2020. They are compostable and compatible with existing sewage treatment facilities so that they can be safely washed down the drain or even flushed down the toilet.

Poly(vinyl alcohol) is one polymer that poses no litter problem, because it dissolves in water and is readily biodegradable. It makes ideal packaging for such things as swimming pool chemicals, detergents, descalers, seed strips, sanitary items, and even oxygen tenting.

The polymer polyhydroxybutyrate (PHB), a polyester of 3-hydroxy-butanoic acid, is completely biodegradable as it is an ideal food for microbes. By copolymerization with another hydroxyacid, such as 3-hydroxypentanoic acid, the polymer can be tailored to take it suitable either for molded articles such as shampoo bottles or thin films for plastic envelopes or carrier bags. PHB is tradenamed Biopol by ICI. Most of the polymer that the company makes is used for packaging, agricultural products, and items of personal hygiene. Wella, the German hair-care company, sells its Sanara brand of shampoo in bottles made of Biopol and has reported increased sales as a result.

Poly(butylene adipate-*co*-terephthalate) (PBAT) is a new type of biodegradable polyester supplied by BASF [6] and Eastman Company [7]. PBAT is more flexible and has higher elongation at break than other biodegradable polyesters, and therefore is more suitable for food packaging and agricultural films.

3.2.12 Pharmaceutics Packaging and Nanomedicines

Packaging for pharmaceutical products can be either internal (integral) or external. External packaging is that which surrounds the pharmaceutical product, such as a plastics container or a blister pack, while integral packaging is that which is part of the actual drug delivery, for example, tablet coating. In both these categories, packaging can be described as either active or passive. Active materials respond to external factors (e.g., temperature and pH triggers for drug release), while passive materials have predefined inherent functionalities independent of external factors (e.g., binding the components of a drug, suitability for food contact, and conforming to environmental and health legislation for packaging and products).

With regard to passive materials for external packaging there has been continuous improvement in traditional properties, such as strength, durability, clarity, moisture resistance, less weight, processability, appearance, and design as well as their new combinations. For example, multilayer films allow a combination of different properties, such as an internal layer with approval for food contact and an external optimized for printing or moisture resistance.

Active polymers for external packaging are being developed that can be applied as thin films in a cost-effective method to act as sensors for biological, chemical or environmental effects and thus help predict shelf life or warn of product contamination or degradation. Similarly, there are polymers with selectivity, for example, to let water in and carbon dioxide out, thus ensuring the survival of the required live microorganisms. Future projections in this field include cheap and flexible carbon-based electronics that will allow the functionality of electronic circuits and displays to be incorporated into packaging, either for promotional purposes or security and tracking. These new developments for the packaging industry and consumer packaging products are being commercialized through FaraPack Polymers Ltd., a joint venture of the Polymer Center, University of Sheffield, with the Faraday Packaging Partnership.

In relation to internal packaging, while passive materials used to aid the handling and delivery of the active ingredients may be inorganic filler materials that make the tablet a sensible size, new developments are looking at methods to attach polymer chains to the surface of such filler particles in order to modify their surface properties for enhanced interaction with other components (particles, binders, colorants) in the tablet or to facilitate attachment of active compounds directly. Passive polymeric materials can also be used in internal (integral) packaging to allow the transport and delivery of substances through hostile environments to specific sites. Several such functional vehicles for drug delivery have been explored at the Polymer Center, University of Sheffield [8], viz., hydrogel dendrimers, hyperbranched polymers, and vesicles (microscale polymer bags). Additional functionality can then be incorporated by using responsive polymers as active materials that can be triggered by a change in pH, pressure, temperature, or light to release the active components carried by them.

Nanotechnology, a newly fashionable field, has added another dimension to internal (integral) packaging of therapeutics. Products, known as nanomedicines, relying on nanotechnology are now in routine clinical use. These include medicines incorporating polyethylene glycol (e.g., *pegfilgrastim*), stealth liposomes (e.g., *Doxil*), polyglutamation (e.g., polyglutamated paclitaxel, *Xyotax*), and antibodies (e.g., *gemtuzumab*, *Mylotarg*). Filgrastim is used to avoid febrile neutropenia in patients undergoing bone marrow transplantation. Attaching polyethylene glycol (PEG) to filgrastim avoids the need for the patient to be injected daily. The product has a linear PEG attached to the N-terminal methionine of granulocyte-colony stimulating factor, so that the bulky PEG component does not interfere with the rest of the molecule or its ability to interact with its cognate receptor. Whereas normal filgrastim is associated with a 48-hour response as it steadily leaks out via the kidneys, this is not the case with pegfilgrastim.

Stealth liposomes differ from conventional liposomes in that they are coated with polyethylene glycol which enhances their hydrophilicity and enables them to evade the reticulo-endothilial system, thereby slowing their clearance from the body. Such stealth liposomes containing doxorubicin are used as nanomedicines, Doxil. Once liposomes reach a permeable tissue, such as tumor, they sneak into the interstitial fluid and release their drug cargo. A significantly high concentration of doxorubicin is thus achieved in tumor tissues by administering Doxil instead of free doxorubicin. Stealth liposomes that contain other drugs for cancer treatment are being developed [9]. They include *cisplatin*, a mitomicin prodrug and a targeted form of Doxil. A polyglutamated form of paclitaxel (*Xyotax*) has made it possible to deliver large doses of the drug without the need for extensive premedication and without the risk of alopecia. Another example is anti-tumor antibiotic *Calicheamycin*, which is licensed in the U.S.A. for the treatment of certain patients with CD33-positive acute myeloid leukamia. Calicheamycin is too toxic for human use on its own, but when linked to a humanized antibody, such as *Gemtuzumab* (Mylotarg), its toxic effects can be targeted.

3.2.13 Wood–Plastic Composites

There is a largely untapped opportunity for wood–plastic composites [10] in packaging and material handling. This is evident from the fact there are nearly 600 million pallets made each year in North America and only about 50 million are plastic. One Canadian manufacturer, Dura-Skid, produces pallets by assembling lineals made from wood-filled HDPE. The main advantage of wood–plastic composites to the pallet manufacturer is the excellent stiffness to weight ratio afforded by wood filler. The greatest opportunity may be in structural foam and twin-sheet thermoformed pallets.

3.3 Polymers in Building and Construction

The building and construction industry is being transformed by economic and demographic changes that are creating increased opportunities for plastics products [11–15]. It is next only to the packaging industry in order of importance as a user of plastics, and has proved to be the fastest growing market for plastics use in the last 20 years with increased sales in roofing, flooring, pipe and conduit, windows and doors, plumbing fixtures, and insulators. With its high potential for replacement of nonplastic materials, this market is expected to grow more rapidly. One of the relatively new areas of building applications has been in sports surfaces, which are discussed in a later section at the end of the chapter.

The use of nonwood materials in building and construction has been stimulated by the escalating prices of wood construction products. However, a key factor in the growth of plastics in residential construction is consumer acceptance. The advantages of plastics over conventional materials, including improved quality, reduced installation cost, better appearance, durability, and low maintenance requirements, have hastened their acceptance.

Poly(vinyl chloride) (PVC) is the predominant plastic material consumed in building and construction; it accounts for almost half of this industry's total consumption of resins. The largest PVC application is in rigid pipe and tubing, where it accounts for some 40% of total industry consumption. Specific PVC piping uses include sanitary sewer lines, storm water lines, and potable water mains. These uses reflect a classic example of new-product acceptance, where a superior quality, moderately priced, and high perceived-value plastic product can supplant existing building materials and achieve a dominant market position. Other major PVC uses include siding, accessories, and profile extrusions for such end products as windows and doors, accounting for some 12% of total consumption of resins in building and construction.

The commodity thermosets of urea, melamine, and phenolics account for about one-third of building and construction industries' resin consumption. Most of this is used for resin-bonded woods such as plywood, particle board, and oriented strand board. High-density polyethylene (HDPE), used primarily for rigid piping, accounts for about 5% of consumption. Other significant resins, accounting for 3–5% of total consumption, include other polyethylenes, unsaturated polyester, and polystyrene.

3.3.1 Roofing

The basic parts of a roof are the deck, the thermal barrier (insulation), and the impervious roofing membrane that seals the roof complex structure. The built-up roofing membrane is made of (1) bitumen (asphalt) or other more modern materials, (2) the roofing felts (reinforcement), and (3) the aggregates for protection of bitumen against ultraviolent light and oxidation.

The molten asphalts used for waterproofing are a mixture of mineral fillers and bitumens. Because of its limited characteristics, bitumen was the subject of many studies aimed at improving its physical and mechanical properties. Several approaches were tried, such as chemical treatment and blending with natural or synthetic rubbers or latices.

So far, in practice, there are various polymer-bitumen mixtures, such as PE-bitumen or elastomer-bitumen, including styrene–butadiene–styrene block copolymer (SBS)-bitumen or styrene–butadiene copolymer (SBR)-bitumen. The use of waterproof roofing membranes based on the thermoplastic elastomer-bitumen mixtures is increasing, especially in northern climate countries.

The most dramatic change in commercial roofing has been the shift from bitumen-based built-up roofing (BUR) to single-ply membrane roofing, particularly in the United States. Pioneered by Carlisle Corporation with its Sureseal ethylene propylene diene copolymer (EPDM) membrane, single-ply roofing became the fastest growing product in U.S. rubber industry history, presently accounting for over 50% of America's nonresidential flat roofing market. The ease of installation and durability of single-ply roofing together with its resistance to weathering, chemicals, and ozone make it a cost-effective roofing system. A distinctive feature of single-ply roofing is that it cures with weathering and increases its elasticity as it ages. Fire-retardant grades of EPDM are also being developed.

Although the material most used for single-ply membranes is EPDM, in competition with it are several other materials—Du Pont's Hypalon (a chlorosulfonated polyethylene), its Neoprene products, and modified PVC. It is believed that future growth in the roofing market will depend on products, such as Hypalon, that are easy to handle and have good ozone resistance, and on PVCs modified with polymeric plasticizers.

3.3.2 Flooring

The range of flooring material increases in direct proportion to the number of new polymers daily added to the market. In addition to the common floor surfacing materials (e.g., ceramic tile, wood block, wood strip and board, marble, granolithic, terrazzo, linoleum, and cork tiles), new materials are used, such as PVC tiles, vinyl chloride–vinyl acetate copolymer, vinyl-asbestos tiles, PVC welded sheet, synthetic fiber-epoxy polymers, PP, and polyurethane.

PVC and vinyl-asbestos tiles are found to be the most economic materials for use in flooring applications. The most widely used synthetic flooring material is produced by calendering compounds based on vinyl chloride–vinyl acetate copolymers filled with asbestos fibers. In general, the copolymers for flooring applications contain about 13%–15% vinyl acetate.

The so-called homogeneous vinyl is processed by laminating precalendered webs produced in a common PVC calendering operation. The laminated product is made of several sheets of varying thickness: the top layer is a clear, relatively stiff formulation; the middle one contains more filler and is usually opaque; and the bottom sheet contains high filler loadings. After laminating, the product is die-cut into tiles and packaged.

PVC flooring materials are widely used in hospitals and schools, and domestic kitchens and bathrooms, where they offer a wide choice of colors and patterns, ease of cleaning, good cushioning, insulation, and reasonable price, the last three features representing advantages over linoleum.

Heavy-duty, lightweight PP duck boarding is claimed to provide a versatile, easily cleaned work platform, increasing operator comfort and safety. PP flooring is noncorroding and resistant to bacteriological attack. The upper surface is ribbed in order to impart nonslip characteristics, with slotted, self-drawing squares allowing free passage for wash solution and contaminants.

Due to its low level of sound insulation and lack of pleasing appearance, epoxy flooring is not used for domestic purposes. But for industrial flooring, epoxy floors are clearly competitive on a performance basis with most of the other floor types. There are three types of epoxy flooring systems: pourable self-leveling seamless floors, troweled floors, and terrazzo floors. These epoxy flooring systems can be used as new floor coverings over a subfloor of concrete, wood, or steel, and can also be used for remedial work and applied over existing floors. Epoxy flooring has a proven durability of over 25 years.

The growth of seamless floors has had an exciting and profound effect on polyurethane flooring industries. This unique flooring concept can produce durable, attractive, and imaginative effects at the job site, by embedding a variety of different colored particles or fillers into a liquid polyurethane resin just poured from the can. When cured, the lay-up is sanded to remove the sharp edges and smooth the surface, and then vacuumed to remove the sanding dust. Depending on the application and the amount and type of traffic, two or more urethane glaze coats are applied as the wear surface. However, for industrial floors where appearance is not critical, sanding and glazing are not necessary.

The features of seamless floors are well established. Not only are these floors easily installed, they are durable, lightweight, flexible, slip and dent resistant, scratch and scuff resistant, stain and dirt resistant, fungus resistant, heel-mark resistant, and have superior chemical resistance compared to many floor materials, there are a number of colored materials that the imaginative mind can use to provide decorative effects for seamless floors.

3.3.3 Windows

The use of PVC window frames is increasing at a steady pace. In Germany nearly 43% of all windows and 70% of all replacement windows are in PVC.

PVC formulations are generally ethylene vinyl acetate (EVA) polymer-based, chlorinated polyethylene (CPE)-based, or acrylic-based. The majority of new grades are acrylic-based and are claimed to exhibit greater impact strength than the other two types.

Grades of PVC are being tailored to meet different requirements of impact strength and weathering resistance for different climates. Solvay, for example, has developed an unmodified material in granular form that is suitable for use in European countries, an acrylic-modified powder for moderate and Mediterranean climates, lead-stabilized grades for extremely hot climates, and an unmodified grade with very high weathering resistance. Other compounds with high weathering resistance include Dow Chemical's Rovel, General Electric's Geloy, and Hul's Vestolit H1 powder.

The majority of window profiles, until fairly recently, were white. Coloring techniques were developed to produce colored window profiles, thus widening the market. Hoechst, for example, uses in-line bonding of acrylic foil to PVC profiles, allowing a wide range of colors and designs including wood-grain finishes. Both production and precision in window extrusion has been improved with the development of new screw designs, better dies, and microprocessor control of production parameters.

In a window technology to produce self-frosting glass, Marvin Windows and Floors, Warroad, Minnesota, relies on a liquid crystal polymer film from 3M. The film allows for windows that can be either frosted or cleared on demand. According to 3M, liquid crystal droplets are dispersed in a polymer matrix that is sandwiched between two clear electrodes (conductive-coated polyester film). The 0.127-mm-thick film is then placed between two panes of glass. To clear the window, one flicks a switch that apples 11 W/m^2, which causes the crystals to "line up." To frost the window, the charge is broken, thus returning the crystals to their random, unaligned state.

A film (Optica) developed by the Reflective company of France is designed to be used in windows to create a reflecting screen capable of returning the image like a conventional mirror, while preserving the transparency and visual properties of glass. A product of space research, Optica was developed as an effective solution to the delicate problem posed by large glazed areas where people can see inside. Optica consists of a single-layer film with a polyester base of high optical quality (treated for UV rays) on which aluminum oxide particles of controlled density are deposited using a complex vaporization process and a second polyester crystal layer to protect the metal coating. The film is coated with a UV-resistant and pressure-sensitive acrylic adhesive that can be reactivated in water. Once applied to the window, the film becomes an integral membrane, forming an authentic laminate.

3.3.4   Pipes

Plastic pipe demand has been growing at a steady rate since 1967. In the United States its share of the total pipe market is more than 60%. Factors contributing to the accelerating use of plastic pipes include their low cost relative to conventional materials, excellent corrosion resistance, ease of installation, and wider building acceptance.

The plastic pipe market is dominated by PVC pipes, which hold 80% of the market. Polyethylene, principally HDPE, is used for about 12% of pipe production. Following polyethylene are other higher priced, specialty pipes made from ABS terpolymer, chlorinated PVC, polypropylene, polybutylene, and fiber-reinforced epoxy and polyester.

3.3.5   Insulation

The major uses of insulation in the construction industry are in roofing, residential sheathing, and walls. In these applications, polymeric foams offer advantages over traditional insulation such as glass fiber, and these include higher insulating value per inch of thickness and lower costs. The use of polymeric foam for insulation increased markedly since the 1970s due to increased awareness of the need for energy conservation.

Foams are available as rigid sheets or slabs (which are used in the majority of roofing systems), as beads and granules (used in cavity wall insulation), and also as spray and pour-in applications. The market is dominated by polyurethane (PU) foams, in particular polyisocyanurate products, expanded polystyrene (PS), and extruded polystyrene.

Polystyrene foam (PS) holds much of the sheathing market. In masonry and brick walls, PS foams are mainly used because of their better moisture resistance. In cavity walls, loose fill PS is used, while exterior wall applications, which have no limitation on space, use low-cost expanded PS.

Polyurethane/polyisocyanurate products have higher insulation value and good flammability ratings and are expected to continue to be the leading products in plastic foam market as sheets and slabs.

A shift toward single-ply roofing as compared to built-up roof systems has an important influence on the type of foam being utilized. Thus the lower cost of expanded PS has promoted its use in preference to PU foam and extruded PS in single-ply applications. This is facilitated by the fact that the problems of damage to expanded PS foam from hot pitch when used in built-up roof systems are not encountered in single-ply systems.

3.3.6   Polymer–Concrete Composites

The mechanical properties, the corrosion stability, and some useful properties are the reasons for the continuous interest shown in polymer–concrete composites by various design, research, and production organizations. The most important types of polymer–concrete composites are polymer-impregnated concrete (PIC), polymer-cement concrete (PCC), and polymer concrete (PC).

PIC is a precast and cured portland cement concrete that has been impregnated with a monomer that is subsequently polymerized in situ. This type of cement composite is the most developed of polymer–concrete products. PCC, on the other hand, is a modified concrete in which a part (10%–15% by weight) of the cement binder is replaced by a synthetic organic polymer. It is produced by incorporating a monomer, prepolymer-monomer mixture, or a dispersed polymer (latex) into a cement-concrete mix. To effect the polymerization of the monomer or prepolymer-monomer, a catalyst (initiator) is added to the mixture. The process technology used is very similar to that of conventional concrete. So, unlike PIC which has to be used as a precast structure, PCC can be cast-in-place in field applications. PC can be described as a composite that contains polymer as a binder instead of the conventional portland cement.

The largest improvement in structural and durability properties has been obtained with the PIC system. With the impregnation of a macromolecular compound, the compressive strength can be increased four times or more, the modulus of elasticity increased at least two times, the water and salt permeability reduced by 99%, the freeze–thaw resistance improved enormously, and—in contrast to conventional concrete—PIC exhibits essentially zero creep properties. Both strength and durability are strongly dependent on the fraction of the porosity of the cement phase that is filled with polymer.

PIC is generally prepared by impregnating dry precast concrete with a liquid monomer and polymerizing the monomer in situ by thermal, catalytic, or radiation methods. Some of the most widely used monomer for PIC systems include methyl methacrylate, styrene, butyl acrylate, vinyl acetate, acrylonitrile, methyl acrylate, and trimethylpropane trimethacrylate (TMPTMA). These monomers may be used alone or in mixtures. TMPTMA is a cross-linking agent used to decrease the time of polymerization. The polymerization of the monomer used for impregnation is realized usually by thermal polymerization (commonly with water or steam) or radio polymerization, especially with Co^{60} gamma radiation. The initiators most commonly used in polymerization processes for producing PIC composites are 2,2′-azobisisobutyronitrile, benzoyl peroxide, lauroyl peroxide, and methyl ethyl ketone peroxide.

With impregnation by an appropriate monomer, the main effect after polymerization is the sealing of the continuous capillary pore system, which reduces the effect of stress concentrations from pores and micro-cracks, thereby increasing the strength. A reduction of porosity by 10% doubles the strength.

PIC must be considered a new complex material with specific characteristics, which place it in a position, from the viewpoint of quality and cost, between traditional concrete and other groups of engineering materials such as metals and ceramics.

The most important applications of PIC are bridge decking, tunnel support-lining systems, pipes (hydrostatic tests show that PIC supported about twice as much hydrostatic pressure as the unimpregnated pipe), desalting plants, beams (ordinary reinforced and post-tensioned), underwater habitats, dam outlets, off-shore structures, underwater oil storage vessels, and ocean thermal energy plants, among others.

Unlike PIC, which requires a precast structure, PCC has the advantage that it can be cast in place for field applications. Most of the PCC composites are based on different kinds of lattices obtained especially by emulsion polymerization. A latex is a stable dispersion of fine polymer particles in water, also containing some nonpolymeric constituents used in emulsion polymerization recipe. The lattices obtained through emulsion polymerization contain small polymer particles of 0.05–5 μm.

Styrene–butadiene rubber latex (SBR, GRS) and acrylonitrile–butadiene rubber latex (NBR) are two of the earliest to arrive on the market. Since then, many other types have appeared, with poly(vinyl acetate) and copolymers, acrylics (generally polymers and copolymers of the esters of acrylic acid and methacrylic acids), and carboxylic-SBR types being the major products. Since latices are aqueous emulsions, less water is usually needed in the production of PIC. The addition rates of SBR, poly(vinyl acetate), and acrylic latices are comparable, ranging from 0.10–0.20 part latex solids to 1 part Portland cement.

The particular characteristics of compatibility of poly(vinyl acetate) with cement have led to wide use of this latex as a main component in polymer-mortar and polymer–concrete composites.

In all cases, the operations for preparing PCC are similar to those used for common concrete and mortar. Curing, however, is different. After only 1 day of moist cure, the surface of PCC work can be uncovered. The film already formed on the surface retains the necessary amount of moisture for the full hydration of the cement. In principle, after only a few days of air cure at around the normal ambient temperature (20°C), the PCC can be put in service. The polymer latex used for making a PCC must be able to form a film under ambient conditions, coat cement grains and aggregate particles, and form a strong bond between the cement particles and aggregates.

Investigations have demonstrated that PCC has a higher corrosion resistance as compared with ordinary concretes and can effectively be used for floor coating in a moderately aggressive atmosphere, at milk processing factories and breweries. However, the presence of cement in PCC is a source of corrosion destruction under the action of more aggressive and concentrated chemical media at sugar refineries and meat processing enterprises, among others. In such cases, PCC may be recommended only for underfloors. The problem may be radically solved by producing floor coatings with a purely polymeric binder. For example, investigations and practice have revealed that the use of epoxy alkyl resorcinol-based polymer concretes (PC) for floor coatings in production shops at food industry enterprises increases the corrosion resistance of the floors to a great extent [16].

Polymer concretes (PC) may be considered as an aggregate filled with a polymeric matrix. The main technique in producing PCs is to minimize void volume in the aggregate mass so as to reduce the quantity of the relatively expensive polymer necessary for binding the aggregate. By carefully grading the aggregate, it is possible to wet the aggregate and fill the voids by the use of as little as 7%–8% polymer. With high degree of packing, high compressive strength can be obtained.

A wide variety of monomers, prepolymers, and aggregates have been used to realize PC. The list includes epoxy polymer, unsaturated polyester–styrene system, methyl methacylate, and furane derivatives, usually in conjunction with cross-linking agents. To obtain the best chemical resistance, complete curing of the polymer is necessary. This usually is achieved by careful heating, or by using an appropriate cross-linking agent. In order to improve the bond strength between the macromolecular matrix and the aggregate, a silane coupling agent is added to the monomer before the polymerization process. The nature of the aggregate influences the hydrothermal stability of PC composites. Aggregates commonly used include quartz, silica, fly-ash, cement, and their combinations.

The PC systems based on unsaturated polyester and wet aggregates of cement and silica result in significant strength improvements. The mechanism of chemical bonding is believed to involve cross-liking reaction between Ca^{2+} irons of cement particles and carboxylate anions of unsaturated polyester brought about by a hydrolytic reaction. The addition of MMA to a polyester–styrene system provides a hard, clear mirror finish, improves the workability without reducing the strength, and enhances durability.

Besides fast curing, PCs offer several other advantages. Thus, because the polymer is virtually impermeable to moisture, there is very little cracking of the concrete caused by freezing and expansion of moisture within the cured mix. Polymer concretes also resists salts and other agents that eat away at the reinforcing steel within the concrete. In Germany, these properties have led to the use of PCs in water-treatment and sewage plants.

Although PCs do cost more initially, they can be put down in thinner layers than Portland cement concrete to give the same strength at lower volume, thus allowing a weight reduction of up to 50%. PC overlays are normally 3/8–1/2 in. thick, while Portland concretes average 2½–3 in. thick. For bridge overlays, large stretches of epoxy and polyester resins are combined with Portland cement. These mixtures have a stress failure point of 10,000 psi (69 MPa) compared to 2,000–5,000 psi (13.84–34.5 MPa) for Portland cement.

Polymer concretes can be mass produced, taking advantage of short curing times. With special processing techniques and molds it is also possible to cast equipment such as wash basins and bath tubs, plain or with a marble or onyx effect, slabs with many different surface effects, and even marbles with two or more colors.

Finished PC products are much lighter than the cement-bonded material because they have thinner walls. The excellent resistance to chemicals allows many applications in the construction of sewer systems, sewer-treatment plants, animals stables, oil separations, chlorine electrolysis, machine foundations, high-resistance floors, and so on.

3.3.7 Wood–Plastic Composites

The construction industry likes wood–plastic composites (WPCs) for their increased stiffness and reduced thermal expansion, as compared to unfilled plastic products, and for greater durability and lower maintenance when compared to wood. For decking application, the most common WPC products (in decreasing order of sales volume) are Trex, Choice Dek by AERT (Advanced Environmental Recycling Technologies), Master Mark, and Duraboard. Both Choice Dek and Trex are also used for industrial flooring. TimberTech, manufactured by Crane Plastics is a 50% wood-filled virgin HDPE composite made with a very precise tongue and groove configuration.

About 50% of AERT's production is sold into the window and door industry. The material is called MoistureShield and is similar to Choice Dek in composition. Crane Plastics also markets its TimberTech product to window and door manufacturers. Profiles made by modified technology of Strandex Corporation with high wood content (typically 70%) composite to tolerances of ±0.001 in. can go directly to vinyl cladding operations, and can be used with other precision components such as extruded aluminum profiles. The Strandex product can also be stained or painted.

Andersen Windows, which dominates the use of wood-filled PVC for window and door lineals, uses the waste PVC generated in their cladding operations as a polymer base for their composites, typically formulated from 60% PVC and 40% wood.

Re-New Wood, Wagoner, Oklahoma, injection molds wood-filled PVC shingles and does its own compounding. The shingles which are made to imitate wood shakes, a traditional roofing material, are wind, rain, and hail resistant.

3.4 Polymers in Corrosion Prevention and Control

World consumption of plastics has significantly increased in the past 30 years because of their numerous advantages, including excellent corrosion resistance, competitive cost, low weight, easiness of installation, and higher specific strength than many other materials. Plastics have extended the choice of corrosion–resistant materials by an order of magnitude. (Note: The mechanism of corrosion of plastics in quite different from that of metal. Unlike metals, plastics deteriorate by loss in mechanical properties, softening, swelling, hardening, spalling, or discoloration.) One special advantage of polymers over metals is the maintenance cost after their installation. This cost is absolutely lower with polymers because they basically do not require any of those maintenance and corrosion-control systems normally used with metals, such as cathodic protection, external coating, or corrosion inhibitors. All these reasons have increased the use of polymers in corrosion prevention and control.

The thermoplastics used for corrosion resistance are fluorocarbons, acrylics, nylon, chlorinated polyether, polyethylenes, polypropylenes, polystyrene, and poly (vinyl chloride); important thermosetters used in this field are epoxies, phenolics, polyesters, silicones, and ureas. All plastics show useful physical properties. One drawback in the use of plastics is their limited tolerance to temperature. Fluorocarbons are the noble metals of plastics; they are corrosion resistant to practically all environments up to 290°C.

Fluoroelastomers, which include elastomeric fluorocarbons, are family of carbon- or silicon-based polymers containing significant amount of fluorine. Fluoroelastomer products are a family of specialty materials valued for their elastic behavior under comparatively harsh conditions involving a broad thermal range and/or aggressive environments. Unique features and benefits afforded by specific fluoroelastomers have led to their diverse industrial applications as in seals, linings, tubing, hose, belting, fabrics, caulks, adhesives, laminated, and vibration absorbers, for service in harsh chemical and thermal conditions. Sound knowledge of all types of candidate materials is fundamental to making the right choice.

3.4.1 Flue Gas Desulfurization

Flue gas desulfurization (FGD) is an extremely important technology enabling power plants to produce electricity in an environmentally acceptable manner. A number of polymers that can provide protection from corrosion are used with varying degrees of success in FGD systems. They include chloroprene, epoxies, fluorocarbons, polyesters, and vinyl ester systems. However, each system has its advantages and disadvantages and should be carefully scrutinized before selection. For example, although a coating formulated utilizing the epoxy technology can provide resistance to corrosion in FGD systems, the condensate formed in a stack or outlet duct generally provides a high concentration of acid, which results in blistering of most epoxy systems. Similarly, a fluorocarbon system can provide sufficient acid and temperature resistance, but has a major disadvantage in its ability to adhere to a substrate, which can result in premature blistering or disbondment of the system.

Vinyl ester systems [17] have proven to provide a protective barrier to the various components in FGD systems. These linings can be cost-effectively applied with standard spray equipment. They exhibit excellent resistance to the reaction of chlorides, fluorides, and sulfur under high temperature and abrasive conditions during the burning of fossil fuels.

Generally, a vinyl ester could be classified as methyl acrylic acid extended bisphenol A epoxy, acrylic acid bisphenol E epoxy, a bisphenol F novolac epoxy, or a urethane modified ester. Each will provide a different degree of resistance. Vinyl ester lining systems have been successfully applied to provide protection to stack liners, ducting, scrubbers, thickener tanks, and other vessels in FGD plants, stacks and ducts, electrostatic precipitators, and bag house environments.

3.4.2 Chemical Resistant Masonry

Chemical resistant masonry (CRM) is used to provide thermal and mechanical protection of the interior of tanks which are normally in a severe corrosive service at elevated temperatures [18]. CRM is made up of three components: acid brickwork (block), acid proof cement (mortar), and a lining or coating system (membrane). The membrane can be either a sheet lining, a coating, a true pinhole-free membrane, or a semimembrane, depending on the actual service conditions. The membrane system is selected on the basis of its ability to prevent any corrosive materials from contacting the substrate. Because of their excellent resistance to a broad range of corrosive conditions at elevated temperatures, the fluoropolymer plastic membranes have become an excellent choice for use as a membrane beneath chemical-resistant masonry. Most fluoropolymers currently being used as membrane system are poly (vinyl fluoride) (PVF), ethylene chlorotrifluoroethylene (ECTFE), ethylene trifluoroethylene (ETFE), fluorinated ethylene propylene (FEP), and perfluoroalkoxy (PFA). Lack of meltflow processability restricts the selection of PTFE as a viable membrane system.

There are basically three types of mortars in use in CRM. They are thermoset resin based mortars (such as furan, phenolic, epoxy, polyester), silicate mortars, and sulfur mortars.

Furan mortar is the most commonly used mortar in the chemical process industry. It is a two-component system containing a liquid binder and a powder that is mixed in a mortar box to produce a material of trowelable consistency. Furan mortar is suitable for a wide range of corrosive service conditions including various concentrations of acids and alkalies at temperatures up to 170°C. It is not recommended for strongly oxidizing services.

Epoxy mortars are less frequently used in CRM than other resin bonded systems due to higher cost, lower maximum service temperature, and narrower range of chemical resistance. However, because of their excellent bond strength to concrete, epoxides are frequently used for acid brick floors over concrete. They also have applications in alkali exposure.

Polyester mortars are used in oxidizing services such as nitric and chromic acid. They are, however, more expensive than furan and also exhibit a narrower range of chemical resistance. They are usually a three-component system consisting of a resin, a powder, and a curing agent.

3.4.3 Piping Systems

The cooling or heating systems in commercial or institutional buildings are frequently fouled by the accumulation of deposits containing organic material, silt and/or iron oxide. A conventional off-line chemical cleaning in these facilities with an acid or other strong solvent is a very complex and expensive process since the piping is usually not designed for such cleaning procedures and it may be enclosed within walls. In many cases, high-molecular-weight water-soluble polymers can be used as an on-line method for removing substantial quantities of deposits from piping systems. While the method will not work in all cases, it is so inexpensive compared to off-line chemical cleaning that it must be considered.

A slow on-line cleaning over a long period of time has been found to be effective in several applications [19]. Such a cleaning uses high-molecular-weight anionic polymers to slowly soften and disperse the deposits over time. The process may take from several months to up to two years. A high-molecular weight anionic polyacrylamide with molecular weights (Mw) in the range of 15–18 million and charge densities between 30 and 100% anionicity have been used at doses as low as 20 parts per million.

The on-line cleaning is a physical process that pries small pieces of the deposit from the pipe surface. The charged portion of the polymer attaches to surface charges on the deposit. The polymer chain acts like a sail and allows the velocity of the water to exert a physical force on the deposit.

3.4.4 Boiler and Cooling Water Treatment

Programs such as phosphate and zinc treatments of boiler and cooling water require the use of appropriate polymer to demonstrate satisfactory corrosion control. These polymers also prevent bulk precipitation of the corrosion additives in the circulating water. Sulfonated copolymers have thus proven to be effective for both boiler and cooling water treatments.

Possibly the most widely used sulfonated copolymers, used both as dispersants in boiler treatment and as cooling water additives, are composed of acrylic acid and 2-acrylamido-2-methylpropane sulfonic acid (AA–PSA). Since the 1980s, water-treatment service companies from the smallest to the largest have phosphate-based cooling water programs incorporating AA–PSA polymers.

A novel sulfonated copolymer, claimed to be superior to the aforesaid sulfonated polymers, has been introduced for boiler and cooling water treatment [20]. The polymer is said to have a unique composition based on multiple sulfonic acid containing monomers which provide hydrophobic character and high charge density [20]. This composition has been shown to provide a range of exceptional properties. The new polymer demonstrates not only equivalent or better performance for calcium phosphate and zinc stabilization, but also superior thermal stability when compared to other sulfonated polymers. A notable property of the new polymer is its ability to disperse and resolubilize pre-existing scale and corrosion product.

3.4.5 Biodegradable Scale Inhibitor

In the area of water treatment, polyacrylate scale inhibitors and dispersants, based on homopolymers of polyacrylic acid, are commonly used. Although generally of low toxicity, these polymers are not biodegradable. In some sensitive areas, such as the North Sea Oil fields, this has led to the search for biodegradable alternative to polyacrylates.

Polyaspartates, as exemplified by the parent polymer, polyaspartic acid, offer an attractive biodegradable alternative to polyacrylates [21]. In addition to their activity as scale inhibitors and dispersants, polyaspartates have also been shown to have corrosion inhibition activity, particularly for the corrosion environments frequently found in oil-field applications. Polyaspartates have very high biodegradability as indicated by OECD guidelines for aquatic biodegradability.

Sodium polyaspartates are readily obtained by thermal polymerization of aspartic acid followed by hydrolysis with sodium hydroxide:

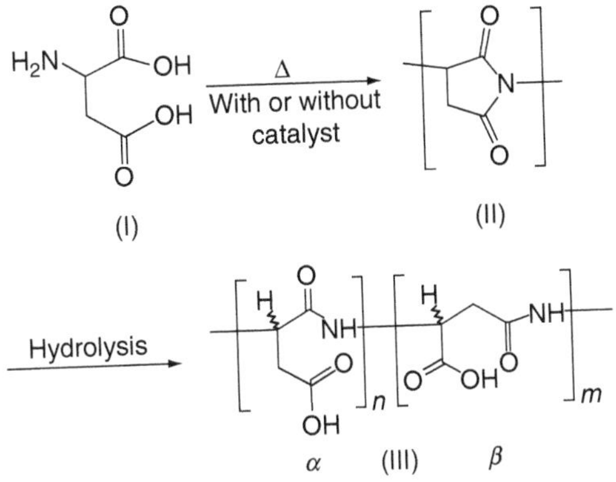

Heating aspartic acid (I) with or without catalyst leads to a linear thermal polycondensation polymer known as polysuccinimide (II), conversions of monomer to polymer of greater than 95% being easily accomplished. Hydrolysis of polysuccinimide with base such as sodium hydroxide, leads to a random copolymer of α and β aspartate units (III), with the β aspartate comprising about 70%–75% of the repeating units.

The performance of polymers as mineral inhibitors is dependent on the ability of the polymer to interact with growth sites on the surface of growing mineral crystals. Generally, the molecular weight of the polymer plays a significant role in this interaction. Polyaspartates have been shown to be very good inhibitors for both calcium and barium scales, the optimum inhibition activity being in the range of 1,000–4,000 Mw. This corresponds to 7–30 mer units in the polymer. For calcium carbonate and barium sulfate scale inhibition, the optimum is in the range of 2,000–4,000 Mw (15–30 mer units). In comparison with commercial scale inhibitors, the polyaspartates show comparable activity, besides being biodegradable, and in many cases better activity than polyacrylates of similar molecular weight.

3.4.6 Reinforcing Steel in Concrete

Corrosion of steel in concrete has been growing at an alarming rate and has become a very costly problem over the past three decades. In attempts to solve this corrosion problem, the successful application of epoxy coatings on underground transmission pipes drew a lot of attention, and it was pronounced by researchers back in 1974 that fusion-bonded epoxy-coated reinforcement (ECR) could significantly extend the time before deterioration of reinforced concrete bridge decks beyond the 5–10 years experienced with uncoated steel bar in area where deicing salts were used. Since then, ECR has been extensively used in the United States, Especially in bridges and viaducts.

For the most part, the performance of ECR in highway bridge decks has been successful as relatively little corrosion-induced concrete deterioration has been found on the great majority of ECR bridge deck structures. However, in the late 1980s, instances of coating disbondment, premature corrosion of ECR and concrete deterioration were recorded in the substructures of our major bridges in the Florida Keys after about 6–10 years of service [22]. Other cases of unsatisfactory performance of ECR were also reported. In several instances of regular maintenance to replace expansion joints in bridge decks containing ECR, the coating was found to be completely disbonded from the steel. These finding raised a lot of concern about the effectiveness of epoxy coatings in preventing corrosion of reinforcing steel in highly corrosive environments and brought into focus the need to study damage morphology in terms of coating characteristics, namely, the adhesion, integrity, and thickness of coating.

Later studies have established the importance of coating defects to the performance of ECR. If there are no defects, the corrosion protection barrier is effective. In the real world, however, there will be defects in placed ECR and the coating's resistance to disbondment from these defects is where the performance can be improved. Several ways have been suggested [23] to improve the long-term adhesion of epoxy coating to steel reinforcing bars. These include: alternative application methods, chemical treatment of blasted steel surface prior to coating application, and evolving a strong quality assurance for coating application plants.

3.5 Plastics in Automotive Applications

The use of plastics in automotive applications is increasing at a significant rate. In Europe and the United States, plastics make up between 7 and 10% of the mass of an automobile. The original impetus for using plastics in automobiles stemmed from the desire to reduce overall weight in order to achieve higher mileage. However, the impetus is now divided more or less equally between weight reduction and other considerations such as performance, parts consolidation, ecology, and processing economics. For example, by replacing a part assembly involving a number of individually cast components by a single molded plastic part, much labor can be saved and, at the same time, weight is reduced and perhaps performance properties are improved.

3.5.1 Exterior Body Parts

The original application of plastics in automobiles, which was for radiator grills (in plated ABS), has been overtaken by the use of plastics body parts to meet styling and aerodynamic requirements, which now represents a major growth area for engineering plastics. New materials have been developed for exterior body parts to fulfill the functional properties of heat stability (140°C–160°C), high impact strength, toughness (no splintering), petrol resistance, and stiffness. The materials being used for body panels include Du Pont's Bexlay range, based on polyamide technology, which replaced reinforced reaction injection molded (RRIM) polyurethane (PU) for Pontiac's Fiero rear quarter panel; XBR alloy, a Celanese product, which is believed to be a thermoplastic polyester alloyed with an acrylate; another Celanese product, Duraloy 2000, which is a polymer bland based on PBT; and Borg Warner's Elemid, which is a nylon/ABS blend.

In 1985 General Electric launched two new resins for car exteriors—Noryl GTX, a polyphenylene oxide (PPO)/nylon blend for fenders and other panels, and Lomod polyether ester elastomers for bumper systems. Noryl GTX combines the heat and dimensional stability and the low water absorption of PPO with the chemical resistance of semicrystalline nylons.

An important requirement for plastic materials in automotive industry is paintability alongside metal components at high temperature. Noryl GTX series has the important requirement of paintability in high-temperature (150°C–160°C) ovens. Several grades of Noryl GTX are available, some of which are glass filled. The unfilled graded are for exterior components such as fenders, spoilers, and wheeltrims, while the glass-reinforced grades are used in under-the-bonnet applications such as cooling fans, radiation end caps, and impeller housings, where, in addition to heat resistance, extra stiffness is needed.

RRIM polyurethanes are widely used for body panels. Bayflex 150 (of Bayer) was chosen by Pontiac for its 1987 Fiero, a well-publicized all-plastic car. The car's panels hang from a metal space frame. Bayflex 150 is used on vertical panels while the bonnet and roof are made of sheet molded components. Bayflex 150 is said to offer a number of advantages (improved thermal stability, stiffness, impact resistance, and dimensional stability) over conventional ureas without their processing limitations.

One problem in using plastics for body panels is the requirement for stiffness under heat along with the need for a high-quality surface finish. To overcome this problem, ICI separated the two requirements and molded tow compatible materials in sandwich form, using coupled glass reinforced polypropylene (PP) for the outer printable skin.

Ford Sierra featured the first all-plastic bumper in Europe, and the Ford Taurus and Sable featured the first of these bumpers in the United States. The bumper is molded from GE's Xenoy polycarbonate (PC)/polybutylene terephthalate (PBT) blend. The high impact strength of PC, combined with the resistance to gasoline and oil and the processing ease of PBT, has boosted the use of PC/PBT blends in car bumpers. Bayer's PC/PBT blends are called Makrolon PR. They are used for several automobile impact-resistant parts.

Formed PP has also been used for car bumpers. Arpro expanded PP bead (of Arco Chemical) is claimed to offer high impact absorption and resilience. It has been used for the front and rear bumper core. Arpro offers a substantial cost saving and a weight reduction of 50% in replacing the steel bumper system.

Toyota, in conjunction with several Japanese resin compounders, recently introduced a super TPO, which consists of polypropylene islands in a rubber matrix. The resulting TPO is softer, ahs better scratch resistance, and is more easily molded for bumper fascia and interior applications than standard TPO products.

TPO-based ionomers have been used in abrasion–resistant automotive exterior applications for a number of years. More recently, Du Pont's Bexloy W, a TPO based on the company's Surlyn ionomer resin, has been employed in competition with conventional TPOs (EPDM/PP compounds) in automotive applications such as fender flares, air dams, and license-plate brackets.

General Motors R&D and Montell U.S.A. have developed thermoplastic olefin (TPO) clay nanocomposites with reduced weight and good dimensional stability for exterior automotive applications [24]. PE nanocomposites have also been made effectively by ethylene homo- and copolymerization in the presence of organoclays (see Chapter 2), catalyzed with methylaluminoxane (MAO)-activated zirconocene, nickel, and palladium catalysts [25].

3.5.2   Interior Components

There is a lot of competition among plastics manufacturers for car interior components although, however, this area is not expected to show much growth [26,27]. ABS plastics continue to dominate the instrument panel market, while newer materials like Dow Chemical's pulse PC/ABS and Bayer's Bayblend PC/ABS alloys have been introduced for use in dashboard housings, interior mirror surrounds, and steering column covers. The alloy's high heat stability allows pigmentation in pastel colors.

Arco Chemical's line of Arpro expanded PP and Arpak expanded PE bead resins are used for the interior components. Polyolefin foams provide a higher degree of chemical resistance, heat resistance, and cushioning as compared to foamed polystyrene. Potential applications include seat backs, head rests, and other critical energy-absorbing areas.

Automotive seat back substrates are among large bow-molded parts made from very-low-melt-flow, high-melt-strength PP. Automotive is a major polypropylene market. The ease of recycling of PP is an important consideration in automotive applications.

A poly (ethylene terephthalate) (PET) product named Curl Lock for car seats has been introduced by Tagaki. In the production of Curl Lock, PET fibers are crimped and curled and then interlocked at their crosspoints with adhesive. The material is formed into car seat upholstery by placing it in a mold.

BP Chemicals have developed a polyurethane (PU) system for the production of integrally-molded sand-insulated carpets. The carpet is laminated with a PU backing to improve shape retention. It is then placed in a mold and PU foam is injected to fill the space between the car floor contours and the carpet. A carpet made with the system has been in use in the Fiat Turbo Uno.

Heterophasic polypropylene copolymers are used in large automotive parts that must withstand high temperatures without distortion. Heterophasic copolymers are formed when a rubber phase, usually ethylenopropylene rubber, is polymerized with the homopolymer phase during manufacture. The addition of rubber increases the impact resistance of the material, while the homopolymer provides stiffness. The primary applications for heterophasic copolymers that take advantage of the high-impact strength include injection-molded consumer products such as child safety seats, luggage and tackle boxes, and appliance and automotive components.

New technologies are broadening the range of available thermoplastic elastomer (TPE) properties. These developments include low-stiffness reactor grades of thermoplastic polyolefin elastomers (TPOs), which are one of the fastest growing TPFs. A factor motivating this growth is the expansion of TPOs into new automotive applications, such as interior skins and airbag covers, in addition to their penetration into existing automotive exterior markets such as bumper fascia, air dams, and trim components. The TPO Product tradenamed Dexflex 756-67 introduced by D&S Plastics International, a major supplier of TPOs to the North American automobile industry, is the first penetration of TPOs into airbags.

Wood–plastic and other natural fiber–plastic composite materials are being increasingly used in automotive industries. Woodstock, made with a mix of about 50% wood flour and 50% PP, is still widely used for automotive interiors, but the industry appears to be especially interested in components made form recycled plastics, and the introduction of wood-filled compounds for molding has spurred some of this interest.

Demand for natural fiber composites has been evident in Europe where environmental issues are prominent. As an example, all BMWs and Mercedes use natural fiber composites for components, such as door liners, boot liners and parcel shelves.

3.5.3   Load-Bearing Parts

Automobile structural components are expected to be a major growth area for advanced composites. Examples include steering column, main door frame, suspension system, and selected engine parts. An early commercial application was Ashland's Arimax 1,000 structural RRIM resin, which was used to produce the spare tire recess cover or the trunk of a number of cars built by GM's Buick–Oldsmobile Cadillac group. Other possible applications of Arimax resins include door trim panels, floor pans, package shelves, radiator supports, and motor side compartments.

The first all-around automobile application of composite leaf-springs was made in the Sherpa 200 series van and mini buses built by Freight Rover. These fiber-reinforced plastic (FRP) components for high stress, high-temperature uses were developed by GKN. ICI's Verton, a long-fiber reinforced thermoplastic is also suitable for such applications.

3.5.4   Under-the-Bonnet (Hood)

The use of plastics in under-the-bonnet applications is expanding at a significant rate. Filled or reinforced polypropylene (PP) or nylon is being used for items such as heater housings. Stampable PP or polyacetal glass mat materials, also known as glass mat thermoplastics, are being used as acoustic shields for diesel engines, chain drives, and timing belts.

From the three basic categories of polypropylene, namely, homopolymers, heterophasic copolymers, and random copolymers (with ethylene), there are specialty resins with enhanced capabilities for specific applications. Producers of large blow-molded or thermoformed parts can thus utilize grades with high melt strength to fabricate heat-resistant under-the-hood automotive parts.

Glass-filled polyether either ketone (PEEK) resins are used in the inlet manifold of a Rover racing engine, while polyether sulfone (Victrex) has been used for piston skirts. Fluoroplastics have found many applications in the sealing rings and bearings of automotive drives.

The Ford Polimotor, which has a virtually all-plastic engine that offers a weight saving of some 60% compared with conventional components, uses polyamide–imide for moving parts with fuel saving advantages.

Blends of polytetrafluoroethylene (PTFE) are replacing metals in applications such as gaskets, bearings, and piston rings in automotive engines. The Polycomp R range of LNP Corporation includes several grades of PTFE filled with polyphenylene sulfide, which show good mechanical properties. Polycomp E combines PTFE with a linear aromatic polyester. It is claimed to offer good electrical, mechanical, and tribological properties. Polycomp A contains a proprietary aromatic polymer developed for reinforcing PTFE.

In the electrical system, glass-reinforced PBTs and PETs find use because of their heat stability and good dimensional stability. An all-plastic headlamp with metallized polyester bulk molding compound or PBT for the housing and reflector is expected to find increasing use. Ford and GM light trucks use an abrasion–resistant coated polycarbonate in headlamps. A new high-intensity lamp introduced in Germany used polyetherimide for the reflector, to resist the higher heat.

3.5.5 Future Trends

Though numerous types of plastics materials are presently used for automotive applications, future vehicles will probably use fewer resin variables. This consolidation trend is well underway and is a result of postconsumer recycling requirements (see Chapter 2 of *Plastics Fabrication and Recycling*), which increasingly emphasize the use of only a few families of materials. Auto manufacturers will rely primarily on four key resin families for their plastics applications: polyolefins, polyamides, polyurethanes, and styrenics. The future use of PVC for automotive applications is uncertain due to pressure of environmental groups who assert that the resin contaminates automotive shredder residue with chlorine.

Polypropylene (PP) in various forms (filled, unfilled, reinforced, rubber blended) finds the greatest usage of all automotive plastics. Its consumption is likely to accelerate as a result of new laws requiring recyclability, and because of the weight and cost reductions it offers. New fabrication techniques may also contribute to growth. For example, the monomaterial sandwich construction techniques developed by fabricators yield rugged and lightweight PP-based instrument and door panels. These parts consist of a sandwich of reinforced or neat PP substrate, a cross-linked PP foam, and a thermoplastic polyolefin elastomer (TPO) cover.

There is considerable competition between materials in automotive markets. For example, a major shift is taking place from polyurethanes to TPO for bumper fascia. The TPOs, with molded-in color capability, light weight, relatively low cost, and ease of recycling, are poised to capture a large share of the fascia market both in North America and Europe. A significant cost saving is possible with a TPO fascia, especially if it is not painted. A polyurethane fascia, by contrast, must always be painted, which adds some 60% to the cost of a fascia.

Many of the factors responsible for shifts in bumper fascia are also behind similar changes in side-body moldings. Various grades of TPO have effectively displaced PVC in this application.

Vertical and horizontal exterior auto body panels have always represented a tremendous opportunity for plastics. In fact, plastics have been very successful in penetrating fender applications in Europe and North America. Although steel remains the dominant material for fenders, the benefits of plastics such as weight reduction, dent resistance, and styling flexibility override cost disadvantages.

The poor temperature resistance and inferior mechanical strength of polyolefins have so far prohibited their use in exterior body panels. However, the advent of metallocene catalyst technology (see Chapter 1) may lead to a new generation of PP materials that will meet the stringent requirements of exterior body panels. This will create yet another automotive market for PP-based products.

3.5.6 Polymer Nanocomposites

PP has great potential for composites and nanocomposites because it can be processed by conventional technologies, such as extrusion and injection molding, to make parts for automotive applications. Reinforcement of PP by micro- and nanofillers yields composites with high rigidity and toughness. PP-clay nanocomposites have been prepared at Toyota Central R&D via melt intercalation (see *Polymer Nanocomposites* in Chapter 2) of montmorillonite organo-clays with PP modified with either maleic anhydride or hydroxyl groups [28,29]. This method was pioneered by Toyota Motor Company to produce nylon-6/clay hybrid, used to make a timing-belt cover, the first practical example of polymeric nanocomposites for automotive applications [30]. The tensile modulus of this nanocomposite is twice that of nylon-6 (2.1 vs. 1.1 GPa), and the coefficeint of linear thermal expansion—a key factor in dimensional stability and an essential factor to manufacture large vehicle parts—is reduced in half (6×10^{-5} vs. 13×10^{-5}) for the nanocomposite containing only 1.6 vol.% clay mineral. The use of polymeric nanocomposites in automotive applications, however, depends on meeting stringent demands on cost and performance. Each application may require variations in the production process to make parts for specific uses.

3.5.7 "Green" Composites

Eco-friendly biocomposites from plant-derived plastics (bioplastic) and fibers (natural/biofiber) are novel materials for the 21st century. They are seen not only as a solution to growing environmental threats but also as an answer to the uncertainty of the petroleum supply. Natural/biofiber composites (biocomposites) are emerging as a viable alternative to glass–fiber reinforced polymer (GFRP) composites, especially in automotive applications [31]. They can be 25%–30% stronger than GFRP for the same weight and can deliver the same performance for lower weight [32]. As compared to GFRP, they also exhibit a favorable nonbrittle fracture on impact, which is an important requirement for use in the compartment of passenger transport.

Autocompanies are also seeking materials that posses sound abatement capability, besides shatter resistance and reduced weight for fuel efficiency. Natural fibers possess excellent sound absorbing efficiency, are more shatter resistant and have better energy economy than glass–fiber based composites. In automotive parts, biocomposites made from natural fibers reduce the mass of the component and can lower the total energy consumed in producing this material by 80% [33]. Natural fibers such as kenaf and hemp used to replace glass fibers have ~1/3rd the cost of E-glass fiber, ~1/2 the density, acceptable specific mechanical properties, and enhanced biodegradability. Interior parts and door trim panels from natural fiber–polypropylene and exterior parts from natural fiber–polyester resins are already in use [34,35].

Composites derived from natural fibers and synthetic polymers such as polyethylene, polypropylene, epoxies and polyesters are, however, not sufficiently eco-friendly because of the synthetic polymers' petroleum-based source, as well as the nonbiodegrdable nature of the polymer matrix. Efforts are therefore on to develop "green" biocomposites by using both fibers and matrix resin based on renewable sources. Thus, by embedding biofibers such as hemp, kenaf, pineapple, leaf fiber, henequen, corn straw fibers, and grasses with renewable-resource-based biopolymers such as cellulose plastic, starch plastic, and soy-based plastic, green biocomposites with acceptable mechanical properties are being developed at Michigan State University [33]. Cellulose esters (cellulosic plastics), such as cellulose acetate (CA), cellulose acetate propionate (CAP), or cellulose acetate butyrate (CAB) are considered as potentially useful biosourced polymers for the future [36]. Possible hydrogen bonds formed by the hydroxyl and carbonyl groups in these esters with hydroxyl groups of cellulose fibers can lead to increased adhesion between fibers and cellulose ester matrix.

Novel biocomposites have been made utilizing two different processing approaches [33]: powder impregnation followed by compression molding (process I) and extrusion followed by injection molding (process II). As an illustration of process I, vacuum dried, chopped hemp fiber was added in the required weight percentage into predried cellulose acetate (CA) powder premixed with 30 wt.% eco-friendly triethyl citrate (TEC) plasticizer and the resulting "green" mixture was compression molded in a press at 195°C under pressure.

For making CA-based green composites by process II, a two-step extrusion was used to make CA-hemp biocomposite pellets which were then injection molded. The first step was to produce CA plastic granules from CA powder and 30 wt.% TEC plasticizer. [Note: CA needs an external plasticizer added to enhance its flow and allow processing below its degradation temperature (230°C), which is near the lower portion of its typical melting range.] In the second step, CA plastic granules were fed into a twin-screw extruder ($L/D=30$) operated at 190°C, while feeding chopped hemp ($\sim 1/4$ in.) in the second-to-last port of the extruder, with the die maintained at 195°C. Cellulose acetate butyrate (CAB), unlike CA, processes easily without the need for a plasticizer. For CAB-hemp biocomposite pellet production, CAB powder (without plasticizers or additives) and chopped hemp fibers were directly fed into the extruder in the requisite amounts. The biocomposite pellets were injection molded at 195°C.

Cellulose acetate plasticized with 30% citrate plasticizer proved to be a better matrix compared to polypropylene for hemp fiber reinforcements in terms of flexural and damping properties [33]. Biocomposites with 30 wt.% of industrial hemp fiber processed through extrusion and injection molding exhibited a flexural strength of ~ 78 MPa and modulus of elasticity of ~ 5.6 GPa. CAB, however, is a better matrix than plasticized CA as far as composite processing and physicochemical properties are concerned. Both CAB and CA matrices have good interaction with hemp fibers as shown by fiber pull-out micrographs, with the former being the better, and thus leading to higher strength.

The ultimate goal of research in green composites is to replace the existing synthetic glass fibers with natural fibers as reinforcements and also to replace petroleum-based polymers with renewable-resource-based bipolymers as matrices in designing and engineering of biocomposite materials [33].

3.6   Polymers in Aerospace Applications

Advanced polymer composites, which are high-performance materials consisting of a polymer matrix resin reinforced with fibers such as carbon, graphite, aramid, boron, or S-glass, have their market in aerospace. This is also expected to be the fastest growing sector of plastics sales, with growth projected at 22% a year.

Much of the original impetus for a changeover from the use of aluminum materials to plastics in the aerospace industry came from the availability of lightweight polymer composites based on reinforcements such as Du Pont's aramid fiber (made under the tradename Kevlar) and honeycomb materials such as Nomex from Ciba–Geigy. By using these reinforcements through impregnation with resins such as epoxies, and with applications of precise tooling, components can be produced that involve complex shapes and meet difficult aerodynamic requirements. Weight reduction is the major advantage of using light-weight composites in place of metals. The incorporation of composite materials in Corcorde afforded a saving of 500 kg in mass. The Airbus 320 is said to have saved about one ton in mass by the use of advanced composites.

Within the aerospace industry, advanced composites are used to greatest extent in military aircraft. Commercial aircraft such as the Boeing 757 and 767 contain less than 5% advanced composites by weight, while military aircraft like the GA-18A contain 10%–20%. Typical applications of composites include aircraft wings, tail sections, pressure vessels, and seat rails.

Advanced composites have been used most extensively in helicopters. Sikorsky's S-75 helicopter, for example, is about 25% composite by weight, mostly graphite–epoxy and aramid–epoxy composite materials. Composites are used in rotors, blades, and tail assemblies. Future military helicopters are likely to comprise up to 80% advanced composites by structural weight. Graphite–epoxy composites are likely to be used in the airframe, bulk-heads, tail bones, and vertical fins, while the less stiff glass–epoxy composites will be used in rotor systems.

3.6.1   Carbon Fibers

The fiber most commonly used in advanced polymer composites is carbon fiber. Research and development of carbon fiber components (CFCs) for aerospace applications gained impetus since 1970 when British Aerospace used them in the Jet Provost rudder trim tabs. Since then CFCs have been used in the foreplanes, wings, access panels and doors, cockpit floor structures, and side panels of a number of aircraft including Jaguars, Tornados, and Harriers. Advantages of CFCs over aluminum materials include an increased strength-to-weight ratio; a reduction of up to 30% in the manufacturing cost, due to reduction in the number of detailed parts; a reduction in the volume of waste materials; and the ability to produce large, complex curved shapes required in advanced military aircraft.

The satellites ECS1 and AMSAT developed by Aerospatiale, France, feature two of the largest carbon–fiber reinforced honeycomb structures made in Europe. A sandwich structure reinforced with high modulus carbon fiber is used for the dual system capsule. The fiber, Grafil HM-S, provides exceptional dimensional rigidity for extremely low weight and matches very specific resonance characteristics required in the design. The fiber in epoxy resin is laminated to form composite skins using prepreg materials. The same grade of carbon fiber is also used in the manufacture of the actuator control struts for the second stage motor of the Ariane launcher.

3.6.2   Resins

There is a growing competition in the market for moderate-temperature materials for aircraft structures such as fuselages and cargo bays in response to the aerospace industry's endeavour to produce an all-composite aircraft, which is currently led by Boeing.

Shell Chemical has produced EPON HPT epoxy resins, which are used in large parts made of graphite-7 reinforced prepreg material for primary aircraft structures. EPON 9000 series of epoxy resins produced by the same company have been used for fabricating, by wet filament winding process, fuselages and other large components that have been used in the prototype of Beech Aircraft's experimental small business aircraft, the Starship 1.

Dow developed an experimental moderate-temperature resin, XV-71794, which is claimed to offer incredibly high toughness and low moisture absorption. Baltimore Aerospace has used Nextel and graphite fibers in a bismaleimide matrix for a hot air duct in a jet engine. The duct is 75% lighter than one made of metal.

More recently, thermoplastics such as polyphenylene sulfide (PPS), polyether sulfone, and polyether ether ketone (PEEK) have entered the primary structures market in competition with thermosets.

PEEK, a relatively new thermoplastic material introduced in 1978 by ICI, is used in aerospace applications. ICI Plastics have produced PEEK reinforced with 68% continuous carbon fiber in narrow tows suitable for braiding. The braid is used to strengthen tubular structures without adding significantly to the weight. One application is in the repair of battle-damaged military equipment, particularly control rods and helicopter rotor shafts. The braid is slipped on where needed and built up in layers to strengthen the component. Another possible application of the braid is for the reinforcement of aluminum air frame struts in the hand-glider.

A 20% glass-filled PEEK introduced by ICI for injection molding was chosen by Boeing to replace aluminum fairing on its 757-200 aircraft. The substitution of PEEK for aluminum reduced weight by 30% and lowered costs considerably. Compared to other materials PEEK also showed greater heat and chemical resistance in addition to the required dimensional stability.

Carbon and glass–fiber reinforced PEEK has been used with advanced plastics technology in the redesigning of the helicopter blades profile, improving aerodynamic properties, and reducing overall weight [37]. The use of PEEK composites in the RAF Linx helicopter enabled it to break the world helicopter speed record.

A good example of large-size fiber-reinforced components in aerospace application is the radome covering the underbelly radar on the Hercules transport aircraft. It is made from very thin polyethylene sulfide (PES) film interleaved with PES-impregnated glass fiber cloth, which is subsequently hot molded in a closed die. The composite radome, nearly 1 m in diameter and 6 mm thick, weighs only 10 kg.

Ryton PPS woven carbon fabric prepreg sheeting (AC31-60), supplied by Phillips Petroleum, is particularly suited for thermoforming into high-strength parts. Boeing Aerospace designed and fabricated a prototype access door using this composite. It resulted in weight reduction of 25% compared with a metal counterpart and a 67% cost saving compared with thermoset composite construction.

To improve on Ryton PPS, Phillips introduced a family of high-temperature performance polyarylene sulfide (PAS) polymers. The crystalline form, PAS1, is said to be suitable for structural components at temperatures up to 120°C and for nonstructural components at temperatures up to 230°C. PAS2, an amorphous form, has temperature limits of 160°C–200°C for structural components and 270°C for nonstructural components.

Raychem developed heat-shrinkable woven fabrics that incorporate heat shrink fibers in the weft and nonshrink protective fibers in the warp. Slipped over a cable harness, for example, they shrink down to a quarter of their size to form a tough outer tube. The problem of damage caused to aircraft radar covers, landing gear wiring, and hydraulic lines by stones and debris thrown up off runways, is being solved by the use of heat shrinkable woven fabric tubes containing high wear and impact resistant fibers such as Kevlar, polyether, and glass. The British Aerospace 146 jet has a Kevlar-based heat shrink fabric protecting its undercarriage cables. Radar umbilical cables on RAF fighter aircraft are similarly protected.

A major problem relating to passenger safety has been the use of epoxy paste, which can emit toxic gas in the event of fire. The epoxy paste compounds are used to fill the honeycomb cells of sandwich panels, such as those used in aircraft wings and flooring, in the region of joints and attachments. A phenolic equivalent, Corfil 792, has been developed by Cyanamid Fothergill. It is a high strength, low density, spreadable thixotropic paste formulated for single stage curing of sandwich constructions and has been accepted by the aerospace industry worldwide.

Cellular materials are used in aircraft for a number of applications including cockpit insulation, duct insulation, and vibration damping for aircraft fuselage. Monsanto's Skybond polyimide foams have been used for sound deadening of jet engines. Solimide polyimide foams produced by Imi-Tech Corporation are said to be lightweight, fire resistant, and show good thermal and acoustic properties. They have been specified by Boeing for several aerospace applications. One space application for Solimide is cryogenic insulation for fuel tanks on major rocket propulsion systems.

3.7 Polymers in Electrical and Electronic Applications

Plastics have played critical roles in every aspect of the ongoing revolution in electronics [38–41]. They are now found in everything from computer housings and floppy discs to encapsulants and connectors. The wide range of electrical and electronic applications of polymers includes wire and cable coverings, housing adhesives, printed circuit boards and insulators (Table 3.1).

One of the key trends in electronic components today is miniaturization, which began in the 1980s and has accelerated. This development is bringing new opportunities for plastics in the thin-walled and

TABLE 3.1 Resins Commonly used for Electrical and Electronic Applications

Resin	Typical Applications
ABS	Consumer electronics, enclosures, telecommunications
Epoxy	Electrical laminates
EVA	Wire and cable
HDPE	Wire and cable
LDPE/LLDPE	Wire and cable
Nylon	Electrical/electronic, wire and cable
Phenolic	Electrical/electronic, electrical laminates
Polycarbonate	Electrical/electronic, enclosures, storage disks
Polyester (unsaturated)	Electrical, enclosures
Polyester (thermoplastic)	Electrical/electronic
PPO-based resin	Electrical/electronic, enclosures
Polypropylene	Wire and cable
Polystyrene	Cassettes, reels, enclosures
PVC	Wire and cable, plugs, connectors, enclosures, electric tapes
Urea	Electrical

downsized products now reaching consumers. In consumer and office electronics, portability is a key design and marketing criterion. Portable computers, printers, copiers, cellular phones, and measuring and monitoring devices are highly plastics intensive. Today's portable (laptop and notebook) computer, for example, has 50%–75% of its weight (and even more of the volume) represented by plastics.

3.7.1   Wire and Cable Insulation

Du Pont's Neoprene synthetic rubber has been used for many years in the cable industry as it provides good low voltage insulation and mechanical toughness as well as good resistance to oil, high temperature, and oxidation. Alcryn, also from Du Pont, is a melt-processable rubber cable sheathing for signal and power transmission. It is claimed to have excellent oil resistance even at high temperatures, and good resistance to severe mechanical treatment. Another key product from Du Pont is Hypalon, a chlorosulfonated polyethylene that exhibits high resistance to oils and a host of chemicals. One application of this product is the jacketing of cables in chemical plants.

The largest single outlet for plastics in electrical and electronic applications—roughly half of the volume consumed—is in wire and cable, where PVC dominates on the basis of its formulating flexibility, ease of processing, and low cost. However, the halogen-containing polymer systems tend to be more corrosive to copper than the non-halogen-based systems. Chlorinated compounds can also result in unacceptable emissions from conventional incinerators. Because of industry and regulatory concern over combustion, corrosivity, toxicity and environmental impact at the points of manufacture and disposal, extensive use of PVC in future thus appears uncertain.

To cover the 1987 NEC (National Electric Code) flame and smoke performance guidelines which took effect in the U.S., industry generally uses a non-flame-retardant polyolefin as the insulation material and a highly flame-retardant PVC as the jacketing. If greater flame retardancy is required, fluoropolymers can be incorporated in the system.

Fluoropolymers such as Teflon and Tefzel that have high mechanical strength and dielectric properties with heat resistance enable thin wall insulation for high current densities. Du Pont's fluoroelastomer Viton is used for insulation and heat shrinkable sleeving. In addition to excellent heat and chemical resistance it has high resistance to fuels, oils, and lubricants. Fluon, a polytetrafluoroethylene (PTFE) product from Fothergill Cables, which specializes in electronics cables for aerospace and defense industries, is applied to thin wall insulation for equipment wires, multicore cables, thermocouple wires, and radio-frequency coaxial cables. Fluon has a service temperature range of -75 to $+260°C$, nonflammability, high chemical resistance, good dielectric properties, and resistance to damage by soldering.

Polyesters in wire and cable applications include Hytrel (Du Pont), a thermoplastic polyester elastomer for the offshore industry; Kapton (Du Pont), a polyimide film offering weight saving and chemical resistance; and Mylar (Du Pont), used in primary insulation, shield isolation, chemical barrier, and mechanical protection.

ICI has developed a large number of products to suit different applications in the wire and cable industry. While for general purpose insulation and sheathing, poly(vinyl chloride) (PVC) is used with chlorinated plasticizers such as the company's Cereclor compounds, its Victrex polyether ether ketone (PEEK) wire and cable covering is designed to operate in high-demand environments. PEEK has high-performance electrical properties that permit the use of smaller diameter cable covering without loss of electrical power or reduction in insulation. Because of the product's good inherent flame retardancy, low smoke emission, and low toxic and corrosive gas emissions it is used in underground mass transit wiring and cable insulation. Victrex has also been used for insulating a coaxial control cable operating in a high radiation environment at a nuclear installation in the United Kingdom.

BP Chemicals have introduced a new technology for silane cross-linking in low-voltage cable insulations to produce more uniform cross-linking and greater shelf-life. The new silane-linked polyethylene compounds can be run on conventional extrusion equipment. The company has also produced a strippable double layer cable shield in cross-linkable polyethylene for medium voltage cables.

The shield has a strippable outer layer formed by a coextrusion of 20% strip material and 50% standard semi-conductive material.

3.7.2 Polymer Insulators

The Ohio Brass Hi*Lite XL insulator, offered by the Ohio Brass Company, is the first nonceramic insulator (NCI) to see widespread service on utility transmission lines throughout North America. It set in motion since 1976, a basic change in line insulation practice. This change which has accelerated since the mid 1980s, has seen the old standard porcelain give way to the more cost-efficient, light-weight FRP designs offered by Ohio Brass and several other manufacturers with plants in North America. Polymer insulators are in use on nearly 80% of all new transmission constructed in North America.

Transmission insulators provide mechanical support and electrical isolation for overhead, open wire, power transmission lines. The primary insulation on these lines is air.

All insulators in the Hi*Lite family use an innovative and unique silicone compound to provide an unbonded interface between insulating rubber housing and load-bearing fiberglass core. This interface has proven to be more reliable than the chemically bonded type of interface.

3.7.3 Printed Circuit Boards

Developments in the electronics industry are focused on two main areas—miniaturization and improved cost/performance ratios—creating the need for polymers that can tolerate higher temperatures for longer periods with good dimensional stability and mechanical strength.

Though the majority of printed circuit boards (PCBs) at present are two-dimensional and made of glass-filled epoxy resins, increasing miniaturization has led to more PCBs being multi-layered with several copper circuits sandwiched between insulating layers. In these applications, because of the added stress due to thermal expansion, specialty materials such as polyamides and fluoropolymers have been increasingly used.

The present trend is towards injection-molded thermoplastic boards in place of thermoset laminate PCBs. The two main reasons for this shift are the difficulty of etching on thermoset laminates, and the greater design flexibility and potential cost saving in high-volume production by injection molding. A further advantage of injection molding is the ability to produce circuit boards with three-dimensional features such as ribs, bosses, heat sinks, and holding devices for snap-parts. Engineering polymers that can be used for molded circuit boards include Du Pont's Rynite polyethylene terephthalate (PET) grades and glass-reinforced grades of ICI's Victrex polyethersulfone (PES).

In the electronics industry, one of the most successful uses of advanced composites has been in PCBs. General Electric's Ultem polyetherimide is used in transport PCBs, especially aerospace an missile guidance systems. Polysulfones and other high-performance thermoplastics are being used in multilayer PCBs for mainframes and supercomputers operating at gigahertz frequencies.

Processes developed for the injection molding of circuit boards include PCK Technology's mold-n-plate process uses two shots, the first being one of catalytic resins, to form paths on plastic parts that are then processed in a copper bath to form circuitry. Special grades of polysulfone, polyethersulfone, and polyarylsulfone are used for the process. In the in-mold metallization process, various conductive, resistive, capacitive, or inductive components are embedded directly into the plastic part in the tool.

3.7.4 Connectors

The miniaturization of PCBs has created the need for connectors to be made from special materials suitable for shaping into thinner parts that can be packed into higher-destiny spaces and that can withstand higher soldering temperatures. Such materials available on the market include tailored grades of polyphenylene sulfide (PPS), PEEK, PET, polyetherimide, sulfone-based alloys, and liquid crystal polymers. Well-known commercial products are Rynite FR-530 and FR-945 modified PET grades of Du

Pont; Valox 9730 developmental alloy of General Electric; Victrex high-performance thermoplastics of ICI; Mindel alloys of Amoco Performance Products, which combine various grades of sulfone polymers; and Ryton PPS grades R-4 and R-7 of Phillips Petroleum Chemicals.

3.7.5 Enclosures

Plastics are increasingly being used for electrical/electronic enclosures. Those used for enclosures of circuit chips should, however, be of high purity and high flow. Since miniaturization reduces the surface area of heat dissipation, the materials used are required to have high heat resistance in addition to high purity. While epoxy resins such as Quatrex 3430 and 3450 of Dow Chemical dominate the market, other materials used include polyurethanes, polyimides, and silicones.

Plastics used in electrical housings are also required to provide protection against electromagnetic interference (EMI), radio frequencies, and electrostatic charges that could damage sensitive equipment circuits. To impart EMI shielding properties to a plastic component, a coating of conductive layer may be provided by post-molding or in-molding processes or polymers may be used that have been metal or carbon black filled to make them conductive (see Chapter 2).

The methods mostly used to produce EMI shielding on plastic moldings are conductive paints, zinc arc spraying, vacuum metallizing, electroplating, and foil application. There are also a range of filled plastic compounds on the market that are based on metal fillers such as aluminum flame, brass fibers, stainless steel fibers, graphite-coated fibers, and metal-coated graphite fibers. The most cost effective conductive compounds are, however, based on carbon black.

The fabrication and electrical properties of carbon nanofiber-polystyrene composites and their potential applications for EMI shielding have been reported [42]. Being lightweight is a key technological requirement for the development of practical EMI shielding systems. Thus the fabrication of foam structures to further reduce the weight of carbon nonofiber-ploymer composites has been recently demonstrated and simple preparation routine has been reported [43] by which this novel foam structure can be prepared.

One of the key determinants of future material selection for enclosures will be the growing global demand for recycling of these units. The recycling of electronics equipment enclosures and other molded electronic parts, however, presents formidable obstacles, since there are many different resins in use, as well as a wide range of performance-enhancing additives and functional coatings for EMI shielding and ESD protection. To overcome these problems, industry is likely to narrow the enclosures resin menu to the usable plastics stream and focus on a well-defined group of additives, coatings, adhesives, and other ingredients.

3.7.6 Optical Fibers

A major advance in communications has been the introduction of optical fiber transmission. In optical fiber transmission, voice and data signals are converted into beams of light that then travel along extremely thin glass fibers. Major advantages of this transmission over the conventional current carrying metal wire are higher message speed and density, lighter weight, and immunity to electromagnetic interference. The main growth in the use of optical fibers is expected to be in telephone networks.

Plastics find extensive use in several areas of fiber optic cables. Buffer tubes, usually extruded from high-performance plastics such as fluoropolymers, nylon, acetal resins, or polybutylene terephthalate (PBT) are used for sheathing optical fibers. A blend o PVC and ethylene vinyl acetate (EVA) polymer, such as Pantalast 1162 of Pantasote Incorporated, does not require a plasticizer, which helps the material maintain stability when in contact with water-proofing materials. PVC and elastomer blends, Carloy 6190 and 6178, of Cary Chemicals are also used for fiber optic applications (Stiffening rods for fiber optics are either pultruded epoxy and glass or steel. Around these is the outer jacketing, which is similar to conventional cable.)

Polymer optical fibers (POFs) combine a lot of very attractive properties like high bandwidth, total EMF immunity, and ease of handling. The focus of POF development has been in Japan, mainly driven by inventions of the group of Koike/Keio University and three companies, namely, Mitsubishi Rayon, Toray, and Asahi, which dominate the world market. The developments in the U.S.A. and Europe, on the other hand, are restricted mainly to the areas of POF compatible emitters and receivers and POF transmission systems, mainly for industrial production lines, automated wiring, and home wiring.

Japan's Mitsubishi Rayon Co. Ltd. Developed a plastic optical fiber cable "Eska Premium" with a wide range of freedom, excellent durability and which can be used for wiring in narrow spaces. It is claimed that the cable can be bent or wound without deteriorating the characteristics of the fiber cable. The cable is lighter and easier to work with than glass-based cables, and is ideal for short communications and transmission wiring between the controllers of machine tools, personal computers, and peripheral equipment.

Toray polymer optical fibers (POFs), developed by Toray Industries, Inc., are of step index type with a core of high purity PMMA and cladding of special fluorinated polymer. These are produced under two series: POFs in PF series for communication use have fiber diameter 0.25–1.5 mm, numerical aperture (NA) 0.46, acceptance angle (AA) 55°, attenuation (ATN) <0.15–0.30 dB/m at 650 nm, and allowable bending radius (ABR) >17–27 mm, main usages being data transmission in the temperature range −40°C to 70°C. POFs in PG series for communication and industrial use have fiber diameter 0.25–3.0 mm, NA=0.5, AA=60°, ATN <0.15–0.35 dB/m at 650 nm, and ABR >9–20 mm, main usages being data transmission, light guide, optical sensor, electric appliances and displays.

While a new POF, Lucina® (120 μm core diameter) made out of CYTOP® [44] offering low transmission loss combined with high band width has reached the market, a new class of optical waveguides, also known as photonic crystal fibers or as holey fibers, have been disclosed since 2001. Structures like small air tubes with diameters in the micron range lead either to so-called photonic band gaps where no photons are allowed to propagate like electrons in solid matter, or to effective index variations shaping an appropriate profile. Polymers are very promising candidates for this new class of fibers because unlike silica photonic crystal fibers, polymer fibers are not restricted to hexagonal symmetry, thus offering a wide variety of new solutions [45,46].

3.7.7 Information Storage Discs

In recent years, information has begun to be stored magnetically, on tape and on disc. Today their position is being challenged by optical discs, because an optical disc can store 1,000 times more information than its magnetic equivalent, has a longer guaranteed lifetime (usually 10 years), and more easily accessible data.

Although magnetic storage discs are beginning to be regarded as archival as compared to be regarded as archival as compared to optical discs, the advantage they have over most optical systems is that information can be added and erased. The two types of magnetic discs on the market are Winchester hard discs and stretched discs.

Winchester discs are molded as blanks in engineering plastics, usually General Electric's Ultem polyetherimide, then coated with an epoxy-ferric oxide layer. The main requirement in manufacturing these discs is absolute flatness because the magnetic read/write head moves only 8–20 microinches above the disc surface and is spinning at a high speed of 3,600 rpm. Any microscopic bump can thus cause damage to both the disc and head.

The stretched disc, developed by 3M Company as an alternative to the Winchester disc, is said to combine the performance characteristics of hard discs with the economy and environmental tolerance of a floppy disc. The discs are 5¼ in. in diameter and are able to store 12-Megabytes of computer data with storage capacity up to 100-Megabytes is thought possible. Unlike Winchester discs, stretched discs consist of a magnetically coated polyester film stretched over an injection-molded disc. The film is bonded to the raised edges of the disc, thus producing a compliant surface with a 250 μm gap between the film and substrate. Since information is stored on the film rather than on the disc, flatness requirements are not critical as with Winchester discs.

The first optical disc to appear on the market was the laser-read video disc, 8 or 12 in. in diameter, which is produced in acrylic resin. The disc is made of two halves sandwiched together, which reduces the problems of warpage and means that the disc can be played on both sides. One example is Phillip's Laservision, which dominates the European market.

The compact audio disc (CD) based on digital recording and playback technology is 4.75 in. in diameter and can store 16 million bits of information in the form of minute pits in the substrate. It is the presence or absence of the pits that is read by the laser. The pits are 0.1 μm deep, 0.5 μm wide, and between 0.833 and 3.56 μm long. Each track comprises a spiral of these pits. The track is laid in polycarbonate that is backed with reflective aluminum and coated with a protective acrylic lacquer. The CD can thus be played only on one side.

The CDs are manufactured either by injection compression molding or injection molding. Injection compression molding minimizes molded-in stress but requires longer cycle times than injection molding and produces flash. For both processes very clean room conditions, up to 1,000 times cleaner than medical molding, are essential.

The processability of the polycarbonate, or any other material used as the substrate, is crucial in the manufacture of all discs. The material must have how melt viscosity and high melt flow to minimize internal stresses caused by molecular orientation of the polymer during filling. Also important are high long-term dimensional stability, resistance to moisture and gas permeability, low birefringence, and superior optical properties. Bayer's polycarbonate grade Makrolon CD-2000 has been specially developed to fit such requirements. The molecular weight of the polycarbonate has been reduced to give an exceptionally high melt flow rate compared to a standard polycarbonate. In addition to polycarbonate, a number of other materials including acrylic resins, polyacrylates, thermoset epoxies, and blends of copolymers have been tested for their suitability in optical discs.

The compact read-only memory disc (CD-ROM) has been developed from the compact audio disc. A 4.75 in. CD-ROM has a storage capacity equivalent to 1,500 floppy discs, or 250,000 typewritten pages. The purity of the material and the absence of internal stresses and composition variation in the discs is even more crucial than for CDs since any birefringence can cause significant loss of data. The main disadvantage of CD-ROM is that material cannot be added or erased.

Two produces, DRAW (Digital Read After Write) and EDRAW (Erasable DRAW), provide the ability to add more information besides the advantages of CD-ROM, thus bridging the gap between optical and magnetic storage discs. While much of the information on these products is proprietary, Daicel in Japan have reportedly produced magneto-optical discs based on a polycarbonate injection-molded substrate involving the application of opto-magnetic-sensitive coatings and more complex metallizing procedures than other formats. To write, a high-power laser is applied to the magnetic field of a disc to change its polarity in a microscopic area, and to read, a low-power laser measures the differences in polarity.

3.7.8   Polymeric FET and LED

The fact that many conducting polymers are semiconductors in the undoped or slightly doped state has prompted attempts by both industrial and academic groups to use them in microelectronics such as semiconductor devices and especially field effect transistors (FETs). For the semiconducting layer between the source and drain electrodes of a FET, micro-meter-thick films are required. Film as thin as this has been prepared by spin coating a solution of a precursor polymer on a substrate having the required electrode pattern and subsequent heat treatment in a stream of gaseous dopant to convert it into a conducting polymer. Improvement of technique has led to carrier mobilities as high as 10^{-1} cm^2/V-s in polymeric FETs [40]. This has opened the way to uses such as flat-panel color displays for computers and possibly flat color televisions when tied to a liquid crystal matrix.

Light emitting diodes (LEDs) have also been fabricated from conjugated polymers such as poly (*p*-phenylene-vinylene) as the emissive layer [40]. The advances in the chemistry of processible conjugated polymers and focused work on the physics of electroluminescence in these materials have led to the development of flexible, almost entirely metal-free LEDs [41]. These polymer-based LEDs could be competitive in display applications because of the potential ease, low cost of fabrication, and large surface area of devices based upon processible polymers.

3.8 Polymers in Agriculture and Horticulture

The application of polymeric materials in agriculture and horticulture has increased considerably in recent years, not only as replacement for traditional materials but also as a means of effecting improvement [47]. For example, the application of plastics has led to a significant improvement in technological processes in the growing and storing of agricultural crops, construction of storages for fertilizers, vegetables and animals, and in agricultural equipment and drainage technology.

3.8.1 Plastic Film

The main uses of plastic films in agricultural and horticultural applications are for covering fodder and cultivation units such as seed-beds, for mulching and warming soil, for the building of canals and reservoirs, and as packaging material for storing and transporting food.

The most widely used plastic film is polyethylene (PE). PE film tunnels and perforated flat PE film allow better use of natural resources such as solar energy, water, and soil. Crops may be made to ripen early making cultivation possible all year round. In addition to covering produce, PE film as a number of other uses. Livestock feeding stuff such as green foliage grass and maize can be preserved by covering the forage in a silo with large gastight black or white PE films. Shrinkable PE films are used for sheet steaming in horticulture.

Polyethylene films for agricultural applications need to have high strength and elasticity, resistance to wind forces, and a long service life. They are mostly produced from low density PE combined with linear, low density PE and ethylene–vinyl acetate copolymer.

As PE films, like all synthetic polymers, suffer from decomposition by environmental influences such as light and atmospheric oxygen, special photoprotective systems are added to delay these effects. Systems that are commonly used are ultraviolet (UV) light absorbers, quenchers, radical scavengers, and hydroperoxide decomposing agents.

Benzophenone and benzotriazole are UV light absorbers that are frequently used in packaging film. While the addition of UV absorbants increase the service life of the film by 100%–200%, a disadvantage of such absorbants is that their effectiveness is dependent on the thickness of the film to be protected. For film less than 100 μm thick, they can only offer a limited protective effect.

The mode of action of quenchers, however, is not dependent on the thickness of the film to be protected. Quenchers are photoprotective compounds that can take up and conduct away energy that has been absorbed by chromophores, such as hydroperoxide, which are present in PE film. Organo-nickel compounds are quenchers that also act as decomposing agents of hydroperoxide.

Hindered amine light stabilizers (HALS), which represent more recent developments in photoreactive compounds and are referred to as scavengers, absorb light above 250 nm and therefore do not act as UV absorbers or quenchers.

Several specialized types of polyethylene film are also available. These include heat-resistant film, heat-retaining film, water-absorbing antistatic film, and photodegradable film.

A heat-resistant PE film for warming the soil will stand heating for 200 h at 120°C and at a pressure of 0.2–0.3 MPa. A heat-retaining PE film with enhanced absorption in the long wave region of the IR spectrum enables the temperature under the film to be 3°C–4°C higher than when under normal PE film.

Soil solarization, or heating soil with solar radiation, involves covering the soils with clean polyethylene sheets during summer or months with bright sunshine and clear skies. The process raises the soil heat and temperature, killing soil-borne pathogens and pests that lower the yield of most of the field crops. This nonchemical management of soil pathogens is an ecofriendly and economical technique to control pests and diseases in the soil for a profitable yield.

Moisture-absorbing antistatic film with enhanced permeability to UV radiation is used primarily for seed beds, as it does not become dusty and therefore creates better conditions for growing plants inside hot-houses. The film also has a surface that prevents the deposition of condensed droplets, increasing the yield of vegetable crops by 15%–20% compared to normal PE film.

Photodegradable PE film is used for mulching the soil in vegetable growing. It improves soil drainage and discourages weed growth. The film breaks down within two weeks to three months as a result of solar radiation, combines with the soil, and is broken down by microorganisms [48].

3.8.2 Plastic Crates

The German company Schoeller International GmbH at Munich has developed a system for production and circulation of collapsible and reusable polypropylene crates to replace traditional cardboard and wooden crates for transporting fruits and vegetables. At the heart of the system is a crate that can collapse inwardly to one-fifth of its original size and still retain its original footprint. The complex design of one piece crates involves molded-in hinges that allow the units to fold and unfold reliably thousands of times. Despite their light weight (740–1,800 g), the containers are capable of holding up to 25 kg of produce, can withstand a load of 500 kg, and can be easily stacked. They are said to be compatible with European pellet standards and meet the ISO standard container specification numbers 600/400 and 400/300.

3.8.3 Building

Plastics have several advantages over metals in the construction of agricultural buildings. For example, in the construction of low-rise farm buildings, metal roof sheeting shows deterioration after 2–3 years due to condensation of water vapor produced by dairy cows. In contrast, glass–fiber reinforced plastic (GRP) polyester sheet has a service life of 15–20 years and is unaffected by condensation of water vapor as well as by most acids, alkalis, grease, and caustic soda solutions. In translucent form it allows a high level of natural light but cuts out glare and excessive solar heating.

GRP sheet has also been used as an overall cladding material. GRP polyester sheet, Filon, produced by BIP Chemicals has a tensile strength of 10^4 psi (69 MPa), better than most other cladding materials of equal thickness and thus allows the wall thickness to be reduced.

3.8.4 Pipes and Hoses

Plastic pipes are being increasingly used in the agricultural industry for drainage and irrigation. The improved strength and greater crack resistance of new grades of polyethylene have allowed the wall thickness of such pipes to be reduced and their diameters increased.

Polyethylene has been irradiated to produce cross-links thus giving it improved strength. Drip irrigation for raw crops uses such irradiated LDPE tubing. For vine and tree crops and drop emitter is made of PE molded in large multicavity molds.

Plastics are now being used in a variety of ways to replace metal components in different irrigation systems. Interestingly, one of the first uses of PVC pipe in the 1960s was in sprinkler irrigation systems as a replacement of brass, gun-metal, or zinc alloys.

Hoses used in irrigation are invariably made of PE. Crimped PE pipes are being more frequently used for drainage and irrigation of soils. Flexible hoses that can be rolled up are used in spraying and sprinkling machines. Hoses reinforced with a rigid spiral are being used with machines to introduce chemicals into the soil for the protection of plants.

3.8.5 Greenhouses

The use of plastics in horticultural industry has grown substantially in recent years. Several plastics, and PVC in particular, are used in applications such as hosepipes, glazing strips, paint, and wiring. However, it has come to light that as greenhouses become airtight, some chemical vapors released by plastics may build up to levels toxic to plants. The plasticizer dibutyl phthalate (DBP) used in PVC has thus been found to be phytotoxic in a greenhouse environment. Research has shown that a fire retardant used in polystyrene trays and granules reduced the growth of some seedlings. Similarly, a silicone joining compound used in greenhouses and fumes from a paint used on an electric boiler were also found to be harmful to test plants.

Responding to the toxicity problem of plastics in greenhouse applications, the Dutch horticultural industry developed a test for screening. Plastics that pass the test qualify as safe horticultural plastics and can be used in a greenhouse environment.

3.9 Polymers in Domestic Appliances and Business Machines

As plastics are more lightweight than glass or metal, are generally impact resistant, lend themselves well to manufacturing techniques, and allow designers greater freedom in design, they are continuing to replace metals in domestic appliances and equipment housings. The use of plastics has been well exploited in both these areas.

Many resins have well-defined niches in the appliance sector. But there is also intense competition among various materials for the same end-use applications (Table 3.2).

TABLE 3.2   Resin Contenders for Key Appliance Applications

Category	Competing Resins
Small appliances	
Clock cases	PS
Kitchen appliances	
Housings	ABS, PP
Mixing bowls	Acrylic, PC
Safety and alarm device	ABS, PS, PVC
housings	
Large appliances	
Air conditioners grills	PC, PP, PS
and covers	
Carpet cleaner housings	ABS, PC, PP, PS
Dishwasher tubs	PP, PVC
Humidifier covers	PE, PP, PS
Laundry units	
Tubs and agitators	PP
Control panels	PP, PVC
Microwave oven covers	PC/ABS, PP, PS
Refrigerators	
Clear interior parts	ABS, acrylic, PC, PETG
Foam insulation	PU
Liners	ABS, PS
Waste compactor and disposer	ABS, PP
housings	
Large and small appliances	
Clear parts	ABS, acrylic, cellulosics, PETG, PC, PS, SAN
Small functional interior parts	Nylon, phenolics, polyacetal
Hose and tubing	Nylon, PE, PP, PU, PVC
Wire and cable insulation	PE, PVC

3.9.1   Large Appliances

While the use of plastics in refrigerators is well established, product development has led to replacement of some of the conventional plastics with newer ones of improved properties. For insulation rigid polyurethane foam has, on the whole, replaced fiberglass in cabinets and doors. In this application, diphenyl-methane diisocyanate (MDI), which has lower vapor pressure than toluene diisocyanate (TDI), is now preferred.

For refrigerators lines low-cost polystyrene (PS) was the first choice, but this was soon replaced by ABS, which has higher stress crack resistance to trichloromethane, a foaming agent that is used to blow polyurethane insulation behind the liners. However, because of the higher cost of ABS there has been a swing back to PS with special grades of PS now being available. Dow Chemical and BASF produce special crack-resistant grades of PS, although the sheets are not as rigid as ABS.

Mobil Chemical Company produces a composite structure consisting of the company's PSMX7100A and an extrusion-grade high-impact PS. The structure is claimed to offer a cost saving of 30%–40% over ABS when equal thicknesses are compared. Mobil's translucent impact grade polystyrene PSMX7800A can be coextruded or laminated as a cap layer to yield a surface glass.

Responding to new PS products, Borg Warner has produced special ABS resins tradenamed CTB, which permit more uniform drawdown and thus allow the use of thinner gauze sheet without the risk of thinning of corners. For interior refrigerator components such as crispers and dairy drawers, ABS grades are commonly used. Lexan PC film has been used for refrigerator consoles.

Because of the limitation of temperature resistance, the use of plastic components in ovens has been mostly limited to exterior trim. Plastics however are gradually being used for semi-structural components. Endcaps with a metallic finish are a typical example. General Electric have used a sputter-coated modified Noryl polyphenylene oxide for a shiny metallic appearance for the Tappan oven. The material provides flame resistance, electrical insulation, and resistance to heat, grease, and abrasion.

Celanex thermoplastic elastomer produced by Celanese Engineering Resins has been used for certain oven components. The 30% glass reinforced grade, offering faster molding cycles, improved impact resistance, and greater flexibility and color range has replaced phenolic resins. Belling and Company Limited used the material for the two oven handles and corner panel supports of its Format Cook Center.

General Electric produces several blow molding grades of polycarbonate for the vertical panels of ovens and dishwashers. Advantages of polycarbonate panels over metallic panels include low cost product, rigidity, and dimensional stability.

In microwave ovens, General Electric's glass-filled Ultem polyetherimide is used for shelf supports and stirrers. Dartco's Xydar liquid crystal polymer is used for turntables, shelf supports, stirrers, and meat probes. Membrane touch switches for microwave ovens are normally made from polyester film such as Du Pont's Mylar.

Plastics are used to make a number of dishwasher components, such as impeller and other pump parts, inner door, tubs, and consoles. Polyphenylene sulfide is used to make dishwasher impellers. The material shows processing ease, good durability, and resistance to high heat and detergents. For dishwasher pump parts, tubs, and door liners, glass-filled polypropylene has all but replaced diecast zinc. Noryl-modified polyphenylene oxide (PPO) is replacing stamped metal for the dryer for blower wheels. Both rigid PVC and PC are used for consoles. Rigid PVC is used because of its inherent flame retardancy, rigidity, and high heat distortion temperature. Polycarbonate (PC) is used for its high heat resistance, dimensional stability, and good electrical insulating properties.

Plastics are replacing metals in washing machines. Procom GS 30H254, a 30% glass reinforced structural foam grade of ICI's polypropylene, has been used by Hotpoint for the oven drums of its washing machine replacing enamelled steel. The Procom drums provide improved insulation and quieter operation. Nonfoamed grades of the product are used for the drum front moldings and detergent dispenser. An acetal copolymer spin gear has been used in place of the traditional belt drive in

Whirlpool's Design 2000 washing machine. The acetal copolymer eliminates the need for secondary finishing operations and affords a significant reduction in weight.

Propathene random copolymer PXC 22406 PP of ICI is used by Electrolux in its vacuum cleaners. The material is used to produce impact-resistant side and rear bumpers to protect the machine during use.

3.9.2 Small Appliances

The market for plastic cookware that can be used in the freezer and microwave is set for a big expansion. Most cookware is made up of fiberglass reinforced thermoset resins. High-barrier multilayer containers that have gained popularity in the market are heat retortable, microwaveable, and provide a shelf life of up to two years without refrigeration for a variety of convenience foods. Liquid crystal polymers have also been used to make cookware as they are transparent to microwave, offer high heat resistance, and give an excellent finish.

The most notable development in small appliances has been that of the electric kettle. Nearly all new kettle bodies, lids, and spouts are molded from plastics such as Kemetal acetal copolymer of Celanese Corporation. Kemetal is chosen for its strength, resilience to impact, smooth surface finish, and wide colorability. Use of the material has also enabled a slimmer design of kettle, suited to the needs of the consumer.

Coffeemakers incorporate plastic in both the jug and housing. The Koffee King decanter from Bloomfield Industries uses for robustness and transparency of its main body Amoco Chemical's Udel polysulfone as a replacement to glass. Krups used a grade of polybutylene terephthalate from Mobay Chemical Company for the housing of their coffeemaker. The material has chemical resistance, good surface quality, high dimensional stability, and odor stability along with high heat deflection temperatures and electrical insulation properties. The same material has also been used for the housing of electric irons.

More recently, wood–plastic composite materials are making forays into consumer goods and housewares [49]. Attractive flower pots and tool handles have been made by Bemis Mfg. and East painter, respectively. Allsop, a molder of Office accessories in Washington State, also uses wood-filled PP to produce a line of products with an environmental spin. Batts, Inc. in Zeeland, Michigan, molds over a million coat hanger each day A small but growing portion of these are made using wood-filled compounds.

3.9.3 Business Equipment

Business machine and equipment housings represent a fast growing area for ABS polymers that are commonly described as tough, hard, and rigid—a combination that is unusual for thermoplastics. It's light weight and the ability to economically achieve a one-step finished-appearance part have contributed to large volume applications of ABS. A variety of special ABS grades are used for equipment and housings. These include high-temperature resistant grades (for instrument panels, power tool housings), fire-retardant grades (for appliance housings, business machines, television cabinets), electroplating grades (for exterior decorative trim), high-gloss, low-gloss, and matte-finish grades (for molding and extrusion applications), and structural foam grades (for molded parts with high strength-to-weight ratio).

Since the time of Noryl, the first commercial blend of two dissimilar polymers introduced by General Electric in the 1960s, a large number of different blends or alloys has been introduced, some of which have been developed specifically for office equipment and housings. A number of polyphenylene ether (PPE) alloys have been introduced by Borg-Warner, Hulls and BASF. The great advantage of these PPE blends is that they can be made flame retardant without halogens.

Borg-Warner has various grades of its Prevex PPE/high-impact polystyrene (HIPS) blend. Prevex offers high impact resistance and high heat distortion temperatures. Applications are for office equipment, business machines, and telecommunication equipment. BASF's Luranyl PPE/HIPS blend includes general purpose types, reinforced grades, impact modified types, and flame-proof types. BASF also has a PPE/nylon blend tradenamed Ultranyl developed for a similar market. Hulls produces a PPE/nylon blend called Vestoblend that uses nylon-12 and nylon-6,12. It is claimed to absorb less water than competing types.

Noryl AS alloys, which are blends of PPO/PS and special fillers, have been developed specifically for office equipment and housings. Their main features are high strength, maximum stiffness, and ability to maintain tight dimensional tolerances over a wide temperature range. They are expected to replace metals in such applications as structural chassis for typewriters and printers.

3.9.4 Air Filters

High-efficiency particulate air (HEPA) filters almost entirely use extrafine glass fibers and generally feature excellent functions, but if the fiber diameters are made finer to improve particle collection, the power consumption rate is increased considerably, clogging occurs, the filter becomes vulnerable to chemicals, and combustion treatment becomes difficult. Japan's Toyobo Company Ltd. has developed for air conditioning units a super-high-performance filter, the Eliton Super, made of a polypropylene polymer fiber. It is to be used as HEPA filter in clean rooms for removing over 99.97% of dust particles of 0.3 μm and as an air-cleaning system. The fibers in this new filter are charged with static electricity to improve the particle-collection efficiency. Compared with glass fiber, it consumes less power, has greater resistance to chemicals, and has a substantially extended service life. The filter can be combusted with ease and reduced into a small volume.

The company also supplies filters for industrial clean rooms, primarily to semiconductor and air-conditioning systems manufacturers.

3.9.5 Solar Systems

Polymers have a share in the development of each major solar technology because they offer not only potentially lower costs, easier processing, and light weight, but also a greater design flexibility [50]. Polymers are used in most of the solar systems and equipment such as covers, thin-film honeycombs and housing for flat-plate (nonconcentrating) collectors, reflecting surfaces and optical lenses for concentrating collectors, reflector shells, structural and support members, insulation, piping, moisture barriers, adhesives, and sealants.

A nonconcentrating solar heat collector consists of a transparent cover, an absorber, tubes or ducts for the heat transport fluid, and an insulated case that limits the thermal losses to the environment. The collector cover serves several important functions: it permits the passage of the solar radiation, prevents loss of heat, and protects the absorber from mechanical and weathering damage. The cover is selected for the highest transmittance of solar radiation and the lowest transmittance of the infrared reradiated from the absorber. Polymeric materials have a low absorptivity for solar radiation and the absorptance can be decreased, compared to glass, by using thinner sheets or films.

The polymers that have high short-wave (<2.8 μm) transmittance and low long-wave (>2.8 μm) transmittance are poly(methyl methacrylate) (PMMA), polycarbonate (PC), poly(vinyl fluoride) (PVF), fluorinated ethylene–propylene (FEP), and polyesters, which contain strong polar groups such as oxygen or halogen atoms. However, except for the first two, infrared transmittance is some five times greater than that of glass.

Transparent honeycomb covers with channels parallel to the sunlight have proven effective in decreasing the convection within the collector and the radiation heat losses, while reducing the amount of light entering less than the additional cover sheet that they replace. Honeycombs have been successfully constructed from PVF, FEP, and PC.

The glazing (cover plate) material for thermal solar collectors has to be able to resist wind, snow, and gravitational loads over the range of temperatures achieved by the collector. This can be in the range of −40 to 120°C in normal operation and as high as 200°C when the fluid is not flowing in the absorber.

The strength properties depend largely upon the difference between the test temperature and the glass transition temperature, T_g; raising the former or lowering the latter has similar effects.

Polyethylene and PVC soften and sag above 60°C; polypropylene, polyester, PMMA, and PVF are only slightly better (above 90°C). The superior plastics with load-bearing ability at high temperatures (50°C–200°C) are the PC and FEP, which have maximum operating temperatures of 130°C and 200°C, respectively. The resistance to natural weathering is very good for PMMA, PVF, and FEP, and there is confirming experience over periods of many years. Weathering of polycarbonates, however, results in discoloration.

While PVF and FEP have good resistance to abrasion, PMMA and PC are subject to scratching and abrasion from dust particles. Their resistance to such damage can be greatly improved by hard coating of polydimethylsiloxane.

The superiority of acrylic polymers used for concentrating collector mirrors or lenses has been underlined [29]. Studies on reactivity of different polymers used as protective coatings or in film form for solar mirrors have established PMMA(3M)/Al/adhesive to be the most durable of the polymer/mirror systems.

Plastic solar heaters have also appeared in the market. An Australian company, Birwick March of Perth, patented the world's first plastic solar heater and began marketing a complete manufacturing system for the plastic units made from UV-stabilized polyethylene. Rotational molding has been perfected to produce a one-piece integrated solar collector and storage tank. The tank is surrounded by a structural polyurethane insulation encased in an elastomer coating, with a high solar transmission acrylic cover over the plastic plate. Birwick claims that the 180-liter system will last more than 15 years because there is no metal in contact with the water in the unit, on joints or gaskets. Using the readily available polyethylene, the system can be made and retailed at a price substantially lower than the existing solar water systems.

3.10 Polymers in Medical and Biomedical Applications

A wide variety of polymeric materials are used in various medical and biomedical applications, which range from medical packaging and products used in medical treatment and diagnosis equipment to items that are actually implanted in the body, e.g., pacemakers, hip joints, and artificial heart valves [51–53]. These products have unique requirements from absolute nontoxicity and compatibility with human tissue to mechanical strength, resistance to sterilization techniques, and, in the case of disposable end products, mass production on a large scale. Due to differing requirements a wide variety of materials are used (Table 3.3). While medical grades of poly(vinyl chloride) (PVC) and polystyrene (PS) are the most commonly used polymers in the medical sector, high-performance engineering resins are now finding an increasing number of uses in this field. The areas in which plastics find use can be divided broadly into the following: medical packaging, nontoxic sterilizable items, appliances, biomedical devices, and disposables.

3.10.1 Medical Packaging

Significant advances have been made in recent years in the packaging of medical products requiring medium and high barrier protection. The demand for sterilizable packaging for hospitals and industry is being met by special paper, plastic films, and other nonwoven materials.

The chlorinated tetrafluoroethylene (CTFE) fluoropolymer Aclar from Allied Corporation is one of the highest barrier materials against moisture transmission. It is used in an antibiotic test kit where even trace amounts of moisture would violate the accuracy of the test.

A large number of small medical containers are now molded in poly(ethylene terephthalate) (PET). Its advantage over glass include no breakage of filling lines and higher output. Tamper-evident closures, droppers, and dispensers are usually molded in polypropylene and high-density polyethylene.

TABLE 3.3 Resins Commonly Specified for Medical Applications

Resin	Primary Features	Typical Applications
Polyethylene	Processing ease, chemical resistance	Caps, needle hubs, medical packaging, waste bags
Polypropylene	Autoclavibility, contact clarity	Syringes, specimen collection cups
Gen-purpose polystyrene	Transparency	Petri dishes, labware, test tubes
High-impact polystyrene	Toughness, opacity	Home test kits, diagnostic equipment, housings
Styrene–acrylonitrile	Chemical resistance, transparency, toughness	Diagnostic components, fluid handling devices, flat plate dialyzers
Acrylic	High light-transmission rates, chemical resistance	I.V. components, specimen-collection containers
Acrylonitrile–butadiene–styrene	Toughness, low and high gloss, good processability	Home test kits, housings, surgical staplers, I.V. connectors
Polycarbonate	Chemical and heat resistance, toughness, transparency	Blood centrifuge bowls, cardiotomy reservoirs, profusion devices, hemodialyzers
Polyester	Chemical resistance	I.V. components, catheters, surgical instruments, housings
Poly(vinyl chloride)	Transparency, good scuff resistance	Blood bags, catheters, cannulae, corrugated tubing, renal care products, transfusion supplies, face masks
Polyurethane	Good chemical resistance, toughness, good processability	Catheters, tubing, I.V. connectors, drug delivery systems
Polyetherimide	Autoclavibility, chemical resistance	Sterilization trays
Polysulfone	High heat resistance	Medical trays

3.10.2 Nontoxic Sterilizable Items

Polymers used in nontoxic sterilizable items, such as tubing, artificial organs, and wound coverings, must be able to withstand sterilization by ethylene oxide, steam autoclave, or gamma radiation. For medical products sterilized by gamma irradiation, only plastics that do not degrade or discolor on exposure to radiation, such as polyester and polycarbonates, can be used.

The biological compatibility of materials used in the nontoxic category is vital and is the subject of continuing research. Since no material is completely inert, it is the level of interaction between an implant and the surrounding tissue that determines the acceptability of the material. Several other facts that are important in determining acceptability include mechanical properties of the polymer, e.g., wear resistance and fatigue, and bulk chemical properties such as resistance to degradation by hydrolysis, sensitivity to enzymes, and the way it reacts to the deposition of protein.

PVC is used for nearly all surgical tubing. Recent developments include flexible PVC compounds with greater resistance to body fluids, and x-ray-opaque PVC compounds that enable PVC catheters and tubing to be traced after insertion into the body. Esmedica-V, a plasticizer-free flexible PVC compound produced by Sekisui, Japan, reportedly meets all requirements for medical use, including ability to stand ethylene oxide sterilization. ICI has produced Welvic VK-2004 for coronary dilation catheters. It is reported to offer high flexibility, high-pressure resistance, and low elongation.

The development of artificial wound coverings has received much attention. The major requirements for such coverings are protection from microbial attack from the environment, optimal water permeability, capability of adhering well to the wound, and ready removability without causing tissue damage. Other important factors are prevention of excessive formation of granulation tissue and optimal elasticity to facilitate an intimate cover of the wound.

Sustained-release wound dressings capable of delivering antibiotics or other biomedical agents in small concentrations have been obtained by microcapsulating a variety of drugs into a UV curable urethane elastomer. Thermedics Incorporated has developed special UV-curable polyurethane elastomers for such applications.

Developments related to medical implants, artificial organs, and prosthesis have improved the quality of life and increased the life expectancy of many individuals.

Silicone rubber is a highly biocompatible thermosetting elastomer that has found applications in prosthesis for ophthalmology, neurology, facial reconstruction, replacement of finger, toe and wrist joints, cardiovascular applications, such as pacemaker coatings and lead wires, and tendon replacements. It is also used in drug delivery systems and tubes for carrying blood, drugs, and nutrients.

The artificial heart, Jarvik-7, which has been successful in keeping a recipient alive for more than a year, has valves made from modified polypropylene and the two ventricles made of polyurethane supported on an aluminum base. For arterial replacements or bypass of clogged blood vessels, polyester fiber such as Du Pont's Dacron remains the preferred material.

Ultrahigh molecular weight polyethylene (UHMW PE) is most commonly used for articulation surfaces in joints. It is also used for prosthesis components in total hip replacement. In the latter application, it may be reinforced with carbon fibers to increase wear properties.

Expanded polytetrafluoroethylene (PTFE) grafts have gained increasing popularity as synthetic or nontextile grafts for reconstructive procedures, such as, above- or below-the-knee bypasses for limb salvage.

3.10.3 Biodegradable Polymers

Generally, biodegradable polymers are those that can be broken down by nature either by hydrolytic processes or enzymatic processes producing nontoxic by-products. ICI offers a biodegradable plastic under the tradename Bipol. It is a polyester made from hydroxybutyric and hydroxyvaleric acids. Because of its much higher price than conventional mass polymers, Biopol will not find wide use as long as purely economic considerations determine the use of plastics. A niche market exists, however, for the medical grade of the polymer. For example, fibers can be used for surgical sutures. The compound is absorbed in the body and does not invoke immune reactions. The molecular weight lies between 30,000 and 750,000. The mixture of the two hydroxyacids is produced by a bacterium of the type *alcaligenes eutrophus*.

Biodegradable polymers are very interesting for tissue engineering (see Chapter 2) applications because they can be absorbed gradually by the human body without permanently retaining traces of residuals in the implantation site, and are compatible with tissue repair. Thus cell transplantation using scaffolds of biodegradable polymer offers the exciting possibility of creating, in vivo, completely natural new tissue with the required mechanical or metabolic features to restore the function of tissues such as cartilage, bone, skin, nerve, kidney, and liver. In this process [54], biodegradable polymer scaffolds act as temporary substrates to which cells can adhere, proliferate, and retain their differential function (see Degradable Polymers, Chapter 2.)

3.10.4 Conducting Polymer Nanotubes

Conducting polymers are of considerable interest for a variety of biomedical applications [55]. Their response to electrochemical oxidation or reduction can produce a change inconductivity, color and volume. A change in the electronic charge is accompanied by an equivalent change in the ionic charge, which requires mass transport between the polymer and electrolyte [56]. When counterions enter a polymer it expands and when they exit it contracts, with extent of expansion or contraction depending on the number and size of ions exchanged. Electrochemical actuators using conducting polymers based on this principle have been developed [57]. They can be doped with bioactive drugs, and can be used in actuators such as microfluidic pumps [58].

Microelectrode neural probes facilitate the functional stimulation or recording of neurons in the central nervous system and peripheral nervous system. Minimizing the electrode impedance is an important requirement for obtaining high quality signals (high signal-to-noise ratio). It has been shown [59,60] that conducting polymers such as polypyrrole (PPy) and poly(3,4-ethylenedioxythiophene) (PEDOT) can decrease the impedance of the recording electrode sites on neural prosthetic devices. It has also been demonstrated [61] that the impedance of the neural microelectrodes can be further decreased significantly (by about two orders of magnitude) and the charge-transfer capacity significantly increased (about three orders of magnitude) by creating conducting polymer nanotubes on the microelectrode surface. The conducting nanotubes that have well-defined internal and external surface texture decrease the electrode impedance by increasing the effective surface area for ionic-to-electronic charge transfer to occur at the interface between brain tissue and the recording site. The drugs can be released from the (drug-loaded) conducting nanotubes at desired points in time by using external electrical stimulation of the nanotubes. This process presumably proceeds by a local dilation of the tube that then promotes mass transport.

In order to produce the nanotubular conducting polymers, nanofibers of a biodegradable polymer, such as poly(L-lactide) (PLLA) or poly(lactide-co-glycolide) (PLGA), containing the desired drug or bioactive species, are first electrospun onto the surface of a neural probe followed by electrochemical deposition of conducting polymers around the electrospun nanofibers. In a final step, the fiber templates can be removed (leaving behind the conducting nanotube) or allowed to slowly degrade, thus providing additional means of controlled delivery of biologically active agents incorporated into fibers themselves [61]. PLLA and PLGA are considered as suitable polymers for the template since they can be readily processed into nanoscale fibers, are stable during electrochemical deposition of the conducting-polymer coating, and can be easily removed under conditions (such as by soaking in a solvent) that leave the wall material intact [62].

3.10.5 Biomimetic Actuators

There is growing interest in biomimetic motions, which imitate the action of natural muscles. Since such motions are difficult to realize using conventional appliances such as mechanical, hydraulic, or pneumatic actuators, research efforts are focused on the development of new muscle-like actuators. Electroactive polymers (EAPs) including polymer gels [63], ionic polymer-metal composites (IMPCs) [64], conductive polymers [56], and carbon nanotubes [65] are candidates to address the performance demands.

Among EAPs, IMPC is considered to be the most suitable for artificial muscle, since it exhibits a large bending displacement in the presence of a low applied voltage. An IMPC consists of an ion-exchange membrane sandwiched between two noble metal plates [66]. The ion-exchange membrane is neutralized with a specific amount of metal ions to balance the anions of the membrane. When a small potential is applied to an IPMC in the hydrated state, the mobile metal ions move toward the cathode and IPMC undergoes bending deformation toward the anode due to volume expansion on the cathode side. Having no moving parts and requiring only a small operating voltage for their actuation, IPMCs have many advantages over conventional mechanical actuators and are considered for application in a wide range of areas including medical, space, robotics, and soft microelectronic machines.

Since an IPMC functions as a pathway for hydrated cations, its properties will be expected to affect the performance of an IPMC actuator. The membrane materials used in IPMCs have so far been limited to a few commercially available perfluorinated ionic polymers, such as Nafion, and the thickness of the IPMC has also been restricted to the available thickness of the commercial membrane [67]. However, IPMC actuators employing new ionic membranes have now been reported [68]. The membranes are prepared from fluoropolymers grafted with polystyrene sulfonic acid (PSSA). IPMCs assembled with these membranes have been shown to exhibit at least several times larger displacements than the Nafion-based IPMC with similar thickness.

To prepare PSSA-grafted fluoropolymers, poly(vinylidene fluoride-co-hexafluoropropylene), poly(ethylene-co-tetrafluoroethylene) and poly(tetrafluoroethylene-co-hexafluoropropylene) were used as base polymers. Each polymer was molded into a film (200–300 µm thick) and irradiated with γ-ray at room temperature at the rate of 6.8 kGy h^{-1} using a cobalt-60 source to obtain a total absorbed dose of 50 kGy. The irradiated film was immersed in nitrogen-purged styrene at 70°C for 8 h for grafting. The PS-grafted film was washed with chloroform, dried, soxhlet-extracted with chloroform for 48 h to remove styrene monomer and homopolymer and then sulfonated in 0.5 M solution of chlorosulfonic acid in 1,2-dichloromethane for 48 h. To prepare IPMC actuators based on PSSA-grafted fluoropolymers, platinum electrodes were created on both surfaces of the grafted membranes by chemical reduction process using $[Pt(NH_3)_4]Cl_2$ and sodium borohydride. The larger displacement exhibited by these new IPMC actuators is considered to be the result of the higher concentration of sulfonyl groups, large ion exchange capacity, and consequent larger volume of water moving.

3.10.6 Dental Resin Composites

Dental composites based on acrylate resin systems are strong, durable and make a very natural looking smile. They provide an alternative to mercury amalgam and allow a relatively easy way of processing. However, these materials generally exhibit low toughness and high polymerization shrinkage. Adding short reinforcing fibers can enhance fracture toughness [69]. It has been shown, for example, that addition of poly(vinyl alcohol) (PVA) fibers (6 mm long, 14 µm diameter) to PMMA leads to an increase of elastic modulus, yield stress and impact toughness [70]. Substantial improvements in flexural strength, toughness, and stiffness have been achieved for dental resin composites reinforced with glass fiber preforms, thereby improving the performance of direct filling resin composites in large restorations with high occlusal loads [71].

Novel nano-silica-fused whiskers have been used as reinforcements to increase the strength, toughness, and wear resistance of dental resin composites, which may be useful in stress-bearing posterior restorations involving cusps [72]. The whiskers are made by fusing silica onto silicon nitride single crystalline whiskers at 800°C to roughen the whiskers and silanizing to improve bonding to the resin.

As the lower molecular weight monomers, viz., bis-DMA, EGDMA, and TEGDMA, commonly used as diluent for the highly viscous Bowen's monomer (see *Dental Materials* in Chapter 1 of *Plastics Fundamentals, Properties, and Testing*), can leach out of the cured composites at high rates, several less viscous, high molecular weight monomers that can serve as replacements for bis-GMA have been evaluated as polymer matrices for dental composites. Two such monomers are urethane dimethacrylate (UDMA) and ethoxylated bisphenol A dimethacrylate (EBPADMA). UDMA (mol. wt. 471) and EBPADMA (mol. wt. 562) have viscosities of 28 and 3 Pa s, compared to bis-GMA (mol. wt. 512) with a viscosity of 1369 Pa s. Having lower viscosities, the monomers are attractive candidates for use as matrix resins without lower molecular weight diluent monomers.

3.10.7 Appliances

Plastics components are now used in a number of appliances such as crutches, walking sticks, and wheelchairs. The comfort of the product is an important consideration in all of these applications. An example is BASF's "comfy crutch," which uses the company's Ultramid KR4412 nylon for the handle component.

Thermoformed high density polyethylene (PE) and low density PE foams that can be molded into a variety of complex shapes are being increasingly used in supporting devices in the treatment of orthopaedic and disabling conditions. Some examples are wheelchair seats, spinal jackets, medical shoe inserts, and splint supports.

Specialized high-temperature plastics such as polyether sulfone and polyetherimide are used for components for medical analysis and diagnosis equipment.

3.10.8 Disposables

Disposable medical products such as syringes, test tubes, petridishes, dialysis products, and drainage devices are proving to be high-volume markets for transparent plastics such as rigid PVC, polystyrene,

styrene acrylonitrile (SAN) copolymer, acrylic resins, and polycarbonate. The possibility of mass production is an important consideration in these applications.

3.11   Polymers in Marine and Offshore Applications

Polymers find a wide range of applications in offshore and marine industry. The properties of materials used in these applications invariably include resistance to marine fouling, sea water, and crude oil. The main applications are in marine cables, coatings, and adhesives.

3.11.1   Cables

Due to the damage caused by smoke-filled or halogen-based fumes from cables in the event of a fire, the offshore industry is showing increased interest in low smoke, zero toxic gas emission cable sheath. Research has centered on obtaining halogen-free polymers having fire retardance as well as oil, chemical, and weathering resistance. One example is a series of formulations called NONHAL-E (elastomeric) and NONHAL-T (thermoplastic) developed by Sterling Greengate specifically for the offshore industry. The elastomeric formulations are based on nitrile rubber and are resistant to oil and other commonly encountered fluids. Cables insulated with ethylene propylene diene terpolymer (EPDM) and sheathed with NONHAL-E are suitable for very tough applications. NONHAL-T was developed to replace PVC in cable sheathing. Halogen-free cables produced with a cross-linked polyethylene and bedded and sheathed with NONHAL-T gives excellent performance in fire conditions.

Polyether ether ketone (PEEK) has been considered for use as insulators and cable jackets. The use of PEEK as an insulator allows thinner overall diameters and permits a higher level of continuous operating temperature. The extreme hardness and toughness of PEEK may also eliminate the need of a metal sheath for cable jacket.

New materials and manufacturing techniques have been used to minimize the problem of drag and friction that occurs with heavy umbilicals. Tufflite, a surface diving umbilical, combines the strength of its rope-like cabled lay-up with kink resistance and neutral buoyancy that allows improved maneuverability and lower drag. Tufflite is based on textile braid reinforced polyurethane (PU) hoses for gas and air supply and has a color-coded outer sheath of abrasion–resistant PU.

Du Pont's thermoplastic polyester elastomer (TPE) Hytrel has been used in an underwater fiber optic cable produced by Shiplex Wire and Cable Corporation. The elastomer is used to position the fiber optics and is chosen because of its modulus properties that help prevent microbends in the fiber optics. The thermoplastic polyester elastomer is extruded over a central steel wire that imparts strength to the cable. Six fiber optic strands with a diameter of 5 mils each are then positioned and covered by a second Hytrel layer, which is added as part of a coextrusion with nylon to obtain greater abrasion resistance. Additional steel layers and a longitudinally formed copper tape are added, and the whole structure is then enclosed in a polyethylene jacket.

3.11.2   Coatings

A wide range of coating materials based on polyurethanes and epoxy resins is available to protect drilling rigs from corrosion. Prodoguard S and Dark Screed are two epoxy resin products available from Prodorite Limited. The products offer good adhesion in conditions requiring chemical and corrosion resistance and have the advantages of nonslip properties, high-impact resistance and noise deadening qualities. Prodoguard S offers a greater weight saving. Both products are used on walkways and other metal deck areas on offshore structures.

Zebron is a polyurethane corrosion–resistant coating material available from ACO-Coatings. It is a solvent-free product that cures in cold wet conditions and can be rapidly applied to a high film thickness in one operation. The material provides a thick abrasion and water-resistant coating to steel or concrete.

An extremely fire-resistant coating material designed for spray applications to onshore and offshore oil and gas facilities has been introduced by Hempel's Marine Paints. The product, called ContraFlam 3810, is easy to apply and is particularly well-suited to the protection of structural steel work, walls, and floors. Another coating of a surface hardener/weather barrier material, such as Contraflam Topclad 3811, can then be applied followed by a color coat.

Most anti-fouling paints incorporate a copper pigment that provides a toxic environment for biological growth. The cuprion anti-fouling system introduced by Cathodic Protection Company Limited, however, differs in that it releases copper into solution by passing direction current through a copper electrode. This produces soluble copper in minute amounts making a hostile environment for the growth of algae, slimes, and mussels. The Cuprion system is used by BP to keep its platforms on its Forties Field in the North Sea free from marine growth.

3.11.3 Other Applications

Hoses for offshore applications are in most cases made of abrasion–resistant materials such as poly(vinyl chloride) or polyurethane. Polytetrafluoroethylene hoses have been developed for use as interconnectors to oil or gas combustion chamber burners or turbines. The inert qualities, ability to withstand the hostile environment, and high-pressure capability suit them to this application.

Special underwater adhesives are available for applying to submerged metal substrates and organic-based substrates. A range of such adhesives are available from Wessex Resins and Adhesives Limited as products of WRA series that show great joint strength and durability. Products such as WR 4301, WR 4302, WRA 4303, WRA 4401, and WRA 4501 are basically two-part underwater adhesives based on epoxy resins and a sacrificial pretreatment solution. The pretreatment solution promotes spontaneous and uniform spreading of the adhesive formulation when applied to metal substrates under water. For applying to organic-based substrates, for example fiber-reinforced plastic, the formulations may however be used without the pretreatment solution.

Special polymeric material is needed to make heavy, finned pistons used in through-the-flow-line well servicing operations. Dowty Seals has developed, in collaboration with Shell, a new type of polymer for such critical service duties. The material produced under Dowty's material reference No. 2471 is able to withstand 130 h of simulated deep-well travel at temperatures intermittently peaking to 130°C, without suffering damage.

Several methods are available for the treatment of oil spillages. The Rigidoil system developed by BP has two components, a polymer and a cross-linking agent. The two components are sprayed simultaneously onto the spill where they mix and rapidly cross-link to produce a rubbery solid in which the oil becomes physically entrapped. The solid assembly is removed easily.

3.12 Polymers in Sport

The properties of polymers—resilience, toughness, and good strength-to-weight ratio—ensured the quick entry of polymers into the sports industry. In most cases, the polymer has been used as a replacement for another material and, being a competitive business, the sports industry has also been very receptive to new developments in polymer formulations and advances in design that contribute to improved sporting performance. Uses of polymers in the sports industry fall into three main categories: synthetic surfaces, footwear, and equipment.

3.12.1 Synthetic Surfaces

In sports such as athletics, traditional materials have been almost completely replaced by synthetic materials. Sports surfaces in modern indoor stadiums are usually made from a porous structure of rubber crumb and a binder such as polyurethane. Other polymeric materials used include flexible PVC or rubber sheet, polyolefins, polypropylene (PP) grasses, and foam laminates.

Following the use of Monsanto's woven PP Astro Turf at the Montreal Olympics in 1976, artificial surfaces have been accepted at World Championships level in hockey. Cricket is often considered a traditional and conservative game, but in fact a great many synthetic surfaces are now used. The acceptance of synthetic surfaces in soccer has been somewhat slower. The first synthetic pitch installed inside a stadium in England was by Queen's Park Rangers who used a PP omniturf surface. This was made of tufted PP rather than woven or knitted, with a pile of about 20 mm. Once laid, graded sand was spread into the carpet to within 1–2 mm of the surface to completely fill the pile.

The requirements of ice-skating surfaces, which are very different from other synthetic surfaces, have been met very successfully by plastics. In 1982 High Density Plastics launched the first full-size synthetic skating floor. The surface is made of interlocking panels of high density polyethylene (tradenamed Hi-Den-Ice), which become an ice-rink when sprayed with a gliding fluid. One big advantage over ice is that the surface only needs to be cleaned off and resprayed once a month. However, in a dry form the panels can also be used for basketball, football, or badminton.

PN Structures, a leading producer of synthetic surfaces in England, produces a range of synthetic Neoturf surfaces for outdoor tennis in a choice of hardwearing Neolast and Recaflex T. For indoor tennis, PNS can adapt its Neolast surface to provide varying area elasticity and shock absorption. Nondirectional sand-filled Neoturf for synthetic golfing greens is also offered. Other new surfaces include Supreme Court from Paw Sport and Leisure designed specifically for tennis, and Uni-Turf from Dunlop, a PVC sheet material with a specially designed embossed "Eldorado" finish used for indoor cricket.

3.12.2 Footwear

There has been significant development in sports footwear in recent years, greatly aided by polymeric materials that allow for better shape, manufacturing techniques, and material performance.

In the design of running shoes, much emphasis has been placed on anatomical comfort leading to the development of toe bars and heel canters, wedges, and midsoles, usually molded in polyethylene foams. Greater attention has been focused on the outsole of sport shoes in order to cope with the increased use of indoor facilities and artificial surfaces. Turntec's Apex 580 sports shoe offers a unique feature of interchangeable soles for different surfaces. The soles are held on by Scotchmate, a fastening system developed by 3M and can be interchanged.

Ski boots use a wide variety of plastics adapted for low temperatures. Thermoplastic polyurethane/ABS blends and plyamides are used for the outer shell of ski boots and modified polyethylene terephthalate for the binding. Polyurethane (PU) foam is often used to line the boots. Microsphere fillers may be incorporated into foam-lined boots to add further thermal insulation to the wax binder. The antivibrational characteristics of PU foams have also led to their use in ski fittings. A sandwich construction of polyurethane elastomer and aluminum alloy has been fitted between the ski and the binding to reduce shock and vibration.

3.12.3 Equipment

Developments in plastics sports equipment are mainly aimed at improving player performance and increasing player safety. Items that fall into the first category include balls, bicycles, tennis rackets, and windsurfing boards. Crash helmets belong in the second category.

Since the maximum performance of sports balls is governed by strict specifications, the scope of using plastics to improve them is rather limited. Improvements can however still be made. For example, microsphere fillers have been used in golf balls and bowling balls. The travel distance of gold balls is decreased by up to 30% by the incorporation of microspheres. In bowling balls, incorporating of microspheres reduces the density of the ball's core enabling lightweight balls to be made for children and women players.

Plastics are being used to make a number of bicycle parts replacing metal. Examples. Include polyamides for structural components such as the bike frames and wheels, and polyacetals for tire valves, brake levers, and gear change and power switches.

Windsurfing is a sport that may be said to have been created by the development of suitable plastics. The sailboards themselves are often made of coextruded ABS sheets as General Electric's Lexan. Sailboard accessories use thermoplastic blends; for example, boom fittings and mast steps that have to withstand very high dynamic loads are made of glass–fiber-reinforced polyamide blends. Polycarbonate/ABS blends are commonly used for daggerboard cases. The most popular type of sails are film sails, which are laminates of a plastic film, usually a polyester such as Du Pont's Mylar, bonded to woven fabric. They are claimed to offer superior tear resistance. A more recent development in windsurfers is a sliding mast molded in Du Pont's Delrin acetal homopolymer. The sliding mast permits adjustment of its position while surfing.

The tennis racket is another example of sports equipment that has been improved significantly due to the availability of high-performance plastics. The carbon fiber-reinforced nylon tennis racket frame first introduced by Dunlop has become the industry's standard for this equipment. An improved design incorporates Atochem's Pebax nylon ether block copolymer into the racket frame where maximum vibrations are concentrated, i.e., between the two profiles of the frame in the handle and across the throat of the frame. For racket strings ICI introduced a high performance yarn called Zyex made from Victrex polyether ether ketone. It is claimed that the Zyex strings have a good recovery and low dynamic modulus like the material originally used (natural gut), but unlike the latter they are not affected by exposure to sunlight or moisture and retain their tension better.

An item of sports equipment that has been improved greatly with regard to player safety is the crash helmet. American football helmets are polycarbonate or ABS shelled, while impact nylons are often used on motorbike crash helmets. Another item of sports equipment in the United States where safety has been improved is a Reduced Injury Factor baseball that has a novel polyurethane core material.

References

1. Moskowitz, P. A. and Kovac, C. A. 1990. Plastics: the sine qua non of electronics packaging. *Plast. Eng.*, 46, 6, 39 (June 1990).
2. Wood, S. A. 1990. Performance polymers are finding greater use in packaging markets. *Mod. Plast.*, 67, 8, 62.
3. Mojo, S. A. 1996. Designing for responsible disposal. In *Biodegradable Polymers and Packaging*, C. Chung, D. L. Kaplan, and E. L. Thomas, eds. Technomic Publishing, Lancaster, Pennsylvania.
4. Ackerman, P., Jagerstad, M., and Ohlsson, T. eds. 1995. *Foods and Packaging Materials: Chemical Interactions*. Royal Society of Chemistry, London, U.K.
5. IOPP (Institute of Packaging Professionals). 2002. *Fundamentals of Packaging Technology, 3rd Ed.*, Naperville, IL.
6. Stärke, D. and Skupin, G., 2001, *Kuntstoffe*, 91, KU100.
7. Pruett, W. P., Hilbert, S. D., Weaver, M. A., and Germinario, L. T., 1995, US Patent 5, 459, 224.
8. Butler, M. 2004. Pack to the Future. *Manuf. Chem.*, 1–3, September, 2004.
9. British Pharmaceutical Conference and Exhibition, *Medicines from Cell to Society*, Manchester International Convention Centre, 27–29 September, 2004.
10. Mali, J., Sarsama, P., Suomi-Lindberg, L., Metsa-Kortelainen, S., Peltonen, J., Viekki, M., Koto, T., and Tisala, S., *Wood Fiber–Plastic Composites*, A report dated 31.12.2003, VTT Building and Transport, Finland.
11. Bares, R. A. Ed. 1982. *Plastics in Material and Structural Engineering*. Elsevier Scientific Pub. Co., Amsterdam, U.K.
12. Feldman, D. 1986. *Polymeric Building Materials*. Elsevier Applied Science, Amsterdam, U.K.
13. Kukacka, L. E. 1976. *Polymers in Concrete*. The Construction Press, Hornby, U.K.

14. Vipulanandan, C. and Paul, E. 1990. Performance of epoxy and polyester polymer concrete. *ACI Mater. J.*, 87, 3, 241 (American Concrete Institute).

15. Davidovits, J., Comrie, D. C., Peterson, H. H., and Ritcey, J. 1990. Geometric concretes for environmental protection. *Concr. Int.: Des. Construction*, 12, 7, 30 (July 1990).

16. Borislavskaya, I. V., Corrosion resistance of polymer concrete, *Proceedings of the ICP/ROLEM/IBK International Symposium, Prague, Czech Republic*, 293, (23–25 June 1981).

17. Hendricks, A. L., de Kreij, A., and Moore, R. E. 1996. *Vinyl ester Linings Protect FGD Systems*, CORROSION 96, Paper No. 463, NACE International, Conferences Div., Houston, Texas.

18. Heffner, D. K. and Tuzla K. 1996. *Plastic Membranes Protected with Chemical Resistant Masonry*, CORROSION 96, Paper No. 402, NACE International, Conferences Div., Houston, Texas.

19. Selby, K. A. and Hess, R. T. 1996. *The Use of Polymers for On-Line Cleaning of Building Water Systems*, CORROSION 96, Paper No. 524, NACE International, Conferences Div., Houston, Texas.

20. Standish, M. L. 1996. *A New Polymeric Materi al for Scale Inhibition and Removal*, CORROSION 96, Paper No. 163, NACE International, Conferences Div., Houston, Texas.

21. Ross, R. J., Low, K. C., and Shannon, J. E. 1996. *Polyaspartate Scale Inhibitors—Biodegradable Alternatives to Polyacrylates*, CORROSION 96, Paper No. 162, NACE International, Conferences Div., Houston, Texas.

22. Powers, R. and Kessler, R. 1987. *Corrosion Evaluation of Substructure, Long Key Bridge*, Corrosion Report No. 87-9A, FL Department of Transportation, Gainesville, Florida.

23. Lempton, R. D. Jr., and Schemberger, D. 1996. *Improving the Performance of Fusion-Bonded Epoxy Coated Steel Reinforcing Bars*, CORROSION 96, Paper No. 323, NACE International, Conferences Div., Houston, Texas.

24. Ottaviani, R., Rodgers, W., Fasulo, P., Pietrzyk, T., and Buehler, C., Global SPE TPO Conference, September 1999.

25. Heinemann, J., Reichert, P., Thomann, R., and Mülhaupt, R. 1999. *Macromol. Rapid Commun.*, 20, 423.

26. Weibner, R., Adler, J. 1987. *Plastics for the Interior Trim of Passenger Cars—Present Situation and Trends for the Future*, SAE Technical Paper Series, SAE, Warrendale, Pennsylvania.

27. Rabe, J. 1990. Plastic elements in and around the engine. *Int. J. Veh. Des.*, 11, 3, 246.

28. Kurauchi, T., Okada, A., Nomura, T., Nishio, T., Saegusa, S., and Deguchi, R. 1991. *SAE Pap. Ser.*, 910, 584.

29. Mapleston, P. 1999. Automakers see strong promise in natural fiber reinforcements. *Mod. Plast.*, 73 (April 1999).

30. Misra, M., Mohanty, A. K., and Drzal, L. T. 2002. Plastics impact on the environment, *Proceedings of 8th Annual Global Plastics Environmental Conference (GPEC 2002), February 13&14, 2002*, Detroit, MI; Society of Plastics Engineers, Environmental Division, Troy, MI.

31. Wibowo, A. C., Mohanty, A. K., Misra, M., and Drzal, L. T. 2004. *Ind. Eng. Chem. Res.*, 43, 4883.

32. Anon, 1999, Grown to fit the part, Daimler-Chrysler High Technology Report, p. 82.

33. Anon 2000. Green door-trim panels use PP and natural fibers. *Plast. Technol.*, 46, 27.

34. Wilkinson, S. L. 2001. *Chem. Eng. News*, 22, 61.

35. Marks, N. G. 1989. Polymer composites for helicopter structures. *Met. Mater.*, 5, 8, 456 (August 1989)(Institute of Metals).

36. Nguyen, L. T., 1988, Issues in molding of electronic packages, *Proceedings of Manufacturing International*, Vol. 4, 119, ASME, New York.

37. Pearson, J. M., Polymers in optical recording, *Proceedings of the ACS Division of Polymeric Materials: Science and Engineering*, Vol. 4, p. 4, ACS, Washington, D.C.

38. Salaneck, W. R., Lundstrom, I., and Ranby, B. eds. 1993. *Conjugated Polymers and Related Materials: The Interconnections of Chemical and Electronic Structure*. Oxford Scientific, Oxford, U.K

39. Gustavsson, G., Cao, Y., Treacy, G. M., Klavetter, F., Kolaveri, N., and Heeger, A. J. 1992. *Nature*, 357, 477.

40. Yang, Y. L., Gupta, M. C., Dudley, K. L., and Lawrence, R. W. 2004. *Nanotechnology*, 15, 1545.

41. Yang, Y. L., Gupta, M. C., Dudley, K. L., and Lawrence, R. W. 1999. *Adv. Mater.*, 17, 2005.

42. Naritomi, M., CYTOP amorphous fluoropolymers for low loss POF, *Proc. POF-Asia Pacific Forum*, Tokyo, December 1996.

43. Large, M. C. J., van Eijkelenborg, M. A., Argyros, A., Zagari, J., Manos, S., Issa, N. A., Bassett, I., et al. Microstructured polymer optical fibers: a new approach to POFs, *10th POF Conference*, Amsterdam, September, 2001.

44. van Eijkelenborg, A., Martijn A., van Eijkelenborg, Alexander Argyros, Geoff Barton, Ian Bassett, Felicity Cox, et al. New possibilities with microstructured polymer optical fibers, *Proc. 11th POF Conference*, pp. 49–52, Tokyo, September 2002.

45. McCormick, C. L. 1987. Polymers in agriculture: an overview. *Polym. Prepr.*, 28, 2, 90 (August 1987) (Published by the Division of Polymer Chemistry Inc., Newark, N.J.).

46. Gilead, D. 1990. Photodegradable films for agriculture. *Polym. Degrad. Stab.*, 29, 1, 65.

47. Chelsea Center for Recycling and Economic Development, Univ. of Massachusetts, Technical Report #19, April 2000.

48. Gebelein, C. G. Williams, D. J., and Deanin, R. D., eds. 1983. Polymers in Solar Energy Utilization, *ACS Symposium Series, Vol. 220*, p. 125.

49. Gebelein, C. G., Tai, C., and Yang, V. 1990. Cosmetic, pharmaceutical and medical polymers: a common theme, *Proceedings of the ACS Division of Polymeric Materials Science and Engineering, Vol. 63*, p. 401.

50. Ellis, J. R. 1990. Plastics/packaging, PVC. New products keep coming. *Med. Device Diagn. Ind.*, 12, 3, 90.

51. Lodge, C. 1987. Tomorrow's plastics may add years to your life. *Plast. World*, 45, 10, 5.

52. Thomson, R. C., Wake, M. C., Yaszemski, M. J., and Mikos, A. G. 1995. Biodegradable polymer scaffolds to regenerate organs. *Advan. Polym. Sci.*, 122, 265.

53. Schmidt, C. E., Shastri, V. R., Vacanti, J. P., and Langer, R. 1997. *Proc. Natl., Acad. Sci. U.S.A.*, 94, 8948.

54. Smela, E. 2003. *Adv. Mater.*, 15, 481.

55. Otero, T. F. and Cortes, M. T. 2003. *Adv. Mater.*, 15, 279.

56. Pernaut, J. M. and Reynolds, J. R. 2000. *J. Phys. Chem. B*, 104, 4080.

57. Kim, D. H., Abidian, M., and Martin, D. C. 2004. *J. Biomed. Mater. Res.*, 71A, 577.

58. Yang, J. Y. and Martin, D. C. 2004. *Sens. Actuators B*, 101, 133.

59. Abidian, M. R., Kim, D. H., and Martin, D. C. 2006. *Adv. Mater.*, 18, 405.

60. Zong, X. H., Kim, K., Fang, D. F., Ran, S. F., Hsiao, B. S., and Chu, B. 2003. *Polymer*, 43, 4403.

61. Shiga, T., Hirose, Y., Okada, A., and Kurauchi, T. 1994. *J. Mater. Sci.*, 29, 5715.

62. Shahinpoor, M. and Kim, K. J. 2001. *Smart Mater. Struct.*, 10, 819.

63. Baughman, R. H., Cui, C., Zakhidov, A. A., Iqbal, Z., Barisci, J. N., Spinks, G. M., Wallace, G. G. et al. 1999. *Science.*, 284.

64. Nemat-Nasser, S. and Li, J. Y. 2000. *J. Appl. Phys.*, 87, 3321.

65. Nemat-Nasser, S. and Wu, Y. 2003. *J. Appl. Phys.*, 93, 5255.

66. Han, M. J., Park, J. H., Lee, J. Y., and Jho, J. Y. 2006. *Macromol. Rapid Commun.*, 27, 219.

67. Agarwal, B. D. and Giare, G. S. 1982. *Mater. Sci. Eng.*, 52, 139.

68. Pavelka, V., Jancar, J., and Nezbedova, E., 2005. *e-Polymer*, no. 6.

69. Xu, H. H., Schumacher, G. E., Eichmiller, F. C., Peterson, R. C., Antonucci, J. M., and Mueller, H. J. 2003. *Dent. Mater.*, 19, 9, 523.

70. Xu, H. H., Quinn, J. B., Giuseppetti, A. A., and Eichmiller, F. C. 2000. *J. Dent. Res.*, 79, 1844.

A1

Trade Names for Some Industrial Polymers

Trade Name	Company	Type of Polymer
Abson	B. F. Goodrich	Acrylonitrile–butadiene–styrene terpolymer
Aclar, Aclon	Allied	Polychlorotrifluoroethylene
Acraldon	Bayer	Ethylene–vinyl acetate copolymers
Acrylan	Monsanto	Acrylic fiber
Acrylite	Cyanamid/Rohm	Acrylic resin
Acrypanel	Mitsubishi	Poly(methyl methacrylate)
Adiprene	Du Pont	Polyurethanes
Afcoryl	Pechiney–Saint-Gobain	Acrylonitrile–butadiene–styrene terpolymer
Aflas	Asahi Glass	Tetrafluorethylene–propylene+cure site monomer terpolymer
Aflon	Ashai Glass	Tetrafluoroethylene–ethylene copolymer
Akulon	Akzo	Nylon-6
Alathon	Du Pont	Low-density polyethylene
Algoflon	Montedison	Polytetrafluoroethylene
Alkathene	ICI	Low-density polyethylene
Alkox	Meisei Chemical Works	Poly(ethylene oxide)
Alloprene	ICI	Chlorinated natural rubber
Alpolit	Hoechst	Unsaturated polyester resins
Amberlite	Rohm & Haas	Ion-exchange resin
Ameripol-CB	B. F. Goodrich	Polybutadiene
Amidel	Union Carbide	Transparent amorphous Polyamide
Amilan, Amilon	Toray	Nylon-6
Antron	Du Pont	Polyamide fiber
Araldite	Ciba–Geigy	Epoxy resin
Ardel	Union Carbide	Polyarylate
Arnite	Akzo	Poly(ethylene terephthalate)
Arnite E	Akzo	Thermoplastic polyester elastomers
Arnite PBTP	Akzo	Poly(butylenes terephthalate)
Arnitel	Akzo	Thermoplastic polyester elastomer
Arotone	Du Pont	Polyaryletherketones (PEEK)
Arylef	Solvay	Polyarylate
Arylon	Du Pont	Polyarylates
Astrel	3M	Polyarylsulfone
Avimid	Du Pont	Polyimide

(continued)

Industrial Polymers, Specialty Polymers, and Their Applications

Trade Name	Company	Type of Polymer
Bayblend	Bayer	Polycarbonate/ABS blend
Baygal, Baymidur	Bayer	Polyurethane casting resins
Baylon	Bayer	Polycarbonate
Baypern	Bayer	Polychloroprene
Beetle	British Industrial Plastics	Urea–formaldehyde resin
Benvic	Solvay	Poly(vinyl chloride)
Bipeau	Ato Chimie	Poly(vinyl chloride) (PVC)
Blendex	G.E.	Acrylonitrile–polybutadiene–styrene graft copolymers
Budene	Goodyear	Polybutadiene
Buna-N	Chem. Werke Hüls	Butadiene–acrylonitrile copolymer
Buna-S	Chem. Werke Hüls	Butadiene–styrene copolymer
Butacite	Du Pont	Poly(vinyl butyral)
Butakon	ICI	Butadiene copolymers
Butaprene	Firestone	Styrene–butadiene copolymers
Butvar	Shawinigan	Poly(vinyl butyral)
BXL	Union Carbide	Polysulfone
Cabelec	Cabot	Ethylene–vinyl acetate copolymers
Caprolan	Allied	Polyamide fiber
Capron	Allied	Polyamide resin
Caradol	Shell	Polyhydroxy compound for isocyanate cross-linking
Carbopol	B. F. Goodrich	Acrylic polyelectrolyte
Carboset	B. F. Goodrich	Acrylic polyelectrolyte
Carbowax	Union Carbide	Poly(ethylene oxide)
Cariflex I	Shell	cis-1,4-polyisoprene
Carina	Shell	Poly(vinyl chloride)
Carinex	Shell	Polystyrene
Cebian	Daicel	Styrene–acrylonitrile copolymers
Celanex	Celanese	Poly(butylene terephthalate)
Celcon	Celanese	Polyacetal
Cellioder B	Bayer	Cellulose acetate–butyrate
Cellit	Bayer	Cellulose acetate
Cellon	Dynamit-Nobel	Cellulose acetate
Cellosize	Union Carbide	Hydroxyethylcellulose
Celluloid	Dynamit Nobel	Cellulose nitrate plasticized with camphor
Chemfluor	Norton/Chemplast	Polyvinylidene fluoride
Chemigum	Goodyear	Polyurethane rubber
Cibamin	Ciba–Geigy	Urea–formaldehyde, melamine-formaldehyde resins
Cibanoid	Ciga-Geigy	Urea–formaldehyde resin
cis-4	Phillips	cis-1,4-polybutadiene
Clariflex TR	Shell	Styrene–diene–styrene triblock elastomer
Cobex	Bakelite Xylonite	Poly(vinyl chloride)
Colcolor	Degussa	Ethylene–vinylacetate copolymers
Coral	Firestone	cis-1,4-polyisoprene
Cordura	Du Pont	Polyamide fiber
Corvic	ICI	Poly(vinyl chloride)
Courtelle	Courtaulds	Polyacrylonitrile
Crastin	Ciba–Geigy	Poly(butylene terephthalate)
Craston	Ciba–Geigy	Polyphenylene sulfide
Creslan	Cyanamid	Acrylic fiber
Crofon	Du Pont	Poly(methyl methacrylate)
Crystic	Scott Bader	Polyester resins

(continued)

Trade Names for Some Industrial Polymers

Trade Name	Company	Type of Polymer
Cyanaprene	American Cyanamid	Polyurethane casting resins
Cycolac	G.E.	Acrylonitrile–butadiene–styrene graft copolymers
Cymel	Cyanamid	Melamine–formaldehyde resin
Cyrolite	Röhm	Styrene–polybutadiene graft copolymers
Dacron	Du Pont	Poly(ethylene terephthalate) fiber
Daltocel	ICI	Rigid polyurethane foams
Daltoflex I	ICI	Polyurethane rubber
Dapon	FMC Corp.	Diallyl phthalate resins
Darvic	ICI	Poly(vinyl chloride)
De-acidite	Permutit Co.	Ion-exchange resin
Degaplast, Deglas	Degussa	Poly(methyl methacrylate)
Delmer, Delpet	Asahi Chemical	Poly(methyl methacrylate)
Delrin	Du Pont	Polyacetal
D.E.R.	Dow	Epoxy resin
Desmobond	Mobay	Epoxide resins
Desmopan	Bayer	Thermoplastic polyurethane elastomers
Desmophen	Bayer	Rigid polyurethane foams
Desmophen A	Bayer	Polyurethane rubber
Diakon	ICI	Poly(methyl methacrylate) molding powder
Diene	Firestone	Polybutadiene
Diofan	BASF	Vinyl chloride–vinylidene chloride–acrylonitrile copolymers
Diolen	ENKA-Glazstaff	Poly (ethylene terephthalate)
Dobeckot	BASF	Epoxide resins
Doctolex	Mitsubishi	Polyaryletherketones (PEEK)
Dorlastan	Bayer	Spandex fiber
Dowex	Dow	Ion-exchange resin
Dowlex	Dow	LLDPE
Duolite	Chemical Process Co.	Ion-exchange resins
Duracon	Daicel Polyplastics	Polyacetal
Durel	Hooker	Polyarylate
Durethan	Bayer	Nylon-6
Durethan A	Bayer	Nylon-6,6
Durethan B	Bayer	Nylon-6
Durethan U	Bayer	Thermoplastic polyurethanes
Durez	Occidental Chemical Corp.	Phenol–formaldehyde resins
Dutral	Montecatini	Ethylene–propylene copolymer
Dycryl	Du Pont	Photopolymer system
Dyflor	Dynamit Nobel	Poly(vinylidene fluoride)
Dylene	Arco Chemical	Styrene homopolymers
Dynel	Union Carbide	Vinyl chloride–acrylonitrile copolymer
Dynyl	Rhone-Poulenc	Elastomeric polyamides, copolyamides
Eccofoam	American Micro	Rigid polyurethane foam
Ecdel	Eastman Chem. Products	Thermoplastic polyester elastomer
Econol	Carborundum	Poly(p-hydroxybenzoic acid ester)
Ekcel	Carborundum	Aromatic polyester
Ekonol	Carborundum	Polycarbonate
Ektar	Eastman Chem. Intern.	Polyphenylenesulfide
Elvacet	Du Pont	Poly(vinyl acetate)
Elvanol	Du Pont	Poly(vinyl alcohol)
Elvic	Solvay	Poly(vinyl chloride)
Encron	Akzo	Polester fiber
Ensolite	Uniroyal	Poly(vinyl chloride)

(continued)

Trade Name	Company	Type of Polymer
Epicote	Dow	Epoxide resins
Epodite	Showa Highpolymer	Epoxide resins
Epon	Shell	Epoxide resins
Epi-Rez	Celanese	Epoxide resins
Escor	Exxon	Ethylene–vinyl acetate copolymers
Escorene	Exxon	Polyethylenes
Eska	Mitsubishi	Poly(methyl methacrylate)
Estamid	Upjohn	Nylon-12 elastomers
Estane	B. F. Goodrich	Thermoplastic polyurethane elastomer
Estar	Eastman Kodak	Polyester film
Etar	Eastman Chem. Intern.	Polyethylene terephthalate
Ethocel	Dow	Ethylcellulose
Evatane	Ato Chimie	Ethylene–vinyl acetate copolymers
Evatate	Sumitomo	Ethylene–vinyl acetate copolymers
Extrel	Exxon	Polypropylene
Fertene	Himont	Polyethylenes (HD, LD)
Fiberloc	B. F. Goodrich	Poly(vinyl chloride)
Flemion	Asahi Glass	Carboxylated fluoropolymer
Flovic	ICI	Poly(vinyl acetate)
Fluon	ICI	Polytetrafluoroethylene
Fluorel	3M	Vinylidene fluoride–hexafluoropropylene copolymer
Foraflon	Ato Chimie	Polytetrafluoroethylene
Formica	Cyanamid	Melamine–formaldehyde resin
Forticel	Celanese	Cellulose propionate
Fortron	Hoechst	Polyphenylenesulfide
Gafite	Hoechst	Poly(butylenes terephthalate)
Gaflex	Hoechst	Thermoplastic polyester elastomers
Gaftuf	Hoechst	Poly(butylenes terephthalate)
Gecet	G.E.	Polyphenylene ether
Geloy	G.E.	Styrene–acrylonitrile copolymers
Gelvatol	Shawinigan	Poly(vinyl alcohol)
Genal	G.E.	Phenol–formaldehyde resins
Genopak, Genotherm	Hoechst	Poly(vinyl chloride)
Geon	B. F. Goodrich	Poly(vinyl chloride)
Grilamid	Ems Chemie	Nylon-12
Grilamid ELY 60	Ems Chemie	Polyamide elastomers
Grilamid TR	Emser Werke	Transparent amorphous polyamide
Grilon	Ems Chemie	Nylon-6
Grilonit	Ems Chemie	Epoxide resins
Grilpet	Ems Chemie	Poly(ethylene terephthalate)
Halar	Ausimont	Ethylene–chlorotrifluoroethylene copolymer
Halon	Ausimont	Polytetrafluoroethylene
Herculoid	Du Pont	Celulose nitrate
H-film	Du Pont	Polyamide from pyromellitic anhydride and 4,4′-diaminodiphenyl ether
Hi-fax	Hitachi	Polyethylene
Hitalex	Hitachi	Polyethylene
Hostadur	Hoechst	Poly (ethylene terphthalate)
Hostaflon ET	Hoechst	Tetrafluoroethylene–ethylene copolymer
Hostaflon FEP	Hoechst	Tetrafluoroethylene–hexafluoropropylene copolymer
Hostaflon TF	Hoechst	Polytetrafluoroethylene
Hostaflon TFA	Hoechst	Perfluoroalkoxy copolymers

(continued)

Trade Name	Company	Type of Polymer
Hostaform	Hoechst	Polyoxymethylene
Hostalen	Hoechst	Polyethylenes (HD, LD)
Hostalen GUR	Hoechst	Ultrahigh-molecular weight polyethylene
Hostalen PP	Hoechst	Polypropylene
Hostalit	Hoechst	Poly(vinyl chloride) and blends
Hostamid	Hoechst	Transparent amorphous polyamide
Hostatec	Hoechst	Polyether ketone
Hycar	B. F. Goodrich	Polyacrylate
Hydrin	B. F. Goodrich	Epichlorohydrin rubber
Hylar	Du Pont	Poly(ethylene terephthalate)
Hypalon	Du Pont	Sulfochlorinated polyethylene
Hytrel	Du Pont	Thermoplastic polyester elastomers
Hyvis	BP Chemicals	Polyisobutylene
Icdal Ti40	Dynamit Nobel	Polyesterimide
Impet	Hoechst	Poly(ethylene terephthalate)
Ionac	Ionac Co.	Ion-exchange resins
Irrathene	G.E.	PE, cross-linked by radiation
Isonol	Dow	Rigid polyurethane foams
Ixan	Solvay	Vinyl chloride–vinylidene chloride–acrylonitrile copolymers
Ixef	Solvay	Aromatic polyamide
Jupilon	Mitsubishi	Polycarbonate
Kadel	Amoco	Polyaryletherketones (PEEK)
Kalrez	Du Pont	Fluoroelastomer
Kamax	Rohm & Hass	Polyacrylic esterimide
Kapton	Du Pont	Polyimide film
Kardel	Union Carbide	Styrene homopolymers
Kel-F	3M	Polychlorotrifluoroethylene, poly(vinyl fluoride)
Kel-F elastomer	3M	Vinylidene fluoride–chlorotrifluoroethylene copolymer
Kematal	Hoechst	Acetal homopolymers
Kermel	Rhone-Poulenc	Polyimide fiber
Kerimid	Rhone-Poulenc	Polimide
Kinel	Rhone-Poulenc	Polybismaleinimide
Kodapak	Eastman Chem. Intern.	Poly(butylenes terephthalate)
Kodar PETG	Eastman Chem. Intern.	Copolyester based on 1,4-cyclohexylene glycol and a mixture of terephthalic and isophthalic acids
Kodel	Eastman Chem. Intern.	Polyester fiber
Kralastic	Uniroyal	Acrylonitrile–butadiene–styrene copolymer
Kraton	Shell	Thermoplastic styrene block copolymer
Kuralon	Kuraray (Japan)	Poly (vinyl alchol) fiber
Kynar	Pennwalt	Poly(vinylidene fluoride)
Laminac	Cyanamid	Polyester resins
Leguval	Bayer	Unsaturated polyester resins
Lekutherm	Bayer	Epoxide resins
Levapren	Bayer	Ethylene–vinyl acetate copolymer
Levasint	Bayer	Ethylene–vinyl alcohol copolymers
Lewatit	Bayer	Ion-exchange resins
Lexan	G.E.	Polycarbonate
Localen	BASF	Ethylene–vinyl acetate copolymers
Lomod	G.E.	Thermoplastic polyester elastomers

(continued)

Trade Name	Company	Type of Polymer
Lucite	Du Pont	Poly(methyl methacrylate) and copolymers
Lucryl	BASF	Poly(methyl methacrylate)
Lupolen	BASF	Ethylene–vinyl acetate copolymers
Luran	BASF	Styrene–acrylonitrile copolymers
Luranyl	BASF	Poly(ethylene oxide) blend
Lustran	Monsanto	Acrylonitrile–butadiene–styrene terpolymer
Lustrex	Monsanto	Polystyrene
Lycra	Du Pont	Spandex fiber
Makrolon	Röhm	Polycarbonate
Maplen	Himont	Polyethylenes (HD, LD)
Maranyl	ICI	Nylon-6
Maranyl A	ICI	Nylon-6,6
Marlex	Phillips	Polyethylene, polypropylene
Marlex TR 130	Phillips	Polyethylene (LLD)
Melinar, Melinite	ICI	Poly(ethylene terephthalate)
Melinex	ICI	Polyester film
Melopas	Ciba–Geigy	Melamine–formaldehyde resins
Merlon	Mobay	Polycarbonate
Methocel	Dow	Methyl cellulose
Minlon	Du Pont	Nylon-6,6
Moltopren	Bayer	Rigid polyurethane foam
Moplen	Himont	Polypropylene
Mowicoll	Hoechst	Poly(vinyl acetate) dispersions
Mowilith	Hoechst	Poly(vinyl acetate)
Mowiol	Hoechst	Poly(vinyl alcohol)
Mowital	Hoechst	Poly(vinyl butyral)
Mylar	Du Pont	Poly(ethylene terephthalate) film
Nafion	Du Pont	Persulfonated fluoropolymer
Nalcite	National Aluminate Corp.	Ion-exchange resins
Napryl	Pechiney–Saint-Gobain	Polypropylene
Natene	Pechiney–Saint-Gobain	Polyethylenes
Natsyn	Goodyear	Polyisoprene
Necofene	Ashland	Polyphenylene ether
Neosepta F	Tokoyama Soda	Ionic membrane (based on fluoropolymer)
Nikalet	Nippon Carbide	Epoxide resins
Nipoflex	Toyo Soda	Ethylene–vinyl acetate copolymer
Nipolon	Toyo Soda	Polyethylenes (HD, LD)
Nitron	Monsanto	Cellulose nitrate
Noblen	Mitsubishi	Polypropylene
Nolimid (adhesives)	Rhone-Poulenc	Polybismaleinimide
Nomex	Du Pont	Poly(m-phenylene isophthalimide)
Nordel	Du Pont	Ethylene–propylene–diene terpolymer
Noryl	G.E	Poly(phenylene oxide)–polystyrene blend
Novadur	Mitsubishi	Poly(butylene terephthalate)
Novarex	Mitsubishi	Polycarbonate
Novatec	Mitsubishi	Polypropylene, polyethylenes (HD, LD)
Novatex	Mitsubishi	Polyethylene (LLD)
Novex	BP Chemicals	Polyethylenes (HD, LD)
Novodur	Bayer	Acrylonitrile–polybutadiene–styrene graft copolymers
Oppanol B	BASF	Polyisobutylene
Oppanol C	BASF	Poly(vinyl isobutyl ether)

(continued)

Trade Name	Company	Type of Polymer
Oppanol O	BASF	Copolymer from 90% isobutylene and 10% styrene
Orgalan	Ato Chimie	Polycarbonate
Orgamide	Ato Chimie	Nylon-6
Orgavyl	Ato Chimie	Poly(vinyl chloride)
Orlon	Du Pont	Acrylic fiber
Oroglas	Rohm & Haas	Poly(methyl methacrylate)
Paraglas	Degussa	Poly(methyl methacrylate)
Paraplex	Rohm & Haas	Epoxide resins
Parapol	Exxon	Polyisobutylene
Parlon	Hercules	Chlorinated rubber
Paxon	Allied	Polyethylene
Pebax	Ato Chimie	Polyamide elastomers
Pelaspan	Dow	Polystyrene (expandable)
Pelprene	Toyobo	Thermoplastic polyurethane elastomer
Perbunan N	Bayer	Butadiene–acrylonitrile copolymers
Perlenka	Akzo	Nylon-6
Permutit	Permutit Co.	Ion-exchange resins
Perspex	ICI	Poly(methyl methacrylate) sheet
Petlon	Mobay	Poly(ethylene terephthalate)
Petra	Allied Signal	Poly(ethylene terephthalate)
Petrothene	USI Chemical	Polyethylenes (HD, LD, LLD)
Pevalon	May & Baker	Poly(vinyl alcohol)
Plastazote	American Micro	Polyethylene foam
Plexiglas	Rohm & Haas	Poly(methyl methacrylate)
Plexigum	Rohm & Haas	Acrylate and methacrylate resins
Pliofilm	Goodyear	Rubber hydrochloride
Pliolite	Goodyear	Styrene–butadiene copolymers
Pliovic	Goodyear	Poly(vinyl chloride)
Pocan	Bayer	Poly(butylenes terephthalate)
Polybond	BP Chemicals	Ethylene–vinyl acetate copolymers
Polycal	Ato Chimie	Poly(vinyl chloride)
Poly-Eth	Gulf Oil	Polyethylene
Polymidal	Raychem	Polyimide
Polymin	BASF	Polyethyleneimine
Polyox	Union Carbide	Poly(ethylene oxide)
PolyPro	Mitsui Petrochemcial	Polypropylene
Polysizer	Showa Highpolymer	Poly(vinyl alcohol)
Polythene	Du Pont	Low-density polyethylene
Polyviol	Wacker Chemie	Poly(vinyl alcohol)
Prevex	Borg Warner	Poly(phenylene oxide) blend
Prodorit	T.I.B. Chemie	Epoxide resins
Profax	Himont	Polypropylene
Propathene	ICI	Polypropylene
Pulse	Dow	(ABS+Polycarbonate) polymer blends
Pyralin	Du Pont	Polyimide
Pyre ML	Du Pont	Polyimide coating
Qiana	Du Pont	Polyamide fiber from bis(*p*-aminocyclohexyl)methane and dodecanedioic acid
Radel	Union Carbide	Polyether sulfone
Ravinil	EniChem	Poly(vinyl chloride)
Resolite	Ciba–Geigy	Urea–formaldehyde resin
Rilsan A	Ato Chimie	Nylon-12
Rilsan B	Ato Chimie	Nylon-11
Rimplast	Petrarch Systems	Silicone resins and molding compounds

(continued)

Trade Name	Company	Type of Polymer
Royalene	Uniroyal	Ethylene–propylene–diene terpolymer
Roylar	Uniroyal	Polyurethanes
Roylar	B. F. Goodrich	Thermoplastic polyurethane elastomers
Rucon	Hooker	Poly(vinyl chloride)
Rucothane	Hooker	Polyurethanes
Rynite	Du Pont	Glass-reinforced poly(ethylene terephthalate)
Ryton	Phillips	Polyphenylenesulfide
Santolite	Monsanto	Polyaryletherketones
Santoprene	Monsanto	Modified polypropylene
Saran	Dow	Copolymers of vinylidene chloride, vinyl chloride and acrylonitrile
Scotchcast	3M	Epoxide resins
Sctochpack	3M	Polyester film
Selar	Du Pont	Ethylene–vinyl alcohol copolymers
Sicralan	Degussa	Silicones
Silbione	Rhone-Poulenc	Silicones
Silastic	Dow	Silicones
Silastomer	Dow	Silicones
Silopren	Bayer	Silicone rubber
Siltemp	G.E.	Silicones
Sirfen	Societa Italiana Resine	Phenol–formaldehyde resin
Sirotherm	ICI	Ampholytic polyelectrolyte
Siretene	Societa Italiana Resine	Polyethylene
Skybond	Monsanto	Polyimide
Solef	Solvay	Poly(vinylidene fluoride)
Solvic	Solvay	Poly(vinyl chloride) and blends
Soreflon	Rohne-Poulenc	Polytetrafluoroethylene
Spanzelle	Courtaulds	Spandex fiber
Sparlux	Solvay	Polycarbonate
Stabar	ICI	Polyarylsulfone, polyaryletherketones
Stylac	Asahi	Acrylonitrile–polybutadiene–styrene graft copolymers
Stylex	Mitsubishi	Styrene homopolymers
Styrodur, Styropor	BASF	Styrene homopolymers
Styrocell	Shell	Polystyrene foam
Styrofoam	Dow	Polystyrene foam
Styrol	Idemitsu	Polystyrene
Styron	Dow	Polystyrene
Sumipex	Sumitomo	Poly(methyl methacrylate)
Suntec	Asahi	Polyethylene (HD, LD)
Supec	G.E.	Polyphenylenesulfide
Superacryl	Dental a.s. (Praha)	Poly (methyl methacrylate) dental resin
Surlyn	Du Pont	Ionomers
Technyl A	Rhone-Poulenc	Nylon-6,6
Technyl C	Rhone-Poulenc	Nylon-6
Tecnoflon	Montecatini	Fluoropolymer
Tecolite	Toshiba	Phenol-formaldehyde resins and molding compounds
Tedlar	Du Pont	Poly(vinyl fluoride)
Tedur	Bayer	Polyphenylenesulfide
Teflon	Du Pont	Polytetrafluoroethylene
Teflon FEP	Du Pont	Tetrafluoroethylene– hexafluoropropylene copolymers
Tenac	Asahi	Acetal homopolymers
Tenite	Eastman Chemical	Polyethylenes (HD, LD, LLD)

(continued)

Trade Name	Company	Type of Polymer
Tenite Acetate	Eastman Chemical	Cellulose acetate
Tenite Butyrate	Eastman Chemical	Cellulose acetate–butyrate
Tenite Propionate	Eastman Chemical	Cellulose acetate–propionate
Tenite PTMT	Eastman Chemical	Poly(tetramethylene terephthalate)
Terlenka	Akzo	Poly(ethylene terephthalate)
Terluran	BASF	Acrylonitrile–polybutadiene–styrene graft copolymers
Terylene	ICI	Poly(ethylene terephthalate)
Texicote	Scott Bader	Poly(vinyl acetate)
Texigels	Scott Bader	Acrylic polyelectrolyte
Texin	Mobay	Polyurethane elastomer
Textolite	G.E.	Silicone resins and molding compounds
Thiokol	Thiokol	Polysulfides
Torlon	Amoco	Polyamide–imide
TPX	ICI	Poly-4-methylpent-1-ene
Trans-4	Phillips	*Trans*-1,4-polybutadiene
Trevira	Hoechst	Polyester fiber
Triax	Bayer	Polyamide/ABS blend
Tricel	Bayer	Cellulose acetate
Trogamid	Dynamit Nobel	Transparent polyamide
Trolen	Dynamit Nobel	Polyethylene
Trolit F	Dynamit Nobel	Cellulose nitrate
Trolitan	Dynamit Nobel	Phenol–formaldehyde resin
Trolitul	Dynamit Nobel	Polystyrene
Trosiplast	Dynamit Nobel	Poly(vinyl chloride) and blends
Trovidur	Dynamit Nobel	Polypropylene, poly(vinylchloride)
Trymer	American Micro	Polyisocyanurate rigid foam
Tybrene	Dow	Acrylonitrile–butadiene–styrene terpolymer
Tynex	Du Pont	Nylon-6,6
Tyril	Dow	Styrene–acrylonitrile copolymer
Udel	Union Carbide	Polysulfone
Ultem	G.E.	Polyetherimide
Ultradur	BASF	Poly(butylenes terephthalate)
Ultradur A	BASF	Poly(ethylene terephthalate)
Ultraform	BASF	Acetal homopolymer
Ultramid	BASF	Transparent polyamide, copolyamides
Ultrapas	Dynamit Nobel	Melamine–formaldehyde resins
Ultrapek	BASF	Polyaryletherketones
Ultrason E	BASF	Polyarylethersulfones
Ultrax	BASF	Liquid crystal polymers
Ultryl	Phillips	Poly(vinyl chloride)
U-Polymer	Unitika	Polyarylate
Urafil	Akzo	Thermoplastic polyurethane elastomers
Urecoll	BASF	Urea–formaldehyde
Urepan	Bayer	Polyurethane
Uthane	Urethanes India	Thermoplastic polyurethane elastomers
Valox	G.E.	Poly(butylenes terephthalate)
Valtec	Himont	Polypropylene
Vandar	Hoechst	Poly(butylene terephthalate)
Vectra	Celanese, Hoechst	Liquid crystal polymers
Vedril	Montedison	Poly(methyl methacrylate)
Versamid	General Mills	Fatty polyamides
Versicol	Allied Colloids	Acrylic polyelectrolyte
Verton	ICI	Polyphenylenesulfide
Vespel	Du Pont	Polyimide

(continued)

Trade Name	Company	Type of Polymer
Vestamid	Chem. Werke Hüls	Nylon-12
Vestamid E	Chem. Werke Hüls	Polyamide elastomers
Vestiform	Chem. Werke Hüls	Poly(methyl methacrylate)
Vestodur	Chem. Werke Hüls	Poly(butylene terephthalate)
Vestolen	Chem. Werke Hüls	Polyethylene (HD, LD)
Vestolen P	Chem. Werke Hüls	Polypropylene
Vestoran	Chem. Werke Hüls	Polyphenyleneether
Vestyron	Chem. Werke Hüls	Polystyrene
Vibrathane	Uniroyal	Polyurethane casting resins
Victrex	ICI	Liquid crystal polymers
Vidyne R	Monsanto	Transparent polyamide
Vinacel	Goodyear	Poly(vinyl chloride)
Vinavil	Montedison	Ethylene–vinyl acetate copolymers
Vinnapas	Wacker Chemie	Ethylene–vinyl acetate copolymers
Vinnol	Wacker Chemie	Poly(vinyl chloride)
Vinovil	Montedison	Chlorinated polyethylene
Vipla, Viplast	EniChem	Poly(vinyl chloride)
Vistalon	Exxon	Ethylene–vinyl acetate copolymers, Ethylene–propylene–diene terpolymer
Vistanex	Exxon	Polyisobutylene
Vithane	Goodyear	Polyurethanes
Viton A	Du Pont	Vinylidene fluoride–hexafluoropropylene copolymer
Viton B	Du Pont	Vinylidene fluoride–hexafluoropropylene–tetrafluoroethylene terpolymer
Vulcaprene	ICI	Polyurethanes
Vulkollan	Bayer	Polyurethanes
Vydyne	Monsanto	Polyamide resin
Vyrene	U.S. Rubber	Spandex fiber
Welvic	ICI	Poly(vinyl chloride)
Wofatit	VEB Farbenfabrik	Ion-exchange resins
Xydar	Dartco Mfg.	Wholly aromatic copolyester injection-molding resin
Zeo-karb	Permutit Co.	Ion-exchange resins
Zetabon, Zimek	Dow	Ethylene–vinyl acetate copolymers
Zetafin	Dow	Ethylene–methyl acrylate copolymers
Zytel	Du Pont	Nylon-6, nylon-6,6

This selection confers no priorities and is not exhaustive.

A2

Commonly Used Abbreviations for Industrial Polymers

AAS	Copolymer of acrylonitrile, acrylate, and styrene
ABR	Acrylate–butadiene rubber
ABS	Acrylonitrile–butadiene–styrene terpolymer
ACS	Thermoplastic blend of a copolymer from acrylonitrile and styrene with chlorinated polyethylene
AES	Thermoplastic quaterpolymer from acrylonitrile, ethylene, propylene, and styrene
ASA	Copolymer of acrylonitrile, styrene, and acrylates
BR	Butadiene rubber
CA	Cellulose acetate
CAB	Cellulose acetate butyrate
CAP	Cellulose acetate propionate
CMC	Carboxymethyl cellulose
CN	Cellulose nitrate
CPE	Chlorinated polyethylene
CPVC	Chlorinated poly(vinyl chloride)
CR	Polychloroprene
CTA	Cellulose triacetate
CTFE	Chlorotrifluoroethylene polymer
EC	Ethyl cellulose
ECTFE	Ethylene–chlorotrifluoroethylene copolymer
EEA	Elastomeric copolymer from ethylene and ethyl acrylate
EMA	Ethylene–methyl acrylate copolymer
EP	Epoxy resin
E/P	Ethylene–propylene copolymer
EPDM	Terpolymer from ethylene, proplene, and a (nonconjugated) diene
EPM	Ethylene–propylene copolymer
EPR	Elastomeric copolymer of ethylene and propylene
EPT, EPTR	Elastomeric copolymer of ethylene, propylene, and a diene
ETFE	Ethylene–tetrafluoroethylene copolymer
EVA	Copolymer from ethylene and vinyl acetate
EVOH	Ethylene–vinyl alcohol copolymer
FEP	Fluorinated ethylene–propylene copolymer
HDPE	High-density polyethylene
HIPS	High-impact polystyrene

IIR	Butyl rubber (isobutylene–isoprene copolymer)
IPN	Interpenetrating polymer network
IR	Synthetic *cis*-1,4-polyisoprene rubber
LCP	Liquid crystal polymer
LDPE	Low-density polyethylene
LLDPE	Linear low-density polyethylene
MBS	Methacrylate–butadiene–styrene copolymer
MF	Melamine–formaldehyde resin
NBR	Acrylonitrile–butadiene rubber (nitrile rubber)
NC	Nitrocellulose (cellulose nitrate)
NR	Natural rubber
PA	Polyamide
PAA	Poly(acrylic acid)
PAE	Polyarylether
PAEK	Polyaryletherketone
PAES	Polyarylethersulfone
PAI	Polyamide–imide
PAMS	Poly-α-methylstyrene
PAN	Polyacrylonitrile
PAr	Polyarylate
PAS	Polyarylsulfide
PB	Polybutdaiene
PBT	Poly(butylene terephthalate)
PC	Polycarbonate
PCTG	Poly(cylohexane terephthalate-glycol)
PDMS	Polydimethylsiloxane
PE	Polyethylene
PEBA	Polyether-block amide
PEC	Polyestercarbonate
PEEK	Polyetheretherketone
PEG	Polyethylene glycol
PEI	Polyetherimide
PEO	Poly(ethylene oxide)
PES	Polyethersulfone
PET	Poly(ethylene terephthalate)
PF	Phenol–formaldehyde resin
PFEP	Copolymer from tetrafluoroethylene and hexafluoropropylene
PI	Polyimide
PIB	Polyisobutylene
PIR	Polyisocyanurate foam
PMMA	Poly(methyl methacrylate)
PO	Polyolefin
POM	Polyoxymethylene (Acetal)
PP	Polypropylene
PPE	Polyphenylether
PPG	Polypropylene glycol
PPO	Poly(phenylene oxide)
PPS	Polyphenylenesulfide
PS	Polystyrene
PSO	Polysulfone
PSU	Polyphenylenesulfone
PTFE	Polytetrafluoroethylene
PTMG	Polyoxytetramethyleneglycol

PTMT	Poly(tetramethylene terephthalate)
PU	Polyurethane
PVA	Poly(vinyl alcohol), Poly(vinyl acetate)
PVAc	Poly(vinyl acetate)
PVAL	Poly(vinyl alcohol)
PVB	Poly(vinyl butyral)
PVC	Poly(vinyl chloride)
PVDC	Poly(vinylidene chloride)
PVDF	Poly(vinylidene fluoride)
PVF	Poly(vinyl fluoride)
PVFM	Poly(vinyl formal)
PVME	Poly(vinyl methyl ether)
PVOH	Poly(vinyl alcohol)
PVP	Poly(vinyl pyridine), Poly(vinyl pyrrolidone)
RTV	Room temperature vulcanizing silicone rubber
SAN	Styrene–acrylonitrile copolymer
SBR	Styrene–butadiene rubber
SBS	Styrene–butadiene–styrene block copolymer
SEBS	Styrene–ethylene–butylene–styrene block copolymer (hydrogenated SIS)
SIN	Simultaneous interpenetrating network
SIS	Styrene–isoprene–styrene block copolymer
SMA	Styrene–maleic anhydride copolyemr
SMS	Styrene-α-methylstyrene copolymer
TPE	Thermoplastic elastomer
TPEs	Thermoplastic polyesters, e.g., PBT and PET
TPO	Thermoplastic polyolefin elastomers
TPU	Thermoplastic polyurethane
UF	Urea–formaldehyde resin
UHMWPE	Ultrahigh-molecular weight polyethylene (mol. wt. $>3\times10^6$)
VLDPE	Very-low-density polyethylene (density ca. 0.890–0.915 g/cm^3)

A3

Typical Properties of Polymers Used for Molding and Extrusion

		Polyethylene		
	ASTM Test Method	Low Density	High Density	Polypropylene
1. Specific gravity	D792	0.91–0.925	0.94–0.965	0.900–0.910
2. Tensile modulus $(\text{psi} \times 10^{-5})$	D638	0.14–0.38	0.6–1.8	1.6–2.25
3. Compressive modulus $(\text{psi} \times 10^{-5})$	D695	—	—	1.5–3.0
4. Flexural modulus $(\text{psi} \times 10^{-5})$	D790	0.08–0.6	1.0–2.6	1.7–2.5
5. Tensile strength $(\text{psi} \times 10^{-3})$	D638, D651	0.6–2.3	3.1–5.5	4.5–6.0
6. Elongation at break (%)	D638	90–800	20–130	100–600
7. Compressive strength $(\text{psi} \times 10^{-3})$	D695	2.7–3.6	12–18	5.5–8.0
8. Flexural yield strength $(\text{psi} \times 10^{-3})$	D790	—	1.0	6–8
9. Impact strength, notched Izod, (ft-lb/in.)	D256	No break	0.5–20	0.4–1.0
10. Hardness, Rockwell	D785	D40-51(Shore)	D60-70 (Shore)	R80-102
11. Thermal conduct. $(\text{cal/s–cm–K} \times 10^4)$	C177	8.0	11–12	2.8
12. Specific heat (cal/g–K)	—	0.55	0.55	0.46
13. Linear therm. exp. coeff. $(\text{K}^{-1} \times 10^5)$	D696	10–22	11–13	8.1–10.0
14. Continuous-use temperature (°C)	—	80–100	120	120–160
15. Deflection temp. (°C at 0.45 MPa)	D648	38–49	60–88	107–121
16. Volume resistivity, ohm cm	D257	$>10^{16}$	$>10^{16}$	$>10^{16}$
17. Dielectric constant at 1 kHz	D150	2.25–2.35	2.30–2.35	2.2–2.6
18. Dielectric strength (kV/in.)	D149	450–1000	450–500	500–660

(continued)

Industrial Polymers, Specialty Polymers, and Their Applications

	ASTM Test Method	Polyethylene		Polypropylene
		Low Density	High Density	
19. Dissipation factor at 1 kHz	D150	<0.0005	<0.0005	<0.0018
20. Deleterious media	D543	Oxidizing acids	Oxidizing acids	Strong oxidizing acids
21. Solvents (room temperature) (Cl.H.=chlorinated hydrocarbons)		None	None	None

	ASTM Test Method	Polystyrene		Poly (methyl Methacrylate)
		Gen. Purpose	Impact-Resistant	
1. Specific gravity	D792	1.04–1.05	1.03–1.06	1.17–1.20
2. Tensile modulus ($psi \times 10^{-5}$)	D638	3.5–4.85	2.6–4.65	3.8
3. Compressive modulus ($psi \times 10^{-5}$)	D695	—	—	3.7–4.6
4. Flexural modulus ($psi \times 10^{-5}$)	D790	4.3–4.7	3.3–4.0	4.2–4.6
5. Tensile strength ($psi \times 10^{-3}$)	D638, D651	5.3–7.9	3.2–4.9	7–11
6. Elongation at break (%)	D638	1–2	13–50	2–10
7. Compressive strength ($psi \times 10^{-3}$)	D695	11.5–16	4–9	12–18
8. Flexural yield strength ($psi \times 10^{-3}$)	D790	8.7–14	5–12	13–19
9. Impact strength, notched Izod, (ft. lb/in.)	D256	0.25–0.40	0.5–11	0.3–0.5
10. Hardness, Rockwell	D785	M65-80	M20-80	M85-105
11. Thermal conduct. (cal/s-cm-K $\times 10^4$)	C177	2.4–3.3	1.0–3.0	4–6
12. Specific heat (cal/g-K)	—	0.32	0.32–0.35	0.35
13. Linear therm. exp. coeff. ($K^{-1} \times 10^5$)	D696	6–8	3.4–21	5–9
14. Continuous-use temperature (°C)	—	66–77	60–79	60–88
15. Deflection temp. (°C at 0.45 MPa)	D648	75–100	75–95	80–107

(continued)

		Polystyrene		Poly (methyl Methacrylate)
	ASTM Test Method	Gen. Purpose	Impact-Resistant	
16. Volume resistivity, ohm cm	D257	10^{16}	10^{16}	10^{14}
17. Dielectric constant at 1 kHz	D150	2.4–2.65	2.4–4.5	3.0–3.6
18. Dielectric strength (kV/in.)	D149	500–700	300–600	400
19. Dissipation factor at 1 kHz	D150	0.0001–0.0003	0.0004–0.002	0.03–0.05
20. Deleterious media	D543	Strong oxidizing acids	Strong Oxidizing acids	Strong bases and strong, oxidizing acids
21. Solvents (room temperature) (Cl.H.= chlorinated hydrocarbons)		Aromatic and Cl.H.	Aromatic and Cl.H.	Ketones, esters, aromatic, and Cl.H.

		Poly(Vinyl Chloride)		ABS Medium Impact
	ASTM Test Method	Rigid	Plasticized	
1. Specific gravity	D792	1.30–1.58	1.16–1.35	1.03–1.06
2. Tensile modulus $(\text{psi} \times 10^{-5})$	D638	3.5–6	—	3–4
3. Compressive modulus $(\text{psi} \times 10^{-5})$	D695	—	—	2.0–4.5
4. Flexural modulus $(\text{psi} \times 10^{-5})$	D790	3–5	—	3.7–4.0
5. Tensile strength $(\text{psi} \times 10^{-3})$	D638, D651	6–7.5	1.5–3.5	6–7.5
6. Elongation at break (%)	D638	2–80	200–450	5–25
7. Compressive strength $(\text{psi} \times 10^{-3})$	D695	8–13	0.9–1.7	10.5–12.5
8. Flexural yield strength $(\text{psi} \times 10^{-3})$	D790	10–16	—	11–13
9. Impact strength, notched Izod, (ft-lb/in.)	D256	0.4–20	—	11–13
10. Hardness, Rockwell	D785	D65-85(Shore)	A40-100(Shore)	R107-115
11. Thermal conduct. $(\text{cal/s-cm-K} \times 10^4)$	C177	3.5–5.0	3.0–4.0	4.5–8.0
12. Specific heat (cal/g-K)	—	0.2–0.28	0.3–0.5	0.3–0.4

(continued)

	ASTM Test Method	Poly(Vinyl Chloride)		ABS Medium Impact
		Rigid	Plasticized	
13. Linear therm. exp. coeff. $(K^{-1} \times 10^5)$	D696	5–10	7–25	8–10
14. Continuous-use temperature (°C)	—	65–80	65–80	71–93
15. Deflection temp. (°C at 0.45 MPa)	D648	57–82	—	102–107
16. Volume resistivity, ohm cm	D257	$>10^{16}$	10^{11}–10^{15}	2.7×10^{16}
17. Dielectric constant at 1 kHz	D150	3.0–3.3	4–8	2.4–4.5
18. Dielectric strength (kV/in.)	D149	425–1300	300–1000	350–500
19. Dissipation factor at 1 kHz	D150	0.009–0.017	0.07–0.16	0.004–0.007
20. Deleterious media	D543	None	None	Conc. oxidizing acids, organic solvents
21. Solvents (room temperature) (Cl. H. = chlorinated hydrocarbons)		Ketones, esters, swelling in aromatic, and Cl.H.	Plasticizer may be extracted. Otherwise like rigid PVC	Ketones, esters, some Cl.H.

	ASTM Test Method	Cellulose Acetate	Cellulose Acetate Butyrate	Fluoropolymers	
				–CF_2–CF_2–	–CF_2–CHCl–
1. Specific gravity	D792	1.22–1.34	1.15–1.22	2.14–2.20	2.1–2.2
2. Tensile modulus $(psi \times 10^{-5})$	D638	0.65–4.0	0.5–2.0	0.58	1.5–3.0
3. Compressive modulus $(psi \times 10^{-5})$	D695	—	—	—	—
4. Flexural modulus $(psi \times 10^{-5})$	D790	—	—	—	—
5. Tensile strength $(psi \times 10^{-3})$	D638, D651	1.9–9.0	2.6–6.9	2–5	4.5–6.0
6. Elongation at break (%)	D638	6–70	40–88	200–400	80–250
7. Compressive strength $(psi \times 10^{-3})$	D695	3–8	2.1–7.5	1.7	4.6–7.4
8. Flexural yield strength $(psi \times 10^{-3})$	D790	2–16	1.8–9.3	—	7.4–9.3
9. Impact strength, notched Izod, (ft-lb/in.)	D256	1–7.8	1–11	3.0	2.5–2.7

(continued)

	ASTM Test Method	Cellulose Acetate	Cellulose Acetate Butyrate	Fluoropolymers	
				$-CF_2-CF_2-$	$-CF_2-CHCl-$
10. Hardness, Rockwell	D785	R34-125	R31-116	D50-55(Shore)	R75-95
11. Thermal conduct. (cal/ s-cm-K$\times 10^4$)	C177	4–8	4–8	6.0	4.7–5.3
12. Specific heat (cal/g-K)	—	0.0–0.4	0.3–0.4	0.25	0.22
13. Linear therm. exp. coeff. ($K^{-1}\times 10^5$)	D696	8–18	11–17	10	4.5–7.0
14. Continuous-use temperature (°C)	—	60–105	60–105	290	175–200
15. Deflection temp. (°C at 0.45 MPa)	D648	50–100	54–108	121	126
16. Volume resistivity, ohm cm	D257	$10^{10}-10^{14}$	$10^{11}-10^{15}$	10^{18}	1.2×10^{18}
17. Dielectric constant at 1 kHz	D150	3.4–7.0	3.4–6.4	2.1	2.3–2.7
18. Dielectric strength (kV/ in.)	D149	250–500	250–400	480	500–600
19. Dissipation factor at 1 kHz	D150	0.01–0.07	0.01–0.04	0.002	0.023–0.027
20. Deleterious media	D543	Strong acids and bases	Strong acids and bases	None	None
21. Solvents (room temperature) (Cl.H.= chlorinated hydrocarbons)		Ketones, esters, Cl.H	Ketones, esters, Cl.H	None	Swells in Cl.H.

	ASTM Test Method	Nylon-6,6 (Moisture Conditioned)	Nylon-6 (Moisture Conditioned)	Acetal	Polycarbonate
1. Specific gravity	D792	1.13–1.15	1.12–1.14	1.42	1.2
2. Tensile modulus (psi$\times 10^{-5}$)	D638	—	1.0	5.2	3.5
3. Compressive modulus (psi$\times 10^{-5}$)	D695	—	2.5	6.7	3.5
4. Flexural modulus (psi$\times 10^{-5}$)	D790	1.75–4.1	1.4	3.8–4.3	3.4

(continued)

	ASTM Test Method	Nylon-6,6 (Moisture Conditioned)	Nylon-6 (Moisture Conditioned)	Acetal	Polycarbonate
5. Tensile strength (psi$\times 10^{-3}$)	D638, D651	11	10	9.5–12	9.5
6. Elongation at break (%)	D638	300	300	25–75	110
7. Compressive strength (psi$\times 10^{-3}$)	D695	—	—	18	12.5
8. Flexural yield strength (psi$\times 10^{-3}$)	D790	6.1	5.0	14	13.5
9. Impact strength, notched Izod, (ft-lb/in.)	D256	2.1	3.0	1.3–2.3	16
10. Hardness, Rockwell	D785	R120	R119	M94 to R120	M70
11. Thermal conduct. (cal/s-cm-K$\times 10^{4}$)	C177	5.8	5.8	5.5	4.7
12. Specific heat (cal/g-K)	—	0.4	0.38	0.35	0.3
13. Linear therm. exp. coeff. ($K^{-1}\times 10^{5}$)	D696	8.0	8.0–8.3	10	6.8
14. Continuous-use temperature (°C)	—	80–150	80–120	90	121
15. Deflection temp. (°C at 0.45 MPa)	D648	180–240	150–185	124	138
16. Volume resistivity, ohm cm	D257	10^{14}–10^{15}	10^{12}–10^{15}	1.0×10^{15}	2×10^{16}
17. Dielectric constant at 1 kHz	D150	3.9–4.5	4.0–4.9	3.7	3.02
18. Dielectric strength (kV/in.)	D149	385–470	440–510	500	400
19. Dissipation factor at 1 kHz	D150	0.02–0.04	0.011–0.06	0.004	0.0021
20. Deleterious media	D543	Strong acids	Strong acids	Strong acids, some other acids and bases	Bases and strong acids
21. Solvents (room temperature) (Cl.H. = chlorinated hydrocarbons)		Phenol and formic acid	Phenol and formic acid	None	Aromatic and Cl.H.

Typical Properties of Polymers Used for Molding and Extrusion

	ASTM Test Method	Ionomers	Poly(Phenylene Oxide)	Polysulfone
1. Specific gravity	D792	0.93–0.96	1.06	1.24
2. Tensile modulus $(psi \times 10^{-5})$	D638	0.2–0.6	3.55	3.6
3. Compressive modulus $(psi \times 10^{-5})$	D695	—	—	3.7
4. Flexural modulus $(psi \times 10^{-5})$	D790	—	3.6–4.0	3.9
5. Tensile strength $(psi \times 10^{-3})$	D638, D651	3.5–5.0	9.6	10.2 (yield)
6. Elongation at break (%)	D638	350–450	60	50–100
7. Compressive strength $(psi \times 10^{-3})$	D695	—	16.4	13.9 (yield)
8. Flexural yield strength $(psi \times 10^{-3})$	D790	—	13.5	15.4 (yield)
9. Impact strength, notched Izod, (ft-lb/in.)	D256	6.0–15	5.0	1.2
10. Hardness, Rockwell	D785	D50-65 (Shore)	R119	M69, R120
11. Thermal conduct. $(cal/s\text{-}cm\text{-}K \times 10^4)$	C177	5.8	5.2	2.8
12. Specific heat (cal/g-K)	—	0.55	—	0.31
13. Linear therm. exp. coeff. $(K^{-1} \times 10^5)$	D696	12	3.3–5.9	5.2–5.6
14. Continuous-use temperature (°C)	—	70–95	—	150–175
15. Deflection temp. (°C at 0.45 MPa)	D648	38	—	180
16. Volume resistivity, ohm cm	D257	$> 10^{16}$	10^{18}	5×10^{16}
17. Dielectric constant at 1 kHz	D150	2.4	2.6	3.13
18. Dielectric strength (kV/in.)	D149	900–1100	400–500	425
19. Dissipation factor at 1 kHz	D150	0.0015	0.00035	0.001
20. Deleterious media	D543	Acids, esp. strong Oxidizing acids	None	None
21. Solvents (room temperature) (Cl.H. = chlorinated hydrocarbons)		None	Aromatic and Cl.H.	Aromatic hydrocarbons

	ASTM Test Method	Phenol– Formaldehyde (Cellulose Fill)	Melamine– Formaldehyde (Cellulose Fill)	Cast Epoxy Glass Fiber Fill
1. Specific gravity	D792	1.37–1.46	1.47–1.52	1.6–2.0
2. Tensile modulus $(psi \times 10^{-5})$	D638	8–17	11–14	30

(continued)

	ASTM Test Method	Phenol–Formaldehyde (Cellulose Fill)	Melamine–Formaldehyde (Cellulose Fill)	Cast Epoxy Glass Fiber Fill
3. Compressive modulus (psi $\times 10^{-5}$)	D695	—	—	—
4. Flexural modulus (psi $\times 10^{-5}$)	D790	10–12	—	20–45
5. Tensile strength (psi $\times 10^{-3}$)	D638, D651	5–9	5–13	5–20
6. Elongation at break (%)	D638	0.4–0.8	0.6–1.0	4
7. Compressive strength (psi $\times 10^{-3}$)	D695	25–31	33–45	18–40
8. Flexural yield strength (psi $\times 10^{-3}$)	D790	7–14	9–16	8–30
9. Impact strength, notched Izod, (ft.lb/in.)	D256	0.2–0.6	0.2–0.4	0.3–10
10. Hardness, Rockwell	D785	E64-95	M115-125	M100-112
11. Thermal conduct. (cal/s-cm-K $\times 10^4$)	C177	4–8	6.5–10	4–10
12. Specific heat (cal/g-K)	—	0.35–0.40	0.4	0.19
13. Linear therm. exp. coeff. ($K^{-1} \times 10^5$)	D696	3.0–4.5	4.0–4.5	1–5
14. Continuous-use temperature (°C)	—	150–175	99	150–260
15. Deflection temp. (°C at 0.45 MPa)	D648	—	43	—
16. Volume resistivity, ohm cm	D257	10^9–10^{13}	10^{12}	$>10^{14}$
17. Dielectric constant at 1 kHz	D150	4.4–9.0	7.8–9.2	3.5–5.0
18. Dielectric strength (kV/in.)	D149	200–400	270–300	300–400
19. Dissipation factor at 1 kHz	D150	0.04–0.20	0.015–0.036	0.01
20. Deleterious media	D543	Strong bases and oxidizing acids	Strong acids and bases	None
21. Solvents (room temperature) (Cl. H. = chlorinated hydrocarbons)		None	None	None

Conversion Factors: 1000 psi = 6.895 MPa; 1 ft.lb/in. = 53.4 J/m; 1 cal = 4.187 J; 1 kV/in. = 0.0394 MV/m, data collected from *Modern Plastics Encyclopedia*

A4

Typical Properties of Cross-Linked Rubber Compounds

A. Diene-Based Polymers and Copolymers

	Styrene–Butadiene Random Copolymer, 25% (wt) Styrene (SBR)	Styrene–Butadiene Block Copolymer, about 25% Styrene (YSBR)	Cis-1,4- Polyisoprene (Natural Rubber NR, Also Made Synthetically IR)	Cis-1,4 Polybutadiene (BR)	Polychloroprene (CR), Neoprene	Butadiene–Acrylonitrile Random Copolymer, Variable % Acrylonitrile (NBR)	Reclaimed Rubber (Whole Tires) (Mainly NR and SBR)
Gum stock (cross-linked, Unfilled)							
Density (g/cm^3)	0.94	0.94–1.03	0.93	0.93	1.23	1.00	1.2 (compd'd)
Tensile strength (psi)[a]	200–400	1700–3700	2500–3000	200–1000	3000–4000	500–1000	
Resistivity (ohm cm, log)	15	13	15–17	—	11	10	
Dielect. const. at 1 kHz	3.0	3.4	2.3–3.0	2.3–3.0	9.0	13	
Dielect. str. (kV/in.)[b]	—	485	—	—	150–600		
Diss. factor, at 1 kHz	0.003	0.01	0.002–0.003	0.002–0.003	0.03	0.055	
Reinforced stock							
Tensile strength (psi)[a]	2000–3500	1000–3000	3000–4000	2000–3500	3000–4000	3000–4000	500–1000
Elong. at break (%)	300–700	500–1000	300–700	300–700	300–700	300–700	300–400
Hardness, Shore A	40–100	40–85	20–100	30–100	20–100	30–100	50–100
Cont. high-temp. limit (°C)	110	65	100	100	120	120	100
Stiffening temp. (°C)	−20 to −45	−50 to −60	−30 to −45	−35 to −50	−10 to −30	0 to −30	−20 to −45
Brittle temp. (°C)	−60	−70	−60	−70	−40 to −55	−15 to −55	−60
Resilience	Good	Excellent	Excellent	Excellent	Good	Fair	Good
Resistance to							
Acid	Good	Good	Good	Good	Good	Good	Good
Alkali	Good	Good	Good	Good	Good	Good	Good
Gasoline and oil	Poor	Poor	Poor	Poor	Good	Good	Good
Aromatic hydrocarbons	Poor	Poor	Poor	Poor	Fair	Excellent	Good
Ketones	Good	Poor	Good	Good	Poor	Good	Poor
Chlorinated solvents	Poor	Poor	Poor	Poor	Poor	Poor	Poor
Oxidation	Good	Good	Good	Good	Excellent	Fair	Good
Ozone	Good	Good	Good	Good	Poor	Fair	Good

Abbreviations are according to ASTM.
[a] 1000 psi = 6.895 MPa.
[b] 1 kV/in. = 0.0394 MV/m.

Typical Properties of Cross-Linked Rubber Compounds

B. Saturated, Carbon–chain Polymers

	Polyisobutylene (Butyl Rubber, Copolymer with 0.5–2% Isoprene) (IIR)	Chloro-Sulfonated Polyethylene (CSM)	Ethylene–Propylene Random Copolymer, 50% Ethylene (EPM)	Ethylene–Propylene Random Terpolymer 50% Ethylene (EPDM)	Poly(Ethyl Acrylate), Usually a Copolymer (ACM)	Vinylidene–Fluoride–Chlorotrifluoro Ethylene Random Copolymer (FKM)	Vinylidene–Fluoride–Hexafluoropropylene Random Copolymer (FKM)
Gum stock (cross-linked, unfilled)							
Density (g/cm^3)	0.92	1.12–1.28	0.86	0.86	1.10	1.85	1.85
Tensile strength (psi)[a]	2500–3000	2500	500	200	200–400	200–2500	2000
Resistivity (ohm cm, log)	17	14	16	16	—	14	13
Dielect. const. at 1 kHz	2.1–2.4	7–10	3.0–3.5	3.0–3.5	—	6	—
Dielect. str. (kV/in.)[b]	600	500	900	900	—	600	250–750
Diss. factor, at 1 kHz	0.003	0.03–0.07	0.004–0.008	0.004–0.008	—	0.05	0.03–0.04
Reinforced stock							
Tensile strength (psi)[a]	2000–3000	3000	1000–3000	1000–3500	1500–2500	1500–2500	1500–2500
Elong. at break (%)	300–700	300–500	200–300	200–300	250–350	300–400	300–400
Hardness, Shore A	30–100	50–100	30–100	30–100	40–100	50–90	50–90
Cont. high-temp. limit (°C)	120	160	150	150	175	200	250
Stiffening temp. (°C)	−25 to −45	−10 to −30	−40	−40	—	−35	—
Brittle temp. (°C)	−60	−40 to −55	−50 to −75	−50 to −75	−30	−50	−45
Resilience	Fair	Good	Good	Good	Fair	Fair	Fair
Resistance to							
Acid	Excellent	Good	Excellent	Excellent	Fair	Excellent	Excellent
Alkali	Excellent	Good	Excellent	Excellent	Poor	Good	Good
Gasoline and oil	Poor	Good	Poor	Poor	Good	Excellent	Excellent
Aromatic hydrocarbons	Poor	Fair	Fair	Fair	Good	Good	Excellent
Ketones	Excellent	Poor	Good	Good	Poor	Poor	Poor
Chlorinated solvents	Poor	Poor	Poor	Poor	Poor	Good	Good
Oxidation	Excellent	Excellent	Excellent	Excellent	Excellent	Good	Excellent
Ozone	Excellent	Excellent	Excellent	Excellent	Excellent	Good	Excellent

Abbreviations are according to ASTM.
[a] 1000 psi = 6.895 MPa.
[b] 1 kV/in. = 0.0394 MV/m.

C. Heterochain Polymers

	Poly(Dimethyl Siloxane) Silicone Rubber, Usually Copolymer with Vinyl groups (VMQ)	Poly(Dimethyl Siloxane) Copolymer with Phenyl-Bearing Siloxane and Vinyl Groups (PVMQ)	Room Temperature Vulcanizing Silicone	Polysulfide (ET and EOT)	Polyurethane (AU and EU)
Gum stock (cross-linked, unfilled)					
Density (g/cm³)	0.98	0.98	1.0–1.3 (compd'd)	1.35	1.25
Tensile strength (psi)[a]	50–100	50–100	—	100–200	2000–4000
Resistivity (ohm cm, log)	11–17	11–17	15	12	11–14
Dielect. const. at 1 kHz	3.0–3.5	3.0–3.5	2.8	7–9.5	5–8
Dielect. str. (kV/in.)[b]	100–600	100–600	500	250–600	350–525
Diss. factor, at 1 kHz	0.001–0.010	0.001–0.010	0.003	0.001–0.005	0.02–0.09
Reinforced stock					
Tensile strength (psi)[a]	500–1200	500–1500	400–800	1300–1800	3000–10,000
Elong. at break (%)	200–700	200–700	100–200	200–500	200–600
Hardness, Shore A	30–80	30–80	30–50	25–85	20–100
Cont. high-temp. limit (°C)	250	300	200–250	120	120
Stiffening temp. (°C)	−50	−100	−50 to −100	−25	−25 to −35
Brittle temp. (°C)	−50	−120	−50 to −100	−50	−50 to −60
Resilience	Fair	Fair	Fair	Fair	Poor
Resistance to					
Acid	Fair	Fair	Fair	Fair	Fair
Alkali	Fair	Fair	Fair	Fair	Poor to fair
Gasoline and oil	Poor	Poor	Poor	Good	Excellent
Aromatic hydrocarbons	Poor	Poor	Poor	Excellent	Good
Ketones	Excellent	Excellent	Excellent	Good	Poor
Chlorinated solvents	Poor	Poor	Poor	Good	Poor
Oxidation	Excellent	Excellent	Excellent	Good	Poor
Ozone	Excellent	Excellent	Excellent	Excellent	Excellent
					Excellent

Abbreviations are according to ASTM.
[a] 1000 psi = 6.895 MPa.
[b] 1 kV/in. = 0.0394 MV/m.

A5

Typical Properties of Representative Textile Fibers

Typical Properties of Representative Textile Fibers

Fiber/Chemical Name	Specific Gravity	Breaking Tenacity[a] (g/denier)		Elongation at Break (%)		Water Absorbed at 70°F, 65% rel. Humidity (%)	Thermal Stability
		Standard	Wet	Standard	Wet		
1. Acetate/cellulose acetate							
(a) Diacetate	1.32	1.2–1.4	0.8–1.0	25–45	35–50	6.4	(a) Sticks at 175–205°C; softens, 205–230°C; Melts, 260°C
(b) Triacetate	1.3	1.1–1.3	0.8–1.0	26–40	30–40	3.2	(b) Melts at 300°C
2. Acrylic/polyacrylonitrile	1.17	2.0–2.7	1.6–2.2	34–50	34–60	1.5	Shrinks 5% at 253°C
3. Aramid/aromatic polyamide							
(a) Kevlar (Du Pont)	1.44	21.7	21.7	2.5–4	2.5–4	4.5–7	(a) Decomposes at 500°C
(b) Nomex (Du Pont)	1.38	4.0–5.3	3.0–4.1	22–32	20–30	6.5	(b) Decomposes at 370°C
4. Cotton/α-cellulose	1.54	3.0–4.9	3.0–5.4	3–10		7–8.5	Decomposes at 150°C
5. Fluorocarbon/poly(tetrafluoroethylene)	2.1	0.9–2.0	0.9–2.0	19–140	19–140	Nil	Melts at about 288°C
6. Glass/silica, silicates	2.49–2.55	9.6–19.9	6.7–19.9	3.1–5.3	2.2–5.3	Nil	Softens at 730–850°C; does not burn
7. Nylon/aliphatic polyamide							
(a) Nylon-6	1.14	4.0–9.0	3.7–8.2	16–50	19–47	2.8–5.0	(a) Melts at 216°C; decomposes, 315°C
(b) Nylon-6,6	1.14	3.0–9.5	2.6–8.0	16–66	18–70	4.2–4.5	(b) Sticks at 230°C melts, 250–260°C
8. Olefin							
(a) Polyethylene (branched)	0.92	1.0–3.0	1.0–3.0	20–80	20–80	Nil	(a) Softens 105–115°C; melts 110–120°C; shrinks 5% at 75°C
(b) Polyethylene (linear)	0.95	3.5–7.0	3.5–7.0	10–45	10–45	Nil	(b) Softens 115–125°C; melts 125–138°C; Shrinks 5% at 75°C
(c) Polypropylene	0.90	3.0–8.0	3.0–8.0	14–80	14–80	0.01–0.1	(c) Softens 140–175°C; melts 160–177°C; shrinks 5% at 100–130°C
9. Polyester/poly(ethylene terephthalate)	1.38	2.2–9.5	2.2–9.5	12–55	12–55	0.4–0.8	Sticks at 230°C; melts, 250°C
10. Spandex/segmented polyurethane	1.21	0.7–0.9	—	400–625	—	1.3	Sticks at 215°C
11. Viscose rayon/regenerated cellulose							
(a) Regular	1.46–1.54	0.7–3.2	0.7–1.8	15–30	20–40	11–13	(a) Loses strength at 150°C
(b) High tenacity		3.0–5.7	1.9–4.3	9–26	14–34	—	(b) Decomposes at 175–240°C
12. Wool/protein	1.32	1.0–2.0	0.8–1.8	20–40		11–17	Decomposes at 130°C

[a] Tensile strength (MPa) = tenacity (g/denier) × density (g/cm^3) × 88.3.

Source: *Textile World*, 128(8), 57, 1978.

Index

Index